Multistate Models for the Analysis of Life History Data

MONOGRAPHS ON STATISTICS AND APPLIED PROBABILITY

Editors: F. Bunea, P. Fryzlewicz, R. Henderson, N. Keiding, T. Louis, R. Smith, and W. Wong

For more information about this series please visit:
https://www.crcpress.com/Chapman--HallCRC-Monographs-on-Statistics--Applied-Probability/book-series/CHMONSTAAPP

Monographs on Statistics and Applied Probability 158

Multistate Models for the Analysis of Life History Data

Richard J. Cook
University of Waterloo
Canada

Jerald F. Lawless
University of Waterloo
Canada

CRC Press
Taylor & Francis Group
Boca Raton   London   New York

CRC Press is an imprint of the
Taylor & Francis Group, an **informa** business

A CHAPMAN & HALL BOOK

CRC Press
Taylor & Francis Group
6000 Broken Sound Parkway NW, Suite 300
Boca Raton, FL 33487-2742

First issued in paperback 2020

Version Date: 20180126

ISBN-13: 978-0-367-57172-6 (pbk)
ISBN-13: 978-1-4987-1560-7 (hbk)

Visit the e-resources at: https://www.crcpress.com/9781138578333

Visit the Taylor & Francis Web site at
http://www.taylorandfrancis.com

and the CRC Press Web site at
http://www.crcpress.com

To Eric and Graham.

To Aidan, Callum, Sarah and Sabrina.

Contents

Preface

Multistate models provide a natural and powerful framework for characterizing life history processes. They can play a central role in medical decision making in complex diseases where the prediction of outcomes is necessary under different treatment strategies, and in health economics where the specification of health policies may be based on projections of long-term costs. Conceptualizing processes in terms of multistate models is also useful when considering the impact of sample selection schemes and complex observation processes, both of which must be addressed in observational studies to avoid biased inferences.

The purpose of this book is to provide readers with an understanding of multistate models and to demonstrate their usefulness for the analysis of life history data in public health research. We emphasize the role of intensity functions when interest lies in factors influencing process dynamics. The interpretation of various intensity-based models is stressed, and their role in understanding complex disease processes is explored. Marginal methods for estimating features such as state occupancy probabilities are also considered. Special features include a review of competing risk analysis and connections between various approaches. The use of multistate models is also highlighted as a framework for understanding the impact of study selection conditions and assumptions necessary for observation schemes to be ignorable for both right-censored data and data arising from intermittent observation of a continuous-time process. Issues arising in the design and analysis of longitudinal studies are considered, along with the importance of dealing appropriately with initial conditions, strategies for dealing with heterogeneity and dependence, and prediction. Numerous examples are provided for illustration and comparison of approaches to analysis. We provide sample code and output within some examples and give detailed code for others in Appendix C.

This book has been written in a style similar to our 2007 book *The Statistical Analysis of Recurrent Events* in that we present models and discuss the construction of likelihoods or estimating functions in a rigorous but accessible way. We emphasize ways of dealing with selection and observation schemes that commonly arise in the collection and analysis of life history data. Familiarity with survival analysis is helpful, and we refer readers to Kalbfleisch and Prentice (2002) and Lawless (2003) for background reading with a similar presentation style to this book. Klein and Moeschberger (2003) give a nice introduction to survival analysis with an emphasis on applications, and introduce examples involving multistate modeling. Fleming and Harrington (1991) give a more mathematical presentation of the topic. Andersen et al. (1993) provide a comprehensive and mathematically rigorous account of statistical methods for event history models and counting processes. Aalen et al. (2008) provide excellent coverage of survival and event history analysis and discuss related issues such as random effect modeling, marker processes and causal analysis. Therneau and Grambsch (2000) is an important reference that highlights the remarkably broad range of problems than can be handled with the `coxph` function in R. Martinussen and Scheike (2006) give an authoritative account of modern methods for survival and event history analysis with a good deal of attention to semiparametric additive models and the `timereg` package.

Relatively few books have been devoted to multistate analysis. Crowder (2001, 2012) gives a textbook treatment of competing risks analysis. Beyersmann et al. (2012) and Geskus (2016) focus on this topic as well as more general multistate models with an emphasis on computing. Willekens (2014) deals with multistate life history analysis with an emphasis on demographic and economic applications and software available in R. Commenges and Jacqmin-Gadda (2016) survey a wide range of event history models, with discussions of random effects, causality and software. The recent book by Van den Hout (2017) covers several of the topics in this book, but in less detail. Sun (2006) gives a comprehensive account of interval-censored data on failure times and recurrent events.

The data in our examples are analyzed using R. In many cases there exist analogous procedures in other software packages. Datasets that are available to the public are listed in Appendix D and are posted on the website for this book (`http://www.math.uwaterloo.ca/~rjcook/cook-lawless-multistate.html`) along with some sample R code.

Much of our research on multistate models has been motivated by challenges arising in public health research and ongoing collaborations with health scientists. We would like to acknowledge Dafna Gladman, Vinod Chandran, Lihi Eder and Murray Urowitz of the Centre for Prognosis Studies in Rheumatic Diseases at the University of Toronto, and thank them for stimulating collaborations that have led to some of the methodological developments in this book. We also wish to acknowledge rewarding collaborations with Robert Coleman, Nancy Heddle, Pierre Major and Matthew Smith.

We would like to thank Bayer Canada Inc., GlaxoSmithKline Inc. and Novartis Pharmaceuticals Inc. for permission to use data from some of the studies in examples. Mouna Akacha is gratefully acknowledged for stimulating interactions regarding problems with recurrent events and competing risks data in oncology trials. We thank John Lachin and the DCCT/EDIC investigators for permission to use data from the Diabetes Control and Complications Trial (DCCT), which was sponsored by the Division of Diabetes, Endocrinology, and Metabolic Diseases of the National Institute of Diabetes and Digestive and Kidney Diseases, National Institutes of Health, as well as the National Heart, Lung, and Blood Institute, the National Eye Institute, and the National Center for Research Resources. We are grateful to Andrew Paterson for introducing us to the DCCT and EDIC studies and for helpful discussions about them. We thank Janet Raboud for access to data from the CANOC project, which was supported by funding from the Canadian Institutes of Health Research (Emerging Team Grant 53444, Dr. R. Hogg, P.I.), and for helpful discussion.

Many ideas in this book originated in research with Jack Kalbfleisch beginning in the 1980s and we thank him for continuing discussions on life history analysis since that time. We are grateful to the faculty, graduate students, and staff at the University of Waterloo who help make ours a stimulating and supportive environment for research. In particular, we would like to acknowledge enjoyable and fruitful collaborations involving multistate models with colleagues Vern Farewell, Robert Gentleman, Grace Yi and Leilei Zeng. Much of the work in multistate analysis has been carried out with the involvement of exceptional graduate students and fellows at the University of Waterloo. We gratefully acknowledge Jean-Marie Boher, Audrey Boruvka, Baojiang Chen, Liqun Diao, Daniel Fong, Dagmar Mariaca Hajducek, Shu Jiang, Lajmi Lakhal-Chaieb, Jooyoung Lee, Nathalie Moon, Narges Nazeri Rad, Edmund Ng, Rhonda Rosychuk, Hua Shen, Ying Wu, Ping Yan, Yildiz Yilmaz, Yujie Zhong and Yayuan Zhu.

A special thanks is expressed to Ker-Ai Lee whose expertise in statistical computing has been instrumental in our research on multistate models, the completion of empirical studies and applications, and the overall preparation of this monograph.

Much of the work here was developed while the first author held a Tier I Canada Research Chair in Statistical Methods for Health Research and while the second author held an Industrial Research Chair co-sponsored by General Motors Canada and the Natural Sciences and Engineering Research Council of Canada. Support from the Natural Sciences and Engineering Research Council of Canada and the Canadian Institutes of Health Research is gratefully acknowledged.

Finally we would like to thank our families for their support and understanding during the writing of this book and always.

Richard Cook
Jerry Lawless
University of Waterloo

Glossary

c.d.f. – cumulative distribution function

p.d.f. – probability density function

$F(t)$ is a c.d.f. for T; $dF(t) = f(t) \, dt$ where $f(t)$ is the p.d.f. of T; $\mathcal{F}(t) = 1 - F(t) = P(T > t)$

$N_p(\mu, \Sigma)$ – a p-variate normal distribution with mean μ and covariance matrix Σ

χ_p^2 – a chi-square random variable on p degrees of freedom

$\mathcal{A} \cup \mathcal{B}$ is the union of sets $\mathcal{A}$ and $\mathcal{B}$

θ is a $p \times 1$ parameter vector

$X^{\otimes 2} = X X'$ where X is a $p \times 1$ vector

$A \perp B$ means the events A and B are independent; $A \perp B \mid C$ means the events A and B are conditionally independent given C

EST – estimate

SE or s.e. – standard error

RR – a generic notation for a relative risk obtained typically by exponentiating a regression coefficient in a multiplicative intensity model

$\emptyset$ – the null set

0^0 is taken to be 1

$x. = \sum_{i=1}^{n} x_i$ – a "dot" subscript indicates summation over the corresponding index

$L(\theta) \propto P(\text{Data}; \theta)$ is a likelihood function; $\ell(\theta) = \log L(\theta)$

$\widehat{\theta}$ maximizing $\ell(\theta)$ is the MLE

$U(\theta) = \partial \log L(\theta) / \partial \theta$ is a score function

$I(\theta) = -\partial U(\theta) / \partial \theta'$ is the observed information matrix; $\mathcal{I}(\theta) = E(I(\theta))$ is the expected or Fisher information matrix

$LRS(\theta) = 2 \left[\log L(\widehat{\theta}) - \log L(\theta) \right]$ is a likelihood ratio statistic

$\{1, 2, \ldots, K\}$ is a collection of labels for states in a K-state process; sometimes they are labeled $\{0, 1, \ldots, K\}$

$Z(t)$ is the state occupied at time t for a multistate process where t is the time since the origin of a process

$\{Z(t), t \geq 0\}$ is the multistate process

$\mathcal{H}(t) = \{Z(s), 0 \leq s \leq t\}$ is the history of the process over $[0, t]$

$\mathcal{A}$ is a set of absorbing states in a multistate process

C^A is an administrative censoring time, typically independent of the processes of interest

C^R is a random censoring time

CHAPTER 1

$N(t)$ is the number of events over $[0, t]$

$\Delta N(t) = N(t + \Delta t^-) - N(t^-)$ and $dN(t) = \lim\limits_{\Delta t \downarrow 0} \Delta N(t)$

$\{N(t), t \geq 0\}$ is a counting process

$X(t)$ is the value of a time-dependent covariate at time t; $x(t)$ denotes the realized value of $X(t)$

$\{X(t), t \geq 0\}$ is a covariate process

$\mathcal{H}(t) = \{Z(s), X(s), 0 \leq s \leq t\}$ is the history for the multistate and covariate processes; if $X(t) = X$ then $\mathcal{H}(t) = \{Z(s), 0 \leq s \leq t; X\}$. It is also used to represent related counting process histories such as $\mathcal{H}(t) = \{N(s), 0 \leq s \leq t\}$ or $\mathcal{H}(t) = \{N(s), 0 \leq s \leq t; X\}$

$B(t)$ is the time since entry to the state occupied at t

$P_{kl}(s, t | \mathcal{H}(s)) = P(Z(t) = l | Z(s) = k, \mathcal{H}(s))$ is a transition probability; for a Markov process $P_{kl}(s, t | \mathcal{H}(s)) = P(Z(t) = l | Z(s) = k) = P_{kl}(s, t)$

$P_k(t) = P(Z(t) = k)$ is the prevalence (or state occupancy probability) for state k at time t

$S_k(t)$ is the total time spent in state k over $[0, t]$; $\psi_k(t) = E(S_k(t))$ and $\psi_{jk}(t) = E(S_k(t) | Z(0) = j)$

CHAPTER 2

$N_{kl}(t)$ is the number of $k \to l$ transitions over $[0, t]$

The intensity for $k \to l$ transitions is defined as

$$\lim_{\Delta t \downarrow 0} \frac{P(Z(t + \Delta t^-) = l \mid Z(t^-) = k, \mathcal{H}(t^-))}{\Delta t} = \lambda_{kl}(t \mid \mathcal{H}(t^-))$$

or equivalently as

$$\lim_{\Delta t \downarrow 0} \frac{P(\Delta N_{kl}(t) = 1 \mid \mathcal{H}(t^-))}{\Delta t} = Y_k(t^-) \lambda_{kl}(t \mid \mathcal{H}(t^-))$$

where $Y_k(t^-) = I(Z(t^-) = k)$ and $\Delta N_{kl}(t) = N_{kl}(t + \Delta t^-) - N_{kl}(t^-)$

$H(u_r) = \{Z(u_s), s = 0, 1, \ldots, r\}$ is the history at u_r of a multistate process under a partition $a = u_0 < u_1 < \cdots < u_R = b$ defined for product integration over $[a, b]$ where $H(u_0) = Z(u_0)$

$\mathcal{D}_{kl}$ is the set of all $k \to l$ transition times in a given period of observation

C^A is an administrative censoring time typically assumed to be completely independent of $\{Z(t), t \geq 0\}$ and $\{X(t), t \geq 0\}$

C^R is a random censoring time

$C^R(t) = I(C^R \leq t)$ is right-continuous counting process for random censoring; $\Delta C^R(t) = C^R(t + \Delta t^-) - C^R(t^-)$ and $dC^R(t) = \lim_{\Delta t \downarrow 0} \Delta C^R(t)$

$C = \min(C^R, C^A)$ is the net right censoring time

$Y(t) = I(t \leq C)$ indicates a process is under observation at t

$Y_k(t) = I(Z(t) = k)$ indicates that state k is occupied at t and $\bar{Y}_k(t) = Y(t)Y_k(t^-)$ indicates a transition out of state k may be observed at time t

$d\bar{N}_{kl}(t) = \bar{Y}_k(t) dN_{kl}(t)$ and $\bar{N}_{kl}(t) = \int_0^t \bar{Y}_k(s) dN_{kl}(s)$ is an observed counting process

$d\bar{X}(t) = Y(t) dX(t)$, $\bar{X}(t) = \int_0^t d\bar{X}(s)$; with a partition of $[a, b]$ given by $a = u_0 < u_1 < \cdots < u_R = b$, $\Delta \bar{X}(u_r) = Y(u_r)(X(u_r) - X(u_{r-1}))$

$\bar{\mathcal{H}}(t) = \{Y(s), \bar{N}(s), \bar{X}(s), 0 < s \leq t; Z(0), X(0)\}$ is the observed history of the multistate and covariate processes under right censoring

$\bar{H}(u_r) = \{Y(u_s), \bar{N}(u_s), \bar{X}(u_s), s = 1, \ldots, r; Z(0), X(0)\}$ is the observed history at u_r of censoring, multistate and covariate processes used under a partition $a = u_0 < u_1 < \cdots < u_R = b$ of $[a, b]$ for product integration

A_0 is the time a process begins to be under observation and a_0 is its realized value

$T_k^{(r)}$ denotes the time of the rth entry into state k

$W_k^{(r)}$ denotes the duration of the rth sojourn in state k

$V_k^{(r)} = T_k^{(r)} + W_k^{(r)}$ denotes the time of the rth exit from state k

$F_k^{(r)}(t) = P(T_k^{(r)} \leq t)$ is the cumulative (sub)-distribution function for the rth entry to state k

$Q(t)$ is a $K \times K$ transition intensity matrix for a K-state Markov process with $\lambda_{kl}(t)$ in entry (k, l) with $k \neq l$, and $-\lambda_{k\cdot}(t)$ in the diagonal entries, $k = 1, \ldots, K$

$P(s, t)$ is a $K \times K$ transition probability matrix with (k, l) entry $P_{kl}(s, t) = P(Z(t) = l \mid Z(s) = k)$; with fixed covariates we write $P(s, t \mid x)$ with (k, l) entry $P_{kl}(s, t \mid x) = P(Z(t) = l \mid Z(s) = k, X = x)$

$\lambda_{kl}(t \mid x(t)) = \lambda_{kl0}(t) \, g(x(t); \beta_{kl})$ is the intensity for a modulated Markov process; we typically use $g(x(t); \beta_{kl}) = \exp(x'(t) \beta_{kl})$

$\lambda_{kl}(t \mid \mathcal{H}(t^-)) = h_{kl}(B(t); x(t))$ is the intensity for a modulated semi-Markov process, where $B(t)$ is the time since the most recent entry to state k

---

CHAPTER 3

$0 = b_0 < b_1 < \cdots < b_R = \infty$ are the break-points (or cut-points) that define a piecewise-constant intensity; $\mathcal{B}_r = [b_{r-1}, b_r)$ defines the rth interval

$t^{(1)} < \cdots < t^{(m)}$ are the distinct times at which transitions of any type are observed in a dataset

$\widehat{\Lambda}_{kl}(t) = \int_0^t d\widehat{\Lambda}_{kl}(u)$ is the Nelson-Aalen (NA) estimate of the cumulative intensity for $k \to l$ transitions

$\widehat{P}(s,t) = \prod_{(s,t]} \{I + \widehat{Q}(u)du\}$ is the Aalen-Johansen (AJ) estimate of the transition probability matrix where $\widehat{Q}(t)dt$ is a $K \times K$ matrix with $d\widehat{\Lambda}_{kl}(t)$ in entry (k,l) for $k \neq l$, and $-d\widehat{\Lambda}_{k\cdot}(t)$ in the diagonal entries, $k = 1, \ldots, K$

$\widehat{P}(s,t|x) = \prod_{(s,t]} \{I + \widehat{Q}(u|x)du\}$ is the Aalen-Johansen estimate of the conditional transition probability matrix where $\widehat{Q}(t|x)dt$ is a $K \times K$ matrix with $d\widehat{\Lambda}_{kl}(t|x)$ in entry (k,l) and $k \neq l$, and $-d\widehat{\Lambda}_{k\cdot}(t|x)$ in the diagonal entries, $k = 1, \ldots, K$

$H(t)$ is a cumulative hazard function where $H_{kl}(t)$ is the cumulative hazard for (possibly latent) sojourn time in state k ending in a $k \to l$ transition; $H_{kl}(t) = \int_0^t h_{kl}(u)du$ and $H_{k\cdot}(t) = \sum_{l \neq k} H_{kl}(t)$

$\widetilde{\mathcal{D}}_{ik}$ is a set of all sojourn times in state k for individual i

$\mathcal{H}(t)$ is the aggregated histories $\{\mathcal{H}_i(t), i = 1, \ldots, n\}$ (Section 3.3)

$P_k(t) = P(Z(t) = k)$; in general this is obtained as $P_k(t) = \sum_{l=1}^{K} P_l(0) \, P_{lk}(0,t)$; if the initial state is 1 with probability one then $P_k(t) = P_{1k}(0,t)$

The intensity for random censoring is

$$\lim_{\Delta t \downarrow 0} \frac{P(\Delta C^R(t) = 1 \mid \bar{\mathcal{H}}(t^-))}{\Delta t} = Y(t) \lambda^c(t \mid \bar{\mathcal{H}}(t^-))$$

and $d\Lambda^c(u|\mathcal{H}(u^-)) = \lambda^c(u|\mathcal{H}(u^-))du$

$\widehat{\Lambda}_{kl}^w(u)$ is a weighted NA estimate of the cumulative $k \to l$ transition rate

$\widehat{Q}^w(u)du$ is a matrix of weighted increments to the NA estimates of cumulative transition rates

$\widehat{P}^w(s,t)$ is a weighted AJ estimate of the transition probability matrix

$O_{kl}(t_{j-1}, t_j)$ and $E_{kl}(t_{j-1}, t_j)$ denote observed and expected transition counts

$E_{ik}^{(r)}$ is the exponential residual for the rth sojourn in state k for individual i

$F_k(t|x_i;\theta) = P(T_{ik} \leq t|x_i;\theta)$ is the cumulative distribution function of the time of entry to state k, T_k, given $X_i = x_i$ when each state can be entered at most once

$\widehat{\theta}_{(-g)}$ denotes a parameter estimate based on data excluding those in group g

$BS(t)$ is a Brier score; $BS_{CV}(t)$ is the Brier score via cross-validation; $EBS(t)$ is the expected Brier score

LS is a logarithmic score; LS_{CV} is the logarithmic score obtained under cross-validation

$\mathcal{Z}_i(C_i)$ is a full observed sample path for the multistate process of individual i

---

CHAPTER 4

For competing risk models state 0 represents being event-free and state k represents failure due to cause k, $k = 1, \ldots, K$

$P(0,t) = \prod_{(0,t]} \{I + Q(u)\,du\}$ is the transition probability matrix where $P_{0k}(0,t) = P_k(t)$

is the cumulative incidence function for failure due to cause k; this is sometimes denoted by $F_k(t)$ and viewed as a sub-distribution function since $\lim_{t\uparrow\infty} P_k(t) < 1$ with

$f_k(t) = dF_k(t)/dt$ the sub-density function, $-\log F_k(t) = \Gamma_k(t)$ and $-d\log F_k(t)/dt$ the sub-distribution hazard

ε records the cause of failure, $\varepsilon = 1, 2, \ldots, K$

$N_k(t) = I(T \leq t, \varepsilon = k)$ indicates a failure due to cause k occurred by time t and $dN_k(t) = \lim_{\Delta t \downarrow 0} \Delta N_k(t)$

$\bar{Y}_i(t) = Y_i(t)Y_{i0}(t^-)$ where $Y_i(t) = I(t \leq C_i)$ and $Y_{i0}(t) = I(Z_i(t) = 0)$

$Y_k^{\ddagger}(t) = I(T_k \geq t) = 1 - Y_k(t)$ indicates an individual has not failed from cause k prior to time t

$\bar{C}(t) = I(C \geq \min(T,t))$

$G_i^c(t) = P(C_i > t|x_i)$ denotes the conditional survival function of the net censoring time given fixed covariates x_i

$\tilde{P}_k^{-i}(t)$ denotes an estimate of $P_k(t)$ based on the sample excluding individual i

$w_i(t)$ is a weight; $\widehat{w}_i(t)$ is an estimated weight

---

CHAPTER 5

$A_0 < A_1 < \cdots$ represent random visit times and $a_0 < a_1 < \cdots$ their realized values

$\mathcal{H}^\circ(a_j) = \{a_r, Z(a_r), X(a_r),\ r = 0, 1, \ldots, j\}$; $\mathcal{H}^\circ(a_0) = \{A_0 = a_0, Z(a_0), X(a_0)\}$

$\mathcal{X}^\circ(a_j) = \{(a_r, X(a_r)), r = 1, \ldots, j\}$

$\mathcal{Z}^\circ(a_j) = \{(a_r, Z(a_r)), r = 1, \ldots, j\}$

$O_{kl}^{(j)} = \sum_{i=1}^{n} Y_i(a_j) Y_{ik}(a_{j-1}) Y_{il}(a_j), \ k, l = 1, \ldots$ are the observed transition counts between a_{j-1} and a_j

$E_{kl}^{(j)} = O_{k\cdot}^{(j)} \hat{P}(Z_i(a_j) = l \mid Z_i(a_{j-1}) = k)$ are estimated expected counts

$\{A(t), t \geq 0\}$ is a right-continuous counting process where $A(t)$ records the number of visits over $[0, t]$

$\bar{\mathcal{H}}(t) = \{Y(u), A(u), Z(u), 0 \leq u \leq t\}$ is the (latent) history under an intermittent observation scheme

$\mathcal{Z}(t) = \{Z(s), 0 \leq s \leq t\}$

$\bar{\mathcal{H}}^\circ(t) = \{Y(u), A(u), 0 \leq u \leq t; (a_j, Z(a_j)), j = 1, \ldots, A(t)\}$ represents the observed data history at time t

$\mathcal{Z}^\circ(t) = \{(a_j, Z(a_j)), j = 1, \ldots, A(t)\}$ is the observed history of the multistate process at time t

The intensity for a visit under the CIVP condition is

$$\lim_{\Delta t \downarrow 0} \frac{P(\Delta A(t) = 1 \mid \bar{\mathcal{H}}^\circ(t^-))}{\Delta t} = Y(t) \lambda^a(t \mid \bar{\mathcal{H}}^\circ(t^-))$$

Chapter 6

V_{kl} is a random variable associated with the $k \to l$ transition intensity; $V_k = (V_{kl}, l \neq k, l = 1, \ldots, K)$ and $V = (V_1, \ldots, V_K)'$

$G(v)$ is the joint c.d.f. for V

$h(t|x, v)$ denotes the conditional hazard for a failure time T give $X = x, V = v$

$h^m(t|x)$ is the marginal hazard for failure time T give $X = x$ and $H^m(t|x) = \int_0^t h^m(u|x) du$

$\lambda(t|Y_k(t^-) = 1, \mathcal{H}(t^-), v) = \lambda_k(t|x, v)$ is the conditional Markov intensity given $X = x$ and $V = v$ in a progressive process

$\lambda_k^m(t|Y_k(t^-) = 1, \mathcal{H}(t^-))$ is the marginal intensity in a progressive process

$W_{i1}, \ldots, W_{iK}$ denote the sojourn (gap) times in a K-state progressive process

$\mathcal{C}(u_1, u_2; \phi) = P(U_1 \leq u_1, U_2 \leq u_2; \phi)$ is a bivariate copula function with dependence parameter ϕ

$\mathcal{D}_{ik}$ is the set of all $k \to 3 - k$ transition times for individual i over a period of observation

For a bivariate multistate process, $Z(t) = (Z_1(t), Z_2(t))$

The intensity of a $k_r \to l$ transition for process r is

$$\lim_{\Delta t \downarrow 0} \frac{P(Z_r(t+\Delta t^-) = l \mid Z(t^-) = (k_1, k_2), \mathcal{H}(t^-))}{\Delta t} = \lambda^{(r)}_{k_1 k_2, l}(t \mid \mathcal{H}(t^-))$$

and $\Lambda^{(r)}_{kl}(t|\mathcal{H}(t^-)) = \int_0^t \lambda^{(r)}_{kl}(u|\mathcal{H}(u^-))\,du$, $r = 1, 2$

$P_{rkl}(t|x) = P(Z_r(t) = l|Z_r(0) = k, X = x)$ is the $k \to l$ Markov transition probability over $(s, t]$ for process r

$CL(\theta)$ is a composite likelihood

M records the class of a latent class model with G classes and $P(M = g : X; \gamma)$, $g = 1, \ldots, G$, with $\sum_{g=1}^G P(M = g|X; \gamma) = 1$

$\mathcal{H}^\circ(\infty) = \{(a_r, Z(a_r)), r = 0, 1, \ldots, m; X\}$ and $\mathcal{Z}^\circ(\infty) = \{(a_r, Z(a_r)), r = 0, 1, \ldots, m\}$

$L_C(\theta)$ is a complete data likelihood in an incomplete data problem

$\mathcal{Q}(\theta; \tilde{\theta}) = E\{\log L_C(\theta)|D; \tilde{\theta}\}$ is the estimated expectation of the complete data log-likelihood given observed data D and an estimate $\tilde{\theta}$

$W(t)$ is the state recorded at time t in a hidden Markov model; $\{W(t), t \geq 0\}$ is the observed process

$Z(t)$ is the true underlying state occupied at time t; $\{Z(t), t \geq 0\}$ is the latent or "hidden" Markov process

$\nu_{kh} = P(W(s) = k \mid Z(s) = h, x)$ denote the state misclassification probabilities for a hidden Markov model with $\sum_{k=1}^K \nu_{kh} = 1$

$\bar{\mathcal{H}}(t) = \{Y(u), A(u), Z(u), W(u), 0 \leq u \leq t\}$ denotes the complete history of all processes including the latent process

$\bar{\mathcal{H}}^*(t) = \{Y(u), A(u), 0 \leq u \leq t; (a_j, Z(a_j), W(a_j)), j = 0, 1, \ldots, A(t)\}$ denotes the history incorporating information on the true states and the potentially misclassified states occupied at the assessment times

$\bar{\mathcal{H}}^\circ(t) = \{Y(u), A(u), 0 \leq u \leq t; (a_j, W(a_j)), j = 0, 1, \ldots, A(t)\}$ denote the history of the observed process at time t

---

CHAPTER 7

$\{Y_i, W_i, i = 1, \ldots, N\}$ denote data from a Phase 1 sample with response variable $Y = \min(T, C)$ and covariates W

R_i indicates whether individual i is selected for the subsample

$\pi(Y, \Delta, W) = P(R = 1|Y, \Delta, W)$ is the Phase 2 selection model where $\Delta_i = I(T_i \leq C_i)$

$S_{(1)}, \ldots, S_{(M)}$ denote M strata in the Phase 1 population

CHAPTER 8

$\mathcal{U}(t)$ represents the accumulated costs/utilities over $[0, t]$

$\{\mathcal{U}(t), t \geq 0\}$ is a cost process associated with the multistate process

$\mu(t) = E\{\mathcal{U}(t)\}$ is the mean cost/utility function

$F_{\mathcal{U}(t)}(w) = P(\mathcal{U}(t) \leq w)$ is the distribution function of the cumulative cost/utility

$\mathcal{H}(t) = \{N(s), \mathcal{U}(s), 0 \leq s \leq t\}$

$\bar{\mathcal{H}}^{\circ}(t_0) = \{\mathcal{Z}(t_0), \bar{X}(t_0)\}$

Abbreviations

AECB - Acute exacerbation of chronic bronchitis

AIDS - Acquired immune deficiency syndrome

AJ - Aalen-Johansen

ALZ - Alzheimer's disease

AMD - Age-related macular degeneration

AUC - Area under the curve

BALP - Bone alkaline phosphatase

BS - Brier score

CANOC - Canadian Observational Cohort Collaboration

cART - Combination anti-retroviral therapy

CDVP - Conditionally dependent visit process

CI - Confidence interval

CIVP - Conditionally independent visit process

CL - Composite likelihood

CLSA - Canadian Longitudinal Study on Aging

CV - Cross-validation

DCCT - Diabetes Control and Complications Trial

EBS - Expected Brier score

EKL - Expected Kullback-Leibler score

EM - Expectation-maximization (algorithm)

EST - Estimate

ETDRS - Early Treatment Diabetic Retinopathy Study Research Group

FNR - False-negative rate

FPR - False-positive rate

HIV - Human immunodeficiency virus

HLA - Human leukocyte antigen

HMM - Hidden Markov model

IIVW - Inverse intensity of visit weight

IPC - Inverse probability of censoring

IPCW - Inverse probability of censoring weight

KLS - Kullback-Leibler score

KM - Kaplan-Meier

LOOCV - Leave-one-out cross-validation

LRS - Likelihood ratio statistic

LS - Logarithmic scores

LTF - Loss to follow-up

MLE - Maximum likelihood estimate

NA - Nelson-Aalen

NPV - Negative predictive value

NSAIDS - Non-steroidal anti-inflammatory drug

NYC - New York Criteria

OR - Odds ratio

PL - Profile likelihood, or partial likelihood

PPV - Positive predictive value

PRL - Profile relative likelihood

Ps - Psoriasis

PsA - Psoriatic arthritis

QoL - Quality of life

RBC - Red blood cell

ROC - Receiver operating characteristic

RR - Relative risk

SE or s.e. - Standard error

SRE - Skeletal-related event

TPR - True-positive rate

UTPC - University of Toronto Psoriasis Clinic

UTPAC - University of Toronto Psoriatic Arthritis Clinic

Chapter 1

Introduction to Life History Processes and Multistate Models

1.1 Life History Analysis with Multistate Models

The term *life history data* refers to information about events and other outcomes during people's lifetimes, and life history analysis to the analysis of such data. The collection of data, and well-planned studies for doing that, are important in understanding processes related to education, employment, fertility, health and other aspects of human lives. Life history studies can have various goals, for example, to study genetic and environmental risk factors for certain diseases; to understand patterns of fertility in specific populations; to study age-related cognitive or physical decline; and thereby, to formulate models that may be used to predict health care costs and to design interventions. Likewise, life history studies vary in form, two extremes being randomized intervention trials and observational studies based on administrative data. A common feature, however, is that the data one seeks are longitudinal: they cover certain aspects of an individual's life over some period of time.

Our objective in this book is to describe important features of life history analysis and to show how multistate models can be used to describe and analyze life history processes. This approach has been widely used in medicine, public health and other fields such as economics and the social sciences. In the next section we provide some examples drawn from medicine and public health. In the remainder of this section we will describe some types of data that are typically collected, and some notation for representing them.

We consider a generic individual on whom data may be collected. Some types of data and associated notation are described next for the various types of variables that are common in life history studies.

(i) Events are assumed to occur at a specific instant in time as, for example, is assumed for births or diagnoses with a disease. In fact, the exact times some events occur may be nebulous or not easily determined, but for now we assume that exact times are available. There may also be different types of events of interest. In the formulation of models for analysis, counting process notation is useful: assuming for convenience that an individual can experience a potentially recurrent event beginning at a time origin $t = 0$, we let $N(t)$ denote the number of events occurring up to time t. The process $\{N(t), t \geq 0\}$ is called a counting process. When there are $R \geq 2$ types of events, we extend the notation by letting $N_r(t)$ denote the number of events of type r over the time interval $(0, t]$, for $r = 1, \ldots, R$. Features of interest in the case of counting processes include the numbers of events experienced over specific time periods and the lengths of time between specific events.

(ii) Categorical variables are often used to denote the status or state of an individual at a given time. Suppose that at a given time an individual can be in one of K mutually exclusive states, which we label $1, \ldots, K$, and let $Z(t)$ denote the state occupied at time t. We refer to this as a multistate framework. In many applications the definition of appropriate states is fairly obvious; for example, $Z(t)$ might denote the number of children a woman has given birth to by age t. In other cases there is latitude in how states are defined. For example, in the next section we describe situations where states are used to represent ranges for biomarkers such as viral load or blood glucose level. For a multistate context, features of interest include the probabilities of moving from one state to another, and the duration of spells spent in specific states. Although it appears rather different, the multistate framework is closely connected to the counting process framework, since a transition from one state to another can be considered as a type of event.

(iii) Fixed variables such as year of birth or sex are also associated with individuals; we will typically use a vector x to represent such factors.

(iv) Time-dependent variables aside from event counts and states occupied are also important; examples include blood pressure, weight, medication and external factors such as air pollution counts. We will use a vector $x(t)$ to represent the values of such variables at time t.

The focus of this book is on multistate models, but we will have occasion to use counting process terminology in dealing with certain topics. In addition, covariates will feature prominently in many models. Before introducing illustrations, we mention some important aspects of multistate modeling and life history processes. A first set of considerations is the definition of states, what transitions between states are allowed, and how we should model transition probabilities. In most contexts the choice of states has some degree of arbitrariness, but we should seek to specify states that are meaningful, and that allow the objectives of modeling and analysis to be met. Ideally, it should also be possible to determine accurately the state an individual is in at a given time. For example, in the next section we describe studies of the progression of retinopathy for persons with Type 1 diabetes; retinopathy is measured on an ordinal scale with a fairly large number of levels, and for various reasons it is best to group adjacent levels into a smaller number of states. This can of course be done in various ways; moreover, measurement for the original ordinal retinopathy scale is based on photos of the eye and is subject to some degree of variation across raters. More generally, in other contexts, we often have to decide whether to distinguish between different but similar states, or to consider them as a single state.

A second set of considerations is associated with study design and the collection of data. Generally there is a group of individuals on whom longitudinal data on state occupancy and covariates are collected. In some settings the study group is referred to as a cohort or a panel; this is often the case when the individuals are part of a planned study with clear selection criteria. For example, the Canadian Longitudinal Study on Aging (CLSA) is a study of adult development and aging, and recruitment of an initial cohort of about 50,000 persons aged 45−85 began in 2009; see Raina et al. (2009). Individuals were selected according to a stratified random sampling plan, and are to be followed for 20 years or to death, with primary data collection every 3 years. Multistate modeling and analysis is also frequently applied to observational data from sources such as disease registries and administrative databases. In such cases the process by which an individual appears in a data base may not be completely known, and the collection of data can occur sporadically at times that vary from individual to individual. The level of detail about states, events, and covariates can also vary widely across studies; for example, in some cases the exact times of transitions or events may be obtainable whereas in others it is known only that they occurred within some time interval. The period of time over which a person's data are available is also subject to various factors, and can vary across individuals.

Features like those mentioned affect what kinds of models are feasible and what types of questions can be answered. In Section 1.4.2 we will provide a more detailed breakdown of study designs and data but now we turn to some examples that illustrate points made so far and motivate subsequent development.

1.2 Some Illustrative Studies

1.2.1 Disease Recurrence Following Treatment in a Clinical Trial

In randomized trials associated with cancer and other diseases, we are often interested in comparing two or more treatments with respect to their ability to slow or prevent disease recurrence. Death is also an outcome in some trials and in many cases it must be considered explicitly in the design and analysis of a trial.

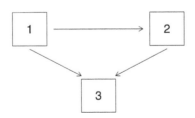

Figure 1.1: Diagram of an illness-death model.

Figure 1.1 shows a multistate model where state 1 represents a disease-free state achieved following treatment, state 2 represents a recurrence of disease and state 3 is death. In some contexts this is called an illness-death model. States 1 and 2 are termed transient since they can be left; state 3 is called an absorbing state since no exit from it is possible. Even if the main focus is on transitions from state 1 to state 2 it will be necessary to consider deaths that occur before recurrence in planning a trial, because they will tend to decrease power for testing treatment differences on the basis of recurrence. Moreover, to increase power, investigators often decide instead to plan and analyze a trial on the basis of time spent in state 1 as the primary outcome; this outcome is known as recurrence-free survival. Treatment comparisons based on recurrence-free survival can have enhanced power if treatments tend to have the same type of effect (an increase or a decrease) on transitions from state 1 to 2 and from state 1 to 3. Later in this book we will consider the implications of doing this and compare analyses based on recurrence-free survival with ones based on other features of the model in Figure 1.1.

1.2.2   Complications from Type 1 Diabetes

The Diabetes Control and Complications Trial (DCCT) was a randomized study that ran from 1983 to 1993. Its main purpose was to assess the effect of an intensive approach to maintain near-normal blood glucose levels and to compare it to conventional therapy that was designed to prevent hyperglycemic symptoms (Diabetes Control and Complications Trial Research Group, 1993). The primary response was retinopathy, damage to the blood vessels in the retina, which is a major complication of diabetes and can lead to visual impairment and blindness. The trial had two cohorts: a Primary Prevention cohort consisting of persons who had no retinopathy at the time of study entry, and a Secondary Intervention cohort, whose members had some degree of retinopathy at entry. The trial showed that the intensive therapy was associated with a significant reduction in the onset and progression of retinopathy as well as nephropathy (nephropathy is the damage of the micro vessels in the kidney which compromises the ability of the kidney to filter waste). After termination of the trial in 1993, 1375 of the 1441 subjects joined an observational study called the Epidemiology of Diabetes Interventions and Complications study (EDIC), which began in 1994. The extent of retinopathy is measured on an ordinal ETDRS scale developed by the Early Treatment Diabetic Retinopathy Study Research Group (1991), which ranges from 1 to 23; level 1 represents no retinopathy and roughly, levels 2 and 3 represent mild retinopathy, levels 4 to 9 represent moderate retinopathy of increasing severity, and levels 10 and over represent increasing degrees of severe retinopathy. Assessment of nephropathy is based on renal function measurements that include urinary albumin excretion rates (AERs) and two states, persistent albuminaria and severe nephropathy, are based on this. Eye and kidney measurements were taken every 6 months in the DCCT and are taken every 2 years in the EDIC study. Numerous other variables are also measured on each individual; the most important is the biomarker glycosylated hemoglobin (HbA1c), which measures average blood glucose over the previous 2 to 3 months; it was measured every 3 months in the DCCT.

For now let us consider retinopathy. In order to study progression over time it is helpful to use multistate models related to ETDRS measurements. It is possible to let each of the levels 1 to 23 on the ETDRS scale represent a state, but a model with 23 states is unwieldy and difficult to fit when individuals are observed at 6-month or 2-year intervals. A simpler model that has proven useful involves five states, defined as follows: State 1 − ETDRS = 1; State 2 − ETDRS = 2 or 3; State 3 − ETDRS = 4−6; State 4 − ETDRS = 7−9; State 5 − ETDRS = 10 or higher (Cook and Lawless, 2014). This balances the need for clinically relevant states with information in the data; for example, there are relatively few ETDRS measurements over 10 in the data and so subdividing state 5 is not feasible. In addition, the primary endpoint for treatment comparisons in the DCCT for persons in the Primary Prevention cohort was progression to an ETDRS score of 4 or higher; subjects in this cohort started in state 1 and this represents an entry to state 3. A decision also has to be made concerning allowable transitions between states. It makes sense to assume that instantaneous transitions are between adjacent states; Figure 1.2 shows state diagrams for two such processes. Model M2 allows transitions only in one direction, which is appropriate if we consider retinopathy as a progressively worsening condition. However, we find many instances in the data where a subject's ETDRS score decreases over two or more successive measurements and so transitions to states representing lower degrees of retinopathy are observed. In this case model M1 is a more accurate representation of the observed data. As we discuss later, it would still be possible to adopt model M2 as representing the true underlying degree of retinopathy and to consider the observed states as resulting from measurement or classification error. The decision as to whether to consider unobserved latent states as part of the model should be based on a careful assessment of the disease and measurement processes. We consider this data in applications in Sections 5.2.4 and 6.4.2 where we compare these approaches and the inferences one can draw.

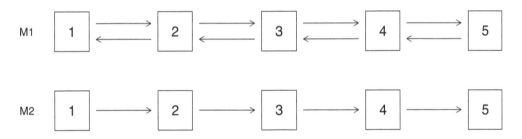

Figure 1.2: State space diagram for a reversible (M1) and a progressive (M2) multistate model for ETDRS score of the degree of retinopathy.

1.2.3 Joint Damage in Psoriatic Arthritis

Psoriatic arthritis (PsA) is an immunological disease in which persons experience pain, inflammation and ultimately destruction of joints in the body. The University of Toronto Psoriatic Arthritis Cohort is associated with a tertiary referral centre at the Toronto Western Hospital that treats patients with various rheumatic diseases, and since 1976 has maintained a clinic registry of persons with PsA (Gladman and Chandran, 2011). Persons undergo a detailed examination and provide serum samples upon entry to the clinic. They are then nominally assessed annually with respect to levels of joint inflammation, joint damage, functional ability, biomarker levels and other factors. Radiological examination of joints is undertaken every second year. Data that we will discuss later in the book are based on about 1200 patients with median follow-up of about 5 years.

Multistate models can be used to address a wide range of issues related to chronic diseases and their effects, and they have been used to study various aspects of PsA. Data in the Toronto PsA cohort are recorded on a total of 64 joints in the body, 28 of which are in the two hands, and one type of analysis is based on classifications of the total number of joints (0−64) with at least some specified level of damage according to an established scoring system (Rahman et al., 1998). For example, we might define states 1−6 to correspond to 0, 1, 2, 3, 4 and 5 or more damaged joints, respectively. Individuals with five or more severely damaged joints are considered to have an aggressive form of the disease called arthritis mutilans. We consider the use of a 6-state model for this process in Section 5.4.5.

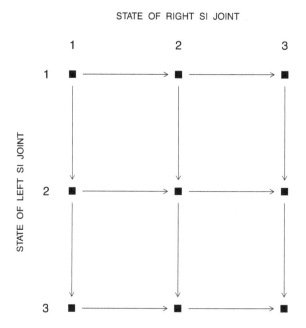

Figure 1.3: State space diagram for the joint process for damage in left and right sacroiliac (SI) joints among patients with psoriatic arthritis; states 1, 2 and 3 represent no damage, mild damage and moderate or more severe joint damage, respectively.

Models are also used to represent progression in the severity of damage for individual joints. Sutradhar and Cook (2008) consider four states of increasing severity of damage based on the modified Steinbrocker score (Rahman et al., 1998). Each of the 28 hand joints were then classified into one of these states at each assessment time, and the data were analyzed to model the development of joint damage. When multiple joints are considered, we can consider association between levels of damage in different joints. Figure 1.3 shows a model used by Cook and Lawless (2014) for modeling damage in the left and right sacroiliac (SI) joints; the three states for each joint again represent levels of severity as determined from radiological examination. Using this model we can assess whether increased damage in one joint is associated with increased damage in the other, and whether any such relationship is symmetrical. As a final illustration, multistate models can combine two or more types of factors. Tom and Farewell (2011), for example, base states on the joint levels of physical disability for an individual patient (none, moderate or severe) and their number of actively inflamed joints (none, 1–5 and 6 or more). We consider the analysis of damage in the sacroiliac joints in Section 6.2.4.

1.2.4 Viral Load Dynamics in Individuals with HIV Infection

The Canadian Observational Cohort on HIV (CANOC) is composed of several Canadian cohorts of HIV-positive persons who initiated combination antiretroviral therapy (cART) since January 1, 2000 (Palmer et al., 2011). Individuals who achieved initial viral suppression (non-detectable presence of virus in the blood) are then nominally followed up at visits approximately every 3 or 4 months, although the times between visits can vary substantially. Upon each individual's entry to the cohort personal information and biomarker measurements are recorded and at the repeat visits, biomarkers such as CD4, CD8 and viral load (VL) counts and blood lipid levels are recorded. Information on clinical events such as AIDS-defining illnesses, cardiovascular events, diagnoses of cancer and death is also obtained. The CANOC data provide insight into disease processes related to HIV infection, associated factors, and the effectiveness of treatment and patient management strategies.

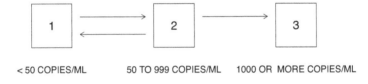

Figure 1.4: Diagram of a model used in an HIV viral load study.

Multistate models are often used to study the dynamics of biomarkers such as CD4 or VL counts. Such markers can be considered effectively as time-varying continuous variables, and one option is to model their paths using stochastic processes with continuous state spaces. For example, Gaussian processes have been used as models for log CD4 counts over time. In many cases, detailed modeling of raw biomarker values is difficult because of short-term variations that are hard to predict, the presence of measurement error and the fact that biomarkers are measured only at intermittent visits. In such contexts it is often helpful to partition the range of biomarkers and to associate a state with each interval in the partition. Figure 1.4 shows a model that has been used to study viral load fluctuations (Lawless and Nazeri Rad, 2015): the three states represent VL counts of less than 50 (deemed as the virus being undetectable), 50–999 and 1000 or more copies per mL. This model has been used to study the occurrence of viral "rebounds" (VR) following viral suppression. In this case we define a viral rebound as occurring when a person's viral load hits 1000 or higher and in order to study times to VR we make state 3 an absorbing state. Models with more than three states could of course also be used and should be considered, but because of short-term fluctuations and the presence of measurement error for VL, it is hard to fit models with too many states when we have widely spaced intermittent measurements. Multistate models are also used for other aspects of HIV disease; for example, we can base states on a combination of biomarker measurements and clinical events in order to study the association between the biomarkers and the events.

1.3   Introduction to Multistate Processes

1.3.1   Counting Processes and Multistate Models

Event history analysis deals with the occurrence of events over time. For a generic individual, suppose that R types of events labeled $r = 1, \ldots, R$ are of interest. It is standard for modeling and analysis to use counting process notation; assuming for convenience that events can occur at times $t \geq 0$, we let $N_r(t)$ denote the number of type r events occurring up to time t. The processes $\{N_r(t), t \geq 0\}$ are called counting processes and together they give a multivariate counting process $\{N(t), t \geq 0\}$ where $N(t) = (N_1(t), \ldots, N_R(t))'$. Models for events and their counting processes in continuous time are specified through intensity functions. Let $\mathcal{H}(t) = \{N(s), 0 \leq s \leq t\}$ denote the history of all events (that is, their types and occurrence times) over $[0, t]$. There may also be covariates or initial conditions at $t = 0$ but we ignore these for now. The intensity function for events of type r is then defined as

$$\lambda_r(t \mid \mathcal{H}(t^-)) = \lim_{\Delta t \downarrow 0} \frac{P(\Delta N_r(t) = 1 \mid \mathcal{H}(t^-))}{\Delta t} \tag{1.1}$$

for $t \geq 0$ where $\Delta N_r(t) = N_r(t + \Delta t^-) - N_r(t^-)$ and $\mathcal{H}(0^-) = \emptyset$. For continuous-time processes we assume that two or more events cannot occur simultaneously, and then the intensity functions for $r = 1, \ldots, R$ fully specify the multivariate event process. It should be noted that (1.1) allows past events of any type to affect new occurrences of type r events, and that the conditional probability of a type r event in the short interval $[t, t + \Delta t)$, given past event history, is approximately $\lambda_r(t|\mathcal{H}(t^-))\Delta t$.

Multistate models in continuous time with state space $\{1, 2, \ldots, K\}$ are specified similarly. With $Z(t)$ denoting the state occupied at time t for a generic individual, transition intensity functions between states are defined as

$$\lambda_{kl}(t \mid \mathcal{H}(t^-)) = \lim_{\Delta t \downarrow 0} \frac{P(Z(t + \Delta t^-) = l \mid Z(t^-) = k, \mathcal{H}(t^-))}{\Delta t} \tag{1.2}$$

for $k \neq l$, where now $\mathcal{H}(t) = \{Z(s), 0 \leq s \leq t\}$ and $\mathcal{H}(t^-)$ represents the history of state occupancy over $[0, t)$. Multistate models can be represented as counting processes by considering the types of transitions allowed as different types of events; we will use notation that reflects this in Chapter 2 and elsewhere. As for general event processes, we assume that only one event (transition) can occur at a given instant.

Multistate models in discrete time can also be considered. We will focus mainly on continuous-time models in this book but mention the discrete case for future reference. Assume that $\{Z(t), t = 0, 1, 2, \ldots\}$ is a process where states are observed only at discrete times $t = 0, 1, 2, \ldots$, and let $\mathcal{H}(t) = \{Z(0), \ldots, Z(t)\}$ for $t \geq 0$. Transition probabilities (or discrete intensities) are now defined as

$$\pi_{kl}(t \mid \mathcal{H}(t-1)) = P(Z(t) = l \mid Z(t-1) = k, \mathcal{H}(t-1)) \tag{1.3}$$

for $k = 1, \ldots, K$, $l = 1, \ldots, K$. We note that $\sum_{l=1}^{K} \pi_{kl}(t \mid \mathcal{H}(t-1))$ must equal one for each $k = 1, \ldots, K$.

Multistate models in continuous time are formulated by specifying the transition intensity functions for allowable transitions. Thus, for the process in Figure 1.3 we need to specify $\lambda_{12}(t \mid \mathcal{H}(t^-))$, $\lambda_{21}(t \mid \mathcal{H}(t^-))$ and $\lambda_{23}(t \mid \mathcal{H}(t^-))$. Certain types of multistate process have been thoroughly studied as stochastic processes. These include Markov processes, for which the dependence on the history is only through the state currently occupied so $\lambda_{kl}(t \mid \mathcal{H}(t^-)) = \lambda_{kl}(t)$, and semi-Markov processes, for which $\lambda_{kl}(t \mid \mathcal{H}(t^-)) = h_{kl}(B(t))$ with $B(t)$ the time since entry to the current state, k. In a given context these or other models can be considered for each of the allowable transitions. Intensities can also be specified in such a way that they depend on covariates, in which case we expand the meaning of $\mathcal{H}(t)$ to include all relevant covariates; in the case of fixed covariates represented by X then $\mathcal{H}(t) = \{Z(s), 0 \leq s \leq t; X\}$. Intensities in which covariates act in a multiplicative or additive fashion are both useful. For Markov models the former take transition intensities of the form

$$\lambda_{kl}(t \mid \mathcal{H}(t^-)) = \lambda_{kl}(t \mid x) = \lambda_{kl0}(t)\, g(x; \beta_{kl}), \tag{1.4}$$

where x and β_{kl} are vectors of covariates and regression coefficients, respectively, and the function $g(x; \beta_{kl})$ is constrained to be positive. A common approach is to use $g(x; \beta) = \exp(x'\beta)$, in which case the $\lambda_{kl0}(t)$ are termed *baseline intensities* which apply for an individual with $x = 0$. Additive regression models, on the other hand, take a form such as

$$\lambda_{kl}(t \mid \mathcal{H}(t^-)) = \lambda_{kl0}(t) + g(x; \beta_{kl}).$$

A mild inconvenience in the use of additive models is the need to constrain the two model components so that the intensity is non-negative.

1.3.2 Features of Multistate Processes

Various features of a multistate process may be of interest. Transition intensities are fundamental; they describe the instantaneous risk of a change in the process by specifying how the probability of a transition occurring over a short time interval depends on the process history up to that time. Transition probabilities over longer periods of time are also important; they take the form

$$P_{kl}(s, t \mid \mathcal{H}(s^-)) = P(Z(t) = l \mid Z(s) = k, \mathcal{H}(s^-)) \tag{1.5}$$

for k and $l = 1, \ldots, K$ and $s \leq t$. When individuals must be in state 1 at $t = 0$, the probabilities $P_{1l}(0, t \mid \mathcal{H}(0))$ for $t > 0$ and $l = 1, \ldots, K$ are often called prevalence or occupancy probability functions. The durations of sojourns in certain states or the time until a specific state is first entered are also often of interest. In the context of Section 1.2.2 and Figure 1.2 for individuals in state 1 (no diabetic retinopathy) at the initiation of intensive therapy for blood glucose control, the prevalence functions, for example, give the probability a person will have no, mild, moderate or severe retinopathy t years later. Similarly, we can consider the length of time until a person experiences some level of retinopathy (that is, leaves state 1). The effects of fixed or time-dependent treatments or covariates on such probabilities are the focus of many studies.

In Chapter 2 we present ways to obtain process features such as transition probabilities and sojourn time distributions. These features can in principle be determined from the intensity functions for a process, but for some processes direct mathematical calculation is untractable. In such cases simulation is a feasible alternative. The following examples illustrate the types of features discussed for some simple processes.

Example 1.3.1:   The Illness-Death Model

In Section 1.2.1 we introduced the 3-state illness-death model (Figure 1.1). Markov models are mathematically very tractable and if we denote the transition intensities as $\lambda_{12}(t)$, $\lambda_{13}(t)$, $\lambda_{23}(t)$, then the following can be obtained:

$$P_{11}(0,t) = \exp\left(-\int_0^t (\lambda_{12}(u) + \lambda_{13}(u))\, du\right).$$

This result follows from the relationship between hazard and survivor functions in survival models. In this case we let T denote the time of exit from state 1 and note that T has hazard function $\lambda_{12}(u) + \lambda_{13}(u)$.

Transition probabilities such as $P_{12}(s,t)$ or $P_{13}(s,t)$ are slightly more complicated and will be described in Chapter 2. Sojourn time probabilities for state 2 are also straightforward if we exercise a little care. We simply need to condition on the time s of entry to state 2: thus, we find

$$P(\text{still in state 2 at time } t \mid \text{entry at time } s) = \exp\left(-\int_s^t \lambda_{23}(u)\, du\right) \qquad (1.6)$$

by considering exit from state 2 as a "failure". The expression (1.6) gives us the probability a sojourn in state 2 exceeds $t - s$, given that it started at time s.

Example 1.3.2:   Competing Risks Models

Competing risks is the term used to describe situations in which an individual may die or fail from different causes, or simply have different modes of failure. Figure 1.5 portrays a model with three modes; for convenience we label the states here as $0, 1, 2, 3$. This figure would describe, for example, a setting where a person with end-stage renal disease (ESRD) is wait-listed for a kidney transplant (Gaston et al., 2003) and where they can leave the wait list in three ways: by receiving a transplant (mode 1), by death (mode 2), or by being de-listed for some reason (mode 3).

As in the preceding example, let us consider a Markov model and assume that $t = 0$ corresponds to the time the person is put on the wait list. Covariates such as age and factors related to the person's ESRD would be relevant but for simplicity we suppress their notation and denote the three transition intensities as $\lambda_{01}(t)$, $\lambda_{02}(t)$ and $\lambda_{03}(t)$. By the same reasoning as in Example 1.3.1, we regard the time of exit

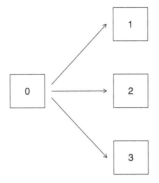

Figure 1.5: A multistate diagram for a competing risks model with three modes of failure.

from state 0 as a failure time. Its hazard function is $\lambda_{01}(u) + \lambda_{02}(u) + \lambda_{03}(u)$, and therefore

$$P_{00}(0,t) = \exp\left(-\int_0^t (\lambda_{01}(u) + \lambda_{02}(u) + \lambda_{03}(u))\, du\right) \tag{1.7}$$

corresponds to the survivor function. We also note that

$$P_{0k}(0,t) = \int_0^t P_{00}(0,u^-)\,\lambda_{0k}(u)\, du\,, \quad k = 1,2,3 \tag{1.8}$$

since if a person is in state k at time t, there must be some time $u \leq t$ for which they were in state 0 at u^- and then made the transition to state k at u. In the competing risks literature the functions (1.8) are called cumulative incidence functions, since they give the cumulative probabilities of failures of each mode, up to time t.

1.3.3 Marginal Features and Partial Models

Given a complete specification of a multistate process through its intensity functions, we can at least in principle determine features such as transition probabilities or average sojourn time in a state. In some cases, however, we may prefer to avoid modeling the entire process and just to specify a "partial" model that addresses a particular feature of interest, as we discuss later in Section 1.4.2. We sometimes refer to features that are not conditional on previous life history as marginal features, and they are a common target for this approach. For example, suppose that in a given process we are interested in prevalence or occupancy probabilities $P_k(t) = P(Z(t) = k)$ for certain states k. This might apply, for example, to the diabetic retinopathy process in Figure 1.2 or the competing risks process in Figure 1.5. We note that

$$P_k(t) = \sum_{l=1}^K P(Z(0) = l)\, P_{lk}(0,t)\,.$$

However, it is also possible to estimate $P_k(t)$ without using a specific multistate model; for example, if all individuals in a cohort are in state 1 at $t = 0$ and if none are lost to follow-up before time t, then we can simply use the proportion of persons in state k at time t as our estimate. A number of methods of estimating prevalence probabilities have been developed and will be discussed later in Section 3.4; these methods can deal with settings where individuals may be prematurely lost and therefore censored.

More generally, features of interest may not be strictly marginal, but more accurately described as partially conditional. In the case of prevalence probabilities, for example, we may want to condition on certain baseline covariates $x(0)$, and estimate $P_k(t|x(0)) = P(Z(t) = k|x(0))$. This can also be applied to transition probabilities. For example, we might want to consider $P(Z(t) = l|Z(s) = k, x(s))$ for a specified set of states l and covariates x, without conditioning on the full process history up to time s. This approach to modeling is often adopted when individuals are under intermittent observation and hence state transitions and covariate paths are incompletely observed.

The durations of sojourns in the states of a progressive process such as model M2 in Figure 1.2 are other marginal features of possible interest, as is the total time spent in a state k over a period $[0,t]$ given by $S_k(t) = \int_0^t I(Z(u) = k) \, du$. We let

$$\psi_k(t) = E\{S_k(t)\} = \int_0^t P_k(u) \, du,$$

denote the expected total sojourn time in state k over $[0,t]$, which is a quantity that has a role in utility-based analyses we discuss in Sections 4.2.3, 8.1 and 8.4. We stress later that different methods of estimation may involve different assumptions about observation times or censoring processes. For example, some methods require censoring or loss to follow-up to be independent of the life history process, given any covariates that are conditioned upon.

1.4 Some Aspects of Modeling, Analysis and Design

1.4.1 Objectives

Objectives associated with life history studies may vary considerably but frequently they include one or more of: increasing the understanding of individuals' processes and of variation across individuals, groups or populations; identifying and characterizing relationships between processes and covariates, or between two or more processes; identifying risk factors associated with adverse outcomes; assessing the effectiveness of individual or population level interventions; and developing predictive models that can be used for activities such as resource allocation, policy formulation and patient management. In order to achieve objectives it is important that there be due attention to the selection of individuals for a study, to definitions and measurement issues associated with states, event occurrence times and covariates, and to the careful follow-up of individuals over some specified time period. In randomized trials these features are usually controlled and in addition, outcomes are well defined. However, in observational studies there are often measurement error, data completeness or data quality issues, and the selection and follow-up of individuals may be subject to conditions that make it difficult or impossible to realize certain objectives. In some cases such features can be addressed by careful modeling of both the life history and data collection processes, but there is no substitute for high quality data. We comment on some of these issues in the following two sections.

1.4.2 Components of a Model

We deal in this book with settings where dynamics of a life history process can be characterized using a multistate model. The components of a multistate model have already been described, from the definition of states and allowable transitions to the specification of transition intensity functions. The definitions of states and the forms specified for the intensities depend very much on context and to some degree on study objectives. For example, in the study in Section 1.2.2 concerning complications from diabetes, a major concern is to assess rates of progression for complications such as retinopathy and nephropathy. Progression can be described in terms of transition intensities and also, over time, transition probabilities; for example, if a person has a mild degree of retinopathy we can consider the probability they will progress to moderate or severe retinopathy within the next 3 years. It is important that models reflect process dynamics and measurement realistically, but it is also important that the models allow the calculation of transition probabilities, and in many applications, that we be able to fit them using intermittent observations of state occupancy. Markov models have major advantages in this regard and as we will describe subsequently, most of the analysis of intermittently observed data is based on Markov models. In Chapter 2 we describe maximum likelihood methods for model fitting, and ways to assess covariate effects and modeling assumptions. We also stress that although Markov models have advantages in many settings, there are settings where other types of models are needed. For example, the transition intensities out of a transient state may depend strongly on time since entry to the state; for example, this can occur when the state represents diagnosis with an especially lethal disease, or when the state represents a period of physical disability whose duration is not highly variable. In many settings, some transition intensities may depend most strongly on process "age" (that is, on t) and others may depend mainly on elapsed time in the state.

Transition intensity models attempt to account for the effects of time and of prior process history on transitions. The effects of fixed covariates x on the other hand are used to describe differences in processes across individuals; we can think of this as "observed" heterogeneity. Similarly, external time-dependent covariates, which evolve independent of individuals, may affect an individual's multistate process. For example, the occurrence of transitions into states representing episodes of severe asthma in children may depend on local air pollution counts. Internal time-dependent covariates also may appear in models, representing variables such as biomarker values or aspects of a person's prior life history. We sometimes find that even after incorporating covariates, the life histories of individuals are more variable than expected. Models that incorporate "unobserved" heterogeneity by specifying latent (unobserved) random effects can be considered in this case. This can be done in various ways. One is through the frailty idea used in survival models; in the multistate context this might involve continuous random effects for specific transition intensities. This can be complicated because one might expect random effects for different transition intensities to be correlated. Another approach involves finite mixture models; in this case we assume that there are G classes of individuals, with each class having its own set of transition intensities. The number of classes G usually has to be small to facilitate model fitting. One type that has been studied a good deal involves two classes consisting of "movers" and "stayers": the latter are individuals who tend not to make transitions, whereas the former make frequent transitions. A third type of model involves latent processes; in this case we assume there is an underlying life history process that is unobservable and that the observable process is linked to it in some specific way. Hidden Markov models are a well-known example in which the underlying process is Markov. They are often used, for example, in cases where classification errors can occur in determining what state a person is in. Then, it is convenient to let $Z(t)$ represent the true state at time t and to assume that given $Z(t)$, the observed state $W(t)$ is governed by a model for the misclassification errors. Models that accommodate heterogeneity will be discussed in subsequent chapters, beginning with Section 2.3.4.

Another modeling feature arises when multistate processes occur within groups or clusters such that processes within a cluster display association. In Section 1.2.3, for example, we considered pairs of joints on the left and right sides of persons with psoriatic arthritis. It is to be expected that the levels of damage in the left and right joints will be correlated. Another situation arises in family studies where we consider multistate models for each family member under observation. In this case shared genetic or environmental factors may produce association among their processes. We consider ways to model and analyze correlated multistate processes in Chapter 6.

We conclude this section with some remarks about the level of detail in models. Intensity functions provide a complete probabilistic structure for a process, from which any process features of interest can be determined. All models are approximations of reality, of course, and rather different model specifications may agree well with observed data. When continuous data on individuals are available it is possible to assess the adequacy of a model fairly well, though the presence of covariates complicates matters. When data on individuals are incomplete in certain respects, as when observations are intermittent, model checking becomes more difficult. Partially specified models mentioned in Section 1.3.3, in which we describe certain features, but not the full probabilistic structure of a process, are often useful. For example, we might choose to model only prevalence probabilities in a setting like the one in Section 1.2.2, where individuals all start in the same state and disease progression is the primary interest. We discuss estimation of prevalence functions based on partially specified models in Section 3.4 and subsequent Sections 4.1 and 4.2. Model checking techniques are illustrated in various examples.

1.4.3 Study Design and Data

Life history studies can select individuals and collect data in a variety of ways. A crucial distinction is between purely retrospective studies and purely prospective studies. The simplest case of the former is when individuals are selected at some point in time, at which data on past events and other variables are obtained. Frequently the individuals must satisfy some condition in order to be selected. For example, Andersen and Green (1985) describe a study designed to estimate the incidence function for diabetes mellitus (DM) in children up to the age of 20. The study members were selected from the Danish National Conscript Registry of 20 year olds who were born in the years 1949–1956. For the 20 year olds who were sampled, it was determined whether they had DM and if so, at what age it was diagnosed. To be in the registry, however, a person had to have survived to the age of 20 and to have not emigrated from Denmark. The process of interest here can be represented by an illness-death model as in Figure 1.1, with state 2 representing DM and state 3 death or emigration. In terms of this model, the selection criterion is thus that a person had to be in states 1 or 2 at age 20.

A purely prospective study is one in which study members are followed for some time period after their selection. This time period may in principle be fixed, but random events such as deaths or losses to follow-up can occur. In some studies the individuals are a random sample from a population of interest. For example, the Canadian Longitudinal Study on Aging mentioned in Section 1.1 is a national longitudinal study of adult life histories and aging. Recruitment for a stratified random sample of 50,000 persons aged 45–85 years began in 2009 (Raina et al., 2009); the plan is to follow individuals for 20 years or to death, with formal assessments every 3 years. Some prospective studies involve randomized interventions or treatment assignment; clinical trials described in Section 1.2.1 are an example of this. Observational studies can also be based on administrative data, disease clinic data and so on. For example, individuals who enroll in the psoriatic arthritis clinic described in Section 1.2.3 are followed prospectively until death or loss to follow-up, with data collected at intermittent visits.

Although the studies just described involve prospective follow-up of participants, the selection of individuals may depend on aspects of their prior life history, even in randomized intervention studies. For example, Cook and Lawless (2007, Section 6.7.2) describe a randomized trial on persons with chronic bronchitis, in which a 2-state model was involved. One state represented an acute bronchial exacerbation and the other represented the exacerbation-free state. The purpose of the trial was to compare an experimental therapy with standard care for the treatment of exacerbations. An eligibility criterion was that at the time of selection for randomization of treatment, a person had to be in the exacerbation state. More complex forms of selection can occur in family studies of disease, where one or more family members may be required to satisfy certain conditions.

Studies can also involve both retrospective and prospective outcomes. For example, in studies of the onset and progression of dementia we might select individuals in a certain age range and follow them for a period of time, as in the Canadian Longitudinal Study on Aging. However, if a person has already experienced onset and perhaps some progression before their time of selection, we might attempt to use this information as well, assuming that there are records from which the age at onset and data on progression prior to selection can be ascertained. The need for reasonably accurate determination of past events and other relevant data must be kept in mind when we consider using retrospective outcomes. Other conditions such as the requirement that a study member be alive at their time of selection can also affect how retrospective data must be treated.

A final key issue in a life history study concerns the collection of data. In most prospective studies data are collected at intermittent "visits" for an individual; in the case of the HIV cohort described in Section 1.2.4, for example, individuals are in principle seen every 3 or 4 months. Variables that require special measurement or diagnostic techniques such as radiological examination can only be obtained at visit times. In some cases the times of certain events or state changes occurring between two visit times can be obtained; for example, exact times of entry to states such as death, disease diagnosis or organ transplantation will usually be known. The exact times of transitions between states such as those in Figure 1.2 are not ascertainable, however; in this case observations on the level of retinopathy are based on photographs of the eyes which are taken at visit times. In studies where the visit times vary considerably both across and within individuals, we may also have to consider the possibility that visit times depend in some way on previous life history, and to allow for this in modeling and analysis. Losses to follow-up can also be dependent on previous life history.

We consider aspects of the selection and follow-up of individuals throughout the book. In Chapters 3 and 4 we develop methods for situations where complete prospective data on transitions and on covariates are available; this allows for a wide range of models to be investigated conveniently. Chapter 5 deals with situations in which occupancy for some states is only known with certainty at visit times for individuals; we will show how this restricts the models that can be handled. Chapter 6 deals with heterogeneous, clustered, or otherwise dependent multistate processes, as well as models that accommodate one or more latent process features. Chapter 7 deals with retrospective or partially retrospective studies in which the sampling schemes for individuals involve selection criteria. Two-phase studies in which certain variables are observed only for a selective subset of a cohort are also introduced. Chapter 8 covers additional miscellaneous topics related to the use of multistate processes in specific contexts.

1.5   Software

We will soon see a strong connection between the approaches to modeling and inference for survival analysis and those for multistate and more general life history analyses. This connection means that many multistate models can be fitted using software that has been developed for survival analysis. We focus on software in R and TIBCO Spotfire S+ in this book; many new packages have recently been developed in R. In S-PLUS the key function for parametric survival analysis is `censorReg` which is based on the location-scale (accelerated failure time) parameterization. The `eha` library of functions in R written by Göran Broström handles parametric proportional hazards models, including flexible models with piecewise-constant baseline hazard functions. The function `coxph` in R and S-PLUS handles semiparametric proportional hazards (i.e. Cox) models, but the full breadth of problems it can handle is well illustrated in Therneau and Grambsch (2000). The R package `timereg` enables one to fit additive regression models; see Martinussen and Scheike

(2006) for a comprehensive account of the associated models, theory and illustrative applications.

Several packages have been written in R specifically to handle multistate data. We mention some of them here; a more complete list and description is given in Appendix A. Specific packages will be used throughout the book, and their capabilities will be discussed. The cmprsk library contains a suite of functions for competing risks analysis including semiparametric models. The prodlim package provides nonparametric estimates in competing risks settings. The package mvna can be used to obtain the nonparametric (Nelson-Aalen) estimates of cumulative transition rates in a Markov model and the etm package can be used to obtain nonparametric estimates of the transition probability matrix. The mstate package handles right-censored data and permits nonparametric estimation as well as fitting semiparametric Cox models for intensity functions. The msSurv package estimates transition probabilities as well as state entry and exit time distributions.

When multistate processes are subject to intermittent observation, parametric models are typically adopted. Jackson (2011) describes the breadth of problems that can be tackled with the msm package, which include Markov models with piecewise-constant baseline transition rates and hidden Markov models. Methods of estimation and model assessment are available, and msm will also fit Markov models to continuously observed data. Other packages address specific types of models. For example, Joly et al. (2002) describe a package SmoothHazard, which enables one to fit parametric or "semiparametric" intensity models for survival data or data from the 3-state illness-death model, in which the transition time to the intermediate state may be interval censored. The semiparametric approach is based on M-splines and estimation is carried out based on penalized likelihood.

When planning an analysis of continuously observed data from a multistate process in R or S-PLUS, dataframes are typically organized in a "counting process" format (Therneau and Grambsch, 2000) similar to that used for survival data. For a multistate process there are multiple lines (rows) in the dataframe that correspond to separate "at-risk" periods for separate types of events (transitions). The columns in the dataframe include identifiers for the types of events, and start and stop (end) times for the period at risk. A column usually labeled "status" indicates whether the event in question occurred at the end of the time period (status = 1) or did not (status = 0). There are also columns for identifying individuals and for covariate values, which we assume are fixed over at-risk periods. The following examples include some sample dataframes, and more will be encountered throughout the book.

1.6   Introduction to Some Studies and Dataframes

1.6.1   A Trial of Breast Cancer Patients with Skeletal Metastases

Here we consider data from an international double-blind phase III randomized trial of patients with breast cancer metastatic to bone (Hortobagyi et al., 1996). Patients were randomized to receive monthly infusions of a bisphosphonate (pamidronate) or a placebo and followed for up to 2 years in order to record the occurrence of skeletal complications arising due to the bone metastases.

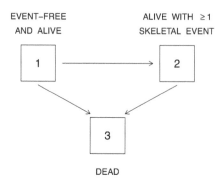

Figure 1.6: A 3-state diagram for a joint model of the onset of a skeletal event and death in a trial of breast cancer patients with bone metastases.

Bisphosphonates are a class of drugs that strengthen bone and are thought, therefore, to reduce the occurrence of skeletal complications. These skeletal complications include symptomatic or asymptomatic fractures, bone pain requiring radiotherapy, or surgical intervention to treat or prevent fracture. We focus here on the occurrence of the first skeletal complication since randomization. Since the patients have metastatic cancer, they are at high risk of death; the mortality rate is approximately 10% over 2 years in the control arm.

The 3-state illness-death model discussed in Section 1.2.1 offers a suitable representation of the possible outcomes for individuals, which we show explicitly in Figure 1.6. Here the initial state 1 represents the condition of being alive without having experienced a skeletal complication; individuals enter state 2 upon experiencing a skeletal complication and enter state 3 upon death. The time origin $t = 0$ corresponds to the time an individual is assigned to a treatment, and subsequent times are recorded in days.

The data for this study can be recorded in a variety of forms. We write it below in a dataframe called mbc with the acronym standing for "metastatic breast cancer".

```
> mbc
  id start stop from to status trt gtime
   1     0   68    1  2      0   0    68
   1     0   68    1  3      1   0    68
   2     0  749    1  2      0   0   749
   2     0  749    1  3      0   0   749
   5     0  127    1  2      1   1   127
   5   127  365    2  3      0   1   238
   5     0  127    1  3      0   1   127
  24     0   83    1  2      1   0    83
  24    83  639    2  3      0   0   556
  24     0   83    1  3      0   0    83
  69     0   39    1  2      1   0    39
  69    39  637    2  3      1   0   598
```

69   0   39   1   3   0   0   39

The variable id is the label associated with each individual, denoted by i in the notes. As will be seen in Section 2.2, the sample path for an individual may be broken down into segments defined by the start and stop times for periods when the individual is at risk of specific transitions. The variables start and stop contain the left and right endpoints of these time intervals, and the status variable indicates whether or not this interval ends with a transition out of the state recorded in the from column to the state in the to column. Note that because individuals can be at risk for transitions from a given state into two or more states at any given time, there may be multiple lines in the dataframe corresponding to each at-risk period for an individual. The variable trt is the treatment indicator which is 1 for individuals assigned to receive pamidronate and 0 otherwise. Finally the variable gtime is the observed sojourn time for the state in the from column; when from = 1, the gtime variable is equal to the (possibly censored) exit time from state 1, but when from = 2, it is the sojourn time W_2 in state 2 or its censored value. Consider the data for individual 1, for example, who begins in state 1 at $t = 0$ at which point they are at risk of both a $1 \rightarrow 2$ transition and a $1 \rightarrow 3$ transition; there are two lines of data reflecting this. The status column in the row with to = 3 and the fact that stop = 68 indicate that at $t = 68$ days the individual made a $1 \rightarrow 3$ transition (that is, they died). Subject 5, on the other hand, made a $1 \rightarrow 2$ transition at $t = 243$ days. They were then at risk of a $2 \rightarrow 3$ transition from $t = 243$ to $t = 365$ days; at $t = 365$ they made a $2 \rightarrow 3$ transition upon death.

1.6.2 An International Breast Cancer Trial

Here we consider a randomized clinical trial of adjuvant chemotherapy for breast cancer that was conducted by the International Breast Cancer Study Group (IBCSG). This study investigated the effectiveness of short-duration (1 month) and long-duration (6 or 7 months) chemotherapy (Gelber et al., 1995). A total of 1229 patients were randomized to treatment with 413 allocated to the short-duration treatment and 816 to the long-duration treatment. The median follow-up time was about 7 years. These data have been the subject of various quality-of-life analyses, based on the 4-state progressive model displayed in Figure 1.7.

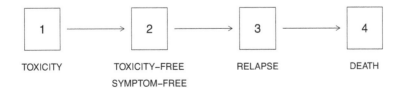

Figure 1.7: A 4-state progressive model for the breast cancer trial.

State 1 is occupied by individuals experiencing treatment-related toxicity, state 2 by those who are both toxicity-free and symptom-free, state 3 is entered upon relapse, and death is represented by a transition to state 4. For the following discussion we exclude 16 individuals for whom one or more state transition times were missing, leaving 411 and 802 subjects in the short- and long-duration groups, respectively.

A portion of the dataframe containing the information on transitions is given below. Time is in months since the initiation of treatment; all individuals are in state 1 at $t = 0$ since the chemotherapy they receive induces toxicities.

```
> bc
  id start   stop status from to trt
   1  0.00   7.00      1    1  2   1
   1  7.00 114.84      0    2  3   1
   2  0.00   3.00      1    1  2   0
   2  3.00  72.76      1    2  3   0
   2 72.76 113.82      0    3  4   0
   6  0.00   8.00      1    1  2   1
   6  8.00 113.03      0    2  3   1
   7  0.00   8.00      1    1  2   1
   7  8.00  32.50      1    2  3   1
   7 32.50  33.91      1    3  4   1
```

This dataframe has a counting process structure with a patient identification number id, a treatment indicator trt, and variables start and stop giving the left and right endpoints of the periods at risk. However, since individuals can only enter state $k + 1$ from state k, $k = 1, 2, 3$, it is only necessary to include either a from variable or a to variable, but we include both here for clarity and for consistency with how dataframes for more complex models are constructed.

We consider this study further in Section 4.2.3 where Figure 4.5 displays Kaplan-Meier estimates of the cumulative distribution functions for the times of entry T_k ($k = 2, 3, 4$) to states 2, 3 and 4 for the short- (left panel) and long- (right panel) duration chemotherapy groups, respectively. It is apparent, and not surprising, that the median duration of toxicity is higher in the arm where the chemotherapy is given for a longer time. There is also a trend apparent for a longer time to relapse and death in this arm.

1.6.3   Viral Rebounds in HIV-Positive Individuals

The CANOC collaboration mentioned in Section 1.2.4 involves the follow-up of HIV-infected individuals from $t = 0$, the point of viral suppression from the initial administration of combination antiretroviral therapy (cART). The states are defined in Section 1.2.4; all individuals begin in state 1, and time is measured in years. In this case, individuals are observed only intermittently at visits that vary in frequency both within and across individuals. The vtime variable contains the times at which individuals are observed, and blood measurements taken. At a given visit time, a variable vload records the HIV viral load in copies/mL; this variable determines the variable state that records the actual state membership. The covariates shown here are fixed and therefore do not change value across the multiple lines provided by an individual. These include gender (1 = male, 2 = female), cat.age (1 if ≥ 45 years, 0 if < 45 years), cart.year recording the year cART began, and part, an indicator of whether the individual engaged in sexual relations with men (1 if yes, 0 if no).

1.6.4 Viral Shedding in HIV Patients with CMV Infection

In the AIDS Clinical Trials Group study ACTG 181, blood and urine samples were periodically provided by individuals with cytomegalovirus infection (Betensky and Finkelstein, 1999; Goggins and Finkelstein, 2000). Laboratory tests were performed to determine whether there was any evidence of cytomegalovirus (CMV) in the samples, which would mean that the infection had progressed to the point of viral shedding. Urine samples of patients were to be drawn every four weeks, while blood samples were to be drawn every twelve weeks at clinic visits.

We can associate the failure time T_1 with the onset of viral shedding in the urine and another time T_2 with the onset of viral shedding in the blood. Because the urine and blood samples were only available occasionally, the times of each of these events are interval censored and known only to lie between the last time a sample yielded a negative test result and the first time a sample yielded a positive result; for individuals with a negative final test, the respective time to viral shedding is right censored at the time of this test.

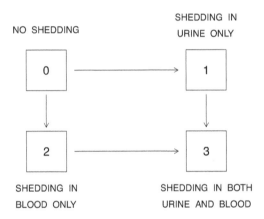

Figure 1.8: A 4-state model for joint consideration of the time to viral shedding in the bloodstream and urine.

Interest lies in modeling the joint distribution of the times to viral shedding in the urine and blood in order to (i) help characterize the onset time distribution for viral shedding of either type, (ii) learn whether evidence of viral shedding is detectable in the urine or blood first and (iii) learn about how shedding in one type of sample is related to the risk of shedding in the other type of sample. These questions are conveniently studied through specification of a 4-state model as depicted in Figure 1.8 where the initial state is labeled 0 to correspond to the absence of any viral shedding. A $k \to k+1$ transition represents the appearance of CMV virus in the urine ($k = 0$ or 2), and a $k \to k+2$ transition represents the appearance of CMV virus in the blood ($k = 0$ or 1). We can address objective (i) by considering $F(t) = P(\min(T_1, T_2) \le t) = 1 - P_{00}(0, t)$. Objective (ii) can be addressed by removing state 3, and then comparing $P_{01}(t)$ and $P_{02}(t)$ for $t > 0$. Objective (iii) can be considered by comparing the transition intensities (a) $\lambda_{01}(t|\mathcal{H}(t^-))$ and $\lambda_{23}(t|\mathcal{H}(t^-))$, and (b) $\lambda_{02}(t|\mathcal{H}(t^-))$ and $\lambda_{13}(t|\mathcal{H}(t^-))$.

1.7   Bibliographic Notes

Life history analysis is a fascinating field that involves the selection, fitting and assessment of stochastic models to data on the occurrence of events over a period of observation. As such it is helpful to have familiarity with the basics of stochastic processes; Cox and Miller (1965) give an excellent introduction to stochastic modeling in discrete and continuous time, as do Bartlett (1978) and Ross (1996). As pointed out in this chapter and as will be more evident shortly, models and methods for survival analysis play a central role in the analysis of multistate and more general life history processes. Cox and Oakes (1984) provide an introduction to the field. The most recent editions of Kalbfleisch and Prentice (2002) and Lawless (2003) have integrated much material on more recent developments including methods for competing risks analysis, multistate modeling and recurrent event analysis. The third edition of Collett (2015) has likewise been greatly expanded and offers similar coverage. Klein and Moeschberger (2003) discuss issues related to truncated data that are relevant for the analysis of multistate processes. The common structure of likelihoods for many types of problems in life history analysis can be exploited from a computational standpoint and this is well illustrated in the book by Therneau and Grambsch (2000). Counting processes are the basis of the presentation of survival analysis in Fleming and Harrington (1991), who use martingale theory for the derivation of asymptotic results. Andersen et al. (1993) likewise give a presentation in terms of martingales for a broad range of problems and methods including recurrent event and multistate processes. Aalen et al. (2008) cover a breadth of topics pertinent to life history analysis including causal inference, time-dependent covariates, study design and the effects of heterogeneity on model features. Keiding (2014) gives a brief but authoritative account of the general field of event history analysis. Martinussen and Scheike (2006) emphasize the utility of additive models developed by Aalen (1989) in the survival and more general life history context.

Multistate analysis has a long history; see, for example, references in Kalbfleisch and Lawless (1985), Andersen et al. (1993) and Yang (2013). Hougaard (1999) and Andersen and Keiding (2002) give reviews of multistate analysis and the journal issue containing the latter paper has some additional articles. A tutorial article by Putter et al. (2007) illustrates how to fit models with right-censored data in R. A more recent paper by Andersen and Keiding (2012) gives important guidance on ensuring that functionals of fitted multistate models are interpretable. Crowder (2001) gives a full treatment of the competing risks problem, and Hougaard (2000) and Crowder (2012) cover multivariate survival analysis and some multistate analyses. Cook and Lawless (2007) deal with recurrent events, but there are numerous connections with multistate models. More recently, there have been some textbooks devoted exclusively to multistate analysis including Beyersmann et al. (2012) and Willekens (2014), who illustrate how to fit competing risks and multistate models using R.

While the literature on frailty models is extensive in the survival context (Hougaard, 2000; Duchateau and Janssen, 2008; Wienke, 2011), less work has been done on modeling heterogeneity in the context of multistate models. We review and extend methodology in this area in Chapter 6, but for examples of random effects modeling see Ng and Cook (1997), O'Keeffe et al. (2013) and Lange and Minin (2013).

When event or transition times are not observed in the survival context they may be interval censored. Sun (2006) gives a textbook treatment of this area, and recent developments for more specialized problems are given in Chen et al. (2013). Gómez et al. (2009) give a tutorial on fitting models to interval-censored data in R. The book by Sun and Zhao (2013) considers recurrent event processes that are intermittently observed, yielding interval-count data. Van den Hout (2017) considers multistate models with interval-censored transition times. Jackson (2011) highlights the utility of the `msm` package in R for analyzing multistate processes under intermittent observation. This suite of functions continues to be developed, and it accommodates a number of complications including misclassification of states via the use of hidden Markov models (Jackson and Sharples, 2002).

1.8   Problems

Problem 1.1   Consider the illness-death model of Figure 1.1 and Example 1.3.1 where $Z(0) = 1$.

(a) For the Markov case with intensities $\lambda_{12}(t), \lambda_{13}(t)$ and $\lambda_{23}(t)$, derive the occupancy probabilities $P_k(t) = P(Z(t) = k)$ for $k = 1, 2, 3$.

(b) Repeat this for the case where $\lambda_{23}(t|\mathcal{H}(t^-), T_2 = t_2) = h(t - t_2)$ for $t > t_2$.

(c) For each case calculate $P(Z(t) = 3, T_2 < t)$ and $P(Z(t) = 3, T_2 > t)$, which are the probabilities of being in state 3 with and without passing through state 2.

(d) Consider the time-homogeneous case where $\lambda_{13} = 1, \lambda_{12} = 0.2$ and $\lambda_{23} = \alpha$. Obtain the two probabilities in part (c) and compare them over $0 \le t \le 3$ when (i) $\alpha = 1$, (ii) $\alpha = 2$ and (iii) $\alpha = 5$.

(Section 1.3)

Problem 1.2   Consider the 2-state homogeneous Markov chain in continuous time with transition intensity matrix

$$Q = \begin{pmatrix} -\lambda_{12} & \lambda_{12} \\ \lambda_{21} & -\lambda_{21} \end{pmatrix}.$$

In Section 2.3.1 we give the expression (2.24),

$$P(s,t) = \exp(Q \cdot (t - s)), \quad 0 \le s \le t$$

for time-homogeneous processes. The matrix exponential function for a square matrix A is defined as $\exp(A) = \sum_{r=0}^{\infty} A^r / r!$, where $A^0 = I$ is the identity matrix.

(a) Show that the eigenvalues of Q are $\nu_1 = 0$ and $\nu_2 = -(\lambda_{12} + \lambda_{21})$. Obtain the right eigenvectors of Q and thus find a matrix A such that

$$Q = ADA^{-1}, \quad \text{with} \quad D = \text{diag}\,(\nu_1, \nu_2).$$

(b) Prove that $P(0,t) = P(t) = A\,\text{diag}(e^{\nu_1 t}, e^{\nu_2 t})\,A^{-1}$ and hence find an explicit expression for $P(t)$.

(c) Find $\lim_{t \to \infty} P(t)$ and hence the limiting distribution of the process. Rewrite $P(t)$ in terms of the new parameters $\pi = \lambda_{21}/(\lambda_{12} + \lambda_{21})$ and $\psi = \lambda_{12} + \lambda_{21}$.

(Section 1.3)

Problem 1.3   Consider the competing risks model with two modes of failure, where state 0 represents persons diagnosed with a form of cancer, state 1 represents death from that cancer, and state 2 represents death from other causes.

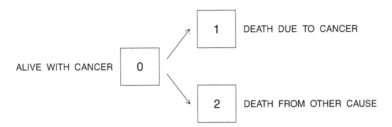

Let t represent age and suppose a person is diagnosed at age a. If a treatment reduces $\lambda_{01}(t|\mathcal{H}(t^-)) = \lambda_{01}(t,a)$ for $t \geq a$, but does not affect $\lambda_{02}(t|\mathcal{H}(t^-)) = \lambda_{02}(t)$, discuss how you could quantify the benefits of treatment.

(Sections 1.3, 1.4)

Chapter 2

Event History Processes and Multistate Models

In this chapter we demonstrate how intensity functions characterizing the instantaneous risks of transitions can be used to compute the probability of a sample path for a multistate process. This also allows us to construct likelihood functions for the typical setting in which processes are subject to right censoring due to limited follow-up. We highlight the close relation between the form of the resulting likelihoods and those for survival times subject to left truncation and right censoring. In Section 2.2 we also discuss marginal process features of common interest such as the entry time distribution for states, and the probability a particular state is occupied at a specified time. In Section 2.3 we introduce some important types of multistate models. In Section 2.4 we consider how process features such as transition probabilities and sojourn time distributions can be obtained, and simulation of multistate processes is discussed in Section 2.5.

2.1 Intensity Functions and Counting Processes

Consider a multistate process with K states. As illustrated in the examples of Chapter 1, it is often appropriate to include one or more absorbing states to represent reasons the process may terminate. We let $Z(t)$ denote the state occupied at time t and $\{Z(t), t \geq 0\}$ represent the associated stochastic process. A $p \times 1$ vector of left-continuous, time-dependent covariates is represented by $X(t)$, and the history for the two processes is denoted by $\mathcal{H}(t) = \{Z(s), X(s), 0 \leq s \leq t\}$. The intensity functions introduced in (1.2) govern the stochastic movement between states and are defined by

$$\lambda_{kl}(t \mid \mathcal{H}(t^-)) = \lim_{\Delta t \downarrow 0} \frac{P(Z(t + \Delta t^-) = l \mid Z(t^-) = k, \mathcal{H}(t^-))}{\Delta t} \tag{2.1}$$

for all $k \neq l$; if k is an absorbing state $\lambda_{kl}(t|\mathcal{H}(t^-)) = 0$. Note that when we write the history $\mathcal{H}(t) = \{Z(s), X(s), 0 \leq s \leq t\}$ at time t the path of the multistate and covariate processes are realized over $[0, t]$ and as a result it may be natural to write the history in terms of $z(s)$ and $x(s)$, $0 \leq s \leq t$. It is conventional, however, to use a capital N for $N(t)$ over $0 < s \leq t$, and moreover we often take expectations with respect to the process history, in which case the history is treated as random. We therefore use capital letters for the elements of the history. When fitting regression models we typically use lower-case x to reflect the fact that model fitting is based on observed covariates.

27

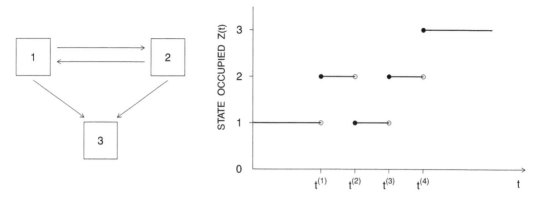

Figure 2.1: A reversible illness-death process (left panel) and a hypothetical realization of a process path $\{Z(t), t \geq 0\}$ (right panel).

Figure 2.1 displays the multistate diagram for a reversible illness-death process in the left panel; it is often useful for characterizing the dynamics of relapsing and remitting conditions (e.g. infections or exacerbations of symptoms). A realization of the sample path is depicted in the right panel for a hypothetical individual where the right-continuous nature of this process is reflected by the closed and open circles on the left and right endpoints of the line segments. The states change at the four transition times $t^{(1)} < \cdots < t^{(4)}$ and the process terminates upon entry to the absorbing state 3 at $t^{(4)}$.

It is often helpful to express the data and models in terms of counting processes. We let $N_{kl}(t)$ denote a right-continuous function that counts the number of instantaneous transitions from state k to l over $[0,t]$. Then $\Delta N_{kl}(t) = N_{kl}(t + \Delta t^-) - N_{kl}(t^-)$ is the number of $k \to l$ transitions over $[t, t + \Delta t)$ and $dN_{kl}(t) = \lim_{\Delta t \downarrow 0} \Delta N_{kl}(t)$ indicates whether a $k \to l$ transition occurred at time t; $dN_{kl}(t) = 1$ if so, and $dN_{kl}(t) = 0$ otherwise. The vector $dN_k(t) = (dN_{kl}(t), l \neq k, l = 1, \ldots, K)'$ contains all elements $dN_{kl}(t)$ for $l \neq k$ and hence the information on whether a transition occurs out of state k at time t, and the nature of any such transition. Specifically if $\sum_{l \neq k} dN_{kl}(t) = 1$ then a transition occurred, and the non-zero element in $dN_k(t)$ indicates which state was entered. The corresponding vector $N_k(t) = (N_{kl}(t), l \neq k, l = 1, \ldots, K)'$, where $N_{kl}(t) = \int_0^t dN_{kl}(s)$, gives the cumulative number and types of transitions out of state k of each type over $[0,t]$. For the multistate model depicted in the left panel of Figure 2.1 four counting processes are needed, which we can represent in $\{N(t), t \geq 0\}$ where $N(t) = (N_1'(t), N_2'(t))'$ with $N_1(t) = (N_{12}(t), N_{13}(t))'$ and $N_2(t) = (N_{21}(t), N_{23}(t))'$. The realizations of the four respective counting processes corresponding to the path in the right panel of Figure 2.1 are depicted in Figure 2.2.

The full vector $N(t) = (N_1'(t), \ldots, N_K'(t))'$ records the nature and number of all transitions occurring over $[0,t]$, and hence when viewed as a process, $\{N(t), t \geq 0\}$ offers an alternative way of representing $\{Z(t), t \geq 0\}$. Note that the history $\{Z(s), 0 \leq s \leq t\}$ contains the information on the state occupied at $t = 0$ so to write the equivalent history in terms of the counting process, in the absence of covariates we must write $\{N(s), 0 \leq s \leq t; Z(0)\}$.

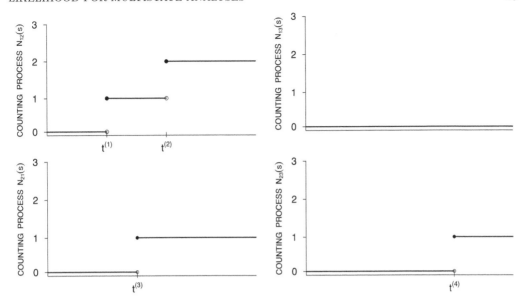

Figure 2.2: The four realized counting processes corresponding to the path depicted in the right panel of Figure 2.1.

It is convenient to define $Y_k(t) = I(Z(t) = k)$ as the indicator that an individual is in state k at t. The intensity function in (2.1) can be then expressed in counting process notation as

$$\lim_{\Delta t \downarrow 0} \frac{P(\Delta N_{kl}(t) = 1 \mid \mathcal{H}(t^-))}{\Delta t} = Y_k(t^-)\lambda_{kl}(t \mid \mathcal{H}(t^-)), \qquad (2.2)$$

which reduces to zero if the process is not in state k at t^-.

2.2   Likelihood for Multistate Analyses

2.2.1   Product Integration and Sample Path Probabilities

Under mild conditions the intensity functions fully specify a multistate stochastic process. While it is convenient to specify models in terms of intensity functions, it is necessary to obtain expressions for the probability of particular sample paths in terms of these intensities. To do so, we introduce the concept of product integration.

Suppose $g(u)$ is a continuous function that is integrable over the interval $[a,b]$. Let $a = u_0 < u_1 < \cdots < u_R = b$ define a partition of $[a,b]$ as depicted in Figure 2.3. This partition creates R sub-intervals $[u_{r-1}, u_r)$ of lengths $\Delta u_r = u_r - u_{r-1}$, $r = 1, \ldots, R$. We now consider a sequence of partitions as $R \to \infty$, and assume that as R increases, $\max_r \Delta u_r \to 0$. The product integral of $g(u)$ over the interval $[a,b]$ is defined as

$$\prod_{(a,b]} \{1 + g(u)\, du\} = \lim_{R \to \infty} \prod_{r=1}^{R} \{1 + g(u_r)\Delta u_r\}, \qquad (2.3)$$

where the left-hand side of the equation is the notation representing product integration, and the right-hand side is the definition. Noting that $\log(1 + g(u_r)\Delta u_r) =$

$g(u_r)\Delta u_r + o(\Delta u_r)$, we see that (2.3) can be written as

$$\lim_{R \to \infty} \prod_{r=1}^{R} \exp(g(u_r)\Delta u_r + o(\Delta u_r)) = \exp\left(\int_a^b g(u)\, du\right), \tag{2.4}$$

where the exponent is a Riemann integral.

Figure 2.3: A timeline representing a partition of the interval $[a,b]$.

Suppose an individual is observed over $[0, C^A]$ where C^A is a fixed administrative censoring time. Consider a partition of $[0, C^A]$ defined by $0 = u_0 < u_1 < \cdots < u_R = C^A$. Let $\Delta N_{kl}(u_r) = N_{kl}(u_r^-) - N_{kl}(u_{r-1}^-)$ denote the number of $k \to l$ transitions over $[u_{r-1}, u_r)$, $\Delta N_k(u_r)$ be a $(K-1) \times 1$ vector containing elements $\Delta N_{kl}(u_r)$, $l = 1, \ldots, K$ excluding the case $l = k$, and $\Delta N(u_r) = (\Delta N_1'(u_r), \ldots, \Delta N_K'(u_r))'$. We also let $H(u_r) = \{Z(u_s), s = 0, 1, \ldots, r\}$ denote the history of the process as recorded at the points of the partition, $r = 0, 1, \ldots, R$ where $H(u_0) = Z(u_0)$.

With this partition we can write the probability of a realization of $\Delta N(u_1), \ldots, \Delta N(u_R)$ given $Z(0)$ as

$$\prod_{r=1}^{R} P(\Delta N(u_r) \mid H(u_{r-1})). \tag{2.5}$$

If we take the limit as $R \to \infty$ and all the Δu_r approach zero, the probability of two or more transitions occurring within $[u_{r-1}, u_r)$ becomes vanishingly small. With R large enough, we can consider the different possible contributions depending on the state occupied at u_{r-1}, and write

$$P(\Delta N(u_r) \mid H(u_{r-1})) = \prod_{k=1}^{K} [P(\Delta N_k(u_r) \mid H(u_{r-1}))]^{Y_k(u_{r-1})}, \tag{2.6}$$

with the convention that 0^0 is taken to be one. We let $N_{k\cdot}(t) = \sum_{l \neq k=1}^{K} N_{kl}(t)$ and $\Delta N_{k\cdot}(t) = N_{k\cdot}(t + \Delta t^-) - N_{k\cdot}(t^-)$. Since $\Delta N_k(u_r)$ is a multinomial random variable written in vector form (i.e. there are K elements to $\Delta N_k(u_r)$ and at most one element is non-zero), given $Y_k(u_{r-1}) = 1$ and $H(u_{r-1})$, we can write $P(\Delta N_k(u_r) \mid H(u_{r-1}))$ as

$$\prod_{l \neq k=1}^{K} \left(Y_k(u_{r-1})\lambda_{kl}(u_r \mid H(u_{r-1}))\Delta u_r + o(\Delta u_r)\right)^{\Delta N_{kl}(u_r)} \tag{2.7}$$

$$\times \left(1 - Y_k(u_{r-1})\lambda_{k\cdot}(u_r \mid H(u_{r-1}))\Delta u_r + o(\Delta u_r)\right)^{1 - \Delta N_{k\cdot}(u_r)}$$

where $\lambda_{k\cdot}(u_r|H(u_{r-1})) = \sum_{l\neq k=1}^{K} \lambda_{kl}(u_r|H(u_{r-1}))$. Making use of (2.6) and (2.7), we rewrite (2.5) as

$$\prod_{r=1}^{R} \prod_{k=1}^{K} \left[\left\{ \prod_{l\neq k=1}^{K} (Y_k(u_{r-1})\lambda_{kl}(u_r \mid H(u_{r-1}))\Delta u_r + o(\Delta u_r))^{\Delta N_{kl}(u_r)} \right\} \right. \tag{2.8}$$

$$\left. \times \; (1 - Y_k(u_{r-1})\lambda_{k\cdot}(u_r \mid H(u_{r-1}))\Delta u_r + o(\Delta u_r))^{1-\Delta N_{k\cdot}(u_r)} \right]^{Y_k(u_{r-1})} .$$

We can take the product in (2.8) in the order $\prod_{k=1}^{K} \prod_{l\neq k=1}^{K} \prod_{r=1}^{R}$ and divide it by $\prod_{k=1}^{K} \prod_{l\neq k=1}^{K} \prod_{r=1}^{R} (\Delta u_r)^{\Delta N_{kl}(u_r)}$, and then take the limit as $R \to \infty$. For a given pair of indices (k,l), the terms $o(\Delta u_r)/\Delta u_r$ in curly brackets vanish as $R \to \infty$, and only the terms $Y_k(\cdot)\lambda_{kl}(\cdot|H(\cdot))$ remain for times at which $k \to l$ transitions occur, since it is at these instants the indicator $dN_{kl}(u) = 1$. The intervals not containing transition times will have contributions of the sort represented outside the curly brackets in (2.8) which, from (2.4), gives

$$\exp\left(-\int_0^{C^A} Y_k(s)\lambda_{kl}(s \mid \mathcal{H}(s^-))\, ds \right)$$

in the limit as $R \to \infty$. So (2.8) leads to the probability density of a sample path over $[0, C^A]$ given the initial state $Z(0)$ as

$$\prod_{k=1}^{K} \prod_{l\neq k=1}^{K} \left[\prod_{t_j \in \mathcal{D}_{kl}} \lambda_{kl}(t_j \mid \mathcal{H}(t_j^-)) \; \exp\left(-\int_0^{C^A} Y_k(u^-)\lambda_{kl}(u \mid \mathcal{H}(u^-)) \right) \right], \tag{2.9}$$

where $\mathcal{D}_{kl}$ denotes the set of $k \to l$ transition times over $[0, C^A]$.

2.2.2   Time-Dependent Covariates and Random Censoring

We next consider the more general problem of constructing the likelihood in the presence of possible time-dependent covariates and random censoring.

Consider the setting where interest lies in the process $\{Z(t), t \geq 0\}$ over $[0, C^A]$ where C^A is an administrative censoring time. We suppose that observation may be incomplete due to random censoring at C^R, so ultimately we observe the process over $[0, C]$, where $C = \min(C^R, C^A)$. The left-continuous indicator function $Y(t) = I(t \leq C)$ is used to indicate that the process is under observation at time t in which case a transition can be observed. We let $C^R(t) = I(C^R \leq t)$ be the indicator of whether random censoring occurred by time t, and denote the corresponding counting process as $\{C^R(t), t \geq 0\}$. Henceforth, we assume that C^A is conditionally independent of the event process given fixed covariates, and condition on the values of C^A for all individuals in a sample.

We let $X(t)$ denote a $p \times 1$ vector of possibly time-dependent covariates and $\{X(t), t \geq 0\}$ denote the covariate process. We suppose that the vector of covariates can be partitioned as $X(t) = (X_1'(t), X_2'(t))'$, where the $p_1 \times 1$ vector $X_1(t)$ reflects endogenous features and hence is an internal covariate, and $X_2(t)$ contains exogenous features and so is an external covariate. Information on covariates may or may not be subject to censoring in the same way the process of interest is. Internal co-variate processes are typically censored at C since that is when the individual ceases to be under observation, but external covariates may be observable beyond C. We focus here on the case where information is only collected up until the censoring time C.

In light of random censoring, we define the observable quantities $d\bar{N}_{kl}(t) = Y(t) dN_{kl}(t)$ that is one if a $k \to l$ transition occurs *and is observed* at t, and let $\bar{N}_{kl}(t) = \int_0^t Y(s) dN_{kl}(s)$ denote the total number of $k \to l$ transitions ob-served to have occurred over $[0, t]$. The corresponding vectors are denoted $\bar{N}_k(t) = (\bar{N}_{kl}(t), l \neq k, l = 1, \ldots, K)'$ and $\bar{N}(t) = (\bar{N}_1'(t), \ldots, \bar{N}_K'(t))'$. We let $\Delta\bar{X}(t) = Y(t + \Delta t) \cdot (X(t + \Delta t) - X(t))$ represent an increment in the covariate vector over $[t, t + \Delta t]$ which is observed because the individual was not censored prior to time $t + \Delta t$; we then let $d\bar{X}(t) = \lim_{\Delta t \downarrow 0} \Delta\bar{X}(t)$ and $\bar{X}(t) = \int_0^t d\bar{X}(s)$. The history of the ob-served multistate, covariate, and random censoring processes is denoted by $\bar{\mathcal{H}}(t) = \{Y(s), \bar{N}(s), \bar{X}(s), 0 \leq s \leq t; Z(0), X(0)\}$. The intensity for random censoring is then defined generally as

$$\lim_{\Delta t \downarrow 0} \frac{P(\Delta C^R(t) = 1 \mid \bar{\mathcal{H}}(t^-))}{\Delta t} = Y(t^-) \lambda^c(t \mid \bar{\mathcal{H}}(t^-)), \qquad (2.10)$$

where $\Delta C^R(t) = C^R(t + \Delta t^-) - C^R(t^-)$. Note that the $Y(t^-)$ term in (2.10) plays a similar role to the $Y_k(t^-)$ term in (2.2) in that $Y(t^-)$ in (2.10) ensures that the cen-soring intensity is zero for all $t > C$. It is worth emphasizing that this formulation accommodates the situation in which the risk of random withdrawal from a study at any time t is influenced by the observed history of the multistate and covariate processes. In particular, once a process enters an absorbing state the random cen-soring intensity may become zero. We can think of C^R as infinite in this case, so that $Y(t) = 1$ as long as $t \leq C^A$. We remark that in some situations an administrative censoring time is not given and then we may assume follow-up ceases upon entry to an absorbing state. We also may encounter situations where a censoring time is present in the data, but with no indication of whether it is an administrative cen-soring time or due to premature loss to follow-up. In that case it may be advisable to treat C as a random censoring time for protection. Censoring is discussed further in Sections 3.4, 4.1.2 and 7.2.5.

To construct the full likelihood we consider a partition of $[0, C^A]$ defined by the points $0 = u_0 < u_1 < \cdots < u_R = C^A$ as before. We again consider the likelihood based on data available over the sub-intervals $[u_{r-1}, u_r)$, $r = 1, \ldots, R$, but here we consider a likelihood based on joint consideration of the multi-state, covariate and censoring processes. To this end we let $\Delta\bar{X}(u_r) = Y(u_r) \cdot (X(u_r) - X(u_{r-1}))$ and $\Delta\bar{N}(u_r) = Y(u_r) \cdot (N(u_r^-) - N(u_{r-1}^-))$ and define $\bar{H}(u_r) = \{Y(u_s), \Delta\bar{N}(u_s), \Delta\bar{X}(u_s), s = 1, \ldots, r; Z(0), X(0)\}$. We consider the following likeli-hood contribution for the interval $[u_{r-1}, u_r)$ for the case in which the individual was not censored over $[u_{r-1}, u_r)$ and so $Y(u_r) = 1$:

$$\left\{ \left[P\left(\Delta \bar{X}(u_r) \mid \Delta \bar{N}(u_r), Y(u_r) = 1, \bar{H}(u_{r-1})\right) \; P\left(\Delta \bar{N}(u_r) \mid Y(u_r) = 1, \bar{H}(u_{r-1})\right) \right]^{Y(u_r)} \right.$$

$$\left. \times \; P\left(\Delta C^R(u_r) = 0 \mid \bar{H}(u_{r-1})\right)^{1-\Delta C^R(u_r)} \right\}^{Y(u_{r-1})I(Z(u_{r-1})\notin \mathcal{A})} \tag{2.11}$$

where $\mathcal{A}$ is the set of absorbing states. This factorization is natural if the covariate affects the future risk of transitions since the value for $\Delta X(u_{r-1})$ is contained in $\bar{H}(u_{r-1})$ in $P(\Delta \bar{N}(u_r) \mid Y(u_r) = 1, \bar{H}(u_{r-1}))$.

At this point there are several issues to consider. First, note that a likelihood contribution is only made over $[u_{r-1}, u_r)$ by an individual if they have not been censored and the multistate response process is not in an absorbing state at time u_{r-1}. Second, note that there is a contribution related to the multistate and covariate processes only if the individual is not censored by u_r. Third, by adopting the particular factorization here the stochastic model for the increment in the covariate process is conditional not just on $\Delta \bar{X}(u_{r-1}), \ldots, \Delta \bar{X}(u_1), X(0)$, but also on $\Delta \bar{N}(u_r), \ldots, \Delta \bar{N}(u_1)$, $Z(0)$ and $I(C > u_r)$. This accomodates the setting in which covariates may cease to be defined when certain (usually absorbing) states are entered in the multistate process; see Section 8.3.2.

The full likelihood based on data over $[0, C^A]$ arising from this partition can be written as

$$L \propto L_X \cdot L_N \cdot L_C \tag{2.12}$$

where

$$L_X = \prod_{r=1}^{R} P\left(\Delta \bar{X}(u_r) \mid \Delta \bar{N}(u_r), Y(u_r) = 1, \bar{H}(u_{r-1})\right)^{Y(u_r)I(Z(u_{r-1})\notin \mathcal{A})} \tag{2.13}$$

pertains to the covariate process,

$$L_N = \prod_{r=1}^{R} P\left(\Delta \bar{N}(u_r) \mid Y(u_r) = 1, \bar{H}(u_{r-1})\right)^{Y(u_r)I(Z(u_{r-1})\notin \mathcal{A})} \tag{2.14}$$

pertains to the multistate process, and

$$L_C = \prod_{r=1}^{R} P\left(\Delta C^R(u_r) \mid \bar{H}(u_{r-1})\right)^{Y(u_{r-1})I(Z(u_{r-1})\notin \mathcal{A})} \tag{2.15}$$

pertains to the random censoring process.

The censoring process and covariate process are said to be noninformative if there is no information to be gained about the parameters of primary interest (i.e. those indexing the multistate process) by modeling the censoring or covariate processes. Under the assumption that the censoring and covariate processes are noninformative, it is customary to restrict attention to the partial likelihood L_N; an exception to this is when interest lies in joint modeling of covariate (often termed "marker") and multistate processes (see Section 8.3).

This likelihood requires an intensity for the observable counting process. Proceeding as we did in Section 2.2.1, we write the probability in (2.14) as

$$
P\left(\Delta\bar{N}(u_r) \mid Y(u_r) = 1, \bar{H}(u_{r-1})\right) = \prod_{k=1}^{K}\left[P\left(\Delta\bar{N}_k(u_r) \mid Y(u_r) = 1, \bar{H}(u_{r-1})\right)\right]^{Y_k(u_{r-1})},
$$

(2.16)

which neglecting terms of order $o(\Delta u_r)$, can be written more explicitly as

$$
\prod_{k=1}^{K}\left[\prod_{l\neq k=1}^{K} P\left(\Delta\bar{N}_{kl}(u_r) = 1 \mid Y(u_r) = 1, \bar{H}(u_{r-1})\right)^{\Delta\bar{N}_{kl}(u_r)}\right.
$$

$$
\left. \times \ P\left(\Delta\bar{N}_{k\cdot}(u_r) = 0 \mid Y(u_r) = 1, \bar{H}(u_{r-1})\right)^{1-\Delta\bar{N}_{k\cdot}(u_r)}\right]^{Y_k(u_{r-1})}.
$$

(2.17)

To proceed further it is necessary to define the intensity for the *observable* counting process

$$
\lim_{\Delta t\downarrow 0} \frac{P(\Delta\bar{N}_{kl}(t) = 1 \mid Y(t) = 1, \bar{\mathcal{H}}(t^-))}{\Delta t}.
$$

Under the assumption that the random censoring time is conditionally independent of the multistate process given the history $\bar{\mathcal{H}}(t^-)$, the probability in the numerator is equal to $P(\Delta N_{kl}(t) = 1 | \mathcal{H}(t^-))$, where $\mathcal{H}(t) = \{Z(s), X(s), 0 \leq s \leq t\}$, and we can then write

$$
\lim_{\Delta t\downarrow 0} \frac{P(\Delta\bar{N}_{kl}(t) = 1 \mid Y(t) = 1, \mathcal{H}(t^-))}{\Delta t} = \bar{Y}_k(t)\,\lambda_{kl}(t \mid \mathcal{H}(t^-)), \qquad (2.18)
$$

where $\bar{Y}_k(t) = Y(t)Y_k(t^-)$ indicates the process is under observation at t and at risk of a transition out of state k. The assumption that censoring is conditionally independent of the multistate process given the full process history is crucial for the methodology in this book. It is discussed further in Section 7.2.4.

Following the same steps as in Section 2.2.1 we replace the probabilities in (2.17) with functions of the intensities in (2.18), scale the discrete likelihood contributions, and take the limit as $R \to \infty$ to obtain

$$
L_N \propto \prod_{k=1}^{K}\prod_{l\neq k=1}^{K} L_{kl}, \qquad (2.19)
$$

where

$$
L_{kl} \propto \left[\prod_{t_r\in\mathcal{D}_{kl}} \lambda_{kl}(t_r \mid \mathcal{H}(t_r^-))\right]\exp\left(-\int_0^{\infty} \bar{Y}_k(u)\,\lambda_{kl}(u \mid \mathcal{H}(u^-))\,du\right), \qquad (2.20)
$$

with $\mathcal{D}_{kl}$ the set of $k \to l$ transition times observed over $[0, C]$.

Individuals are often sampled for inclusion in a study at some time A_0 after the process of interest has begun. Prevalent cohort studies, for example, involve screening a population for a disease of interest, sampling those individuals found to have the disease, and following them over time to collect information on the disease process. In this setting the time origin $t = 0$ is often taken to be the time of disease onset, and $A_0 \geq 0$ is the time an individual is selected and their follow-up starts. The time A_0 since the onset of the process is often called a delayed entry time, or left-truncation time. The former term is more natural in this setting because A_0 represents the delay from the onset of the process to the time of recruitment. The latter term is most natural in the survival analysis setting since those individuals whose survival time is lower than (i.e. "left of") A_0 are excluded (or truncated) out of the sample. Here we will use these terms interchangeably.

The construction of the likelihood (2.19) can be modified to accommodate delayed entry by redefining $Y(u)$ as $Y(u) = I(A_0 \leq u \leq C)$, which indicates whether the individual is under observation at time u. Then, in (2.20) $\bar{Y}_k(u) = Y(u)Y_k(u^-)$ as before.

We note here that (2.20) can be factored into terms that correspond to likelihoods from failure times that are possibly left truncated and right censored. Let $t_k^{(r)}$ denote the time at which state k is entered for the rth time and let $w_k^{(r)}$ denote the observed sojourn time such that $v_k^{(r)} = t_k^{(r)} + w_k^{(r)}$ is either the time a transition to another state occurs or the censoring time, whichever is smaller. In addition, we let $\delta_{kl}^{(r)} = 1$ if a transition to state $l \neq k$ is observed, and let $\delta_{kl}^{(r)} = 0$ otherwise. We then see that (2.20) factors into pieces of the form

$$\left[\lambda_{kl}(v_k^{(r)} \mid \mathcal{H}(v_k^{(r)-})) \right]^{\delta_{kl}^{(r)}} \exp\left(- \int_{t_k^{(r)}}^{v_k^{(r)}} \lambda_{kl}(u \mid \mathcal{H}(u^-)) \, du \right). \qquad (2.21)$$

This is exactly the form of the likelihood for a failure time T that is left truncated at $t_k^{(r)}$ with $\delta_{kl}^{(r)}$ the failure status indicator. In terms of the notation used for the dataframes in Section 1.6, the row of data corresponding to this piece of the likelihood would have $\texttt{start} = t_k^{(r)}$, $\texttt{stop} = v_k^{(r)}$ and $\texttt{status} = \delta_{kl}^{(r)}$.

2.3 Some Important Multistate Models

2.3.1 Modulated Markov Models

When modeling life history processes with transition intensity functions a formative decision involves specification of the basic time scale. In progressive processes it is often natural to use the age of the process ("global" time) as the time scale. The canonical multistate model with this time scale is the Markov model which in the absence of covariates involves transition intensities of the form

$$\lambda_{kl}(t \mid \mathcal{H}(t^-)) = \lambda_{kl}(t). \qquad (2.22)$$

For Markov models it is particularly straightforward to express the transition probability function in terms of the transition intensities through product integration. If we let $\Lambda_{kl}(t) = \int_0^t \lambda_{kl}(s) ds$ denote the cumulative transition intensity, then $d\Lambda_{kl}(t) = \lambda_{kl}(t) dt$. We may construct a $K \times K$ matrix $Q(t)$ with $\lambda_{kl}(t)$ in the off diagonal entries ($k \neq l$) and $-\lambda_{k\cdot}(t) = -\sum_{l \neq k} \lambda_{kl}(t)$ in the diagonal entries, $k = 1, \dots, K$. Consider an interval $[s, t]$ and let $s = u_0 < u_1 < \cdots < u_R = t$ denote a partition with $\Delta u_r = u_r - u_{r-1}$, $r = 1, \dots, R$, and let $\Delta Q(u_r) = Q(u_r) \Delta u_r$. If I is a $K \times K$ identity matrix then the $K \times K$ transition probability matrix $P(s,t)$ is obtained by product integration as

$$P(s,t) = \prod_{(s,t]} \{I + Q(u)\,du\} = \lim_{R \to \infty} \prod_{r=1}^{R} \{I + \Delta Q(u_r)\}\,; \qquad (2.23)$$

$P(s,t)$ has (k,l) element $P(Z(t) = l | Z(s) = k)$. This result is easily shown by using the fact that $P(s,t) = \prod_{r=1}^{R} P(u_{r-1}, u_r)$. For time-homogeneous models where $\lambda_{kl}(u) = \lambda_{kl}$ and $Q(u) = Q$, the formula above gives

$$P(s,t) = \exp(Q \cdot (t-s))\,, \qquad (2.24)$$

where the matrix exponential function for a $K \times K$ matrix B is defined as $\exp(B) = \sum_{r=0}^{\infty} B^r / r!$. Kalbfleisch and Lawless (1985) and Jackson (2011) discuss ways to compute (2.24) and the R function $\mathtt{expm}$ implements them.

In the presence of time-dependent covariates $x(t)$, models with intensities of the form

$$\lambda_{kl}(t \mid \mathcal{H}(t^-)) = Y_k(t^-)\,\lambda_{kl}(t \mid x(t)) \qquad (2.25)$$

are often referred to as modulated Markov models. Such processes offer a natural basis for modeling disease processes such as the occurrence of joint damage in arthritis, the decline in bone density in osteoporosis, and the decline in kidney function in diabetics; in these settings biomarkers are often used as covariates to explain and predict disease progression.

There are several frameworks one can consider for modeling fixed or time-dependent covariate effects. The most widely adopted are multiplicative models in which

$$\lambda_{kl}(t \mid x(t)) = \lambda_{kl0}(t)\,g(x(t); \beta)\,,$$

where $\lambda_{kl0}(t)$ is a baseline transition intensity applicable to individuals for whom $x(t) = 0$, and $g(\cdot; \beta) > 0$ is a parametric function indexed by β. Typically we let $g(x(t); \beta) = \exp(x'(t)\beta)$ in which case $\exp(\beta_j)$ is interpretable as a multiplicative factor that increases (if $\exp(\beta_j) > 0$) or decreases (if $\exp(\beta_j) < 0$) the intensity for each unit increase in $x_j(t)$ while all other elements of $x(t)$ are fixed. The exponentiated regression coefficients in this class of models are referred to as relative risks and the short-form RR is used when reporting estimates. We will see in Chapter 3 that multiplicative models offer an appealing framework for estimation and inference based on the partial likelihood.

Additive hazard models are another formulation often used in survival analysis. The additive intensity model for multistate Markov processes, in its most general form, is written

$$\lambda_{kl}(t \mid x(t)) = \alpha_{kl0}(t) + \alpha_{kl1}(t)x_1(t) + \cdots + \alpha_{klp}(t)x_p(t)\,.$$

Semiparametric versions of both additive and multiplicative models are widely used, and will be discussed in Section 3.3. It should be noted that calculation of transition probabilities when time-dependent covariates are present is usually infeasible except in special cases, and the main use of the models here is in identifying factors affecting the multistate process.

We consider models that facilitate calculation of transition probabilities in Section 8.3.

2.3.2   Modulated Semi-Markov Models

Semi-Markov models play a useful role in settings where it is natural to focus on sojourn times in particular states. In the absence of covariates, transition intensities for semi-Markov models take the form

$$\lambda_{kl}(t \mid \mathcal{H}(t^-)) = h_{kl}(B(t)), \tag{2.26}$$

where $B(t)$ is the time since the most recent entry to state k. With time-dependent covariates we use the term modulated semi-Markov model and write

$$\lambda_{kl}(t \mid \mathcal{H}(t^-)) = h_{kl}(B(t) \mid x(t)). \tag{2.27}$$

Covariate effects are routinely expressed using multiplicative models but additive models and other formulations can also be used. An illustration we discuss later concerns studies of chronic bronchitis where interest lies in the duration of transient recurrent infections. Other situations where sojourn times in states are of interest are the duration of hospitalizations among individuals with psychiatric disorders, studies of the duration of quit attempts in smoking cessation trials, and the duration of periods of exacerbations of symptoms in episodic disorders such as lupus.

2.3.3   Models with Dual Time Scales

In some settings there is a scientific rationale for considering more than one time scale. In chronic obstructive pulmonary disease, for example, individuals are at risk of recurrent infections. The risk for, and duration of, infections may increase the longer an individual has had the condition. One may specify models with dual time scales through intensity functions of the form

$$\lambda_{kl}(t \mid \mathcal{H}(t^-)) = \lambda_{kl}(t) \, g(B(t); \gamma), \tag{2.28}$$

where here the basic time scale is Markov but the time since entry to the current state modulates the risk of transition. Alternatively, if the semi-Markov time scale is most natural one can specify

$$\lambda_{kl}(t \mid \mathcal{H}(t^-)) = h_{kl}(B(t)) \, g(t; \gamma), \tag{2.29}$$

where $h_{kl}(\cdot)$ is baseline hazard function that is modulated by the defined covariate "t". In addition, some multistate processes may have intensities for some transitions of Markov form and others of semi-Markov form. For example, in studies of chronic obstructive pulmonary disease the risk of exacerbations may be modeled on a Markov scale (time since onset of the chronic condition), but the duration of exacerbations may naturally be modeled using the time since the onset of the exacerbation as the primary scale.

2.3.4 Models Accommodating Heterogeneity

When data from multiple processes are available, it is common for variation between processes to exceed what can be explained by available covariates and the stochastic variation that is compatible with the adopted model. In a sample of n individuals, we may, for example, observe realizations of each individual's processes which exhibit quite different patterns. This unexplained variation is often called heterogeneity, and must be dealt with in some way to ensure valid statistical inference. A common approach is to define intensities conditional on latent (unobservable) variables or processes that are associated with each individual. When the latent variables are constant they are often called random effects and in the survival analysis literature the evocative term *frailty* is often used to reflect the fact that individuals with physical frailties are often at higher risk of death, for example. Since we are dealing with general multistate processes we will tend to use the term *random effect*.

Let V_{kl} denote a random effect associated with the $k \to l$ transition intensity, $V_k = (V_{kl}, l \neq k, l = 1, \ldots, K)'$ and $V = (V_1', \ldots, V_K')'$ have a multivariate cumulative distribution function $G(\cdot)$. With multiplicative models, conditional transition intensities are expressed given random effects

$$\lambda_{kl}(t \mid V = v, \mathcal{H}(t^-)) = v_{kl}\, \lambda_{kl}(t \mid \mathcal{H}(t^-)), \quad k \neq l,$$

where v_{kl} is the realized valued of V_{kl}. Marginal likelihoods, obtained from (2.19) as

$$L \propto \int_0^\infty \cdots \int_0^\infty \prod_{k=1}^K \prod_{l \neq k=1}^K \left\{ \prod_{t_r \in \mathcal{D}_{kl}} v_{kl}\, \lambda_{kl}(t_r \mid \mathcal{H}(t_r^-)) \right.$$
$$\left. \times \, \exp\left(-\int_0^\infty Y_k(u^-)\, v_{kl}\, \lambda_{kl}(u \mid \mathcal{H}(u^-))\, du\right) \right\} dG(v)$$

can be challenging to work with since the factorization into separate functions for each (k, l) combination is not possible in general and so high dimensional integration may be required. If $v_{kl} = v_k$, for all $l \neq k$ the dimension of the integration is reduced considerably, and if $v_k = v$ a common scalar random effect is employed and only one-dimensional integration is required. This is a highly specialized model, however, which is only appealing in a few situations.

The introduction of random effects accommodates heterogeneity, but it also has the effect of generating a marginal model with a different history dependence than is reflected by the conditional intensities. If, for example, a conditional intensity is Markov and $\lambda_{kl}(t|v, \mathcal{H}(t^-)) = v_{kl}\, \lambda_{kl}(t)$, then after integrating out the random effects the intensity is $E\{V_{kl}\lambda_{kl}(t) \mid \mathcal{H}(t^-)\}$ and the process will not satisfy the Markov property; here the expectation is taken with respect to V_{kl} given the history $\mathcal{H}(t^-)$. It is therefore difficult to distinguish the need to accommodate heterogeneity in risk between individuals with the need to accommodate a residual history dependence not accounted for in the baseline intensity. We comment on this further in Chapter 6.

When individuals are sampled at some time $A_0 > 0$ after the process began, then as discussed in Section 2.2 it is important to account for this in the construction of the likelihood. With delayed entry at $A_0 = a_0$ and under right censoring, we write

$$L \propto \int_0^\infty \cdots \int_0^\infty \prod_{k=1}^K \prod_{l \neq k=1}^K \left\{ \prod_{t_r \in D_{kl}} v_{kl} \lambda_{kl}(t_r \mid \mathcal{H}(t_r^-)) \right.$$

$$\left. \times \exp\left(-\int_0^\infty \bar{Y}_k(u) v_{kl} \lambda_{kl}(u \mid \mathcal{H}(u^-)) \, du \, dG(v \mid \mathcal{H}(a_0)) \right) \right\}$$

where $\bar{Y}_k(u) = Y(u) Y_k(u^-)$ with $Y(u) = I(a_0 \leq u \leq C)$. This likelihood can be challenging, and perhaps impossible, to compute if data on the transition times before sampling are unreliable or completely unavailable since such data are required to obtain $dG(v \mid \mathcal{H}(a_0))$.

Another approach to dealing with heterogeneity involves conceptualizing the population as comprised of a number of sub-populations (or classes) where individuals in the same class make transitions according to the same process. Which class an individual belongs to is generally unknown and so methods are based on latent class modeling; we refer to these models as finite mixture models since at the population level the process is generated by a mixture of a finite number of distinct processes. Suppose there are G classes of individuals such that individuals in class g have transition intensity functions

$$\lambda_{kl}(t \mid \mathcal{H}(t^-), M = g) = \lambda_{kl}^{(g)}(t \mid \mathcal{H}(t^-); \theta_g),$$

where M is an unobservable random variable taking values in $\{1, \ldots, G\}$ according to the probability model

$$P(M = g \mid X; \gamma) = \pi_g(X; \gamma), \quad g = 1, \ldots, G,$$

and we assume that $G \geq 2$ is a specified value. As indicated above we may allow the π_g to depend on baseline covariates X. Mover-stayer models, in which $G = 2$ and one group of individuals never moves from their initial state are often useful but we focus primarily on the more general finite mixture models in Chapter 6.

2.3.5 Linked Models and Local Dependence

Sometimes there may be parallel processes operating for an individual, for example, in studies involving paired or multiple organ systems. In diabetes studies, for example, deterioration in vision may occur in both left and right eyes. Interest may lie in characterizing the rate of progression in both eyes, but also in understanding how deterioration in one eye may be correlated with deterioration in the other eye. Interest may also lie in jointly modeling two distinct progressive features such as retinopathy and nephropathy, where the latter represents a deterioration in kidney function. Both of these conditions arise due to vascular disease which, in diabetics, is believed to be driven by poor glucose control.

Let $\{Z_1(t), t \geq 0\}$ and $\{Z_2(t), t \geq 0\}$ denote two processes of interest; when they arise from paired organs the parameters governing the marginal processes may be the same, but for distinct correlated processes they would typically differ. We assume for convenience the processes have the same number of states. Interest lies in jointly modeling the processes to estimate the association between them, to consider the extent to which one process can be used to improve efficiency in estimation of another, and to enhance prediction.

Random effects are often used to link two or more processes. Let $\mathcal{H}_r(t) = \{Z_r(s), 0 \leq s \leq t; X\}$ denote the history for process r in the case where there are fixed covariates and let V_r be a vector of random effects for process r, $r = 1, 2$. Let $Y_{rk}(t) = I(Z_r(t) = k)$, $r = 1, 2$, and

$$\lim_{\Delta t \downarrow 0} \frac{P(Z_r(t + \Delta t^-) = l \mid v_r, \mathcal{H}_r(t^-))}{\Delta t} = \lambda_{rkl}(t \mid v_r, \mathcal{H}_r(t^-)) \qquad (2.30)$$

denote the conditional intensity of a $k \rightarrow l$ transition for process r given the random effect $V_r = v_r$. A conditional Markov model for process r has intensity of the form

$$\lambda_{rkl}(t \mid v_r, \mathcal{H}_r(t^-)) = v_{rkl} \, \lambda_{kl}(t \mid X),$$

where v_{rkl} is the component of the vector v_r that influences the $k \rightarrow l$ transitions; we assume $E(V_{rkl}) = 1$ and $\text{var}(V_{rkl}) = \phi_{rkl}$. If $V_1 \perp V_2 \mid X$, then the two processes are independent given X and if $V_1 = V_2$ the vector of random effects is shared between the processes. If $V = (V_1', V_2')'$ follows a multivariate distribution with $\text{cov}(V_1, V_2) \neq 0$, then a more general model is obtained with association accommodated between the two processes. A major issue is that this formulation does not typically lead to easily interpreted measures of association, and as in Section 2.3.4 the resulting marginal processes do not have simple forms.

Another framework for joint models arises by specification of an expanded state space defined by combinations of states from the two processes; Section 1.2.3 contains an example involving left and right sacroiliac joints. To this end we let $Z(t) = (Z_1(t), Z_2(t))$ denote states for this joint process $\{Z(t), t \geq 0\}$ and define transition intensity

$$\lim_{\Delta t \downarrow 0} \frac{P(Z(t + \Delta t^-) = (l, k_2) \mid Z(t^-) = (k_1, k_2), \mathcal{H}(t^-))}{\Delta t} = \lambda^{(1)}_{k_1 k_2, l}(t \mid \mathcal{H}(t^-))$$

for process 1, where $\mathcal{H}(t) = \{Z(s), 0 \leq s \leq t; X\}$; the intensity $\lambda^{(2)}_{k_1 k_2, l}(t \mid \mathcal{H}(t^-))$ is analogously defined; see Section 6.2.2. While this retains the general form and interpretability of intensity-based models, we pay the price of losing direct interpretation of covariate effects on features of the marginal processes.

Copula functions offer an appealing framework for constructing multivariate survival models since the marginal distributions retain their simple interpretation. In the multistate setting these methods can be useful for modeling dependencies in certain features of the processes, such as sojourn times in semi-Markov models or state entry times in Markov models. Finally, in some contexts interest may lie in simultaneous inferences regarding two or more processes, but not in the association between processes. In the DCCT study, for example, the effectiveness of the intensive glucose control program in delaying progression of nephropathy and retinopathy was of interest. In this case a working independence assumption for the retinopathy and nephropathy processes can furnish estimates of effects for each, and robust sandwich-type variance estimates can be used to ensure valid simultaneous inferences regarding the two processes (Lee and Kim, 1998). Chapter 6 discusses both heterogeneity and models for associated processes in some detail.

2.4   Process Features of Interest

While intensity functions fully characterize processes and are fundamental to their understanding, often there is interest in descriptive analyses or inferences about other features of a process. Transition probabilities $P_{kl}(s,t)$ are often of considerable interest, and we reiterate that for non-Markov models these cannot be easily obtained in terms of process intensities. Thus, in later chapters we will consider partial process models. We now mention some other features of interest.

Entry Time Distributions

Let $T_k^{(r)}$ denote the rth time of entry to state k, and $F_k^{(r)}(t) = P(T_k^{(r)} \leq t)$ be the cumulative (sub)-distribution function. In this case nonparametric (e.g. Kaplan-Meier) estimates can be computed for the $F_k^{(r)}(\cdot)$ functions. If state k is not ultimately visited with probability one, then even the distribution of $T_k^{(1)}$ is improper so $\lim_{t \to \infty} F_k^{(1)}(t \mid x) < 1$; in this case $F_k^{(1)}(t)$ can be estimated using nonparametric estimates of this cumulative incidence function; this is discussed in Sections 3.2 and 4.1.

State Occupancy Probabilities

Another feature of common interest is a state occupancy or prevalence probability. If all processes are in the common initial state $Z(0) = 1$, the state occupancy probabilities are then

$$P_k(t) = P(Z(t) = k \mid Z(0) = 1), \quad k = 1, \ldots, K. \tag{2.31}$$

The function $P_k(t) = P(Z(t) = k)$ may be defined more generally if there is a multinomial distribution for the initial state with probabilities $P(Z(0) = l)$, $l = 1, \ldots, K$. In this case,

$$P_k(t) = \sum_{l=1}^{K} P(Z(t) = k \mid Z(0) = l) \, P(Z(0) = l).$$

Estimation of state occupancy probabilities via estimation of transition probability functions is discussed in Section 3.4. Other approaches are described in Sections 4.1 and 4.2. For example, for a progressive K-state process with $Z(0) = 1$ and transitions are only allowed from state $k \to k+1$, $k = 1, \ldots, K-1$,

$$P_k(t) = F_k(t) - F_{k+1}(t), \quad k = 1, \ldots, K \tag{2.32}$$

for $t \geq 0$, where $F_1(t) = 1$ and $F_{K+1}(t) = 0$.

Sojourn Time Distributions

We let $W_k^{(r)}$ denote the sojourn time in state k, $k = 1, \ldots, K$, on the rth occasion it is entered. Such sojourn times are infinite for absorbing states so the following remarks pertain to non-absorbing states. In general (see Problem 2.11),

$$P(W_k^{(r)} > w \mid t_k^{(r)}, \mathcal{H}(t_k^{(r)-})) = \exp\left(-\int_{t_k^{(r)}}^{t_k^{(r)}+w} \sum_{l \neq k} \lambda_{kl}(u \mid \mathcal{H}(u^-))\,du\right). \qquad (2.33)$$

In some models it may be possible to conceive of a marginal sojourn time distribution for a particular state, for example, in the case of a pure semi-Markov process.

Total State Occupancy Times

Another quantity of interest is the cumulative mean time spent in a particular state over a particular interval of time. If

$$S_k(t) = \int_0^t I(Z(u) = k)\,du$$

is the total time spent in state k over $[0, t]$ for an individual starting in state 1 at $t = 0$,

$$E\{S_k(t) \mid Z(0) = 1\} = \int_0^t P(Z(u) = k \mid Z(0) = 1)\,du.$$

These quantities are useful when the states represent different disease conditions and interest lies in summarizing and comparing the total number of days with specific conditions; we give an application of this in Section 4.2.3. They are also often useful to consider when designing a study, which we consider in Section 3.6. In Section 8.1 we consider applications to health economic analyses where there may be fixed costs per day depending on the state occupied (e.g. hospitalized or discharged).

2.5   Simulation of Multistate Processes

Simulation often plays an important role in multistate analysis. First, as in any other area involving statistical modeling, simulation offers a way of assessing the properties of statistical methods. Second, as noted in Section 2.4, simple marginal features are often complex (even intractable) functions of the process intensities. Model-based estimates of some features may be obtained most easily by simulation using estimated parameter values. Third, it can be challenging computing expressions for asymptotic variances which, in simple settings, can help guide sampling strategies, the planned duration of follow-up, or other aspects of study design, and simulation is often used to inform such choices. Fourth, prediction is often of scientific interest, or of interest for health management. With complex models prediction is often most easily done via simulation. For example, multistate processes may be designed to model the dynamics of population members in terms of the onset and progression of some disease, in conjunction with individuals' utilization of health care services. Such models are, for example, useful in modeling surgical waiting times, or the need for home care for those with physical disabilities. Simulation can be used to explore the consequences of altering resources so as to inform, or examine the consequences of, health policy decisions.

Simulation of data from multistate processes can proceed in a stepwise fashion by using (2.33) and separately dealing with the sojourns in successive states. The process is in effect a series of competing risks simulations. We consider first the setting involving only fixed covariates. If the process begins in state 1, the first thing to simulate is the initial transition time out of state 1 and the state entered upon the first transition. The survivor function for the first transition time is, from (2.33),

$$P(T^{(1)} > t^{(1)} \mid x) = \exp\left(-\sum_{l>1} \int_0^{t^{(1)}} \lambda_{1l}(s \mid \mathcal{H}(s^-)) \, ds\right). \qquad (2.34)$$

With fixed covariates the history $\mathcal{H}(s^-)$ in (2.34) only contains $X = x$ and the information that no transition out of state 1 has occurred over $[0, s)$. To simulate $T^{(1)}$ we can generate a standard uniform random variable $U^{(1)} \sim \text{unif}(0,1)$, set the realized value $u^{(1)} = P(T^{(1)} > t^{(1)} \mid x)$ in (2.34) and solve for $t^{(1)}$. To determine the state entered, we then consider a multinomial experiment where the transition at time $t^{(1)}$ is made to state k with probability

$$P(dN_{1k}(t^{(1)}) = 1 \mid T^{(1)} = t^{(1)}, x) = \frac{\lambda_{1k}(t^{(1)} \mid \mathcal{H}(t^{(1)-}))}{\sum_{l>1} \lambda_{1l}(t^{(1)} \mid \mathcal{H}(t^{(1)-}))}. \qquad (2.35)$$

We now repeat this process. If we let $Z^{(1)} = k$ denote the state entered, then the exit time from it, $T^{(2)}$, has conditional survivor function

$$P(T^{(2)} > t^{(2)} \mid T^{(1)} = t^{(1)}, Z^{(1)} = k, x) = \exp\left(-\sum_{l \neq k} \int_{t^{(1)}}^{t^{(2)}} \lambda_{kl}(s \mid \mathcal{H}(s^-)) \, ds\right) \qquad (2.36)$$

which has the form of a survival function for a left-truncated failure time with hazard function $\sum_{l \neq k} \lambda_{kl}(s \mid \mathcal{H}(s^-))$. If $U^{(2)} \sim \text{unif}(0,1)$, then $t^{(2)}$ can be obtained by solving $P(T^{(2)} > t^{(2)} \mid T^{(1)} = t^{(1)}, Z^{(1)} = k, x) = u^{(2)}$, where $u^{(2)}$ is the realized value. The new state entered is again determined by a multinomial experiment with state l being entered with probability

$$\frac{\lambda_{kl}(t^{(2)} \mid \mathcal{H}(t^{(2)-}))}{\sum_{j \neq k} \lambda_{kj}(t^{(2)} \mid \mathcal{H}(t^{(2)-}))},$$

$l \neq k$, $l = 1, \ldots, K$. This process can be repeated until entry to an absorbing state.

Time-dependent covariates are harder to handle. Defined external covariates are in principle straightforward to incorporate, but stochastic external covariates have to be generated independently according to a suitable model. Then, at time s their histories are incorporated in $\mathcal{H}(s)$ in (2.34), (2.35) and subsequent expressions. Internal time-dependent covariates are of two types. If they are defined in terms of the process history then they are implicitly recorded in $\mathcal{H}(s)$. Other types of internal covariates must be generated jointly with the response process. Some code for generating sample paths along with remarks on available software is given in Appendix B.

2.6   Bibliographic Notes

In settings where inference is to be based on censored data through the likelihood function, product integration offers a convenient way to construct the likelihood. Gill and Johansen (1990) survey the ways in which product integration can be viewed and applied with an emphasis on time to event data. Books such as those by Andersen et al. (1993), Kalbfleisch and Prentice (2002), Lawless (2003) and Aalen et al. (2008) illustrate the application of product integration for likelihood construction in general and for particular models. Cox and Hinkley (1979) and Lindsey (1996) cover inference based on the likelihood. Semiparametric theory is presented in Fleming and Harrington (1991), Andersen et al. (1993) and Aalen et al. (2008).

Mixture models offer an important way of accommodating heterogeneity and modeling dependence. For heterogeneity models see, for example, Ng and Cook (1997), Mealli and Pudney (1999) and Lange and Minin (2013).

In some settings the context gives some guidance on how to model the heterogeneity; this is also the case when data arise in clusters and it is envisioned that life histories from individuals in the same cluster are more similar than life histories from individuals in different clusters in some sense. In this case one may assign cluster-level random effects, or frailties, and model association within clusters and the variation between clusters through these random effects (Cook et al., 2004; Sutradhar and Cook, 2008). Alternatively, a marginal model can be conceived in which association can be modeled using copula functions (Diao and Cook, 2014), or robust variance estimation can be carried out following model fits under a working independence assumption; this is done by Lee and Kim (1998) in the case of intermittent observation of processes.

If clusters are not defined by the context, one can envision sub-populations of individuals who have similar process characteristics and develop finite mixture (or latent class) models. The mover-stayer model has received much attention going back to Goodman (1961); maximum likelihood estimation was developed by Frydman (1984). Relatively little work has been done on more general mixture models for multistate processes but O'Keeffe et al. (2013) develop models accommodating a mover-stayer component along with a component characterizing continuous variation in progression rates in the setting of intermittently observed individuals. We consider models using random or latent effects at length in Chapter 6.

The importance of occupancy probabilities and other process features besides the transition intensities has received a good deal of attention in specific applications; see, for example, Pepe and Mori (1993), Keiding et al. (2001), Cook et al. (2003) and Cook et al. (2009).

A variety of software packages have simulation features, typically for quite specific types of processes. The R packages `msm` (Jackson, 2011) and `Epi` (Carstensen and Plummer, 2011) can simulate Markov intensities through functions `simmulti.msm` and `simLexis`, respectively. Discrete event simulation and microsimulation software systems can also be useful. For example, the `MicSim` package in R (Zinn, 2014) will simulate nonhomogeneous Markov processes.

2.7 Problems

Problem 2.1 Discrete-time multistate processes can be modeled using multinomial distributions for the risk of transition. Consider a K-state process for which $Z_i(t)$ records the state occupied at time t, $t = 0, 1, \ldots$ If $X_i(s)$ is a vector of time-dependent covariates at s, the history of the joint process at t is $\mathcal{H}_i(t) = \{Z_i(s), s = 0, 1, \ldots, t; X_i(s)\}$. Let $Y_{ik}(t) = I(Z_i(t) = k)$ indicate individual i is in state k at time t, let $N_{ikl}(t) = I(Z_i(t-1) = k, Z_i(t) = l)$ indicate that a $k \to l$ transition occurred at t, and let $N_{ik}(t) = (N_{ik1}(t), \ldots, N_{ikK}(t))'$. Note that $N_{ik}(t)$ is a multinomial random variable. For a general process $P(Z_i(t) = l | \mathcal{H}_i(t-1))$ is a transition probability that allows a dependence on the process history.

(a) Under a first-order Markov process, we let

$$P(Z_i(t) = l \mid Z_i(t-1) = k, \mathcal{H}_i(t-1)) = P(Z_i(t) = l \mid Z_i(t-1) = k, X_i) = \pi_{ikl}(t),$$

where $\pi_{ikl}(t) = \pi_{ikl}$ if the process is time-homogeneous. If we observe a sample of n independent individuals over $0, 1, \ldots, T$, show that the likelihood under a time-homogeneous process can be written

$$L \propto \prod_{i=1}^{n} \prod_{t=1}^{T} \prod_{k=1}^{K} \left(\prod_{l=1}^{K} \pi_{ikl}^{N_{ikl}(t)} \right)^{Y_{ik}(t-1)}.$$

(b) Let X_{ik} denote a $p \times 1$ vector of covariates which affects the transitions out of state k. Generalized logit models are useful to examine covariate effects on transition probabilities, which we often write as

$$\log(\pi_{ikl} / \pi_{ikk}) = x'_{ik} \beta_{kl}, \quad l = 1, \ldots, K,$$

implying

$$\pi_{ikl} = \frac{\exp(x'_{ik} \beta_{kl})}{\sum_{h=1}^{k} \exp(x'_{ik} \beta_{kh})}$$

with $\beta_{kk} = 0$ and $\sum_{h=1}^{k} \pi_{ikh} = 1$. Derive the log-likelihood for $\beta = (\beta'_1, \ldots, \beta'_K)'$ where $\beta_k = (\beta_{k1}, \ldots, \beta_{kK})'$, $k = 1, \ldots, K$.

(c) Derive the score vector and information matrix for β.

(Section 2.2)

Problem 2.2 If $X_i(s)$ is discrete and time-varying in Problem 2.1, derive the form of the likelihood for the joint processes where we model $P(X_i(t)|Z_i(t), \mathcal{H}_i(t-1))$ as well as $P(Z_i(t)|\mathcal{H}_i(t-1))$.

(Section 2.2)

Problem 2.3 Consider a binary covariate X with $P(X = 1) = p$ and $P(X = 0) = 1 - p$. Next let $\{Z(t), t \geq 0\}$ denote a 2-state process with $P(Z(0) = 1) = 1$ and $\mathcal{H}(t) = \{Z(s), 0 \leq s \leq t; X\}$ denote the history of the process at t. The process is Markov given X, with transition intensities

$$\lim_{\Delta t \downarrow 0} \frac{P(Z(t + \Delta t^-) = 3 - k \mid \mathcal{H}(t^-))}{\Delta t} = Y_k(t^-) \lambda_k(t) \exp(\beta_k X), \quad k = 1, 2,$$

where $Y_k(t) = I(Z(t) = k)$. Suppose now that X is unobservable and therefore cannot be regressed upon as it is above. Derive an expression for the intensity

$$\lim_{\Delta t \downarrow 0} \frac{P(Z(t + \Delta t^-) = 3 - k \mid Y_k(t^-) = 1, \mathcal{Z}(t^-))}{\Delta t}, \quad k = 1, 2,$$

where $\mathcal{Z}(t) = \{Z(s), 0 \leq s \leq t\}$ and show that it depends on the entire process history.

(Section 2.3)

Problem 2.4 Consider a heterogeneous population for which a proportion π of individuals follow a progressive K-state process with only $k \to k+1$ transitions possible $(k = 1, \ldots, K-1)$ and the remaining $1 - \pi$ of the population remain in the initial state 1. For simplicity we assume there are no covariates. Let V denote a binary random variable such that $V = 1$ indicates an individual follows the multistate process and if $V = 0$ no transitions will occur out of state 1. Suppose the process is Markov given $V = 1$, so with $\mathcal{H}(t) = \{Z(s), 0 \leq s \leq t\}$,

$$\lim_{\Delta t \downarrow 0} \frac{P(Z(t + \Delta t^-) = 2 \mid \mathcal{H}(t^-), V = v)}{\Delta t} = Y_1(t^-) v \lambda_1(t),$$

where $Y_k(t) = I(Z(t) = k)$. Show that the observable process with $1 \to 2$ intensity defined by

$$\lim_{\Delta t \downarrow 0} \frac{P(Z(t + \Delta t^-) = 2 \mid Z(t^-) = 1, \mathcal{H}(t^-))}{\Delta t}$$

is not Markov. What are the observable intensities for $k \to k+1$ transitions for $k \geq 2$?

(Section 2.3)

Problem 2.5 Consider a progressive K-state model with state diagram given below.

(a) If

$$\lim_{\Delta t \downarrow 0} \frac{P(Z(t + \Delta t^-) = k + 1 \mid \mathcal{H}(t^-))}{\Delta t} = Y_k(t^-) \lambda_k$$

show that

$$P_{kl}(0, t) = \sum_{j=k}^{l} C(k, j, l; \lambda) \exp(-\lambda_j t)$$

for $k \leq l$ and is zero otherwise, where

$$C(k, j, l; \lambda) = \frac{\prod_{h=k}^{l-1} \lambda_h}{\prod_{h=k, h \neq j}^{l} (\lambda_h - \lambda_j)},$$

and $\lambda = (\lambda_1, \ldots, \lambda_{K-1})'$.

(b) Satten (1999) considers a random effect model in which

$$\lim_{\Delta t \downarrow 0} \frac{P(Z(t + \Delta t^-) = k + 1 \mid V = v, \mathcal{H}(t^-))}{\Delta t} = Y_k(t^-) \, v \, \lambda_k .$$

Derive $P_{1l}(0,t) = E_V\{P_{1l}(0,t \mid V = v)\}$, where

$$P_{1l}(0,t \mid V = v) = P(Z(t) = l \mid Z(0) = 1, V = v)$$

and V is gamma distributed with mean one and variance $\phi > 0$.

(c) Derive the means and variances of the state sojourn times W_k resulting from the model in (b) along with $\operatorname{cov}(W_k, W_l)$ for $1 \le k < l \le K - 1$.

(d) Let $K = 5$, $\lambda = 1.0$ and $\lambda_{k+1} = 1.1\lambda_k$, $k = 1, 2, 3$. Plot $P(Z(t) = K \mid Z(0) = 1)$ from (a) and (b) with $\phi = 0.2$ and comment on the differences.

(e) Derive the form of the $k \to k + 1$ intensity function for the marginal model (i.e. after averaging over the random effect), $k = 1, \dots, K - 1$.

(Sections 2.3, 2.4)

Problem 2.6   Outline procedures for simulating life histories from the models in each of Problems 1.1 and 2.5.

(Section 2.5)

Problem 2.7   Consider the illness-death model of Figure 1.1 for a time-homogeneous Markov process with transition intensities $\lambda_{12} = \lambda_1$, $\lambda_{13} = \mu_1$ and $\lambda_{23} = \mu_2$. Let T represent time to death (entry to state 3) and let $\Delta_2 = I$(individual entered state 2) indicate whether the individual passed through state 2 or not.

(a) Prove that

$$E(T \mid \Delta_2 = 0) = \frac{1}{\lambda_1 + \mu_1} \quad \text{and} \quad E(T \mid \Delta_2 = 1) = \frac{\lambda_1 + \mu_1 + \mu_2}{\mu_2(\lambda_1 + \mu_1)} .$$

(b) Show that $E(T \mid \Delta_2 = 0) < E(T \mid \Delta_2 = 1)$.

(Sections 2.3, 2.4)

Problem 2.8   Consider the multistate diagram in Figure 1.5 for a setting involving three modes of failure. Denote the intensities as $\lambda_{0k}(t)$, $k = 1, 2, 3$, let T be the time of failure, and let ε record the mode of failure.

(a) Give $P(T \ge t)$, the probability density function of T, and $P(\varepsilon = j \mid T = t)$.

(b) Outline how you would simulate realizations of (T, ε) for this type of process. Implement this and generate a sample of size $n = 2000$ for the case where $\lambda_{01}(t) = 1$, $\lambda_{02}(t) = 2$, $\lambda_{03}(t) = 0.5t$. Confirm that the empirical frequencies of entry into the absorbing states are in rough agreement with what you would expect based on your result in (a).

(Sections 2.4, 2.5)

Problem 2.9 Consider a continuous-time Markov process with K states, with transition intensities of the form $\lambda_{kl}(t) = \lambda_{kl}\, g(t)$, where the λ_{kl} are positive constants and $g(t)$ is positive valued. Find the form of the transition probability matrices $P(s,t)$ in terms of $g(t)$ and the matrix Q with (k,l) entry λ_{kl} for $k \neq l$ and diagonal entries $-\sum_{l\neq k}\lambda_{kl}$, for $k = 1,\dots,K$.

(Section 2.3)

Problem 2.10 Consider a K-state process in which individuals begin in state 1 at $t = 0$. It is often of interest to determine the marginal state occupancy probabilities (sometimes called state prevalence functions)

$$P_j(t) = P(Z(t) = j) = P(Z(t) = j \mid Z(0) = 1),$$

as well as the expected total time spent in state j over the time interval $[0, C^A]$, denoted $\psi_j(C^A)$.

(a) Show that

$$\psi_j(C^A) = \int_0^{C^A} P_j(t)\, dt, \quad j = 1,\dots,K.$$

(b) Derive functions $\psi_j(C^A)$, $j = 1,2$, for a progressive 3-state time-homogeneous Markov process with intensities $\lambda_{12} = \alpha$ and $\lambda_{23} = \beta$.

(c) Consider a survival time model as a 2-state process with state 1 representing the condition of being alive and state 2 entered upon death. Use part (a) to prove the well-known result

$$E(T) = \int_0^\infty \mathcal{F}(t)\, dt,$$

where T is the survival time variable with survivor function $\mathcal{F}(t) = P(T > t)$, and $E(T)$ is assumed to exist.

(Section 2.4)

Problem 2.11 Let $T_k^{(r)}$ and $V_k^{(r)}$ be the entry and exit times for the rth visit to state k $(r = 1,2,\dots)$ for a multistate process with intensity functions $\lambda_{kl}(t|\mathcal{H}(t^-))$ as in (2.2). Define $W_k^{(r)} = V_k^{(r)} - T_k^{(r)}$ as the rth sojourn time in state k and

$$E_k^{(r)} = \int_{T_k^{(r)}}^{V_k^{(r)}} \lambda_{k\cdot}(t \mid \mathcal{H}(t^-))\, dt$$

where $\lambda_{k\cdot}(t|\mathcal{H}(t^-)) = \sum_{l\neq k}\lambda_{kl}(t|\mathcal{H}(t^-))$.

(a) Show that $W_k^{(r)}$ has an exponential distribution with mean 1 and that distinct $W_k^{(r)}$, $k = 1,\dots,K$, $r = 1,2,\dots$, are mutually independent. By using product integrals show first that

$$P(W_k^{(r)} > w \mid t_k^{(r)}, \mathcal{H}(t_k^{(r)-})) = \exp\left(-\int_{t_k^{(r)}}^{t_k^{(r)}+w} \lambda(t \mid \mathcal{H}(t^-))\, dt\right).$$

(b) How might the $E_k^{(r)}$ be used for modeling checking?

(Section 2.2)

Chapter 3

Multistate Analysis Based on Continuous Observation

In this chapter we present methods for fitting multistate models for the analysis of data from individuals who are observed continuously over a period of time in a prospective study. We first construct likelihoods and develop methods for estimation and inference for parametric models. Nonparametric methods and semiparametric analyses are then considered. Examples are provided to illustrate the use of software and the interpretation of results.

3.1 Maximum Likelihood Methods for Parametric Models

Let θ be a vector of parameters indexing the transition intensity functions $\lambda_{kl}(t|\mathcal{H}(t^-))$ in (2.1) for a multistate process with K states. In most cases there are functionally independent vectors θ_{kl} $(k \neq l)$ that parameterize each individual intensity and unless stated otherwise we assume this here. The likelihood contribution for a single individual was provided in Section 2.2. We assume that observation of an individual begins at time A_0 $(A_0 \geq 0)$ and stops at time C $(C > A_0)$, which satisfy conditions specified in Section 2.2.2. To extend the notation to allow for a sample of n independent individuals, we introduce the subscript i with $i = 1, \ldots, n$ and from (2.19) and (2.20) the full likelihood for $\theta = (\theta_{kl}, k \neq l = 1, \ldots, K)$ is

$$L(\theta) = \prod_{k \neq l} L_{kl}(\theta_{kl}), \tag{3.1}$$

where

$$L_{kl}(\theta_{kl}) = \prod_{i=1}^{n} \left\{ \prod_{t_{ir} \in \mathcal{D}_{ikl}} \lambda_{kl}(t_{ir} \mid \mathcal{H}_i(t_{ir}^-); \theta_{kl}) \right. \tag{3.2}$$

$$\left. \times \exp\left(-\int_0^\infty \bar{Y}_{ik}(u)\lambda_{kl}(u \mid \mathcal{H}_i(u^-); \theta_{kl}) \, du \right) \right\}$$

with $\bar{Y}_{ik}(u) = I(A_{i0} \leq u \leq C_i) I(Z_i(u^-) = k)$ and $\mathcal{D}_{ikl}$ the set of distinct times t_{ir} at which individual i makes an observed transition from state k to state l. It should be noted that the time origin of the process must be known relative to the start of follow-up in order to determine A_{i0}. We discuss some situations where A_{i0} might not be known in Chapter 7. Finally we note that the history over $[0, A_0)$ must be available to obtain the intensity needed to compute (3.2).

The log-likelihood $\ell_{kl}(\theta_{kl}) = \log L_{kl}(\theta_{kl})$ corresponding to (3.2) can be written in the counting process notation introduced in Section 2.1 as

$$\sum_{i=1}^{n} \int_0^\infty \bar{Y}_{ik}(u) \left\{ \log \lambda_{kl}(u \mid \mathcal{H}_i(u^-); \theta_{kl}) \, dN_{ikl}(u) - \lambda_{kl}(u \mid \mathcal{H}_i(u^-); \theta_{kl}) \, du \right\},$$

where $\{N_{ikl}(u), u > 0\}$ is the counting process for $k \to l$ transitions for individual i. The maximum likelihood estimate (MLE) for θ_{kl}, denoted $\hat{\theta}_{kl}$, is obtained by maximizing $L_{kl}(\theta_{kl})$ or, equivalently, $\ell_{kl}(\theta_{kl})$. If we write $\lambda_{ikl}(u; \theta_{kl})$ for $\lambda_{kl}(u|\mathcal{H}_i(u^-); \theta_{kl})$ for convenience, the likelihood score function $U_{kl}(\theta_{kl}) = \partial \ell_{kl}(\theta_{kl})/\partial \theta_{kl}$ can be expressed as

$$U_{kl}(\theta_{kl}) = \sum_{i=1}^{n} \int_0^\infty \bar{Y}_{ik}(u) \left\{ \frac{\partial \log \lambda_{ikl}(u; \theta_{kl})}{\partial \theta_{kl}} \, dN_{ikl}(u) - \frac{\partial \lambda_{ikl}(u; \theta_{kl})}{\partial \theta_{kl}} \, du \right\}. \tag{3.3}$$

The observed information matrix $I_{kl}(\theta_{kl}) = -\partial U_{kl}(\theta_{kl})/\partial \theta'_{kl}$ is then

$$I_{kl}(\theta_{kl}) = \sum_{i=1}^{n} \int_0^\infty \bar{Y}_{ik}(u) \left\{ \frac{\partial^2 \lambda_{ikl}(u; \theta_{kl})}{\partial \theta_{kl} \partial \theta'_{kl}} \, du - \frac{\partial^2 \log \lambda_{ikl}(u; \theta_{kl})}{\partial \theta_{kl} \partial \theta'_{kl}} \, dN_{ikl}(u) \right\}. \tag{3.4}$$

The normalized Fisher information matrix $\mathcal{I}_{kl}(\theta_{kl})$ is the probability limit of $n^{-1} I_{kl}(\theta_{kl})$ as $n \to \infty$. Under mild regularity conditions that imply the existence of these limits and some other criteria (Andersen et al., 1993, Section 6.1), $\sqrt{n}(\hat{\theta}_{kl} - \theta_{kl})$ has a limiting normal distribution with mean vector 0 and covariance matrix $\mathcal{I}_{kl}^{-1}(\theta_{kl})$. A short derivation (see Problem 3.4) shows that $\mathcal{I}_{kl}(\theta_{kl})$ can be estimated consistently by

$$\frac{1}{n} \hat{I}_{kl}(\hat{\theta}_{kl}) = \frac{1}{n} \sum_{i=1}^{n} \int_0^\infty \bar{Y}_{ik}(u) \left\{ \frac{\partial \log \lambda_{ikl}(u; \theta_{kl})}{\partial \theta_{kl}} \frac{\partial \log \lambda_{ikl}(u; \theta_{kl})}{\partial \theta'_{kl}} \right\} \Bigg|_{\hat{\theta}_{kl}} dN_{ikl}(u). \tag{3.5}$$

Note that since (3.5) involves $dN_{ikl}(u)$, it and similar expressions reduce to sums over observed transition times; specifically (3.5) can be written as

$$\hat{I}_{kl}(\hat{\theta}_{kl}) = \sum_{i=1}^{n} \sum_{t_{ir} \in \mathcal{D}_{ikl}} \left\{ \frac{\partial \log \lambda_{ikl}(t_{ir}; \theta_{kl})}{\partial \theta_{kl}} \frac{\partial \log \lambda_{ikl}(t_{ir}; \theta_{kl})}{\partial \theta'_{kl}} \right\} \Bigg|_{\hat{\theta}_{kl}}.$$

Estimation or hypothesis tests can be based on the asymptotic normality of $\hat{\theta}_{kl}$, or on likelihood ratio or score statistics, which have limiting chi-squared or normal distributions.

3.1.1 Markov Models

For Markov models transition intensities $\lambda_{kl}(t|\mathcal{H}_i(t^-); \theta_{kl})$ take the form $\lambda_{kl}(t; \theta_{kl})$. Time-homogeneous models for which $\lambda_{kl}(t; \theta_{kl}) = \theta_{kl}$ are an important special case, and formulas given above then take simple forms. First, we see that the log-likelihood from (3.2) reduces in this case to

$$\log L_{kl}(\theta_{kl}) = \sum_{i=1}^{n} \int_0^\infty \bar{Y}_{ik}(u) \{ \log \theta_{kl} \cdot dN_{ikl}(u) - \theta_{kl} \, du \} = n_{kl} \log \theta_{kl} - S_k \theta_{kl}, \tag{3.6}$$

where here $S_k = \sum_{i=1}^{n} \int_0^\infty \bar{Y}_{ik}(u)\, du$ is the total person-time at risk of a transition out of state k, and $n_{kl} = \sum_{i=1}^{n} \int_0^\infty d\bar{N}_{ikl}(u)$ is the total number of $k \to l$ transitions observed across individuals in the sample. This log-likelihood has the same form as one for a time-homogeneous Poisson process where n_{kl} is the "count" and S_k is the "exposure time". As a result, maximum likelihood estimates for this model and corresponding regression models described later can be computed using software for generalized linear models. Log-likelihoods of this form also arise when fitting exponential failure time models based on censored data in which case n_{kl} corresponds to the total number of event times observed in the sample and S_k the total person-time at risk. We illustrate this in the analyses that follow.

The maximum likelihood estimate for the $k \to l$ transition rate arising from (3.6) has the natural form

$$\hat{\theta}_{kl} = \frac{n_{kl}}{S_k}. \tag{3.7}$$

For the time-homogeneous model, (3.5) reduces to $\hat{I}_{kl}(\hat{\theta}_{kl}) = n^{-1} \cdot n_{kl}/\hat{\theta}_{kl}^2$, so the estimated normal approximation for $\hat{\theta}_{kl}$ is

$$\sqrt{n}(\hat{\theta}_{kl} - \theta_{kl}) \sim \left(0, \frac{n\hat{\theta}_{kl}^2}{n_{kl}}\right). \tag{3.8}$$

Confidence intervals and Wald tests for θ_{kl} can be based on this. If n and n_{kl} are not large, a better approximation is to assume

$$\sqrt{n}(\hat{\phi}_{kl} - \phi_{kl}) \sim N\left(0, \frac{9n \cdot n_{kl}^{1/3}}{S_k^{2/3}}\right)$$

where $\phi_{kl} = \theta_{kl}^{-1/3}$; see Problem 3.4. Likelihood ratio statistics can also be used for tests or confidence intervals; they take the form

$$\text{LRS}_{kl}(\theta_{kl}) = 2\left\{\log L_{kl}(\hat{\theta}_{kl}) - \log L_{kl}(\theta_{kl})\right\} = 2\left\{n_{kl}\log(\hat{\theta}_{kl}/\theta_{kl}) - S_k(\hat{\theta}_{kl} - \theta_{kl})\right\}$$

and assuming that θ_{kl} is the true parameter value, $\text{LRS}_{kl}(\theta_{kl})$ is asymptotically $\chi^2_{(1)}$. It is apparent from the factorization of the full likelihood (3.1) that the $\hat{\theta}_{kl}$ and associated pivotals are asymptotically independent across distinct (k, l).

The assumption that the transition intensities are constant is strong, so more flexible functional forms are often appealing. Models with piecewise-constant (or piecewise time-homogeneous) intensity functions are very useful and quite easy to fit. In this case, we let $0 = b_0 < b_1 < \cdots < b_R = \infty$ specify a partition of the positive real line, with $\mathcal{B}_r = [b_{r-1}, b_r)$ and $\cup_{r=1}^{\infty} \mathcal{B}_r = [0, \infty)$. The intensities are then taken to have the form

$$\lambda_{kl}(t; \theta_{kl}) = \theta_{klr}, \quad t \in \mathcal{B}_r,$$

for $r = 1, \ldots, R$. Following the steps in the earlier derivation for the time-homogeneous model, we write the log-likelihood for the parameters $\theta_{kl} =$

$(\theta_{kl1}, \ldots, \theta_{klR})'$ for a piecewise-constant model as

$$\log L_{kl} = \sum_{i=1}^{n} \sum_{r=1}^{R} \int_{0}^{\infty} \bar{Y}_{ikr}(u) \{\log \theta_{klr} \cdot dN_{ikl}(u) - \theta_{klr} \, du\}$$

$$= \sum_{r=1}^{R} \{n_{klr} \log \theta_{klr} - S_{kr} \theta_{klr}\}$$

where $\bar{Y}_{ikr}(u) = \bar{Y}_{ik}(u) I(u \in \mathcal{B}_r)$, $n_{klr} = \sum_{i=1}^{n} \int_{0}^{\infty} \bar{Y}_{ikr}(u) \, dN_{ikl}(u)$ is the total observed number of $k \to l$ transitions over $\mathcal{B}_r$, and $S_{kr} = \sum_{i=1}^{n} \int_{0}^{\infty} \bar{Y}_{ikr}(u)$ is the total person-time at risk for a $k \to l$ transition over $\mathcal{B}_r$. The MLE for θ_{klr} is then $\widehat{\theta}_{klr} = n_{klr}/S_{kr}$ and we have the analogous results to (3.8) as

$$\sqrt{n}(\widehat{\theta}_{klr} - \theta_{klr}) \sim N\left(0, \frac{n\widehat{\theta}_{klr}^2}{n_{klr}}\right).$$

Specification of the number and location of the break-points $b_1, \ldots, b_{R-1}$ is best done prior to examination of the data, but it is apparent from the form $\widehat{\theta}_{klr} = n_{klr}/S_{kr}$ that the estimate is undefined if there are no individuals at risk of a transition out of state k in $\mathcal{B}_r$; a practical solution is to remove or relocate one of the break-points b_{r-1} or b_r to ensure there is a positive time at risk and the corresponding estimate is defined. Models with different sets of break-points b_r for different transition intensities can also be used. This flexibility is particularly useful when first entry times vary substantially across states; there may then be little data early in the course of follow-up about certain transition intensities.

Transition probabilities are complicated functions of the parameters θ_{kl}, and several approaches are used to obtain variance estimates or confidence intervals for these. Suppose we are interested in $P_{kl}(s,t;\theta)$ for some specific values (k,l) and (s,t), where θ denotes the $m \times 1$ vector $(\theta_1, \ldots, \theta_m)'$ that contains all of the elements of all θ_{kl} parameters in some order. An estimate for the asymptotic variance of the maximum likelihood estimator $P_{kl}(s,t;\widehat{\theta})$ is given by the delta formula:

$$\widehat{\text{var}}(P_{kl}(s,t;\widehat{\theta})) = \sum_{r=1}^{m} \sum_{u=1}^{m} \left\{\frac{\partial P_{kl}(s,t;\theta)}{\partial \theta_r} \frac{\partial P_{kl}(s,t;\theta)}{\partial \theta_u}\right\}\bigg|_{\widehat{\theta}} \widehat{\text{cov}}(\widehat{\theta}_r, \widehat{\theta}_u), \qquad (3.9)$$

where $\widehat{\text{cov}}(\widehat{\theta}) = m^{-1}I^{-1}(\widehat{\theta})$ is the estimated covariance matrix for $\widehat{\theta}$ with elements $\widehat{\text{cov}}(\widehat{\theta}_r, \widehat{\theta}_u)$. The $m \times m$ matrix $I(\widehat{\theta})$ is obtained from components (3.5) and the fact that the separate $\widehat{\theta}_{kl}$ are asymptotically independent.

For a few Markov models there are simple expressions for $P_{kl}(s,t;\theta)$ so that the derivatives in (3.9) can be determined analytically, but in general numerical approximation is needed. An effective approach is to use numerical differentiation, whereby

$$\frac{\partial P_{kl}(s,t;\theta)}{\partial \theta_r} \doteq \frac{P_{kl}(s,t;\theta + \Delta_r) - P_{kl}(s,t;\theta - \Delta_r)}{2\delta_r}, \qquad (3.10)$$

where Δ_r is an $m \times 1$ zero vector except for the small value $\delta_r > 0$ for the element corresponding to θ_r. The choice of δ_r for this numerical derivative depends on the units of time in the model and the magnitudes of the θ_{kl} but after a little trial and error, satisfactory accuracy can be achieved. It is assumed throughout this discussion that there is an accurate way to compute transition probabilities $P_{kl}(s,t;\theta)$ for given s,t and θ. For time-homogeneous models or piecewise homogeneous models, the exponential formula applies as discussed in Section 2.3.1.

The nonparametric bootstrap offers an alternative approach to variance estimation. It involves choosing several (B) samples of n individuals with replacement from the original data. These B bootstrap samples labeled $b = 1,\ldots,B$ each yield estimates $\widehat{\theta}^{(b)}$ by maximizing (3.1); these give corresponding estimates $P(s,t;\widehat{\theta}^{(b)})$, and from these sample variance estimates or confidence intervals can be obtained. For variance estimation 100 or 200 bootstrap samples is often sufficient but several thousand may be necessary for inferences based on quantiles of the bootstrap distribution.

Regression models can also be fitted. For example, for multiplicative models with an external covariate vector $x(t)$ and transition intensity specifications $\lambda_{kl0}(t;\alpha_{kl})\exp(x'(t)\beta_{kl})$, the general expressions in (3.2)–(3.5) hold with $\theta_{kl} = (\alpha'_{kl},\beta'_{kl})'$. However, even for the case of time-homogeneous baseline intensity functions $\lambda_{kl}(t;\alpha_{kl})$, simple closed-form expressions are typically not available.

Example 3.1.1:   Illness-Death Models

Consider the model portrayed in Figure 1.1 where Markov transition intensities are denoted $\lambda_{12}(t;\theta_{12})$, $\lambda_{13}(t;\theta_{13})$ and $\lambda_{23}(t;\theta_{23})$. For convenience we do not consider covariates, but they could be included in the models. To write the likelihood function and related quantities, we adopt notation given in Section 1.6, where dataframes for multistate life history were discussed. For simplicity, we assume that each individual is observed from $t = 0$ at which point $Z_i(0) = 1$. We define T_{ik} as the entry time to state k, $k = 2,3$; let $V_{i1} = \min(T_{i2},T_{i3})$ denote the exit time from state 1; and assume that observation of an individual concludes at $\min(T_{i3},C_i)$, where C_i is an independent censoring time. The observed data are then represented by $\tilde{t}_{ik} = \min(t_{ik},C_i)$ for $k = 2,3$, $\tilde{v}_{i1} = \min(v_{i1},C_i)$, and we let

$$\delta_{ikl} = I(k \to l \text{ transition is observed for individual } i).$$

Due to the progressive nature of the model in Figure 1.1, the sets $\mathcal{D}_{ikl}$ in (3.2) contain at most one element. The likelihood function (3.1) therefore has components (3.2), which can be written as

$$L_{12}(\theta_{12}) = \prod_{i=1}^{n} \lambda_{12}(\tilde{v}_{i1};\theta_{12})^{\delta_{i12}} \exp\left(-\int_{0}^{\tilde{v}_{i1}} \lambda_{12}(u;\theta_{12})\,du\right)$$

$$L_{13}(\theta_{13}) = \prod_{i=1}^{n} \lambda_{13}(\tilde{v}_{i1};\theta_{13})^{\delta_{i13}} \exp\left(-\int_{0}^{\tilde{v}_{i1}} \lambda_{13}(u;\theta_{13})\,du\right)$$

$$L_{23}(\theta_{23}) = \prod_{i=1}^{n} \left\{\lambda_{23}(\tilde{t}_{i3};\theta_{23})^{\delta_{i23}} \exp\left(-\int_{\tilde{t}_{i2}}^{\tilde{t}_{i3}} \lambda_{23}(u;\theta_{23})\,du\right)\right\}^{\delta_{i12}}.$$

We note that there are four possible life histories for an individual, illustrated by the sample data in Section 1.6.1. These are

(i) Still in state 1 at the end of follow-up (e.g. individual with id $= 2$ in the `mbc` dataframe in Section 3.2.2)

(ii) Transition from state 1 to state 2 observed, and still in state 2 at the end of follow-up (e.g. individual id $= 5$)

(iii) A direct transition from state 1 to state 3 is observed (e.g. id $= 1$)

(iv) Transitions from state 1 to state 2 and from state 2 to state 3 both observed (e.g. id $= 69$)

Each of the four types of data contribute to two or more of L_{12}, L_{13} and L_{23}.

For models with time-homogeneous or piecewise-constant intensities, the closed form results described in Section 3.1.1 apply. We find here that $\widehat{\lambda}_{kl} = n_{kl}/S_k$ for $(k,l) = (1,2)$, $(1,3)$ and $(2,3)$, with $n_{kl} = \sum_{i=1}^{n} \delta_{ikl}$, $S_1 = \sum_{i=1}^{n} \tilde{v}_{i1}$ and $S_2 = \sum_{i=1}^{n} \delta_{i12}(\tilde{t}_{i3} - \tilde{t}_{i2})$. These piecewise-constant intensities can be easily fit using the `phreg` function in the `eha` library developed for survival analysis. We defer an illustration of parametric model fitting to Section 3.2.2 where nonparametric estimates are also considered, but observe that the likelihood contributions in $L_{12}(\theta_{12})$, $L_{13}(\theta_{13})$ and $L_{23}(\theta_{23})$ take the form of failure time likelihoods.

3.1.2 Semi-Markov Models

Semi-Markov models are based on the durations of sojourns in states that are occupied once they are entered, which are sometimes called state sojourn times. For semi-Markov models the transition intensity functions $\lambda_{kl}(t|\mathcal{H}_i(t^-);\theta_{kl})$ take the form $h_{kl}(B_i(t);\theta_{kl})$, where $B_i(t)$ is the time since entry to the current state k for individual i in a sample of size n, $i = 1,\ldots,n$. In settings where states are recurrent (i.e. they can be entered more than once), there may be several sojourn times generated by an individual process. The likelihood contribution $L_{kl}(\theta_{kl})$ for k to l transitions in (3.2) becomes

$$\prod_{i=1}^{n} \left[\prod_{w_{ir} \in \mathcal{D}_{ikl}} h_{kl}(w_{ir};\theta_{kl}) \right] \exp\left(- \int_0^{\infty} \bar{Y}_{ik}(u)\, h_{kl}(B_i(u);\theta_{kl})\, du \right), \qquad (3.11)$$

where $\mathcal{D}_{ikl}$ is the set of sojourn times w_{ir} in state k that ended with a transition to state l for individual i. If a number of states may be entered upon exit from state k, then a competing risks framework is adopted for analysis. In this case the $h_{kl}(w;\theta_{kl})$ may be viewed as cause-specific hazard functions governing both the sojourn time in state k and the nature of the transition ending the sojourn in state k. We discuss competing risks in detail in Section 4.1.

As for Markov models, software for parametric survival analysis software can be used with a dataframe suitably constructed. Note, however, that when applying a competing risks approach for the analysis of a general semi-Markov process in the case of (3.11), state k may be entered several times and so there may be several sojourn times observed. If $T_k^{(r)}$ and $V_k^{(r)}$ are the rth entry and exit times of state k, then we let $W_k^{(r)} = V_k^{(r)} - T_k^{(r)}$ denote the rth sojourn time. These sojourn times, along with information on which state was entered at the end of each sojourn (or the censoring information), are represented in the dataframe. The use of sojourn times as the basis for semi-Markov models makes it difficult to obtain explicit expressions for transition probability matrices in all but the simplest of models. Simulation can, however, be used for empirical inferences regarding transition probabilities or state occupancy probabilities from a fitted model. A semi-Markov process with constant intensities $h_{kl}(w;\theta_{kl}) = \theta_{kl}$ is the same as the time-homogeneous Markov process described in Section 3.1, and in this case the calculations are consequently simplified.

Example 3.1.2:   *Illness-Death Models*

A semi-Markov illness-death model has $1 \to 2$ and $1 \to 3$ transition intensities that are of the same form as in the Markov model of Example 3.1.1, since an individual starts in state 1 at time $t = 0$. However, the $2 \to 3$ transition intensity takes the form $h_{23}(b_i(t);\theta_{23})$, where $b_i(t) = t - t_{i2}$ is the current sojourn duration (i.e. at time t) in state 2 for individuals who entered state 2 at time $t_{i2} \leq t$. We let $W_{i2} = T_{i3} - T_{i2}$ denote the ultimate sojourn time in state 2 which of course is only defined for individuals who enter state 2. When there is an independent right censoring time C_i, then the sojourn time in state 2 is censored by $C_i - t_{i2}$, so we define $\tilde{w}_{i2} = \min(w_{i2}, C_i - t_{i2})$ and let $\delta_{i23} = I(\tilde{w}_{i2} = w_{i2})$ indicate that a $2 \to 3$ transition is observed for individual i, $i = 1,\ldots,n$. The likelihood function (3.1) for the semi-Markov model then has components $L_{12}(\theta_{12})$ and $L_{13}(\theta_{13})$ that are of the same form as in Example 3.1.1, with $h_{12}(t;\theta_{12}) = \lambda_{12}(t;\theta_{12})$ and $h_{13}(t;\theta_{13}) = \lambda_{13}(t;\theta_{13})$. The likelihood function for θ_{23}, however, takes the form

$$L_{23}(\theta_{23}) = \left\{ \prod_{i=1}^{n} h_{23}(\tilde{w}_{i2};\theta_{23})^{\delta_{i23}} \exp\left(-\int_0^{\tilde{w}_{i2}} h_{23}(u;\theta_{23})\,du\right) \right\}^{\delta_{i12}}.$$

We note that the likelihoods L_{12} and L_{13} in Example 3.1.1 and $L_{23}(\theta_{23})$ here are exactly the form of the likelihood for a (possibly censored) survival time observation. If $h_{kl}(u;\theta_{kl})$ takes the form of a hazard function for a parametric model that is supported by existing software, we can therefore use the software to estimate θ_{kl}. For example, if $h_{kl}(u;\theta_{kl}) = \alpha_1 \alpha_2 u^{\alpha_2 - 1}$ with $\theta_{kl} = (\alpha_1,\alpha_2)'$, software that handles Weibull survival time distributions (e.g. Lawless, 2003, Section 5.2) can be used. We defer an illustration of model fitting to Section 3.2.4.

3.1.3   Multistate Processes with Hybrid Time Scales

In some settings it may be best to use Markov intensities for certain types of transitions and semi-Markov intensities for others. We refer to these processes as having "hybrid" time scales and give an illustration involving such a formulation in Section 3.2.4, where we model the onset and resolution of outbreaks of symptoms in individuals infected with the herpes simplex virus. In this setting, it is reasonable to model the risk of outbreak as a function of time since diagnosis but more natural to model the duration of any outbreak in terms of the time since the start of the outbreak.

Markov or semi-Markov models may be modulated through the inclusion of covariates that represent aspects of prior life history, as we described in Sections 2.3.1

and 2.3.2. One or more parts of that history may involve the times of the occurrence of previous transitions. This yields intensities which we refer to in Section 2.3.3 as having dual time scales if only one prior event time is used. As an example, for the illness-death model we could consider $2 \to 3$ transition intensity of the additive form

$$\lambda_{23}(t \mid \mathcal{H}(t^-); \theta_{23}) = g(t; \alpha_1) + h(b(t); \alpha_2),$$

where g and h are specified parametric functions.

3.1.4   Comments on Parametric Models

Nonparametric analysis of data from multistate processes has the advantage of allowing an unrestricted look at intensity functions and other features of multistate models; see the next section. Similarly, semiparametric regression models discussed in Section 3.3 are favoured for assessing covariate effects because they avoid parametric assumptions about baseline transition intensities. Fully parametric models are nevertheless appealing in a number of situations, including the following:

(i) Occasionally scientific background may suggest specific forms for certain intensities.

(ii) Data sparsity in certain time regions can lead to imprecise nonparametric estimates; in such settings a more precise parametric estimate that agrees with observed data may be preferable.

(iii) In some settings where data are incomplete, nonparametric estimation may be difficult or the corresponding estimator may be undefined, whereas parametric modeling may be reasonably straightforward. An example discussed in detail in Chapter 5 is when only the states occupied at intermittent observation times are known, and not the exact transition times.

(iv) Parsimonious parametric models provide useful summaries of analysis. The (approximate) adequacy of specific parametric forms across studies makes their comparison easier, provides a natural basis for meta-analysis, and may suggest new hypotheses about the process in question.

Finally, we note that flexible models involving moderately many parameters provide a bridge between parsimonious parametric models and nonparametric models; spline or piecewise-constant specifications for intensity functions are especially useful.

3.2   Nonparametric Estimation

For Markov and semi-Markov models without covariates, nonparametric estimation of cumulative transition intensities and certain other process features is possible. Nonparametric estimators are useful on their own and also for assessing the fit of parametric models.

3.2.1   Markov Models

As discussed in Section 2.3.1, Markov processes have transition intensity functions (2.1), which take the form $\lambda_{kl}(t)$ for $k \neq l$, where we define $\lambda_{kk}(t) = -\sum_{l \neq k} \lambda_{kl}(t)$

for $k = 1, \ldots, K$. The nonparametric Nelson-Aalen (NA) estimator of the cumulative intensities $\Lambda_{kl}(t) = \int_0^t d\Lambda_{kl}(u) = \int_0^t \lambda_{kl}(u)\, du$ is

$$\widehat{\Lambda}_{kl}(t) = \sum_{i=1}^{n} \sum_{t_{ir} \in \mathcal{D}_{ikl}(t)} \frac{I(t_{ir} \le t)}{\bar{Y}_{\cdot k}(t_{ir})}, \quad k \ne l, \tag{3.12}$$

where $\mathcal{D}_{ikl}$ is defined as in Section 3.1. Here and subsequently we use the convention introduced in Section 2.2 that "dot" subscripts indicate summation, so $\bar{Y}_{\cdot k}(t) = \sum_{i=1}^{n} Y_{ik}(t)$. These estimators are analogous to Nelson-Aalen estimators of the cumulative hazard function in survival analysis and can be thought of as discrete maximum likelihood estimates when $\Lambda_{kl}(t)$ increases only at times for which a $k \to l$ transition is observed. We note that (3.12) can be rewritten as

$$\widehat{\Lambda}_{kl}(t) = \int_0^t d\widehat{\Lambda}_{kl}(u) = \int_0^t \frac{d\bar{N}_{\cdot kl}(u)}{\bar{Y}_{\cdot k}(u)}, \quad k \ne l, \tag{3.13}$$

where by convention we define the integrand to equal 0 if $\bar{Y}_{\cdot k}(u) = 0$. In (3.13) we see that $d\Lambda_{kl}(u)$ is estimated as the number of $k \to l$ transitions observed at time u divided by the number of individuals at risk (that is, under observation and in state k at time u^-).

Andersen et al. (1993, Section 4.1) provide a rigorous derivation of properties for Nelson-Aalen estimators. A careful discussion requires that we define $J_k(u) = I(\bar{Y}_{\cdot k}(u) > 0)$, with $J_k(u)/\bar{Y}_{\cdot k}(u) = 0$ when $\bar{Y}_{\cdot k}(u) = 0$. They obtain the variance estimate

$$\widehat{\mathrm{var}}(\widehat{\Lambda}_{kl}(t)) = \int_0^t \frac{J_k(u)\, d\widehat{\Lambda}_{kl}(u)}{\bar{Y}_{\cdot k}(u)} = \sum_{t^{(r)} \le t} \frac{J_k(t^{(r)})}{\bar{Y}_{\cdot k}(t^{(r)})^2}\, d\bar{N}_{\cdot kl}(t^{(r)}), \tag{3.14}$$

where $t^{(1)} < \cdots < t^{(m)}$ are the distinct times at which m observed transitions (of any type) occur. This variance estimator is based on the continuous-time process, for which two or more transitions can never occur simultaneously. In practice there are occasional "ties" because transition times are recorded on a discrete scale. An alternative estimate is based on a discrete time framework, so that conditional on the past process history, $d\bar{N}_{\cdot kl}(t^{(r)})$ has a binomial distribution. This yields

$$\widehat{\mathrm{var}}(\widehat{\Lambda}_{kl}(t)) = \sum_{t^{(r)} \le t} \frac{J_k(t^{(r)})\, d\bar{N}_{\cdot kl}(t^{(r)})\, (\bar{Y}_{\cdot k}(t^{(r)}) - d\bar{N}_{\cdot kl}(t^{(r)}))}{\bar{Y}_{\cdot k}(t^{(r)})^3} \tag{3.15}$$

$$= \sum_{t^{(r)} \le t} \frac{J_k(t^{(r)})\, d\widehat{\Lambda}_{kl}(t^{(r)})\, (1 - d\widehat{\Lambda}_{kl}(t^{(r)}))}{\bar{Y}_{\cdot k}(t^{(r)})}$$

which is preferred to (3.14) when ties are common. The two variance estimates are typically close in value with moderately large numbers $\bar{Y}_{\cdot k}(t^{(r)})$ at risk. It is generally assumed that, asymptotically as $n \to \infty$, $J_k(u) > 0$ with probability one, but in specific finite samples we should recognize that some transition intensities are inestimable over certain time intervals. Andersen et al. (1993, Section 4.1) show that the asymptotic variance $\sigma_{kl}^2(t)$ of $\sqrt{n}\,(\widehat{\Lambda}_{kl}(t) - \Lambda_{kl}(t))$ is

$$\sigma_{kl}^2(t) = \lim_{n \to \infty} \int_0^t E\left\{ \frac{J_k(u)}{n^{-1}\bar{Y}_{\cdot k}(u)} \right\} \lambda_{kl}(u)\, du, \tag{3.16}$$

prove consistency of $\widehat{\Lambda}_{kl}(t)$ under certain conditions, and show that $\{\sqrt{n}\,(\widehat{\Lambda}_{kl}(t) - \Lambda_{kl}(t)), t > 0\}$ converges in distribution to a mean-zero Gaussian process.

The product integral formula (2.23) in Section 2.3.1 also allows us to obtain nonparametric estimates of transition probabilities. This gives

$$\widehat{P}(s,t) = \prod_{(s,t]} \{I + \widehat{Q}(u)\, du\} \tag{3.17}$$

where $\widehat{Q}(u)\, du$ is the $K \times K$ matrix with entries $d\widehat{\Lambda}_{kl}(u)$ given by the integrand in (3.13) in the off-diagonal and $-\sum_{l \neq k} d\widehat{\Lambda}_{kl}(u)$ in the diagonal, $k = 1, \ldots, K$. The estimator (3.17) is known as the Aalen-Johansen estimator.

Asymptotic theory and variances and covariances for transition probabilities $\widehat{P}_{kl}(s,t)$ can be obtained; see Andersen et al. (1993, Section 4.4.1.3) or Aalen et al. (2008, Section 3.4.5). The form of the variance estimates is complicated, but software that provides both the estimates (3.17) and variance estimates is available. In particular, the R package etm does this, as we illustrate in Section 3.4.3; see Appendix A.

3.2.2   An Illness-Death Analysis of a Metastatic Breast Cancer Trial

In Section 1.6.1 we discussed an international multicenter trial of individuals with breast cancer metastatic to bone reported by Hortobagyi et al. (1996). In this study individuals were randomized to receive pamidronate, a bisphosphonate drug with bone strengthening properties, or a placebo control. The primary goal was to assess the effect of pamidronate on reducing the occurrence of skeletal complications arising due to the skeletal metastatic lesions. The multistate diagram in Figure 1.6 recognizes the fact that individuals with metastatic cancer are at non-negligible risk of death and that, in particular, individuals may die without experiencing a skeletal event.

Table 3.1 lists and describes some key variables in this dataset. Recall from Section 1.6 that an individual contributes multiple rows to the dataframe, each of which correspond to periods at risk for specific types of transitions. The first eight variables were discussed in Section 1.6.1; among the remaining, age contains the patient age at study entry, and prior.frac and prior.rad indicate whether a patient experienced a fracture or required radiotherapy for the treatment of bone pain prior to study entry. We will make use of these last three covariates in Section 3.3.2 when discussing regression analyses. A few lines of the dataframe are displayed below.

```
> mbc
  id start stop from to status trt gtime age prior.frac prior.rad
   1     0   68    1  2      0   0    0   68 59          0         0
```

1	0	68	1	3	1	0	68	59	0	0
2	0	749	1	2	0	0	749	38	0	0
2	0	749	1	3	0	0	749	38	0	0
5	0	127	1	2	1	1	127	45	0	1
5	127	365	2	3	0	1	238	45	0	1
5	0	127	1	3	0	1	127	45	0	1
24	0	83	1	2	1	0	83	50	0	0
24	83	639	2	3	0	0	556	50	0	0
24	0	83	1	3	0	0	83	50	0	0
69	0	39	1	2	1	0	39	62	0	1
69	39	637	2	3	1	0	598	62	0	1
69	0	39	1	3	0	0	39	62	0	1
:	:	:	:	:	:	:	:	:	:	:

Table 3.1: Variables and their meaning for the 3-state analysis of skeletal events and death from a metastatic breast cancer trial.

Variable	Description
id	patient ID
start	the time at the beginning of the period at risk
stop	the time at the end of the period at risk
from	the state occupied over the period at risk
to	a state potentially entered at the end of the period at risk
status	indicator of a transition at stop time
trt	0 if placebo; 1 if pamidronate
gtime	the time between the beginning and end of the period at risk
age	age at study entry (years)
prior.frac	1 if patient had experienced a fracture prior to study entry; 0 otherwise
prior.rad	1 if patient had required radiotherapy prior to study entry; 0 otherwise

We can examine and compare trends in the risk of skeletal events or death over time by examining estimates of the cumulative intensities $\Lambda_{kl}(t) = \int_0^t \lambda_{kl}(u)\,du$ for each of the two treatment groups. We first carry out a parametric analysis based on the assumption that the three transition intensities are of a Weibull form with

$$\lambda_{kl}(t \mid \mathcal{H}(t^-)) = \lambda_{kl}(t; \theta_{kl}) = \theta_{kl1}^{-1} \theta_{kl2} \cdot (t/\theta_{kl1})^{\theta_{kl2}-1},$$

where θ_{kl1} and θ_{kl2} are the scale and shape parameters. Such models are often adequate over fairly short time periods; this will be checked later. As mentioned in Section 2.2.2, this model can be fitted using software for parametric survival analysis accommodating left truncation. Here we use the **phreg** function in the event history analysis package **eha**; it deals with covariates and can also be used for analyses involving piecewise-constant transition intensities as described below. We provide the code for estimation of the $1 \to 2$ transition intensity for the control group by selecting the relevant lines of the dataframe; the other transition intensities are estimated in a similar fashion. The **eha** package mentioned briefly in Section 1.5 and used here is discussed in more detail in Appendix A and in the examples of Appendix B.

```
> library(eha)
> weib0.12 <- phreg(Surv(stop, status) ~ 1,
                data=mbc[(mbc$from == 1) & (mbc$to == 2) & (mbc$trt == 0),],
                dist="weibull", center=FALSE)
> summary(weib0.12)
```

Covariate	W.mean	Coef	Exp(Coef)	se(Coef)	Wald p
log(scale)		5.826		0.088	0.000
log(shape)		0.013		0.070	0.854

```
Events                      126
Total time at risk       42745
Max. log. likelihood    -860.15
```

There are no statistics reported under the heading Exp(Coef) since there are no co-variates in this model, but these entries will be populated in regression applications we consider shortly. The entries in coef[1] and coef[2] are maximum likelihood estimates for $\log\theta_{121}$ and $\log\theta_{122}$, respectively, and estimates of the cumulative transition intensity function $(\theta_{121}^{-1}t)^{\theta_{122}}$ at times $0.01, 0.02, \ldots, 720$ are obtained as follows:

```
> tt <- seq(0.01,720,by=0.01)
> weib0.Q12 <- ( exp((-1)*weib0.12$coef[1])*tt )^exp(weib0.12$coef[2])
```

Before examining this further, we fit an intensity with a piecewise-constant form also using the phreg function. This requires specification of the break-points $b_1 < b_2 < \cdots < b_{R-1}$ at which the intensities change value. Here we allow the intensity to take a new value every 90 days for the first year and again at 450 days; these break-points were chosen to provide flexibility in the shape of the intensity while ensuring there is sufficient data to enable estimation of all pieces. The models with piecewise-constant intensity specification are invoked by the command "dist = pch" (which stands for piecewise-constant hazard) and the break-points are specified in the "cuts" object.

```
> pwc0.12 <- phreg(Surv(stop, status) ~ 1,
                data=mbc[(mbc$from == 1) & (mbc$to == 2) & (mbc$trt == 0),],
                dist="pch", cuts=c(90,180,270,360,450), center=FALSE)
> pwc0.12$hazards
        (.., 90] (90, 180] (180, 270] (270, 360] (360, 450] (450, ...]
  [1,] 0.003017  0.003661   0.003363   0.001956   0.004772   0.001043
```

The reported estimates in the hazards object are the $\hat{\theta}_{12r}$ estimates mentioned in Section 3.1.1 and can be summarized as follows:

```
> lambda12 <- data.frame(start=c(0,90,180,270,360,450),
                    stop=c(90,180,270,360,450,9999),
                    lam12=as.vector(pwc0.12$hazards))
> lambda12
  start stop    lam12
      0   90 0.003017
     90  180 0.003661
    180  270 0.003363
```

```
    270   360 0.001956
    360   450 0.004772
    450 9999 0.001043
```

The cumulative transition intensities are obtained in the code that follows:

```
> pwc0.Q12 <- rep(0, length(tt))
> for (p in 1:nrow(lambda12)) {
    pwc0.Q12 <- ifelse((tt > lambda12$start[p]) & (tt <= lambda12$stop[p]),
                    lambda12$lam12[p], pwc0.Q12)
  }
> pwc0.Q12 <- cumsum(pwc0.Q12*0.01)
```

The nonparametric Nelson-Aalen estimates $\widehat{\Lambda}_{12}(t)$, $\widehat{\Lambda}_{13}(t)$, $\widehat{\Lambda}_{23}(t)$ given by (3.13) are the most flexible estimates and can be obtained by a call to the `survfit` function with the specification that estimation be carried out under the `type = "fh2"` option. From the form of (3.13), it is apparent that this is a step function that jumps each time a transition of the respective kind is observed. The number of individuals making jumps at a given time, as well as the number of individuals in the risk set at that time, determine the magnitude of the increment.

```
> np0.12 <- survfit(Surv(stop, status) ~ 1, data=mbc,
          subset = ((from == 1) & (to == 2) & (trt == 0)), type="fh2")
```

The parametric and nonparametric estimates are plotted in the left column of panels in Figure 3.1 for the control group and the estimates for the pamidronate arm are in the right column. The code for the control arm is below; the analogous code for the treatment arm is omitted.

```
> plot(0, 0, type="n", xlim=c(0,24), ylim=c(0,2),
      xlab="MONTHS  SINCE  STUDY  ENTRY",
      ylab="CUMULATIVE  INTENSITY  FOR  SKELETAL  EVENT")
> legend(0, 2, c("NONPARAMETRIC","WEIBULL","PIECEWISE"),
      lty=c(1,2,3), bty="n", cex=0.8)
> lines(c(0, np0.12$time/30), c(0, -log(np0.12$surv)), type="s", lty=1)
> lines(c(0, tt/30), c(0, weib0.Q12), type="l", lty=2)
> lines(c(0, tt/30), c(0, pwc0.Q12), type="l", lty=3)
```

The plots of the three estimates of $\Lambda_{12}(t)$ for each treatment group (top row of Figure 3.1) show fairly constant intensities for skeletal events, reflected by the roughly linear plots over the 24 months of follow-up. For the parametric estimate based on the Weibull-type intensity, the test of $H_0: \theta_{122} = 1$ (or equivalently $\log \theta_{122} = 0$) is a test of the null hypothesis that there is no trend in risk of developing a skeletal complication; this gives a p-value of 0.854 for the control group. The three types of estimates (Weibull, piecewise-constant intensity and Nelson-Aalen estimate) are in broad alignment for each treatment group, but the slope for $\widehat{\Lambda}_{12}(t)$ is much lower for the individuals treated with pamidronate, suggesting an effect in reducing the risk of skeletal complications. There appears to be less evidence of a treatment effect on event-free death or death following a skeletal event, but this is not unexpected given the palliative intent of the treatment. For the pamidronate-treated individuals, the

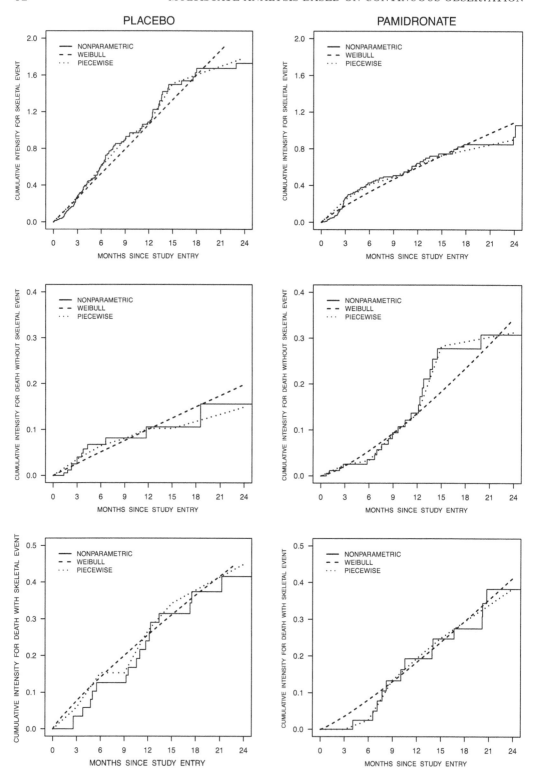

Figure 3.1: Plots of the parametric and nonparametric estimates of the cumulative intensities for a skeletal event $(1 \rightarrow 2)$, skeletal event-free death $(1 \rightarrow 3)$, death following a skeletal event $(2 \rightarrow 3)$ for the placebo arm (left column) and pamidronate arm (right column).

estimate of the $1 \to 3$ intensity based on the piecewise-constant model tracks the Nelson-Aalen estimate much better than the Weibull model in the second year of follow-up. As mentioned following (3.13), when there are no individuals in the risk set at a particular time t, the convention is to define the estimate as zero. As a result the Nelson-Aalen estimate of the cumulative intensity for death following a skeletal event (a $2 \to 3$ transition) may be zero because no individuals have yet entered the risk set, but it will also be zero if individuals have entered the risk set but no transitions have been observed. The fact that the Nelson-Aalen estimate $\widehat{\Lambda}_{12}(t)$ is non-zero at 90 days for the control and pamidronate groups means a reasonable number of individuals are at risk of a $2 \to 3$ transition at some points over the first 90 days. The nonparametric estimate $\widehat{\Lambda}_{23}(t)$ is zero for the first 90 days in the pamidronate group and for about 80 days in the placebo group, indicating that no $2 \to 3$ transitions occurred. It is apparent that the Weibull form of the intensity does not reflect this period of low risk as well as the piecewise-constant estimate for the pamidronate group.

3.2.3 Semi-Markov Models

As noted in Section 3.1.2, the likelihood functions $L_{kl}(\theta_{kl})$ in (3.11) for semi-Markov processes are the same form as for a competing risks setting, except that an individual may contribute more than once if state k can be visited more than once. Competing risks models are described in Section 4.1 in some detail, but we note the forms of nonparametric estimates here. Nonparametric Nelson-Aalen estimates and related survival probability estimates are readily obtained. Let $H_{kl}(w) = \int_0^w h_{kl}(u)du$ denote the cumulative intensity function for $k \to l$ transitions after a sojourn of length w in state k. In addition, we define $\widetilde{\mathcal{D}}_{ik}$ as the set of all sojourn times in state k for individual i including one possibly censored by end of follow-up,

$$ n_k(u) = \sum_{i=1}^{n} \sum_{w \in \widetilde{\mathcal{D}}_{ik}} I(w \geq u) $$

as the size of the risk set for transitions out of state k at time u after entry aggregating across repeated sojourns for each individual and across the entire sample, and

$$ d_{kl}(u) = \sum_{i=1}^{n} \sum_{w \in \widetilde{\mathcal{D}}_{ik}} I(w = u)\, I(\text{transition at sojourn time } w \text{ is to state } l)\,. $$

We also let $w_1^* < w_2^* < \cdots < w_{R(k)}^*$ be the set of distinct sojourn times in state k across all individuals. The Nelson-Aalen estimator for $H_{kl}(w)$ is then

$$ \widehat{H}_{kl}(w) = \int_0^w d\widehat{H}_{kl}(u)\,, \quad w > 0\,, \tag{3.18} $$

where $d\widehat{H}_{kl}(u) = d_{kl}(u)/n_k(u)$. Distribution functions associated with sojourns in a given state can also be estimated. The Nelson-Aalen estimator for the cumulative hazard function for the duration of a spell in state k, no matter which state is visited next, is

$$\widehat{H}_k(w) = \sum_{l \neq k} \widehat{H}_{kl}(w), \quad w > 0. \tag{3.19}$$

The corresponding Kaplan-Meier estimator for the probability a sojourn in state k has duration W_k of length w or greater is

$$\widehat{\mathcal{F}}_k(w) = \prod_{w_r^* < w} \left(1 - \frac{d_{k \cdot}(w_r^*)}{n_k(w_r^*)} \right). \tag{3.20}$$

Sub-distribution functions

$$F_{kl}(w) = P(W_k \leq w, \text{ next state is } l) = \int_0^w \mathcal{F}_k(u) \, dH_{kl}(u)$$

are estimated by

$$\widehat{F}_{kl}(w) = \sum_{w_r^* \leq w} \widehat{\mathcal{F}}_k(w_r^*) \, \frac{d_{kl}(w_r^*)}{n_k(w_r^*)}. \tag{3.21}$$

Variance estimation for the Nelson-Aalen estimators and the sub-distribution functions and software for obtaining them are described in Section 4.1. However, special arguments are needed to prove asymptotics in the semi-Markov setting. Ordinary martingale methods cannot be used because the time scale involved here is not calendar (chronological) time t, but time since a state was entered. Andersen et al. (1993, Sections 10.1, 10.2) discuss this.

3.2.4   Recurrent Outbreaks of Symptoms from Herpes Simplex Virus

Here we consider data from a randomized crossover trial reported in Romanowski et al. (2003) in which 202 patients infected with the herpes simplex virus were randomized to receive a daily 500 mg dose of a drug valacyclovir with an increase to 1000 mg per day when an outbreak of symptoms occurred (suppressive therapy) or to take 1000 mg of valacyclovir only on days when symptoms are present (episodic therapy); those randomized to suppressive therapy switched to episodic therapy after 6 months, and those randomized to episodic therapy switched to suppressive therapy in the second 6-month period.

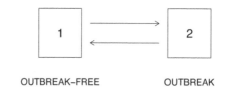

Figure 3.2: An alternating 2-state process characterizing the onset and resolution of outbreaks of symptoms in individuals with herpes simplex virus infection.

The primary purpose of the trial was to obtain data on patient preference of the two treatment regimens, but the data obtained from the study provide an opportunity to study the disease process and associated treatment effects; we restrict attention here to data from the first period during which the average numbers of outbreaks in the episodic and suppressive therapy arms were 3.15 and 0.57, respectively. These averages suggest a very strong prophylactic effect of suppressive therapy; the analyses that follow are directed more at the dynamic aspects of the disease process.

Table 3.2 describes some of the key variables. The first few variables pertain to the onset and resolution of outbreaks, with the **enum** variable conveying the cumulative number of outbreaks that have occurred since randomization for each row. The **gtime** variable is defined as **stop - start** and will serve as the response in analyses of the duration of outbreaks. Additional covariates include **sex** (1 = male, 0 = female), **vtype** (virus type; 1 = HSV type II, 0 otherwise), and **nrecur** which records the number of recurrent outbreaks experienced during the previous year.

Table 3.2: Variables and their meaning for the alternating 2-state analysis of outbreaks from the herpes simplex virus trial.

Variable	Description
id	patient ID
start	the time at the beginning of the period at risk
stop	the time at the end of the period at risk
from[†]	the state occupied over the period at risk (1 = outbreak free; 2 = outbreak)
to[†]	a state potentially entered at the end of the period at risk
status	indicator of a transition at the stop time
gtime	gap time between the beginning and end of the at-risk period
enum	number of the outbreak (since randomization)
	1 = 1st outbreak; 2 = 2nd outbreak; 3 = 3rd outbreak;
	4 = 4th outbreak; 5 = 5th outbreak; 6 = 6th and above outbreak
trt (x_1)	1 if suppressive therapy; 0 if episodic therapy
sex (x_2)	1 if male; 0 if female
vtype (x_3)	1 if infected with Herpes Simplex Virus Type II; 0 otherwise
nrecur (x_4)	number of recurrence in year prior to study entry

[†] The **from** and **to** variables are set to 1 to denote the outbreak-free state and 2 to denote the outbreak state.

A few lines of the dataframe are given below.

```
> hsv
   id start stop from to status gtime enum trt sex vtype nrecur
    1     0   43    1  2      1    43    1   0   0     1      9
    1    43   47    2  1      1     4    1   0   0     1      9
    1    47   72    1  2      1    25    2   0   0     1      9
    1    72   79    2  1      1     7    2   0   0     1      9
    1    79  138    1  2      1    59    3   0   0     1      9
    1   138  143    2  1      1     5    3   0   0     1      9
    1   143  163    1  2      1    20    4   0   0     1      9
    1   163  164    2  1      1     1    4   0   0     1      9
    1   164  166    1 99      0     2    5   0   0     1      9
    2     0    2    1  2      1     2    1   1   1     1      4
    2     2   10    2  1      1     8    1   1   1     1      4
    2    10  176    1 99      0   166    2   1   1     1      4
    :     :    :    :  :      :     :    :   :   :     :      :
```

We initially set aside the covariates and consider a model with intensities of the form

$$\lambda_{i12}(t \mid \mathcal{H}_i(t^-)) = \lambda_{12}(t)$$

reflecting a Markov time scale for the onset of outbreaks, and

$$\lambda_{i21}(t \mid \mathcal{H}_i(t^-)) = h_{21}(B_i(t))$$

reflecting a semi-Markov time scale for the duration of the outbreaks (see Figure 3.2). The Markov time scale for the outbreak process intensity is natural in order to consider risk as a function of time since entry to a randomized study. For the duration of episodic conditions, however, it is more natural to think of the time scale as the time since the start of the outbreak. We proceed with the simple model above as an illustration; in later sections we consider alternating processes of this sort where there may be seasonal effects, more information on the duration of time with the disease, subject heterogeneity, and information on initial conditions, which can be exploited in analysis.

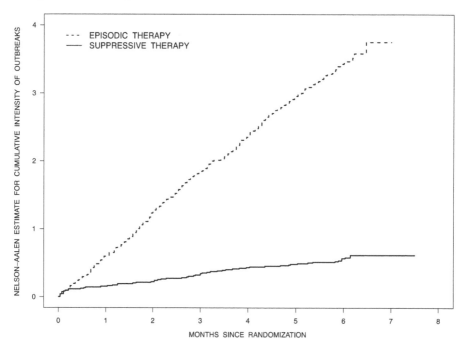

Figure 3.3: Nelson-Aalen estimates of $\Lambda_{12}(t)$, the cumulative intensity for the onset of outbreaks, by treatment group.

Some R code for obtaining the Nelson-Aalen estimates of the cumulative intensity for the onset of outbreaks follows for the episodic therapy arm in period 1, with the estimates of both arms plotted in Figure 3.3. The Markov time scale is reflected by use of the `start` and `stop` variables in the `Surv` object, and the restriction to periods of time when individuals are at risk of an outbreak by selecting the lines with `from == 1`.

```
> library(survival)
> onset0 <- survfit(Surv(start,stop,status) ~ 1,
            data=hsv, subset=((from == 1) & (trt == 0)), type="fh2")

> plot(c(0, onset0$time/28), c(0, -log(onset0$surv)), type="s",
  xlim=c(0,8), ylim=c(0,5), xlab="MONTHS  SINCE  RANDOMIZATION",
  ylab="NELSON-AALEN  ESTIMATE  FOR  CUMULATIVE  INTENSITY  OF  OUTBREAKS")
```

From the estimates plotted in Figure 3.3, we conclude that there is a relatively constant risk of outbreaks over time since randomization, and that the risk of outbreaks is greatly reduced in patients on the suppressive therapy regimen; these inferences are based on the roughly straight lines in Figure 3.3 and the much lower slope of the $\Lambda_{12}(t)$ estimate in the suppressive arm.

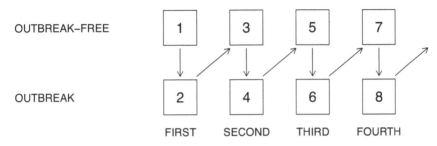

Figure 3.4: A multistate diagram for the analysis of an alternating 2-state process with history-dependent stratification.

Interest often lies in assessing whether there is a trend in other aspects of a process, such as a tendency for the outbreak intensity to increase as the cumulative number of outbreaks increases. This can be conveniently assessed by considering a more general model depicted in Figure 3.4 in which odd and even numbered states represent outbreak-free and outbreak conditions, respectively, and

$$\lim_{\Delta t \downarrow 0} \frac{P(\Delta N_{i12}(t) = 1 \mid Z_i(t^-) = 1 + 2r, \mathcal{H}_i(t^-))}{\Delta t} = \lambda_{12}^{(r)}(t), \quad r = 0, 1, \ldots.$$

Comparison of the slopes of the estimates of $\Lambda_{12}^{(r)}(t)$, $r = 1, 2, \ldots$, enables one to assess how the risk of outbreaks changes according to the number of previous outbreaks. Nelson-Aalen estimates of $\Lambda_{12}^{(r)}(t)$ are easily obtained by stratifying on the cumulative number of events as follows:

```
> onset0.strat <- survfit(Surv(start,stop,status) ~ strata(enum),
            data=hsv, subset=((from == 1) & (trt == 0)), type="fh2")
```

These are plotted in Figure 3.5 for the episodic (left panel) and suppressive (right panel) arms for $r = 1, 2, \ldots, 5$ with an estimate for the sixth stratum obtained by assuming a common rate for the sixth and higher events. We see slight evidence of increasing risk of outbreak with increasing number of previous outbreaks for the episodic therapy arm, but little evidence of this for the suppressive therapy arm where very few patients are observed to experience a third outbreak.

The Nelson-Aalen estimates of the cumulative hazard function $H_{21}(w)$ for the duration of outbreaks are plotted for each treatment arm in Figure 3.6. Here the semi-Markov time scale is reflected by the use of the gap time contained in `gtime`, with relevant lines selected by choosing those with `from == 2`.

```
> res0 <- survfit(Surv(gtime,status) ~ 1,
        data=hsv, subset=((from == 2) & (trt == 0)), type="fh2")

> plot(c(0, res0$time/28), c(0, -log(res0$surv)), type="s", xlim=c(0,2.1),
      ylim=c(0,6), xlab="TIME  SINCE  ONSET  OF  OUTBREAK  (MONTHS)",
      ylab="CUMULATIVE  HAZARDS  FOR  OUTBREAKS")
```

Trends in sojourn time distributions may arise over successive outbreaks due to systematic features of a disease process, or artificially when subject heterogeneity is unaccounted for and an association is induced among sojourn times for an individual. It is difficult to distinguish between these two sources of trend, but more general models involving random effects can help; see Chapter 6. Here we simply explore evidence of trend by outbreak number, by fitting intensities of the form

$$\lambda_{21}(t \mid N_{i12}(t^-) = r, \mathcal{H}_i(t^-)) = h_{21}^{(r)}(B_i(t)).$$

The stratified Nelson-Aalen estimates of $H_{21}^{(r)}(\cdot)$ are plotted in Figure 3.7 and do not reveal evidence of trend for either treatment arm. We therefore return to the unstratified estimates in Figure 3.6, which show comparable rates of resolution in the two arms during the first two weeks but suggest that the distribution of outbreak durations may have a longer tail for individuals on the suppressive arm.

3.3   Semiparametric Regression Models

The flexibility of semiparametric regression models for survival analysis makes them natural tools for the analysis of life histories based on multistate processes. In this section, we describe important multiplicative and additive models for transition intensities, and how they can be implemented for analysis.

3.3.1   Multiplicative Modulated Markov Models

Modulated Markov models were introduced in Section 2.3.1. With $X(t)$ denoting a vector of possibly time-dependent covariates, the $k \to l$ transition intensity functions take the form

$$\lambda_{kl}(t \mid \mathcal{H}_i(t^-)) = \lambda_{kl0}(t)\, g(x_i(t); \beta_{kl}), \qquad (3.22)$$

where $g(x; \beta)$ is a positive-valued function. Note that (3.22) is very flexible, and can accommodate internal and external covariates.

The likelihood methods of Section 3.1 for parametric models in principle deal naturally with regression, although they require full data on time-dependent covariates. For multiplicative Cox-type regression models, we set $g(x; \beta) = \exp(x'\beta)$ and for this specification we may write the log of (3.2) as

$$\log L_{kl}(\theta_{kl}) = \sum_{i=1}^{n} \int_{0}^{\infty} \bar{Y}_{ik}(u) \left\{ \log \lambda_{kl}(u \mid x_i(u)) \cdot dN_{ikl}(u) - \lambda_{kl}(u \mid x_i(u))\, du \right\} \quad (3.23)$$

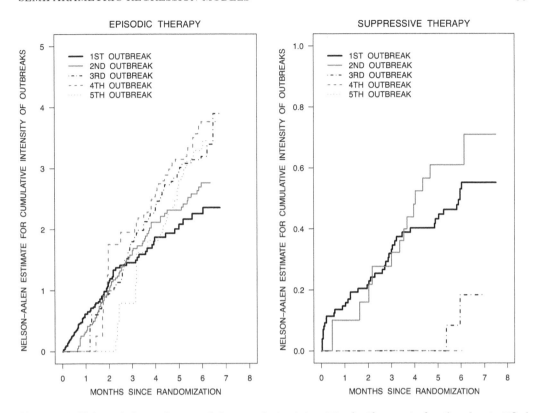

Figure 3.5: Nelson-Aalen estimates of the cumulative intensities for the onset of outbreaks stratified by the cumulative number of outbreaks $(\Lambda_{12}^{(r)}(t),\ r = 1, 2, \ldots)$ and treatment arm.

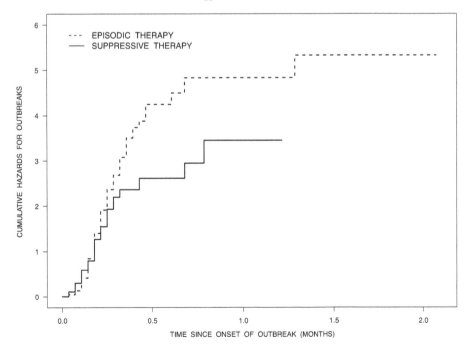

Figure 3.6: Nelson-Aalen estimates of the cumulative hazard functions $(H_{21}(s))$ for the duration of outbreaks by treatment arm.

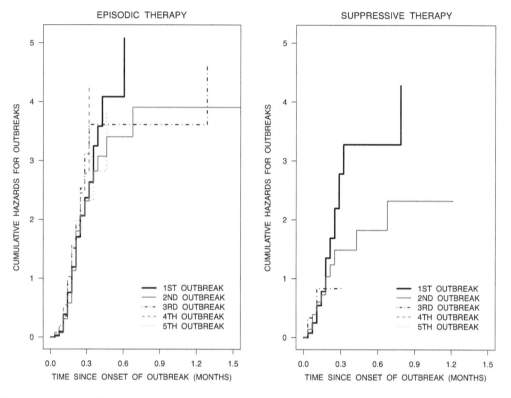

Figure 3.7: Nelson-Aalen estimates of the cumulative hazard functions stratified by the cumulative number of outbreaks $(H_{12}^{(r)}(s),\ r = 1, 2 \ldots)$ and treatment arm.

and maximize it directly. A model of particular interest involves baseline intensities $\lambda_{kl0}(t)$ of a piecewise-constant form. We consider this first before developing semiparametric methods.

As in Section 3.1.1, let $0 = b_0 < b_1 < \cdots < b_{R-1} < b_R = \infty$ denote break-points defining intervals $\mathcal{B}_r = [b_{r-1}, b_r)$, $r = 1, 2, \ldots, R$. We then set

$$\lambda_{kl0}(t) = \lambda_{klr} = \exp(\alpha_{klr}), \quad t \in \mathcal{B}_r,$$

and $x_{1r}(t) = I(t \in \mathcal{B}_r)$, $r = 1, \ldots, R$ and let $x_1(t) = (x_{11}(t), \ldots, x_{1R}(t))'$ be a defined time-dependent covariate. If $x_2(t)$ is a vector of other covariates, we write

$$\lambda_{kl}(t \mid x_2(t)) = \exp(x_1'(t)\alpha_{kl} + x_2'(t)\beta_{kl}) = \exp(x'(t)\theta_{kl})$$

where $\alpha_{kl} = (\alpha_{kl1}, \ldots, \alpha_{klR})$, $x(t) = (x_1'(t), x_2'(t))'$ and $\theta_{kl} = (\alpha_{kl}', \beta_{kl}')'$. We can then write (3.23) as

$$\log L_{kl}(\theta_{kl}) = \sum_{i=1}^{n} \int_0^\infty \bar{Y}_{ik}(u) \left\{ x_i'(u)\theta_{kl}\, dN_{ikl}(u) - \exp(x_i'(u)\theta_{kl})\, du \right\}. \qquad (3.24)$$

The element of the score vector corresponding to α_{kl} is then

$$\frac{\partial \log L_{kl}(\theta_{kl})}{\partial \alpha_{kl}} = \sum_{i=1}^{n} \int_0^\infty \bar{Y}_{ik}(u)\, x_{i1}(u) \left\{ dN_{ikl}(u) - \exp(x_i'(u)\theta_{kl})\, du \right\},$$

and setting this equal to zero and solving for λ_{klr} gives

$$\tilde{\lambda}_{klr}(\beta_{kl}) = \frac{n_{klr}}{\sum_{i=1}^{n} \int_0^{\infty} \bar{Y}_{ikr}(u) \exp(x'_{i2}(u)\beta_{kl})}, \qquad (3.25)$$

where $\bar{Y}_{ikr}(u) = \bar{Y}_{ik}(u) x_{i1r}(u)$ and $n_{klr} = \sum_{i=1}^{n} \int_0^{\infty} \bar{Y}_{ikr}(u) \, dN_{ikl}(u)$. If $x_{i2}(u)$ is fixed this reduces further to

$$\tilde{\lambda}_{klr}(\beta_{kl}) = \frac{n_{klr}}{\sum_{i=1}^{n} S_{ikr} \exp(x'_{i2}\beta_{kl})},$$

where $S_{ikr} = \int_0^{\infty} \bar{Y}_{ikr}(u) \, du$ is the time individual i was at risk of a transition out of state k in $\mathcal{B}_r$.

The score for β_{kl} is likewise given by differentiating (3.24) to obtain

$$\frac{\partial \log L_{kl}(\theta_{kl})}{\partial \beta_{kl}} = \sum_{i=1}^{n} \int_0^{\infty} \bar{Y}_{ik}(u) x_{i2}(u) \Big\{ dN_{ikl}(u) - \exp(x'_i(u)\theta_{kl}) \, du \Big\}, \qquad (3.26)$$

and since $\exp(x'_i(u)\theta_{kl}) = \sum_{r=1}^{R} I(u \in \mathcal{B}_r) \lambda_{klr} \exp(x'_{i2}(u)\beta_{kl})$, substituting (3.25) into (3.26) gives an estimating equation for β_{kl} of the form

$$\sum_{i=1}^{n} \sum_{r=1}^{R} \int_0^{\infty} \bar{Y}_{ikr}(u) x_{i2}(u) \Big\{ dN_{ikl}(u) - \frac{n_{klr} \exp(x'_{i2}(u)\beta_{kl})}{\sum_{i=1}^{n} \int_0^{\infty} \bar{Y}_{ikr}(v) \exp(x'_{i2}(v)\beta_{kl}) \, dv} \, du \Big\}.$$

Again if $x_{i2}(u)$ is fixed, we can write this more compactly as

$$\sum_{i=1}^{n} \sum_{r=1}^{R} n_{iklr} \left\{ x_{i2} - \frac{\sum_{i=1}^{n} S_{ikr} x_{i2} \exp(x'_{i2}\beta_{kl})}{\sum_{i=1}^{n} S_{ikr} \exp(x'_{i2}\beta_{kl})} \right\}.$$

where $n_{iklr} = \int_0^{\infty} \bar{Y}_{ikr}(u) \, dN_{ikl}(u)$. The general theory of maximum likelihood gives the approximation

$$\sqrt{n}(\hat{\theta}_{kl} - \theta_{kl}) \sim \mathrm{MVN}(0, I^{-1}(\hat{\theta}_{kl}))$$

where $I(\theta_{kl}) = -\partial^2 \log L_{kl}(\theta_{kl})/\partial \theta_{kl} \partial \theta'_{kl}$.

When the baseline transition intensities $\lambda_{kl0}(t)$ are not confined to be in a particular parametric family (i.e. they are left unspecified), the model is called semiparametric. There are two ways to approach estimation, both leading to the same result. The first is to consider the full likelihood functions (2.20) and to treat $\Lambda_{kl0}(t)$ as a step function that can jump only at times for which a $k \to l$ transition has been observed in the data. The likelihood function then depends on a finite number of parameters for a given dataset, and it can be shown (see Problem 3.5) that for given β_{kl}, the profile maximum likelihood estimate of $\Lambda_{kl0}(t)$ is

$$\tilde{\Lambda}_{kl0}(t; \beta_{kl}) = \int_0^t d\tilde{\Lambda}_{kl0}(t; \beta_{kl}) = \int_0^t \frac{J_k(u) \, d\bar{N}_{.kl}(u)}{S_k^{(0)}(u; \beta_{kl})}, \qquad (3.27)$$

where $S_k^{(0)}(u; \beta_{kl}) = \sum_{i=1}^{n} \bar{Y}_{ik}(u) g(x_i(u); \beta_{kl})$ and $J_k(u)$ is as defined in Section 3.2.1. When $g(x; \beta_{kl}) = \exp(x'\beta_{kl})$, this can be seen to be the limiting estimate that results from (3.25) if $\Delta b_r = b_r - b_{r-1} \to 0$ as $R \to \infty$. The profile likelihood function for β_{kl} is obtained by replacing $\lambda_{kl0}(u)\,du$ in (3.24) with $d\widetilde{\Lambda}_{kl0}(u; \beta_{kl})$ in (3.27) and exponentiating to give

$$PL_{kl}(\beta_{kl}) = \prod_{i=1}^{n} \left\{ \prod_{u \in \mathcal{D}_{ikl}} \frac{g(x_i(u); \beta_{kl})}{S_k^{(0)}(u; \beta_{kl})} \right\}, \tag{3.28}$$

where $\mathcal{D}_{ikl}$ is the set of all times for which individual i is observed to make a $k \to l$ transition. Estimates $\widehat{\beta}_{kl}$ are obtained by maximizing (3.28) or equivalently solving the partial likelihood score equation for β_{kl} given by

$$U_{kl}(\beta_{kl}) = \sum_{i=1}^{n} \int_0^{\infty} \bar{Y}_{ik}(u) \left\{ x_i(u) - \frac{S_k^{(1)}(u; \beta_{kl})}{S_k^{(0)}(u; \beta_{kl})} \right\} dN_{ikl}(u) = 0,$$

where $S_k^{(1)}(u; \beta_{kl}) = \sum_{i=1}^{n} \bar{Y}_{ik}(u)\, x_i(u)\, g(x_i(u); \beta_{kl})$. This can also be obtained as a limiting case of the piecewise-constant model.

We illustrate in Section 3.3.2 how to obtain estimates of θ_{kl} for the piecewise-constant model at hand using the `phreg` function of the `eha` library with fixed covariates. This function will also handle the case of time-dependent covariates that change at a finite number of times provided the dataframe is appropriately constructed. This is achieved by partitioning the at-risk periods of observation for a given individual into periods of time their covariates are fixed. The data from these sub-intervals can then be treated as if they arose from different individuals under the Markov assumption, and a standard call to the `phreg` function can be made. As suggested in Section 3.1.1 one may also exploit the fact that the likelihood can be partitioned to create a more elaborate dataframe with separate lines for each sub-interval derived by the intersection of intervals of constant baseline intensity and those over which covariates are fixed for a given individual.

The second approach to estimation for semiparametric models uses partial likelihood; the likelihood (3.28) is a partial likelihood (Cox, 1975) whose individual components are based on the conditional probabilities

$$P(d\bar{N}_{ikl}(u) = 1 \mid d\bar{N}_{\cdot kl}(u) = 1, \mathcal{H}(u^-)),$$

where $\mathcal{H}(u)$ represents the aggregated histories $\{\mathcal{H}_i(u), i = 1, \ldots, n\}$ over all individuals. Cox (1972) derived the likelihood function for β in the multiplicative hazards model for a single failure time in this way. The estimator for $\Lambda_{kl0}(t)$ in that case is often called the Breslow estimator, and was originally obtained from an unbiased estimating equation based on the fact that

$$\sum_{i=1}^{n} E\{d\bar{N}_{ikl}(u) - \bar{Y}_{ik}(u)\, g(x_i(u); \beta_{kl})\, d\Lambda_{kl0}(u) \mid \mathcal{H}_i(u^-)\} = 0. \tag{3.29}$$

Solving (3.29) without the expectation and with β_{kl} replaced by $\widehat{\beta}_{kl}$, we find that

$$\widehat{\Lambda}_{kl0}(t) = \int_0^t \frac{J_k(u)\, d\bar{N}_{\cdot kl}(u)}{S_k^{(0)}(u; \widehat{\beta}_{kl})}, \tag{3.30}$$

which is equal to (3.27) evaluated at $\widehat{\beta}_{kl}$. As in (3.12)–(3.14), this integral reduces to the sum

$$\widehat{\Lambda}_{kl0}(t) = \sum_{t^{(r)} \leq t} \frac{J_k(t^{(r)}) \, d\bar{N}_{\cdot kl}(t^{(r)})}{S_k^{(0)}(t^{(r)}; \widehat{\beta}_{kl})}, \qquad (3.31)$$

where $t^{(1)} < \cdots < t^{(m)}$ are the m distinct times at which transitions (of any kind) occur in the dataset. We note that if there are no covariates in the model, or equivalently $\beta_{kl} = 0$, then $S_k^{(0)}(u; 0)$ reduces to $\bar{Y}_{\cdot k}(u)$ under the unrestrictive assumption that $g(x; 0) = 1$, and (3.31) is the Nelson-Aalen estimator (3.13).

For the remainder of this section we focus on Cox models, where $g(x_i(u); \beta_{kl}) = \exp(x_i'(u)\beta_{kl})$. For such models we work with functions of the form

$$S_k^{(r)}(u; \beta_{kl}) = \sum_{i=1}^{n} \bar{Y}_{ik}(u) \, x_i(u)^{\otimes r} \exp(x_i'(u)\beta_{kl}),$$

where $x_i(u)^{\otimes 0} = 1$, $x_i(u)^{\otimes 1} = x_i(u)$, and $x_i(u)^{\otimes 2} = x_i(u)\,x_i'(u)$; we use the function $S_k^{(2)}(u; \beta_{kl})$ in (3.32).

Under mild regularity conditions (Andersen et al., 1993, Section 7.2.2), the estimator $\widehat{\beta}_{kl}$ obtained by maximizing (3.28) can be shown to be asymptotically normal, with $\sqrt{n}(\widehat{\beta}_{kl} - \beta_{kl})$ converging in distribution to a multivariate normal distribution with mean 0. The asymptotic covariance matrix $\mathcal{I}_{kl}^{-1}(\beta_{kl})$ is obtained by applying standard maximum likelihood large sample theory. In particular, some straightforward algebra shows that the observed information matrix $I_{kl}(\beta_{kl}) = -\partial^2 \log PL_{kl}(\beta_{kl})/\partial\beta_{kl}\partial\beta_{kl}'$ is given by

$$I_{kl}(\beta_{kl}) = \sum_{i=1}^{n} \sum_{u \in \mathcal{D}_{ikl}} \left\{ \frac{S_k^{(0)}(u; \beta_{kl}) \, S_k^{(2)}(u; \beta_{kl}) - S_k^{(1)}(u; \beta_{kl}) \, S_k^{(1)}(u; \beta_{kl})'}{S_k^{(0)}(u; \beta_{kl})^2} \right\}, \qquad (3.32)$$

and $\mathcal{I}_{kl}(\beta_{kl})$ is the limit of $n^{-1} I_{kl}(\beta_{kl})$ as $n \to \infty$. It follows that $I_{kl}^{-1}(\widehat{\beta}_{kl})$ can also be used as an asymptotic estimate of the covariance matrix for $\widehat{\beta}_{kl}$.

It can also be shown that the processes $\{\sqrt{n}(\widehat{\Lambda}_{kl0}(t) - \Lambda_{kl0}(t)), \ t > 0\}$ are asymptotically Gaussian, and variance estimates can be based on this (Andersen et al., 1993, Section 7.2.2). These are somewhat complicated, and we will not give the results here. However, the estimation of regression coefficients, baseline cumulative intensity functions and related quantities can be carried out using standard software for fitting Cox models. When the different transition intensities have no common parameters, we may consider each intensity as a separate cause-specific survival model of multiplicative hazards form. Therefore, software for the Cox model can be used to obtain estimators, provided that it allows for left truncation and, if necessary, time-dependent covariates. We use the R/S-PLUS function `coxph` in this book.

3.3.2 Regression Analysis of a Palliative Breast Cancer Trial

We now illustrate some regression analysis using data from the metastatic breast cancer trial considered in Section 3.2.2 with covariates described in Table 3.1. We first fit a multiplicative model for the risk of a skeletal event with covariates for treatment, age, an indicator of whether or not the patient had a fracture prior to study entry, and an indicator of whether they required radiotherapy prior to study entry. We first fit regression models with piecewise-constant baseline intensities, using the **phreg** function in the **eha** library, adopting the same break-points as in Section 3.2.2 at 90, 180, 270, 360 and 450 days. The code below is for fitting the $1 \rightarrow 2$ transition intensity.

```
> pwc12 <- phreg(Surv(stop, status) ~ trt + age + prior.frac + prior.rad,
            data=mbc[(mbc$from == 1) & (mbc$to == 2),],
            dist="pch", cuts=c(90,180,270,360,450), center=FALSE)
> print(pwc12)
```

Covariate	W.mean	Coef	Exp(Coef)	se(Coef)	Wald p
trt	0.549	-0.491	0.612	0.141	0.001
age	57.384	-0.013	0.987	0.006	0.024
prior.frac	0.157	0.205	1.227	0.177	0.249
prior.rad	0.205	0.410	1.507	0.154	0.008

```
Events                   212
Total time at risk        94851
Max. log. likelihood    -1477.8
LR test statistic        29.06
Degrees of freedom       4
Overall p-value          7.60913e-06
```

The column **coef** contains the maximum likelihood estimates of the regression coefficients; the corresponding estimates of the parameters of the baseline intensities are obtainable from the object **pwc12** as illustrated in Section 3.2.2. The **se(Coef)** column contains the standard errors based on the observed information matrix. Additional summary data follow along with results of a 4 degree of freedom (d.f.) likelihood ratio test of H_0: $\beta_{12} = 0$. The results of this fit and those of the other intensities are displayed in the left column of Table 3.3.

The semiparametric model is fitted with analogous code using the **coxph** function. The specification **method = "breslow"** indicates the way that ties in the **stop** times are to be handled; this method ensures that the estimators for $\Lambda_{kl0}(t)$ are of the form given in (3.30).

```
> mbc12 <- coxph(Surv(stop, status) ~ trt + age + prior.frac + prior.rad,
        data=mbc, subset=((from == 1) & (to == 2)), method="breslow")
```

The function **cox.zph** allows us to assess the covariate-specific and global adequacy of the multiplicative assumptions. We discuss this in Section 3.5 in some detail but point out here that **cox.zph** approximates score tests of the null hypothesis $\gamma = 0$ in the expanded model

$$\lambda_{12}(t \mid \mathcal{H}_i(t^-)) = \lambda_{kl0}(t) \exp(x_i' \beta_{12} + g(t) x_i' \gamma)$$

where $g(t)$ is a scalar defined function of time; most commonly we use $g(t) = t$ (identity) or $g(t) = \log t$. The appeal of a score test, or the approximation of one in **cox.zph**, is that one does not have to fit the model under the alternative. The parameter γ is of the same dimension as β_{12} and **cox.zph** gives p-values for tests of the null hypothesis for each component of γ as well as a global test of H_0: $\gamma = 0$. What follows is a call to the `cox.zph` function with the first argument the object from the model fit and the second specifying that the identity function $g(t) = t$ be used. Details on the interpretation of the output will be given later, so here we restrict attention to the significance of the results.

```
> cox.zph(mbc12, transform="identity")
               rho   chisq      p
trt        -0.08645 1.6316 0.2015
age         0.04034 0.3282 0.5667
prior.frac -0.08708 1.6312 0.2015
prior.rad   0.02741 0.1612 0.6880
GLOBAL          NA  3.9058 0.4189
```

None of the p-values provide significant evidence against the assumption of multiplicative covariate effects, nor did the call based on the "log" transform (not shown). Therefore, we proceed with the interpretation of the fitted model with output as follows:

```
> mbc12
              coef exp(coef) se(coef)     z      p
trt       -0.47929   0.61922  0.14146 -3.39 0.0007
age       -0.01307   0.98702  0.00572 -2.29 0.0223
prior.frac 0.22127   1.24765  0.17759  1.25 0.2128
prior.rad  0.43087   1.53860  0.15472  2.78 0.0054

Likelihood ratio test=29.78 on 4 df, p=5.433e-06 n=380, number of events= 212
```

Here the entries under **coef** are the estimates that maximize (3.28) and **se(coef)** contain the entries from the inverse of the information matrix (3.32). The summary statistics that follow include a 4 d.f. likelihood ratio test based on the partial likelihood (3.28); note the similar value to the likelihood ratio statistic based on the piecewise-constant fits with both statistics leading to rejection of H_0: $\beta_{12} = 0$.

Table 3.3 summarizes this output and the corresponding results from the piecewise-constant model. There is very close agreement between the estimates of covariate effects for the piecewise-constant and semiparametric analyses. For the semiparametric model, we find strong evidence of a reduction in the risk of a skeletal event among treated individuals (RR = 0.62; 95% CI: 0.47, 0.82; $p < 0.001$), a lower risk among older individuals (RR = 0.99; 95% CI: 0.98, 1.00; $p = 0.022$), and an increased risk among individuals with need for radiotherapy prior to entry into study (RR = 1.54; 95% CI: 1.14, 2.08; $p = 0.005$); there was little evidence of an effect of prior fracture ($p = 0.213$). The corresponding analyses for $1 \to 3$ transitions and $2 \to 3$ transitions did not yield any evidence of covariate effects.

The occurrence of skeletal complications reflects bone health and so it may be reasonable to assume this has a negligible effect on mortality, supporting the use of a Markov time scale in the application. However, the Markov assumption (conditional on fixed covariates) may be implausible in some applications. One way of assessing model adequacy is by model expansion; we consider an illustration here but discuss this in more detail in Section 3.5.1. Since the time of the skeletal event is a left-truncation time in the analysis of the $2 \to 3$ intensity, it can be introduced as a

Table 3.3: Summary of estimates obtained from fitting multiplicative Markov regression models for the palliative breast cancer trial data.

Covariate	Comparison/Units	Piecewise-Constant			Semiparametric		
		RR	95% CI	p	RR	95% CI	p
		$1 \to 2$ Transition					
Treatment	Pamidronate vs. Placebo	0.61	(0.46, 0.81)	< 0.001	0.62	(0.47, 0.82)	< 0.001
Age	Per 10 years	0.88	(0.79, 0.98)	0.024	0.88	(0.78, 0.98)	0.022
Prior Fracture	Yes vs. No	1.23	(0.87, 1.74)	0.249	1.25	(0.88, 1.77)	0.213
Prior Radiation	Yes vs. No	1.51	(1.11, 2.04)	0.008	1.54	(1.14, 2.08)	0.005
		$1 \to 3$ Transition					
Treatment	Pamidronate vs. Placebo	1.35	(0.66, 2.77)	0.410	1.33	(0.65, 2.72)	0.436
Age	Per 10 years	1.23	(0.92, 1.65)	0.158	1.23	(0.92, 1.65)	0.162
Prior Fracture	Yes vs. No	0.56	(0.17, 1.86)	0.347	0.58	(0.17, 1.91)	0.366
Prior Radiation	Yes vs. No	0.43	(0.13, 1.42)	0.168	0.44	(0.13, 1.45)	0.177
		$2 \to 3$ Transition					
Treatment	Pamidronate vs. Placebo	0.87	(0.42, 1.83)	0.718	0.87	(0.42, 1.83)	0.718
Age	Per 10 years	1.27	(0.94, 1.72)	0.122	1.26	(0.93, 1.71)	0.130
Prior Fracture	Yes vs. No	0.52	(0.18, 1.51)	0.229	0.53	(0.18, 1.54)	0.244
Prior Radiation	Yes vs. No	0.53	(0.21, 1.34)	0.180	0.54	(0.22, 1.35)	0.185

covariate in the $2 \to 3$ intensity in order to assess whether the risk of death among individuals who have had a skeletal event is related to the time of the skeletal event; this would be a violation of the Markov property. Fitting

$$\lambda_{23}(t \mid \mathcal{H}_i(t^-)) = \lambda_{230}(t) \exp(x_i' \beta + \gamma t_{i2})$$

where t_{i2} is the entry time to state 2 and testing H_0: $\gamma = 0$ yields $p = 0.423$, so there is no evidence to contradict the Markov assumption.

3.3.3 Multiplicative Modulated Semi-Markov Models

The semiparametric version of the modulated semi-Markov model (2.27) can be handled in essentially the same way as a Markov model, although the mathematical development of large sample properties is somewhat different. We begin by setting up some notation. As in Section 3.2.3, we let the set $\tilde{\mathcal{D}}_{ik}$ consist of the durations of all sojourns that individual i spends in state k, whether censored or uncensored. We let $w_{ik}^{(r)}$, $r = 1, \ldots, n_{ik}$ denote the $n_{ik} \geq 0$ times in $\tilde{\mathcal{D}}_{ik}$, where $\tilde{\mathcal{D}}_{ik}$ is understood to be empty when $n_{ik} = 0$. We assume that covariates are fixed or remain constant over each sojourn for each state and let $z_{ik}^{(r)}$ denote the covariate vector for the rth sojourn in state k. Time-dependent covariates that change values only at a finite set of times within sojourns can be handled, as in survival analysis, by splitting the sojourn into sub-intervals within which the covariates are fixed. Finally, we let $\delta_{ikl}^{(r)} = 1$ if the rth sojourn ends with a transition to state l and $\delta_{ikl}^{(r)} = 0$ otherwise.

With this notation, the intensity functions (2.27) for the rth sojourn in state k take the form $h_{kl0}(w)\exp(\eta_{ikl}^{(r)})$, where $\eta_{ikl}^{(r)} = [z_{ik}^{(r)}]'\beta_{kl}$, and the likelihood function based on all sojourns in state k is, from (3.11),

$$L = \prod_{i=1}^{n}\left\{\prod_{r=1}^{n_{ik}}\left[\left(h_{kl0}(w_{ik}^{(r)})e^{\eta_{ikl}^{(r)}}\right)^{\delta_{ikl}^{(r)}}\exp\left(-\int_{0}^{w_{ik}^{(r)}}h_{kl0}(u)e^{\eta_{ikl}^{(r)}}\,du\right)\right]\right\}. \qquad (3.33)$$

By assuming that $H_{kl0}(w)$ is a step function that changes value only at times $w = w_{ik}^{(r)}$ with $\delta_{ikl}^{(r)} = 1$, we find that for given β_{kl}, (3.33) is maximized with respect to $H_{kl0}(w)$ by

$$\widehat{H}_{kl0}(w;\beta_{kl}) = \sum_{i=1}^{n}\sum_{w_{ik}^{(r)}\leq w}\frac{\delta_{ikl}^{(r)}}{\widetilde{S}_{k}^{(0)}(w_{ik}^{(r)};\beta_{kl})}, \qquad (3.34)$$

where $\widetilde{S}_{k}^{(0)}(w;\beta_{kl}) = \sum_{j=1}^{n}\sum_{w_{jk}^{(r)}\geq w}\exp(\eta_{jkl}^{(r)})$. Substituting (3.34) into (3.33), we find that the profile likelihood function for β_{kl} is

$$\mathrm{PL}_{kl}(\beta_{kl}) = \prod_{i=1}^{n}\prod_{r=1}^{n_{ik}}\frac{\exp(\delta_{ikl}^{(r)}\eta_{ikl}^{(r)})}{\widetilde{S}_{k}^{(0)}(w_{ik}^{(r)};\beta_{kl})}. \qquad (3.35)$$

We see that (3.35) is the same as a partial likelihood in a Cox model when there are survival times $\{w_{ik}^{(r)}; r = 1,\ldots,n_{ik}, i = 1,\ldots,n\}$ that may be uncensored ($\delta_{ikl}^{(r)} = 1$) or censored ($\delta_{ikl}^{(r)} = 0$). Cox model software can therefore be used to obtain estimates $\widehat{\beta}_{kl}$ and estimates $\widehat{H}_{kl0}(w) = \widehat{H}_{kl0}(w;\widehat{\beta}_{kl})$ of the cumulative baseline intensities. Standard large sample variance estimates and associated inference procedures for $\widehat{\beta}_{kl}$ and $\widehat{H}_{kl0}(w)$ based on the Cox survival model can also be used. Proofs of these results are a little more complicated than for the case of a modulated Markov model. The estimate (3.34) and profile likelihood (3.35) are based on state sojourn times and cannot be represented in terms of martingales as the Markov estimates can. However, alternative methods (Andersen et al., 1993, Sections 10.1, 10.2; Dabrowska et al., 1994) show that the usual methods apply.

To summarize, when a specific transition intensity $\lambda_{kl}(t|\mathcal{H}(t))$ is in the form of a semiparametric semi-Markov model, we can once again use Cox model software for analysis. For an rth sojourn in state k for individual i, the contribution to the dataframe for estimation of β_{kl} and $H_{kl0}(w)$ are as follows: the **start** time is 0, the **stop** time is $w_{ik}^{(r)}$ and the observation status indicator is $\delta_{ikl}^{(r)}$.

3.3.4 Regression Analysis of Outbreaks from Herpes Simplex Virus

Here we revisit the example of Section 3.2.4 involving the trial of patients with herpes simplex virus infection. We now consider regression modeling for the recurrent outbreaks and their respective durations using the four covariates described in Table 3.2, namely, the treatment variable (x_1), sex (x_2), virus type (x_3) and number of outbreaks in the previous year (x_4).

We consider a series of models for the risk of outbreaks and the duration of outbreaks. For the onset of outbreaks we let $x_i = (x_{i1}, \ldots, x_{i4})'$ and consider the following models:

$$\lambda_{i12}(t \mid \mathcal{H}_i(t^-)) = \lambda_{120}(t) \exp(x_i'\beta) \qquad (3.36)$$

$$\lambda_{i12}(t \mid \mathcal{H}_i(t^-)) = \lambda_{120}(t) \exp(x_i'\beta + \gamma N_{i12}(t^-)) \qquad (3.37)$$

$$\lambda_{i12}(t \mid \mathcal{H}_i(t^-)) = \lambda_{12}^{(r)}(t) \exp(x_i'\beta) \quad \text{where } N_{i12}(t^-) = r, \quad r = 0, 1, \ldots \qquad (3.38)$$

Model (3.36) is a standard Markov model, while (3.37) incorporates a parametric dependence on the cumulative number of previous outbreaks. In model (3.38), the baseline intensity is stratified on $N_{i12}(t^-)$. Note that while we have used $\lambda_{120}(t)$ and β in (3.36) and (3.37), their interpretation is different between the two models.

We begin here with a model (3.36) for the onset of outbreaks including the four variables in Table 3.2. The specification `subset(from==1)` restricts the data to the lines containing information about the risks of outbreaks.

```
> onset1 <- coxph(Surv(start, stop, status) ~ trt + sex + vtype + nrecur,
                  data=hsv, subset=(from == 1), method="breslow")
```

We first assess the assumption that the covariates act multiplicatively on the baseline intensity by a call to `cox.zph` with the identity transform.

```
> cox.zph(onset1, transform="identity")
            rho      chisq        p
trt    -0.054927 1.132716 0.2872
sex    -0.034350 0.447757 0.5034
vtype   0.002047 0.001564 0.9685
nrecur -0.027182 0.270233 0.6032
GLOBAL            NA 1.821850 0.7685
```

The p-values do not provide significant evidence against the multiplicative assumptions, and so we interpret the results from the model (3.36).

```
> onset1
            coef exp(coef) se(coef)      z       p
trt    -1.7276    0.1777   0.1437 -12.02 <2e-16
sex    -0.2642    0.7678   0.1169  -2.26 0.0238
vtype   0.1141    1.1208   0.1065   1.07 0.2840
nrecur  0.0655    1.0677   0.0230   2.84 0.0045

Likelihood ratio test=237 on 4 df, p=0 n= 570, number of events= 373
```

We conclude that there is a highly significant effect of treatment ($p < 0.001$) while controlling for the sex of the patient, the virus type and the number of previous outbreaks. Specifically, there is an 82% reduction on the risk of an outbreak associated with the use of the suppressive therapy. Moreover males are at a significantly lower risk of outbreak ($p = 0.024$) with a relative risk of 0.77. While there is insufficient evidence to suggest the virus type matters, there is a significant effect of the number of previous outbreaks; this may reflect heterogeneneity in the susceptibility of individual subjects to outbreaks.

When controlling for the cumulative number of on-study outbreaks by regressing on the internal covariate $N_{i12}(t^-)$ as in (3.37), the results are broadly similar (see below). The effect of $N_{i12}(t^-)$ (enum) is highly significant, as one might expect from Figure 3.5, and the effect of the number of outbreaks in the previous year (nrecur) is somewhat attenuated compared to the previous model. This attenuation is intuitive since both the time-dependent covariate nrecur and the covariate enum reflect an individual's propensity for outbreaks, so some of the effect of nrecur in (3.36) is explained by enum in (3.37).

```
> onset2 <- coxph(Surv(start, stop, status) ~ trt + sex + vtype + nrecur + enum,
               data=hsv, subset=(from == 1), method="breslow")
> onset2
           coef exp(coef) se(coef)     z       p
  trt   -1.5076    0.2214   0.1536 -9.82 < 2e-16
  sex   -0.2335    0.7917   0.1169 -2.00   0.046
  vtype  0.0861    1.0899   0.1068  0.81   0.420
  nrecur 0.0474    1.0485   0.0235  2.02   0.044
  enum   0.1925    1.2123   0.0444  4.33 1.5e-05

Likelihood ratio test=255 on 5 df, p=0 n= 570, number of events= 373
```

The model (3.38) is fitted by stratifying on the time-dependent variable enum. The results below show that the effect of treatment and sex remain significant, and the effect of the number of outbreaks in the previous year is further attenuated.

```
> onset3 <- coxph(Surv(start, stop, status) ~ trt + sex + vtype + nrecur +
               strata(enum), data=hsv, subset=(from == 1), method="breslow")
> onset3
           coef exp(coef) se(coef)     z      p
  trt   -1.4929    0.2247   0.1577 -9.46 <2e-16
  sex   -0.2619    0.7696   0.1182 -2.21  0.027
  vtype  0.0429    1.0438   0.1109  0.39  0.699
  nrecur 0.0445    1.0455   0.0239  1.86  0.063

Likelihood ratio test=120 on 4 df, p=0 n= 570, number of events= 373
```

Multiplicative semi-Markov regression models for the duration of outbreaks are considered next. We consider models analogous to (3.36)−(3.38), but with, for example, $\lambda_{i21}(t|\mathcal{H}_i(t^-)) = h_{210}(B_i(t))\exp(x_i'\alpha)$, where $\alpha = (\alpha_1, \alpha_2, \alpha_3, \alpha_4)'$,

$$\lambda_{i21}(t \mid \mathcal{H}_i(t^-)) = h_{210}(B_i(t))\exp(x_i'\alpha + \xi N_{i12}(t^-)) \tag{3.39}$$

and

$$\lambda_{i21}(t \mid \mathcal{H}_i(t^-)) = h_{21}^{(r)}(B_i(t))\exp(x_i'\alpha), \quad r = N_{i21}(t^-) = 1, 2, \ldots . \tag{3.40}$$

The results of fitting all of three models are summarized in the bottom half of Table 3.4. The code for fitting the first model using the gap time variable gtime follows. Note that subset(from==2) specifies the lines corresponding to the duration of outbreaks are used.

```
> res1 <- coxph(Surv(gtime, status) ~ trt + sex + vtype + nrecur,
               data=hsv, subset=(from == 2), method="breslow")
```

```
> cox.zph(res1, transform="identity")
            rho   chisq      p
  trt   -0.09536 3.4695 0.06251
  sex   -0.03817 0.5484 0.45895
  vtype  0.05707 1.2361 0.26622
  nrecur -0.01114 0.0444 0.83311
  GLOBAL      NA 6.1066 0.19133

> res1
            coef exp(coef) se(coef)      z      p
  trt   -0.07344   0.92919  0.14831 -0.495 0.6205
  sex   -0.01397   0.98613  0.11786 -0.119 0.9056
  vtype  0.13994   1.15020  0.10877  1.287 0.1982
  nrecur 0.05246   1.05386  0.02335  2.247 0.0247

Likelihood ratio test=6.91 on 4 df, p=0.1407 n= 372, number of events= 369
```

Table 3.4: Summary of estimates obtained from fitting multiplicative Markov and semi-Markov regression models for the herpes simplex virus data.

	Markov (3.36)			Modulated Markov (3.37)			Stratified Markov (3.38)		
	RR	95% CI	p	RR	95% CI	p	RR	95% CI	p
Onset of Outbreaks									
Treatment	0.18	(0.13, 0.24)	< 0.001	0.22	(0.16, 0.30)	< 0.001	0.22	(0.16, 0.31)	< 0.001
Sex	0.77	(0.61, 0.97)	0.024	0.79	(0.63, 1.00)	0.046	0.77	(0.61, 0.97)	0.027
Virus type	1.12	(0.91, 1.38)	0.284	1.09	(0.88, 1.34)	0.420	1.04	(0.84, 1.30)	0.699
Prev. yr # recur.	1.07	(1.02, 1.12)	0.005	1.05	(1.00, 1.10)	0.044	1.05	(1.00, 1.10)	0.063
$N_{i12}(t^-)$				1.21	(1.11, 1.32)	< 0.001			

	Semi-Markov			Modulated Semi-Markov (3.39)			Stratified Semi-Markov (3.40)		
	RR	95% CI	p	RR	95% CI	p	RR	95% CI	p
Duration of Outbreaks									
Treatment	0.93	(0.69, 1.24)	0.620	1.01[†]	(0.75, 1.38)	0.925	0.99[†]	(0.73, 1.35)	0.959
Sex	0.99	(0.78, 1.24)	0.906	1.00	(0.79, 1.26)	0.985	1.02	(0.81, 1.28)	0.885
Virus type	1.15	(0.93, 1.42)	0.198	1.14	(0.92, 1.41)	0.237	1.11	(0.90, 1.38)	0.330
Prev. yr # recur.	1.05	(1.01, 1.10)	0.025	1.05	(1.00, 1.10)	0.040	1.05	(1.00, 1.10)	0.053
$N_{i12}(t^-)$				1.08	(1.00, 1.16)	0.039			

[†] Evidence of model violation by non-multiplicative effect of treatment.

The tests of the multiplicative effect of the covariates using coxph do not suggest serious problems with the fitted model, but in models (3.39) and (3.40) there is evidence ($p = 0.029$ and 0.013, respectively) that the treatment does not have a simple multiplicative effect. The result of the cox.zph call for (3.39) is as follows:

```
> res2 <- coxph(Surv(gtime, status) ~ trt + sex + vtype + nrecur + enum,
                data=hsv, subset=(from == 2), method="breslow")

> cox.zph(res2, transform="identity")
               rho   chisq       p
 trt      -0.11282  4.7971  0.0285
 sex      -0.03674  0.5078  0.4761
 vtype     0.06091  1.4110  0.2349
 nrecur   -0.00811  0.0237  0.8775
 enum     -0.05584  1.2051  0.2723
 GLOBAL         NA  7.6822  0.1746
```

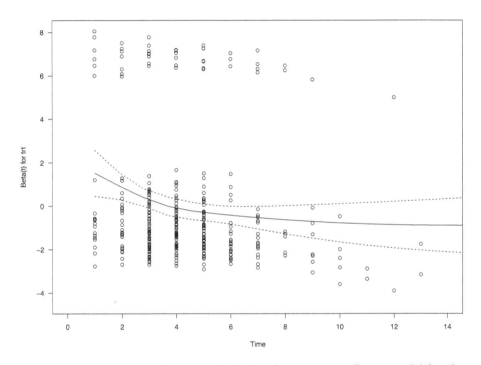

Figure 3.8: A Schoenfeld residual plot for the treatment effect in model (3.39).

The accompanying plot of the standardized Schoenfeld residuals based on (3.39) is given in Figure 3.8, which reveals a mild trend in the effect of treatment over time. To address this, we fit a model with a time-varying effect of treatment by defining the time-dependent indicator $D_i(t) = I(B_i(t) > 3.5)$ and generalize (3.39) to

$$\lambda_{i21}(t \mid \mathcal{H}_i(t^-)) = h_{210}(B_i(t)) \exp(x_i' \alpha + \xi N_{i12}(t^-) + \zeta D_i(t) x_{i1}) \qquad (3.41)$$

such that the treatment effect during the first half-week following onset is reflected by the relative risk $\exp(\alpha_1)$ and after the outbreak has lasted more than 3.5 days, by the relative risk $\exp(\alpha_1 + \zeta)$. This provides only a rough approximation of the pattern seen in Figure 3.8, but allows us to consider the time-varying effect explicitly. The contributions to the revised dataframe from the individual with id = 1 are given below, and the call to the coxph function follows:

```
> hsv2TD
  id gstart gstop gstatus enum trt sex vtype recur Dt
   1    0.0   3.5       0    1   0   0     1     9  0
   1    3.5   4.0       1    1   0   0     1     9  1
   1    0.0   3.5       0    2   0   0     1     9  0
   1    3.5   7.0       1    2   0   0     1     9  1
   1    0.0   3.5       0    3   0   0     1     9  0
   1    3.5   5.0       1    3   0   0     1     9  1
   1    0.0   1.0       1    4   0   0     1     9  0
   :      :     :       :    :   :   :     :     :  :
```

```
> res2TD <- coxph(Surv(gstart, gstop, gstatus) ~ trt + sex + vtype +
                  nrecur + enum + trt:Dt, data=hsv2TD, method="breslow")
```

The tests of the assumptions of multiplicative covariate effects for this extended model are then carried out using `cox.zph`:

```
> cox.zph(res2TD, transform="identity")
               rho      chisq       p
trt     -0.028060 0.2906334 0.5898
sex     -0.041922 0.6558761 0.4180
vtype    0.062159 1.4658586 0.2260
nrecur   0.001218 0.0005348 0.9815
enum    -0.055772 1.2076353 0.2718
trt:Dt   0.010465 0.0404880 0.8405
GLOBAL         NA 3.3730708 0.7608
```

These test results and the associated Schoenfeld residual plots (not shown) do not provide any evidence against the model (3.41) assumptions regarding covariate effects, and so we examine the output and summarize the estimates from fitting models (3.39) and (3.41) in Table 3.5.

```
> res2TD
            coef exp(coef) se(coef)     z      p
trt      0.54161   1.71878  0.22674  2.39 0.0169
sex      0.00581   1.00583  0.11766  0.05 0.9606
vtype    0.13010   1.13894  0.10896  1.19 0.2324
nrecur   0.05202   1.05339  0.02357  2.21 0.0273
enum     0.08036   1.08368  0.03653  2.20 0.0278
trt:Dt  -0.84678   0.42879  0.29657 -2.86 0.0043

Likelihood ratio test=19.01 on 6 df, p=0.004139 n= 612, number of events= 369
```

Overall there is insufficient evidence to claim a systematic effect of the suppressive therapy strategy on the duration of outbreaks. Estimates of effects from model (3.41) suggest that during the first half-week of an outbreak they resolve more rapidly in the suppressive arm, but among outbreaks not resolving within 3.5 days the duration tends to be longer under suppressive therapy. This is consistent with the plots of the marginal Nelson-Aalen estimates of the cumulative hazards in Figure 3.6. Among the other variables, there is insufficient evidence to suggest an effect of sex on the duration of the outbreaks, or of virus type. The greater the number of outbreaks in the year prior to study entry and the greater the number of prior outbreaks on study, the higher the intensity for the resolution of outbreaks. It is difficult to identify the mechanism associated with this from the analysis here. It may be, for example, that individuals tending to have long outbreaks will necessarily have fewer of them in a given period of time, and conversely those individuals having a large number will tend to have shorter outbreaks. We explore this in Chapter 6 where we discuss frailty models, which allow association among outbreak durations within individuals.

Table 3.5: Summary of estimates obtained from fitting modulated semi-Markov regression models (3.39) and (3.41) for the duration of outbreaks for the herpes simplex virus trials.

	Model (3.39)			Model (3.41)		
	RR	95% CI	p	RR	95% CI	p
Treatment	1.01	(0.75, 1.38)	0.925			
Treatment						
$\quad 0 < t \leq 3.5$ days				1.72	(1.10, 2.68)	0.017
$\quad 3.5$ days $< t$				0.74	(0.49, 1.10)	0.135
Sex	1.00	(0.79, 1.26)	0.985	1.01	(0.80, 1.27)	0.961
Virus type	1.14	(0.92, 1.41)	0.237	1.14	(0.92, 1.41)	0.232
Prev. yr # recur.	1.05	(1.00, 1.10)	0.040	1.05	(1.01, 1.10)	0.027
$N_{i12}(t^-)$	1.08	(1.00, 1.16)	0.039	1.08	(1.01, 1.16)	0.028

Finally we remark that given the defined time-dependent covariate $D_i(t)$, it is possible to assess the need for expansion of (3.39) to accommodate a time-varying effect of treatment using coxph only. This can be done by carrying out a partial likelihood score test of H_0: $\zeta = 0$ in (3.41) as follows. Let $\psi = (\alpha', \xi)'$, so $\omega = (\psi', \zeta)'$ are the regression coefficients in (3.41). The estimates obtained from fitting (3.39) may now be viewed as the maximum partial likelihood estimates $\tilde{\psi}_0 = \tilde{\psi}(0)$ for (3.41) under the hypothesis H_0: $\zeta = 0$; we let $\tilde{\omega}_0 = (\tilde{\psi}_0, 0)'$ where the values for $\tilde{\psi}_0$ are saved in rec2$coef:

```
> rec2$coef
        trt        sex       vtype      nrecur        enum
  0.014638  -0.002252   0.128862    0.048229    0.075176
```

The score statistic for testing H_0: $\zeta = 0$ in (3.41) is obtained as

$$U'(\tilde{\omega}_0)\, I^{-1}(\tilde{\omega}_0)\, U(\tilde{\omega}_0) \tag{3.42}$$

where $U(\omega)$ is the partial score vector of (3.41), and $I(\omega)$ is the associated observed information matrix given in (3.4) with the (k, l) notation suppressed here. Note that the only non-zero entry in vector $U(\tilde{\omega}_0)$ is the last one corresponding to the ζ term. Asymptotically, (3.42) has a χ_1^2 distribution under H_0: $\zeta = 0$.

This can be implemented using the coxph function as follows. If the initial values of the parameters in the expanded model (3.41) are specified as the values in rec2$coef with the initial value of 0 for ζ, the coxph function can be called with the dataframe `hsv2TD` specified. If the maximum number of iterations is set to 0, the score test statistic provided from the "fit" of `coxph` is the score statistic (3.42) for testing the null hypothesis $H_0: \zeta = 0$.

```
> res2TD.score <- coxph(Surv(gstart, gstop, gstatus) ~ trt + sex + vtype +
                nrecur + enum + trt:Dt, data=hsv2TD, method="breslow",
                init=init=c(rec2$coef,0), iter.max=0)
```

The p-value (0.004) based on this test (see below) is smaller than the p=0.029 from `cox.zph` reflecting stronger evidence against the multiplicative effect of treatment in (3.39), but this is based on a model under the alternative hypothesis of a particular form that was selected in part by the results from `cox.zph`. We provide this example simply to illustrate how to compute score statistics using the `coxph` function, and stress that tests guided by the preliminary examination of data are problematic. We discuss the use of score tests further in Section 3.5.2.

```
> res2TD.score$score
  [1] 8.45027

> 1 - pchisq(res2TD.score$score, 1)
  [1] 0.00365
```

3.3.5 Additive Markov and Semi-Markov Models

Aalen et al. (2001) discuss the use of modulated Markov intensity functions in which covariates have additive effects. Specifically if $\mathcal{H}_i(t)$ is the history containing time-dependent covariates, then let

$$\lambda_{kl}(t \mid \mathcal{H}_i(t^-)) = \beta_{kl0}(t) + \beta_{kl1}(t)x_{i1}(t) + \cdots + \beta_{klp}(t)x_{ip}(t). \qquad (3.43)$$

With right-censored data, we can write the intensity for the observed process as $\lambda_{kl}(t \mid \bar{\mathcal{H}}_i(t^-)) = \bar{Y}_{ik}(t)x_i'(t)\beta_{kl}(t)$, where $x_i(t) = (1, x_{i1}(t), \ldots, x_{ip}(t))'$ is a $(p+1) \times 1$ covariate vector including a 1 in the first entry, $\beta_{kl}(t) = (\beta_{kl0}(t), \beta_{kl1}(t), \ldots, \beta_{klp}(t))'$, and $\bar{Y}_{ik}(t) = Y_i(t)Y_{ik}(t^-)$.

Estimation can be conveniently carried out based on a least squares approach adapted for censored data. With an $n \times (p+1)$ matrix,

$$\bar{X}_k(t) = \begin{bmatrix} \bar{Y}_{1k}(t) & \bar{Y}_{1k}(t)x_{11}(t) & \cdots & \bar{Y}_{1k}(t)x_{1p}(t) \\ \vdots & \vdots & & \vdots \\ \bar{Y}_{nk}(t) & \bar{Y}_{nk}(t)x_{n1}(t) & \cdots & \bar{Y}_{nk}(t)x_{np}(t) \end{bmatrix},$$

a $(p+1) \times 1$ estimate of $dB_{kl}(t) = \beta_{kl}(t)dt$ is obtained as

$$d\widehat{B}_{kl}(t) = \left(\bar{X}_k'(t)\bar{X}_k(t) \right)^{-1} \bar{X}_k'(t) \, d\bar{N}_{kl}(t),$$

where $d\bar{N}_{kl}(t) = (d\bar{N}_{1kl}(t), \ldots, d\bar{N}_{nkl}(t))'$ is an $n \times 1$ vector of increments in the $k \to l$ counting processes for each individual given by $d\bar{N}_{ikl}(t) = \bar{Y}_{ik}(t) \, dN_{ikl}(t)$, $i = 1, \ldots, n$.

The vector of cumulative regression functions $B_{kl}(t) = \int_0^t dB_{kl}(u) = \int_0^t \beta_{kl}(u)\, du$ can then be estimated as

$$\widehat{B}_{kl}(t) = \int_0^t R(s)\, d\widehat{B}_{kl}(s)$$

where $R(s)$ indicates that the matrix $\bar{X}_k(s)$ has full column rank.

If covariates are fixed, denoted by $x_i = (1, x_{i1}, \ldots, x_{ip})'$, and common for all (k, l), then for a generic individual we can rewrite (3.43) simply as $\lambda_{kl}(t|x) = x'\beta_{kl}(t)$ and let

$$\widehat{\Lambda}_{kl}(t \mid x) = \int_0^t x'd\widehat{B}_{kl}(s) = x'\widehat{B}_{kl}(t)\,,$$

be the estimate of the cumulative transition rate for an individual with covariate vector x. An estimate of the conditional transition probability matrix is then obtained as

$$\widehat{P}(s, t \mid x) = \prod_{(s,t]} \left\{ I + \widehat{Q}(u \mid x)\, du \right\}$$

where as in Section 3.2.1 $\widehat{Q}(u|x)\, du$ has $d\widehat{\Lambda}_{kl}(u|x)$ in the off-diagonals ($k \neq l$) and $-d\widehat{\Lambda}_{k\cdot}(u|x)$, $k = 1, \ldots, K$, in the diagonal entries.

Lin and Ying (1994) consider a simplified version of (3.41) with constant coefficients. Here we redefine the fixed covariate vector for individual i as the $p \times 1$ vector $x_i = (x_{i1}, \ldots, x_{ip})'$ by omitting the leading 1 and write the model for the $k \to l$ transition for individual i as

$$\lambda_{kl}(t \mid \mathcal{H}_i(t^-)) = \beta_{kl0}(t) + x_i'\beta_{kl}\,,$$

where $\beta_{kl0}(t)$ is a baseline transition rate and $\beta_{kl} = (\beta_{kl1}, \ldots, \beta_{klp})'$. If we let

$$\bar{x}_k(u) = \frac{\sum_{i=1}^n \bar{Y}_{ik}(u)\, x_i}{\sum_{i=1}^n \bar{Y}_{ik}(u)}$$

denote the empirical mean covariate vector for $k \to l$ transitions among those under observation and at risk of a transition out of state k at time u, then $\widehat{\beta}_{kl}$ is given by

$$\left[\sum_{i=1}^n \int_0^\infty \bar{Y}_{ik}(u)\, (x_i - \bar{x}_k(u))^{\otimes 2}\, du \right]^{-1} \left[\sum_{i=1}^n \int_0^\infty (x_i - \bar{x}_k(u))\, dN_{ikl}(u) \right]$$

with a covariance matrix estimated by

$$\widehat{\mathrm{cov}}(\widehat{\beta}_{kl}) = n^{-1} \widehat{\Omega}_{kl}^{-1} \widehat{V}_{kl} \widehat{\Omega}_{kl}^{-1}$$

where

$$\widehat{\Omega}_{kl} = \frac{1}{n} \sum_{i=1}^n \int_0^\infty \bar{Y}_{ik}(u)\, (x_i - \bar{x}_k(u))^{\otimes 2}\, du$$

and

$$\widehat{V}_{kl} = \frac{1}{n} \sum_{i=1}^n \int_0^\infty (x_i - \bar{x}_k(u))^{\otimes 2}\, dN_{ikl}(u)\,.$$

Let $d\bar{N}_{kl}(u) = dN_{\cdot kl}(u)/Y_{\cdot k}(u)$, where $dN_{\cdot kl}(u) = \sum_{i=1}^{n} \bar{Y}_{ik}(u)\,dN_{ikl}(u)$ and $\bar{Y}_{\cdot k}(u) = \sum_{i=1}^{n} \bar{Y}_{ik}(u)$. The cumulative baseline intensity is then naturally estimated as

$$\widehat{B}_{kl0}(t) = \int_0^t d\bar{N}_{kl}(u) - \int_0^t \bar{x}_k' \widehat{\beta}_{kl}(u)\,du$$

which may alternatively be denoted by $\widehat{\Lambda}_{kl0}(t)$. We refer readers to Martinussen and Scheike (2006, Chapter 5) for a rigorous discussion of the regularity conditions and asymptotic theory of additive hazards models along with several worked examples.

3.3.6   Analysis of Herpes Outbreaks with Additive Models

Here we fit additive intensity models to data from the herpes simplex virus trial of Section 3.3.4. We adopt (3.43) with fixed covariates treatment, sex, virus type and number of previous outbreaks, and $N_{i12}(t^-)$ as a time-dependent covariate, for the onset of outbreaks. The syntax for the **aalen** function in the **timereg** package is similar to that of **coxph**. With the covariate effects expressed as nonparametric estimates, one may carry out tests of covariate effects based on $\widehat{B}_{klj}(t)$. The supremum test of H_0: $\beta_{klj}(t) = 0$ is based on

$$\sup_{s \in [0, C_{max}]} |B_{klj}(s)|,$$

where C_{max} is some maximum time of interest; here we use $C_{max} = \max_i\{C_i\}$. A resampling approach is used to obtain an approximation to the distribution of the test statistic under the null, so a random number seed is specified. Tests may also be carried out regarding whether covariates effects are constant, i.e., H_0: $\beta_{klp}(t) = \beta_{klp}$. Tests here are based on Kolmogorov-Smirnov or Cramer von Mises statistics.

```
> library(timereg)
> hsv1 <- hsv[hsv$from == 1,]
> hsv1$id <- 1:nrow(hsv1)

> set.seed(2000)
> onset1a <- aalen(Surv(start, stop, status==1) ~ trt + sex + vtype + nrecur + enum,
            data=hsv1, id=hsv1$id, start.time=0, max.time=max(hsv1$stop))
> onset1a
  Additive Aalen Model


  Test for nonparametric terms


  Test for non-significant effects
            Supremum-test of significance  p-value H_0: B(t)=0
  (Intercept)                        3.80                0.004
  trt                                9.24                0.000
  sex                                2.45                0.209
  vtype                              1.58                0.772
  nrecur                             3.27                0.025
  enum                               3.05                0.039


  Test for time invariant effects
            Kolmogorov-Smirnov test  p-value H_0:constant effect
  (Intercept)             0.6590                            0.149
```

```
trt                                    0.2800                        0.185
sex                                    0.1720                        0.610
vtype                                  0.1560                        0.821
nrecur                                 0.0996                        0.073
enum                                   0.2440                        0.024
                      Cramer von Mises test  p-value H_0:constant effect
(Intercept)                             5.720                        0.637
trt                                     3.990                        0.153
sex                                     0.545                        0.828
vtype                                   0.621                        0.853
nrecur                                  0.548                        0.070
enum                                    2.790                        0.042
```

We find significant evidence of a treatment effect ($p < 0.001$), and effects of the number of outbreaks in the previous year ($p = 0.025$) and the cumulative number of outbreaks on study ($p = 0.039$). Only the effect of the cumulative number of on-study outbreaks appears to be time dependent according to the Kolmogorov-Smirnov ($p = 0.024$) or Cramer von Mises ($p = 0.042$) statistics; the nonparametric estimates of the cumulative coefficients are plotted in Figure 3.9. The interpretation of these plots is best made by examination of the slopes, similar to the way slopes of Nelson-Aalen estimates provide insight into trends in hazard or intensity function. The roughly linear plots (best exemplified by the effect of treatment) are what one would expect with constant effects on the additive scale. We next fit the model with constant coefficients for all variables except **enum** through use of the **const()** operator.

```
> set.seed(2000)
> onset1b <- aalen(Surv(start, stop, status==1) ~ const(trt) + const(sex) +
            const(vtype) + const(nrecur) + enum, data=hsv1, id=hsv1$id,
            start.time=0, max.time=max(hsv1$stop))
> onset1b
  Additive Aalen Model

  Test for nonparametric terms

  Test for non-significant effects
                Supremum-test of significance  p-value H_0: B(t)=0
(Intercept)                             4.28                         0.003
enum                                    4.24                         0.002

  Test for time invariant effects
                Kolmogorov-Smirnov test  p-value H_0:constant effect
(Intercept)                             0.254                        0.153
enum                                    0.199                        0.050
                      Cramer von Mises test  p-value H_0:constant effect
(Intercept)                             3.15                         0.158
enum                                    1.03                         0.163

  Parametric terms :
                   Coef.       SE Robust SE      z      P-val
const(trt)      -0.01190 0.001370   0.001430  -8.34 0.0000000
const(sex)      -0.00158 0.001170   0.001180  -1.34 0.1800000
const(vtype)     0.00175 0.001280   0.001240   1.42 0.1560000
const(nrecur)    0.00117 0.000302   0.000278   4.20 0.0000264
```

The treatment is associated with a highly significant reduction in the intensity for outbreaks ($p < 0.001$) and a higher number of outbreaks in the previous year is associated with a significant increase in the risk of outbreaks. The time-varying effect of $N_{i12}(t^-)$ remains significant according to the supremum test ($p = 0.002$), and the Kolmogorov-Smirnov test suggests this should be treated as a time-varying effect; the effect is conveyed graphically in the lower right panel of Figure 3.9.

We next consider the semi-Markov model for the duration of outbreaks. A preliminary fit for the duration of outbreaks suggests the need to accommodate time-varying effects of treatment, virus type and the number of outbreaks in the previous year. We therefore focus on the model resulting from the call:

```
> hsv2 <- hsv[hsv$from == 2,]
> hsv2$id <- 1:nrow(hsv2)

> set.seed(2000)
> res1b <- aalen(Surv(gtime, status==1) ~ trt + const(sex) + vtype + nrecur +
               const(enum), data=hsv2, id=hsv2$id,
               start.time=0, max.time=max(hsv2$gtime))
> res1b
  Additive Aalen Model

  Test for nonparametric terms

  Test for non-significant effects
              Supremum-test of significance  p-value H_0: B(t)=0
  (Intercept)                         2.89                 0.062
  trt                                 3.29                 0.016
  vtype                               2.97                 0.078
  nrecur                              3.41                 0.029


  Test for time invariant effects
              Kolmogorov-Smirnov test  p-value H_0:constant effect
  (Intercept)                    1.06                        0.090
  trt                            1.00                        0.061
  vtype                          1.39                        0.012
  nrecur                         0.24                        0.014
              Cramer von Mises test  p-value H_0:constant effect
  (Intercept)                  7.910                        0.058
  trt                          3.170                        0.184
  vtype                       13.000                        0.007
  nrecur                       0.256                        0.042


  Parametric terms :
               Coef.      SE Robust SE       z      P-val
  const(sex) -0.00214 0.02200    0.02310 -0.0926 0.926000
  const(enum) 0.02140 0.00736    0.00614  3.4900 0.000481
```

We find a significant effect of treatment ($p = 0.016$) and the number of previous outbreaks ($p = 0.029$) with their effects displayed in Figure 3.10. The constant effect of sex is not significant but the time dependent cumulative number of outbreaks on study is $(p < 0.001)$; the greater the value of $N_{i12}(t^-)$, the greater is the intensity for the resolution of outbreaks.

The multiplicative model (3.37) fitted in Section 3.3.4 yielded strong evidence of a large reduction in risk of outbreaks among individuals on suppressive therapy $(p < 0.001)$ and a mildly significant lower risk of outbreaks among males $(p = 0.046)$. The number of outbreaks in the previous year as well as the time-dependent variable $N_{i12}(t^-)$ were both positively associated with increased risk of outbreaks. The corresponding additive model with nonparametric estimates of covariate effects also suggested a significant effect of suppressive therapy in reducing the rate of outbreaks, as well as the number of outbreaks in the previous year; the effects of both were consistent with constant coefficients. The time-dependent variable $N_{i12}(t^-)$ was highly significant in the additive model, but there was significant evidence against the assumption of a constant effect; interestingly, the Schoenfeld residual analysis and associated test did not suggest a time-varying effect of this variable in the multiplicative model $(p = 0.272)$. It is also noteworthy that the effect of sex was not significant in the additive model when modeled nonparametrically or as a constant effect.

The multiplicative and additive models are thus in broad, though not total, agreement concerning covariate effects. The appeal of multiplicative models is in part based on the widespread familiarity with relative risk type measures of effect. For either multiplicative or additive models, with fixed covariates and fixed effects, covariate effects are readily interpreted in terms of transition probabilities or sojourn time distributions. Time-varying effects are problematic in this sense and going beyond intensities is harder. The additive formulation has appeal in the sense that when the coefficients are estimated nonparametrically, they represent raw data summaries concerning intensities, which are most effectively reported on graphically.

3.4   Nonparametric Estimation of State Occupancy Probabilities

The probability an individual occupies particular states at a particular time is often of interest. In the context of longitudinal studies where individuals are recruited at $t = 0$ in a common initial state, there is an implicit conditioning on $Z(0)$ and the state k occupancy probability is $P(Z(t) = k | Z(0))$ which may be written as $P(Z(t) = k)$ for short. In the study of progression in diabetic retinopathy considered in Section 1.2.2, for example, the probabilities $P_k(t) = P(Z(t) = k)$ for states $k = 1, \ldots, 5$ convey the prevalence of different levels of retinopathy at different times on study. Estimates of prevalence or occupancy probabilities can be obtained by fitting specific types of models, but also in more direct and less model-dependent ways. In Section 3.4.1, we focus on nonparametric approaches based on Markov processes, and we consider alternative robust methods in Sections 3.4.2 and 4.2.

3.4.1   Aalen-Johansen Estimates

The nonparametric Aalen-Johansen (AJ) estimator (3.17) of the transition probability matrix $P(s, t)$ for a Markov process was described in Section 3.2.1. We assume

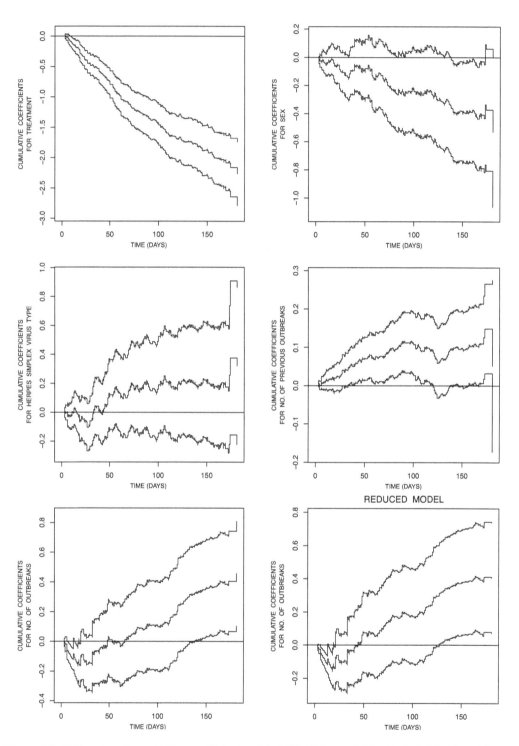

Figure 3.9: Estimates of cumulative coefficients with 95% CI for variables included in the additive intensity for the onset of outbreaks in the herpes simplex trial; lower right figure is obtained from the model with constant coefficients for all covariates except for $N_{i12}(t^-)$ (**enum**).

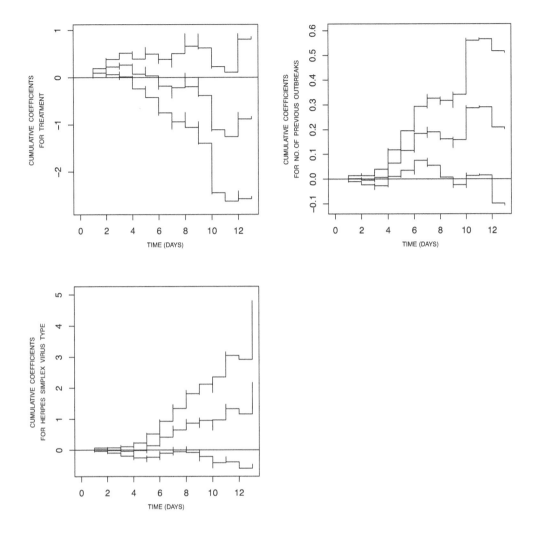

Figure 3.10: Estimates of cumulative coefficients with 95% CI for variables included in the additive intensity for the duration of outbreaks in the herpes simplex trial.

for simplicity that all individuals start in state 1, so that $P_k(t) = P_{1k}(0,t)$. The AJ prevalence estimators for $k = 1, \ldots, K$ are given by (3.17) as $\widehat{P}_k(t) = [\widehat{P}(0,t)]_{1k}$, with

$$\widehat{P}(0,t) = \prod_{(0,t]} \{I + \widehat{Q}(u)\, du\}, \tag{3.44}$$

where $\widehat{Q}(u)\, du$ is the $K \times K$ matrix containing the increments $d\widehat{\Lambda}_{kl}(u)$ of the Nelson-Aalen estimators given by (3.12) in the off-diagonal and $-\sum_{l \neq k} d\widehat{\Lambda}_{kl}(u)$ in the diagonal entries.

The assumption that all individuals begin in the same initial state can be easily relaxed. If individuals may be in different states at $t = 0$, we define the initial distribution through $P_j(0) = P(Z(0) = j)$ and let $P_k(t) = \sum_{j=1}^{K} P_j(0)\, P_{jk}(0,t)$. We may then estimate $P_j(0)$ under a multinomial model and $P_{jk}(0,t)$ using the Aalen-Johansen estimate to give a more general estimate of $P_k(t)$.

Although these estimates are derived from a Markov process, it turns out that they are also valid for non-Markov processes provided that the follow-up (or censoring) times for individuals are independent of their multistate processes. This can be seen by noting that for Δt small enough that at most one transition is observed in $(t, t+\Delta t]$, the AJ estimators in (3.44) satisfy

$$\widehat{P}_l(t+\Delta t) = \sum_{k=1}^{K} \widehat{P}_k(t) \, \Delta\widehat{\Lambda}_{kl}(t), \qquad (3.45)$$

where $\Delta\widehat{\Lambda}_{kl}(t) = \widehat{\Lambda}_{kl}(t+\Delta t) - \widehat{\Lambda}_{kl}(t)$. Moreover, if more than one transition in $(t, t+\Delta t]$ is impossible, then $P_l(t+\Delta t) = \sum_{k=1}^{K} P_k(t) \Delta\Lambda_{kl}(t)$, where $\Delta\Lambda_{kl}(t) = P(Z(t+\Delta t) = l | Z(t) = k)$ for $k \neq l$, and $\Delta\Lambda_{kk}(t) = 1 - \sum_{l \neq k} \Delta\Lambda_{kl}(t)$, no matter what the process intensities are. It should be noted that the $\Lambda_{kl}(t)$ should now be viewed as cumulative rate functions $\int_0^t \lambda_{kl}(u)du$ ($k \neq l$), where

$$\lambda_{kl}(u) = \lim_{\Delta t \downarrow 0} P(Z(t+\Delta t^-) = l \mid Z(t^-) = k)/\Delta t;$$

they are the same as the intensities only for Markov models. The Nelson-Aalen estimator $\Delta\widehat{\Lambda}_{kl}(t)$ in Section 3.2.1 estimates $\Delta\Lambda_{kl}(t)$ consistently if follow-up is independent of the multistate process and $\bar{Y}_{\cdot k}(t)/n$ converges in probability to a positive limit as $n \to \infty$. It can thus be seen that $\widehat{P}_k(\Delta t) = \Delta\widehat{\Lambda}_{1k}(0)$ is consistent and then by recursion that the $\widehat{P}_l(r\Delta t)$ are consistent for $r = 2, 3, \ldots$. Consistency of the $\widehat{P}_l(t)$ at all $t > 0$ follows from this under mild additional conditions.

In some cases censoring times might not be independent of the multistate processes. For example, this could happen if censoring depends on covariates that also affect the processes, or if the censoring intensity depends on the state an individual occupies. In the latter case the AJ transition probability and prevalence estimators remain consistent if the multistate process is Markov, but not more generally. We can see this by noting that the increments of the Nelson-Aalen estimate (3.12) are based on the fact that in a Markov process,

$$E\{\bar{Y}_{ik}(t)\,[dN_{ikl}(t) - d\Lambda_{kl}(t)]\} = 0 \qquad (3.46)$$

provided that the conditional independence conditions on censoring given in Section 2.2.2 are satisfied. For a Markov process they are satisfied, since

$$E\{\bar{Y}_{ik}(t)\,dN_{ikl}(t) \mid \bar{\mathcal{H}}_i(t^-), \bar{Y}_{ik}(t) = 1\} = d\Lambda_{kl}(t)$$

where $\bar{\mathcal{H}}_i(t) = \{Y_i(s), \bar{N}_i(s), \bar{X}_i(s), 0 \leq s \leq t; Z_i(0)\}$ with $\bar{N}_i(s)$ and $\bar{X}_i(s)$ defined as in Section 2.2.2. For a non-Markov process, however, this expectation would in general depend on other features of $\bar{\mathcal{H}}_i(t^-)$ besides the fact that $\bar{Y}_{ik}(t) = 1$ and will not equal $d\Lambda_{kl}(t)$, and the left side of (3.46) will not equal zero. In Section 3.4.2, we describe ways to adjust the AJ estimate for process-dependent censoring. Dependent censoring schemes and methods of analysis which adjust for these are discussed more fully in Section 4.2.

3.4.2 Adjustment for Process-Dependent Censoring

When random censoring is related to previous history of the multistate process, the estimating functions that produce specific estimators may no longer be unbiased. It is possible to correct for this in many cases by using "inverse probability of censoring weights" (IPCWs) to adjust the estimating functions. The key to doing this is to consider random censoring (premature loss to follow-up) as a separate event and to model its intensity. Let $X(t)$ be a vector of fixed or time-dependent covariates that may include features of the multistate process history. We let $\bar{\mathcal{H}}_i(t) = \{Y_i(s), \bar{N}_i(s), \bar{X}_i(s), 0 \le s \le t; Z_i(0)\}$ as before and define histories $\mathcal{X}_i(t) = \{X_i(s), 0 \le s \le t\}$ and $\mathcal{Z}_i(t) = \{Z_i(s), 0 \le s \le t\}$ for the covariate and multistate processes alone, along with their counterparts in the censored data setting denoted $\bar{\mathcal{X}}_i(t) = \{Y_i(s), \bar{X}_i(s), 0 \le s \le t\}$ and $\bar{\mathcal{Z}}_i(t) = \{Y_i(s), \bar{N}_i(s), 0 \le s \le t; Z_i(0)\}$, respectively.

We assume that administrative censoring is completely independent of all other processes. As in Section 2.2.2, let $C^R(t) = I(C^R \le t)$ and $\Delta C^R(t) = C^R(t + \Delta t^-) - C^R(t^-)$. We first assume that the random censoring intensity is independent of the full path of the multistate and covariate processes given the (observed) history of the covariate process at t^-. This is reflected in the condition:

$$\lim_{\Delta t \downarrow 0} \frac{P(\Delta C_i^R(t) = 1 \mid C_i \ge t, \mathcal{X}_i(\infty), \mathcal{Z}_i(\infty))}{\Delta t} = \lim_{\Delta t \downarrow 0} \frac{P(\Delta C_i^R(t) = 1 \mid C_i \ge t, \mathcal{X}_i(t^-))}{\Delta t}. \tag{3.47}$$

We further assume that $X_i(t^-)$ contains the relevant information in the covariate history to model the intensity for random censoring. This is reflected by

$$\lim_{\Delta t \downarrow 0} \frac{P(\Delta C^R(t) = 1 \mid C_i \ge t, \mathcal{X}_i(t^-))}{\Delta t} = \lim_{\Delta t \downarrow 0} \frac{P(\Delta C^R(t) = 1 \mid C_i \ge t, X_i(t^-))}{\Delta t}, \tag{3.48}$$

which we denote by $\lambda_i^c(t|X_i(t^-))$; we also let $d\Lambda^c(t|X_i(t^-)) = \lambda^c(t|X_i(t^-))\,dt$.

We then consider the function $G_i^c(t) = P(C_i^R > t \mid \mathcal{X}_i(t^-))$, which represents the probability of remaining under observation to time t given the covariate path over $[0, t]$. This function plays a crucial role in making adjustments for dependent censoring by reweighting the contributions from individuals who are uncensored so that the resulting weighted pseudo-sample reflects the composition of individuals at risk in the absence of dependent random censoring. The function $G_i^c(t)$ can be computed using product integration via

$$G_i^c(t) = \prod_{u \le t} [1 - d\Lambda^c(u \mid X_i(u^-))] .$$

A weighted contribution to the estimating equation for $d\Lambda_{kl}(t)$ from individual i is then given by

$$\frac{Y_i(t)\, Y_{ik}(t^-)}{G_i^c(t^-)} [dN_{ikl}(t) - d\Lambda_{kl}(t)] . \tag{3.49}$$

To see that this is unbiased, note that $Y_i(t) = \bar{I}(C_i^R \ge t) \cdot I(C_i^A \ge t)$, so (3.49) can be rewritten as

$$I(C_i^A \ge t) \frac{I(C_i^R \ge t)}{G_i^c(t^-)} Y_{ik}(t^-) [dN_{ikl}(t) - d\Lambda_{kl}(t)].$$

Then noting that the expectation of $I(C_i^R \geq t)$ given $\mathcal{X}_i(\infty)$, $\mathcal{Z}_i(\infty)$ and C_i^A is $G_i^c(t^-)$ by (3.47) and (3.48), so the corresponding expectation of (3.49) yields

$$I(C_i^A \geq t)\, Y_{ik}(t^-)\, [dN_{ikl}(t) - d\Lambda_{kl}(t)],$$

which in turn has expectation zero since the administrative censoring time is independent of the process and $E\{dN_{ikl}(t) \mid Y_{ik}(t^-) = 1\} = d\Lambda_{kl}(t)$. As we discuss in Section 3.5.4, the term $d\Lambda_{kl}(t)$ can be viewed as a marginal transition rate, but it is the intensity under a Markov model. It then follows that IPCW estimators of the $\Lambda_{kl}(t)$ are consistent; they are

$$\widehat{\Lambda}_{kl}^w(t) = \int_0^t d\widehat{\Lambda}_{kl}^w(u) = \sum_{u \in \mathcal{D}_{kl}(t)} d\widehat{\Lambda}_{kl}^w(u) \qquad (3.50)$$

where $\mathcal{D}_{kl}(t) = \{u : d\bar{N}_{.kl}(u) = 1, u \leq t\}$ and

$$d\widehat{\Lambda}_{kl}^w(u) = \sum_{i=1}^n \frac{\bar{Y}_{ik}(u)\, dN_{ikl}(u)}{G_i^c(u^-)} \Big/ \sum_{i=1}^n \frac{\bar{Y}_{ik}(u)}{G_i^c(u^-)}. \qquad (3.51)$$

In practice, $\lambda^c(t|X_i(t))$ and $G_i^c(t)$ are unknown, and we must assume that a consistent estimator $\widehat{G}_i^c(t)$ can be obtained.

For processes that start in state 1 with probability one, an estimate of the occupancy probability $P_k(t)$ is obtained as the $(1, k)$ component of the matrix

$$\widehat{P}^w(0, t) = \prod_{u \leq t} \{I + \widehat{Q}^w(u)\, du\},$$

where $\widehat{Q}^w(u)\, du$ is the $K \times K$ matrix with entries $d\widehat{\Lambda}_{kl}^w(u)$ given by (3.51) in the off-diagonal and $d\widehat{\Lambda}_{kk}^w(u) = -d\widehat{\Lambda}_{k.}^w(u)$, where $d\widehat{\Lambda}_{k.}^w(u) = \sum_{l \neq k} d\widehat{\Lambda}_{kl}^w(u)$ in the diagonal, $k = 1, \ldots, K$. The matrix displayed above is called the weighted Aalen-Johansen estimator. By the same reasoning as in Section 3.4.1, the occupancy probability estimators $\widehat{P}_k^w(t)$ are consistent, provided consistent estimators $\widehat{G}_i^c(u^-)$ are used.

We can also use inverse weighting for estimation of the marginal distribution of the time T to some specific event such as the entry time to a given state. Let $dN_i(t) = I(T_i = t)$ indicate the event occurs at time t for individual i, $i = 1, 2, \ldots, n$, and let $\Lambda(t) = -\log \mathcal{F}(t)$, where $\mathcal{F}(t) = P(T > t)$. Following the general steps above, we first obtain the IPCW Nelson-Aalen estimate of the cumulative hazard function for T by solving

$$\sum_{i=1}^n \frac{Y_i(t)}{\widehat{G}_i^c(t^-)} \{dN_i(t) - d\Lambda(t)\} = 0, \qquad (3.52)$$

which gives $d\widehat{\Lambda}^w(t) = \sum_{i=1}^n w_i(t)\, dN_i(t) / \sum_{i=1}^n w_i(t)$, where $w_i(t) = Y_i(t)/\widehat{G}_i^c(t^-)$, and $\widehat{\Lambda}^w(t) = \int_0^t d\widehat{\Lambda}^w(u)$. The IPCW Kaplan-Meier (KM) estimator is then

$$\widehat{\mathcal{F}}^w(t) = \prod_{u \leq t} [1 - d\widehat{\Lambda}^w(u)]. \qquad (3.53)$$

An alternative inverse probability weighted estimator of $\mathcal{F}(t)$ would be $\widehat{\widehat{\mathcal{F}}}^w(t) =$

$\exp(-\widehat{\Lambda}^w(t))$. In cases where $X(t)$ involves both fixed and time-dependent covariates it is preferable to fit a second censoring model that involves only the fixed covariates X_i^*. Letting $\widehat{G}_i^*(t) = \widehat{G}_i^*(t|x_i^*)$ be the resulting estimate of $P(C_i^R > t|x_i^*)$, we then replace $\widehat{G}_i^c(t^-)$ in (3.52) or (3.53) with $\widehat{G}_i^c(t^-)/\widehat{G}_i^*(t^-)$. This process is referred to as weight stabilization and it produces more efficient estimates than the ordinary IPCW method. Robins and Finkelstein (2000) discuss this for IPCW-KM estimation.

Any approach to modeling the censoring process can and should be checked for adequacy. Cox models or semiparametric additive hazards models can be used (Datta and Satten, 2002). Variance estimation for estimators of occupancy probabilities is most conveniently achieved by the bootstrap, although analytical calculations are available in certain cases (Glidden, 2002).

3.4.3 Skeletal Complications and Mortality in Cancer Metastatic to Bone

In Sections 3.2.2 and 3.3.2, we presented analyses of data from a metastatic breast cancer trial. There we restricted attention to the occurrence of the first skeletal complication, and recognized that the non-negligible mortality rate in this population gives rise to a semi-competing risks problem that is naturally handled by an illness-death model. Subjects could, in fact, experience recurrent skeletal events and we address this here.

Table 3.6: Summary statistics on the number of skeletal events experienced by individuals in the placebo and pamidronate arms of the metastatic breast cancer trial.

Number of Events	Placebo		Pamidronate	
	Frequency	Percent	Frequency	Percent
0	69	35.4	99	53.5
1	41	21.0	39	21.1
2	34	17.4	17	9.2
3	18	9.2	13	7.0
≥ 4	33	16.9	17	9.2
	195		185	

Table 3.6 shows the distribution of total events in each treatment group, but it is important to remember that the duration of time at risk varies across subjects due to premature drop-out from the study or death. Moreover, as we discuss below, premature drop-out and death are related to event occurrence.

Multistate models provide a convenient way to consider the analysis of recurrent events when terminal events may also occur. Figure 3.11 displays the state space diagram with event states $E_0, E_1, E_2, \ldots$ and death states $D_0, D_1, D_2, \ldots$. At any time t, an individual is in state E_k if they are alive and have experienced exactly k events over $(0, t]$, and they are in state D_k if they experienced exactly k events prior to dying by time t. The maximum number of events experienced by individuals in the study is not too large so the final states E_K and D_K may be defined so that the maximum number is represented by K.

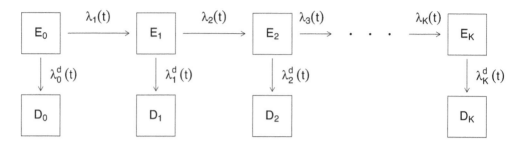

Figure 3.11: Multistate diagram for the joint analysis of recurrent skeletal complications and death for the metastatic breast cancer trial.

The multistate model depicted in Figure 3.11 is quite general and offers a unifying framework for joint consideration of a recurrent event process and death. Specifically it allows us to consider features such as the probability distribution for the cumulative number of (recurrent) events up to time t, the expected number of events, and the probability of death by time t. These can all be approached through nonparametric estimation of state occupancy probabilities through a working Markov model; as discussed in Section 3.4.1, the resulting estimates are often valid for non-Markov processes. Here we let $\lambda_k(t)$ denote the $E_{k-1} \to E_k$ transition rate and $\lambda_k^d(t)$ the $E_k \to D_k$ transition rate, which we may interpret as intensities under a Markov assumption. In addition $\Lambda_k(t) = \int_0^t \lambda_k(s)\, ds$ and $\Lambda_k^d(t) = \int_0^t \lambda_k^d(s)\, ds$. If $Z(t)$ represents the state occupied at time t, we have, for example, that

$$p_k(t) = P(\text{exactly } k \text{ events by time } t) = P(Z(t) = E_k) + P(Z(t) = D_k) \qquad (3.54)$$

and

$$F(t) = P(\text{death by time } t) = \sum_{k=0}^{K} P(Z(t) = D_k). \qquad (3.55)$$

For convenience in the dataframe we number states $E_0, \ldots, E_K$ as 1 to $K+1$ and states D_0 to D_K as $K+2$ to $2K+2$, respectively; here we specify $K = 8$ for the control arm and $K = 10$ for the treated arm. A few lines of the dataframe for this analysis are given in the following text for individuals in the control arm. From this we can see, for example, that individual 1 died at $t = 68$ days without having experienced a skeletal event. Individual 4 experienced three skeletal events on days 61, 277 and 290, respectively, before being censored on day 309 after study entry.

```
> mbc0[1:9,]
  id from   to   time
   1    1   10   68.00
   2    1 cens  749.00
   8    1    2  765.00
   8    2 cens  765.01
   9    1    2   61.00
   9    2    3  277.00
```

```
 9    3    4 290.00
 9    4 cens 309.00
10    1 cens 195.00
```

The general layout of the dataframe is like that of Section 3.2.2 but with the larger state space, individuals may of course contribute many more lines.

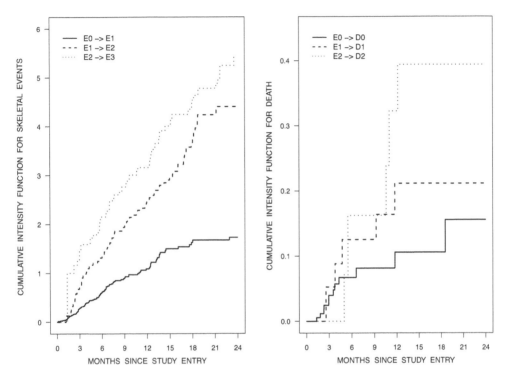

Figure 3.12: Nelson-Aalen estimates of the cumulative intensity functions for skeletal events (left panel) and death (right panel).

The Nelson-Aalen estimates of the cumulative transition rates $\Lambda_k(t) = \int_0^t \lambda_k(s)\, ds$ and $\Lambda_k^d(t) = \int_0^t \lambda_k^d(s)\, ds$ are displayed for the control arm in Figure 3.12 for transitions out of state E_0, E_1 and E_2. The code for obtaining these estimates based on the etm function follows, where the first step is the construction of the matrix of logical variables indicating possible direct transitions.

```
> library(etm)
> tra0 <- matrix(FALSE, ncol=16, nrow=16)
> tra0[1,c(2,10)] <- TRUE
> tra0[2,c(3,11)] <- TRUE
> tra0[3,c(4,12)] <- TRUE
> tra0[4,c(5,13)] <- TRUE
> tra0[5,c(6)]    <- TRUE
> tra0[6,c(7)]    <- TRUE
> tra0[7,c(8,14)] <- TRUE
> tra0[8,c(9,15)] <- TRUE
> tra0[9,c(16)]   <- TRUE
```

```
> ADO <- etm(mbc0, c("1","2","3","4","5","6","7","8","9",
                     "10","11","12","13","14","15","16"),
             tra0, "cens", s=0, t=720)


> ADO.AJ <- data.frame(cbind( ADO$time, t(ADO$est[1,,])))
> dimnames(ADO.AJ)[[2]] <- c("tt", paste("P1",1:16,sep=""))
```

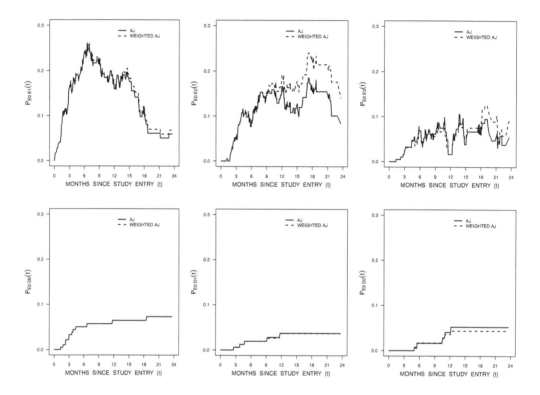

Figure 3.13: Aalen-Johansen estimates of state occupancy probabilities for states E_k and D_{k-1}, $k = 1, 2, 3$.

For the skeletal events process, it is clear that the risk of an event at time t is greater for individuals who have already experienced an event, judged by the fact that the slope of $\widehat{\Lambda}_1(t)$ and $\widehat{\Lambda}_2(t)$ are considerably greater than the slope of $\widehat{\Lambda}_0(t)$; the slopes of $\widehat{\Lambda}_1(t)$ and $\widehat{\Lambda}_2(t)$ are quite comparable. There is also a suggestion that the rate of death is higher at time t among individuals who have experienced an event, but this is based on a modest number of deaths.

Aalen-Johansen estimates of the state occupancy probabilities are also obtainable from the etm function and are plotted in Figure 3.13. These are found as described in Section 3.4.2, using the $(2K+2) \times (2K+2)$ transition rate matrices based on the Nelson-Aalen estimates for $d\widehat{\Lambda}_k(u)$ and $d\widehat{\Lambda}_k^d(u)$. For the transient states, the occupancy probabilities increase and decrease as individuals go on to experience more events or die, but the occupancy probabilities for the absorbing death states are non-decreasing. Recall the cumulative probability of death by time t is $\sum_{k=0}^{K} P(Z(t) = D_k | Z(0) = E_0)$.

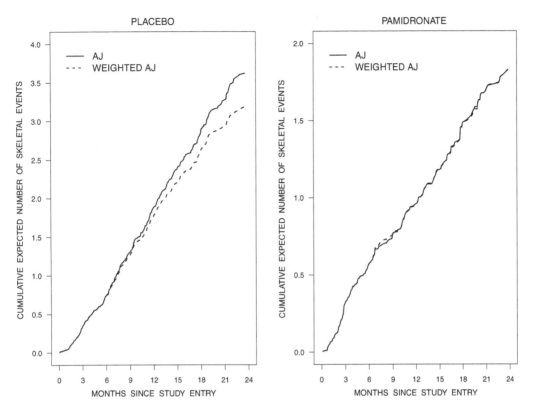

Figure 3.14: Nonparametric estimates of the cumulative expected number of skeletal events based on the Aalen-Johansen estimate of the state occupancy probabilities by treatment arm.

The mean functions for the cumulative number of skeletal events for a given treatment group are defined as

$$\sum_{k=0}^{K} k \cdot P(Z(t) = D_k \cup E_k \mid Z(0) = E_0). \tag{3.56}$$

In this analysis, we fixed $K = 8$ in the control arm and $K = 10$ in the treated arm, because there is little information about transition rates from states E_K over the 2 years of follow-up; this results in (3.56) being slightly under the true expected number of events for larger values of t. With this in mind, the estimates are displayed in Figure 3.14.

The analysis here allows us to compare the probabilities of death and the distributions of skeletal events for the two treatment groups. In particular, Figure 3.14 indicates that subjects on pamidronate experience about half as many skeletal events as those on the placebo, whereas Figure 3.15 indicates that mortality is roughly the same in each group. One limitation of this approach is that we do not have a simple direct way to obtain a measure of treatment effect for the mean number of events. Cook et al. (2009) discuss ways to do this that are related to the methods in Section 4.2.

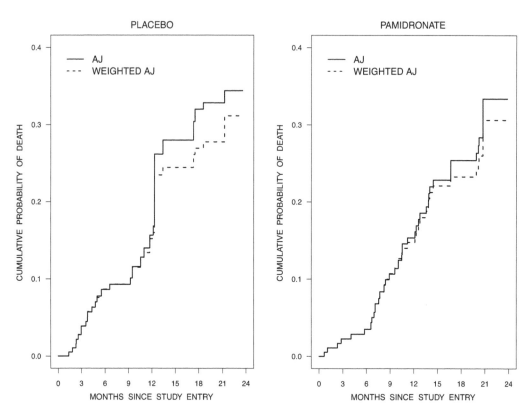

Figure 3.15: Nonparametric estimates of the cumulative probability of death based on the Aalen-Johansen estimate of the state occupancy probabilities by treatment arm.

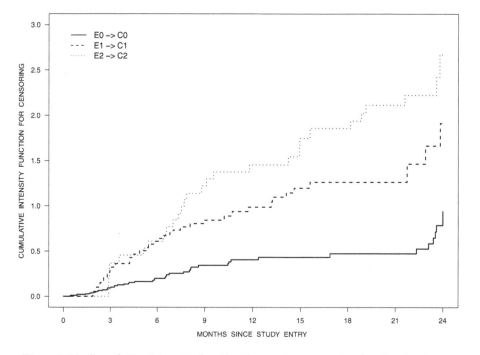

Figure 3.16: Cumulative intensity function for random censoring for the placebo group.

To explore the possibility of dependent random censoring, we consider a censoring model in which the random censoring intensity satisfies (for a given treatment group),

$$\lambda^c(t \mid Z_i(t^-) = E_k, \bar{\mathcal{H}}_i(t^-)) = \lambda_k^c(t), \quad k = 0, 1, \ldots, K.$$

Checks on this assumption, made by introducing covariates representing other process features such as times between previous skeletal events, did not show any significant need for model expansion. Figure 3.16 shows the dependence of the random censoring time on the number of previous skeletal events for the placebo group. This association may arise when individuals who are experiencing events are withdrawn from the study at a higher rate than those with fewer events because of the need for more specialized care.

This suggests the use of IPC weights in (3.51), with

$$\widehat{G}_i^c(u^-) = \prod_{s<u} (1 - d\widehat{\Lambda}^c(s \mid Z_i(s^-)))\,,$$

where

$$d\widehat{\Lambda}^c(s \mid Z_i(s^-) = k) = \frac{\sum_{i=1}^{n} \bar{Y}_{ik}(s)\, dC_i^R(s)}{\sum_{i=1}^{n} \bar{Y}_{ik}(s)}\,, \quad k = 0, 1, \ldots, K. \qquad (3.57)$$

The resulting weighted estimates based on (3.51) are overlaid in the plots of Figures 3.13 to 3.15, where the effect of dependent censoring is most noticeable in the estimated mean function for the individuals receiving placebo treatment (left panel of Figure 3.14). In this arm the number of events is higher so there is a greater opportunity for effects of event-dependent censoring.

Cook and Lawless (2007, Section 6.6) discuss alternative frameworks for the analysis of recurrent events and death, including intensity-based models and robust marginal methods, which can be used for estimation of mean functions. The models considered here are referred to in Cook and Lawless (2007) as partially conditional models, since the Markov assumption can be a working assumption and inferences can be made robust to departures from it.

3.5   Model Assessment

Once the state structure for a process has been given, we can ask questions about the adequacy of specifications for the transition intensity functions or other process features. In this section, we consider ways in which models can be checked against observed data.

3.5.1   Checking Parametric Models

The techniques used to check model assumptions vary according to the model being assessed, but broadly, the main approaches are (i) comparison of parametric model-based estimates with nonparametric (empirical) estimates; (ii) testing a base model against an expanded model that includes it (model expansion); (iii) examination of fitted values, residuals or influence measures across individuals or groups and (iv) predictive assessments, which are important when models are to be used for prediction.

Multistate processes have various features that can be considered when checking a model; they include transition probabilities, state occupancy probabilities, distributions of entry and exit times for states, and distributions of sojourn times in a state. In Section 3.2 we described how to obtain nonparametric estimates of such features for Markov and semi-Markov models, and illustrated how to assess a parametric model without covariates by comparing plots of parametric and nonparametric estimates; see Figure 3.1. The nonparametric estimates used should be robust to departures from the type of process assumed in the model being assessed if we wish to provide the broadest model checks. For example, by using Nelson-Aalen and Aalen-Johansen estimates described in Section 3.2, we can assess parametric specifications $\lambda_{kl}(t;\theta_{kl})$ for Markov transition intensities, but they do not provide checks on the Markov assumption itself. The Markov nonparametric estimates can be made robust under state-dependent censoring by IPC weighting, however, and we consider additional robust estimation procedures in Section 4.2.

Since Markov models are widely used, we consider them further. Checks on parametric Markov models can include comparisons of (i) Nelson-Aalen and parametric estimates of $\Lambda_{kl}(t)$; (ii) Aalen-Johansen and parametric estimates of $P(0,t)$ or $P(s,t)$ and (iii) Kaplan-Meier and parametric estimates of $F_k(t) = P(T_k \leq t)$, where T_k is the time to first entry for state k. We prefer informal comparisons of parametric and nonparametric estimates, but occasionally a formal goodness-of-fit test may be feasible. In the case of transition probabilities, we might consider a set of times $t_0 = 0 < t_1 < \cdots < t_J$ and define observed and expected transition counts

$$O_{kl}(t_{j-1}, t_j) = \sum_{i=1}^{n} Y_i(t_j) I(Z_i(t_{j-1}) = k, Z_i(t_j) = l)$$

$$E_{kl}(t_{j-1}, t_j) = O_{k\cdot}(t_{j-1}, t_j) P_{kl}(t_{j-1}, t_j; \widehat{\theta})$$

for any (k,l) and $j = 1, \ldots, J$. Comparison of the observed and expected counts is helpful. For a formal test, we could consider a Pearson goodness-of-fit statistic

$$W = \sum_{j=1}^{J} \sum_{k=1}^{K} \sum_{l=1}^{K} \frac{[O_{kl}(t_{j-1}, t_j) - E_{kl}(t_{j-1}, t_j)]^2}{E_{kl}(t_{j-1}, t_j)}. \tag{3.58}$$

If $\widehat{\theta}$ is based only on the observed transition counts, then W is asymptotically chi-squared distributed under the null hypothesis that the parametric Markov model is correct (Kalbfleisch and Lawless, 1985), but this is not the case when $\widehat{\theta}$ is based on the likelihood function (3.1). Then, the asymptotic distribution of (3.58) is of the form given by a linear combination of independent $\chi^2_{(1)}$ variables, with coefficients that depend on θ. The R software package msm implements tests based on (3.58) and related extensions to deal with intermittent observation of processes, but p-values provided are questionable in cases where the limiting distribution is complicated and depends on unknown parameter values. Andersen et al. (1993, Section 6.3) describe other types of test statistics based on cumulative intensity functions, some of which also have complicated limiting distributions. Our preference is for tests based on model expansion, which we discuss below.

When a model involves covariates, the model checks just described can in principle be carried out by stratifying individuals according to their covariate values. However, because covariates usually affect different transition intensities in different ways, this approach is not very useful except when we form a small number of strata based on meaningful categorization of covariates. Model expansion, however, can be used as a general technique for assessing parametric assumptions about transition intensities and the effects of covariates. This provides hypothesis tests concerning the base (null) model and, unlike tests based on comparisons of parametric and nonparametric estimators, the test statistics typically have asymptotic normal or chi-square distributions under the null hypothesis, from which p-values are easily obtained.

Some examples of testing based on model expansion are as follows:

(i) Testing the homogeneity of a specific transition intensity with respect to time, for example, by testing the hypothesis H_0: $\beta_{kl} = 0$ within the model

$$\lambda_{kl}(t \mid \mathcal{H}(t^-); \alpha_{kl}, \beta_{kl}) = \exp(\alpha_{kl} + \beta_{kl}\, g_{kl}(t)),$$

where $g_{kl}(t)$ is a specified function. See Problem 3.8 for an illustration.

(ii) Testing for covariate effects within a specified regression model; for example, testing $\beta_{kl} = 0$ within a model

$$\lambda_{kl}(t \mid \mathcal{H}(t^-)) = \lambda_{kl0} \exp(\beta_{kl}x),$$

with x as a fixed covariate.

(iii) Testing whether a Markov intensity is adequate; for example, testing H_0: $\beta_{23} = 0$ in the model

$$\lambda_{23}(t \mid \mathcal{H}(t^-); \alpha_{23}, \beta_{23}) = \exp\{\alpha_{23}(t) + \beta_{23}B(t)\},$$

where $B(t)$ is the time since entry to state 2. In Section 3.3.2, a related convenient test of a Markov assumption for the $2 \to 3$ transition intensity in an illness-death model was carried out by including the entry time to state 2 as a covariate in an expanded model.

Estimation for parametric models described in Section 3.2 can be used to obtain Wald, score or likelihood ratio test statistics. Score statistics are convenient in many settings, and require that only the null model be fitted. In particular, let $\lambda_{ikl}(u; \theta_{kl})$ denote the intensity $\lambda_{kl}(u|\mathcal{H}_i(u^-); \theta_{kl})$ and suppose that $\theta_{kl} = (\alpha'_{kl}, \beta'_{kl})'$ and that we wish to test H_0: $\beta_{kl} = 0$. The likelihood score function (3.3) can be rewritten as

$$U(\theta_{kl}) = \sum_{i=1}^{n} \int_0^\infty \bar{Y}_{ik}(u) \frac{\partial \log \lambda_{ikl}(u; \theta_{kl})}{\partial \theta_{kl}} \left\{ dN_{ikl}(u) - \lambda_{ikl}(u; \theta_{kl})\, du \right\}$$

and with θ_{kl} replaced by the MLE under H_0, denoted by $\widetilde{\theta}_{kl}(\widetilde{\alpha}_{kl}, 0)$, we obtain

$$U_0 = \sum_{i=1}^{n} \int_0^\infty \bar{Y}_{ik}(u)\, g_{ikl}(u; \widetilde{\theta}_{kl}) \left\{ dN_{ikl}(u) - \lambda_{ikl}(u; \widetilde{\theta}_{kl})\, du \right\}, \tag{3.59}$$

where $g_{ikl}(u; \theta_{kl}) = \partial \log \lambda_{ikl}(u; \theta_{kl}) / \partial \beta_{kl}$. Such tests have been developed by Hjort (1990), Pena (1998) and others; see Problem 3.6 for an illustration.

Residuals of various types have also been proposed for multistate and other event history models. Many of them are hard to interpret and require smoothing. Cox-Snell residuals based on sojourn times in different states can be useful. Let $W_{ik}^{(r)} = V_{ik}^{(r)} - T_{ik}^{(r)}$ denote the length of the rth sojourn in state k for individual i. If there are no time-dependent covariates except for functions of process history $\mathcal{H}_i(t^-)$ in the transition intensities $\lambda_{kl}(t|\mathcal{H}(t^-); \theta_{kl})$ for $l \neq k$, then if the model is correct the random variables

$$E_{ik}^{(r)} = \int_{T_{ik}^{(r)}}^{V_{ik}^{(r)}} \lambda_{k \cdot}(t \mid \mathcal{H}_i(t^-)) \, dt \tag{3.60}$$

have exponential distributions with mean one (see Problem 2.11), where $\lambda_{k \cdot}(t|\mathcal{H}_i(t^-)) = \sum_{l \neq k} \lambda_{kl}(t|\mathcal{H}_i(t^-))$. The $\widehat{E}_{ik}^{(r)}$ given by replacing parameters θ_{kl} with consistent estimators are asymptotically independent exponential variables with mean one. The $\widehat{E}_{ik}^{(r)}$ for a given state k can be examined in a probability plot, and plotted against covariates; in the case where the sojourn $E_{ik}^{(r)}$ is censored by the end of follow-up, it is conventional to replace $\widehat{E}_{ik}^{(r)}$ with $\widehat{E}_{ik}^{(r)} + 1$ (Lawless, 2003, Section 6.2). Finally, we mention influence analysis, in which the effects of the data from specific individuals is considered. For convenience, denote a given θ_{kl} as just θ; and let $\widehat{\theta}^{-i}$ be the MLE obtained from the dataset with individual i excluded. The "influence" of individual i can be assessed by considering $\widehat{\theta} - \widehat{\theta}^{-i}$ or the associated likelihood drop (LD) statistics $\mathrm{LD}_i = 2(\ell_{kl}(\widehat{\theta}) - \ell_{kl}(\widehat{\theta}^{-i}))$. It is easily shown (e.g., Lawless, 2003, p. 281) that $\widehat{\theta} - \widehat{\theta}^{-i} \doteq I_{kl}(\widehat{\theta})^{-1} U_{ikl}(\widehat{\theta})$ and $\mathrm{LD}_i \doteq U'_{ikl}(\widehat{\theta}) I_{kl}(\widehat{\theta})^{-1} U_{ikl}(\widehat{\theta})$, where $U_{kl}(\theta)$ is given by (3.3) and $U_{ikl}(\theta)$ is the contribution from individual i. Individuals for whom the approximations for LD_i are large can be investigated, and exact values of $\widehat{\theta}^{-i}$ and LD_i determined.

3.5.2 Semiparametric Models

For semiparametric models such as these described in Section 3.3, we can define residuals and can develop tests based on model expansion in ways analogous to those for parametric models. The examination of transition probabilities is less attractive; aside from Markov models, they are difficult to compute and even for Markov models it is hard to diagnose the source of differences in model-based and empirical transition probabilities. Therefore, we once again emphasize tests based on model expansion and on related plots and summaries.

For a modulated multiplicative Markov model with $k \to l$ transition intensity of the form (3.22) with $g(x_i(t); \beta_{kl}) = \exp(x_i'(t)\beta_{kl})$, we can consider expanded models of the form

$$g(x_i(t), x_i^e(t); \beta_{kl}, \gamma_{kl}) = \exp(x_i'(t)\beta_{kl} + [x_i^e(t)]'\gamma_{kl}),$$

where $x_i^e(t)$ is a predictable vector of fixed or time-varying variables that could include functions of t, $x_i(t)$ or previous process history. From (3.28) the derivative of the profile log-likelihood function with respect to γ_{kl}, evaluated at $\widetilde{\beta}_{kl}$, the estimate of β_{kl} from the base model with $\gamma_{kl} = 0$, is

$$U_0 = \sum_{i=1}^{n} \int_0^\infty \bar{Y}_{ik}(u) x_i^e(u) \left\{ dN_{ikl}(u) - \exp(x_i'(u)\tilde{\beta}_{kl}) \, d\tilde{\Lambda}_{kl0}(u) \right\}, \tag{3.61}$$

with $\tilde{\Lambda}_{kl0}(u)$ the estimator based on (3.30). As described in Section 3.3.1, Cox model software such as the R function `coxph` can be used to fit the full ($\gamma_{kl} \neq 0$) or null ($\gamma_{kl} = 0$) models. Tests of H_0: $\gamma_{kl} = 0$ can be based on Wald or likelihood ratio statistics, or on the score statistic (3.61). For the score statistic some additional calculation is needed to obtain the covariance matrix for U_0 under H_0, but it has two nice features. One is that (3.61) can be partitioned over groups of individuals or time intervals to enable diagnosis of lack of fit. A second is that for some models, tests based on (3.61) or asymptotically equivalent statistics are available in software. In particular, the R function `cox.zph` provides a test of the proportional intensity specification $g(x_i; \beta_{kl}) = \exp(x_i'\beta_{kl})$ for a p-dimensional fixed covariate that is equivalent to tests based on (3.61) with additional covariates $x_{ij}^e(t) = x_{ij}g_j(t)$ for $j = 1, \ldots, p$. The `cox.zph` tests have the advantage that it is not necessary to create a dataframe including the defined covariate. If piecewise-constant time-dependent covariates are added to the dataframe, a score test can be carried out through direct calls to the `coxph` function, as illustrated at the end of Section 3.3.4. Similar tests can be used for additive models.

3.5.3   Predictive Performance of Models

Models can also be assessed in terms of their predictive ability. This is of particular interest when a model is used to assist in decisions about screening, treatment or other interventions. The potential occurrence of various clinical events, and their probability for a specific setting and individual, are key factors in decisions. This can be done in various ways, some of which are closely related to methods used for survival models. For example, we may in some settings be interested in predicting the time T_k until a specific state k is first entered. A common approach is to measure predictive performance for a model that gives a distribution function $F_k(t|x_i;\theta) = P(T_{ik} \leq t|x_i;\theta)$ at a given time t via the Brier score (Gneiting et al., 2007)

$$\text{BS}(t) = \sum_{i=1}^{n} Y_i(t) \left[I(T_{ik} \leq t) - F_k(t \mid x_i; \theta) \right]^2, \tag{3.62}$$

where (T_{ik}, x_i), $i = 1, \ldots, n$ come from a random sample of individuals. Here we assume that x_i is a vector of external covariates; for simplicity we show it as fixed, but x_i could be a history $\bar{x}_i(t)$ of external time-dependent covariates. In practice the model parameters are estimated by $\hat{\theta}$ and if $\hat{\theta}$ is based on data for the individuals $i = 1, \ldots, n$, then we can replace (3.62) with a cross-validated version. If we split individuals into G groups ($g = 1, \ldots, G$) of equal size and if $\hat{\theta}_{(-g)}$ denotes the estimate obtained from the data on all individuals except those in group g, then (3.62) is replaced by

$$\text{BS}_{\text{cv}}(t) = \sum_{g=1}^{G} \sum_{i \in S_g} Y_i(t) \left[I(T_{ik} \leq t) - F_k(t \mid x_i; \hat{\theta}_{(-g)}) \right]^2. \tag{3.63}$$

If the variable $I(T_{ik} \leq t)$ and $Y_i(t)$ are not conditionally independent given x_i, then inverse probability of censoring weights can be incorporated (Gerds et al., 2008). If two models are compared, the one with smaller BS(t) is preferred for prediction, and (3.63) can also be extended to deal with multiple times t.

The Brier score is a type of performance measure for a predictive model; it is also referred to as a scoring rule (Gneiting et al., 2007). In settings where T is an event time and $\tilde{F}(t|x) = \tilde{P}(T \leq t|X = x)$ is a prediction model, we define the Brier score as described above by taking a specific time t and considering the binary response $Y_t = I(T \leq t)$. We term the predictive model $\tilde{F}$ to remind us that it will have been specified through analysis of some data and possibly other considerations. The Brier score for a single Y_t and X is then BS$(t) = (Y_t - \tilde{F}(t|X))^2$, where Y_t and X are potential observations. If the model $\tilde{F}$ is used for a population of individuals for which X has distribution function $G(x)$ and given $X = x$, T has distribution $F(t|x)$, then the average or expected Brier score is

$$\text{EBS}(t) = E\{(Y_t - \tilde{F}(t|X))^2\}. \tag{3.64}$$

The expectation in (3.64) is with respect to (T, X), and it is easily seen that (see Problem 3.7)

$$\text{EBS}(t) = E_X\{F(t \mid X)(1 - F(t \mid X))\} + E_X\{(\tilde{F}(t \mid X)(1 - F(t \mid X)))^2\}. \tag{3.65}$$

The first term in EBS(t) reflects variation in T given X, and the second term reflects model misspecification. These two features are often referred to as sharpness and calibration, respectively (e.g. Gneiting et al., 2007). Given a set of observations (T_i, X_i) from a target population, we can estimate EBS(t) with an estimate of the form (3.62). Smaller Brier scores and average Brier scores indicate better performance for a predictive model, and a main use of such measures is in the comparison of two or more predictive models for a given situation.

A multistate model may not give an easily computed probability $F_k(t|x_i; \theta)$. In that case, it can be estimated by simulating the multistate process, but this can be very time-consuming. A simpler predictive assessment is to consider the probability of the full observed sample path $\mathcal{Z}_i(C_i)$ for an individual. This is typically done by the log probability density, or logarithmic score (Gneiting and Raftery, 2007),

$$\text{LS}_i = -\log P(\mathcal{Z}_i(C_i) \mid X_i; \theta),$$

with smaller LS_i indicating better performance. This score is also often referred to as the Kullback-Leibler score, or KLS. Once again, we need to estimate θ and a cross-validated LS is used when $\hat{\theta}$ is based on individuals $i = 1, \ldots, n$:

$$\text{LS}_{\text{CV}} = -\sum_{g=1}^{G} \sum_{i \in S_g} \log P(\mathcal{Z}_i(C_i) \mid X_i; \hat{\theta}_{(-g)}). \tag{3.66}$$

We return to the topic of prediction in Section 8.2 where we give a more detailed discussion.

3.5.4   Consequences of Model Misspecification and Robustness

In many settings where covariate effects are expressed on marginal means, inferences regarding these first moments can be robust to misspecification of higher-order moments; quasi-likelihood methods, for example, used in fitting certain generalized linear models have this property, and we may use robust variance estimates to provide protection against misspecification of the variance function (McCullagh and Nelder, 1989). In settings with multivariate or clustered data, robust variance estimates provide protection against misspecification of both the variance and dependence structure (Liang and Zeger, 1986). In Section 4.2, we discuss marginal models for state occupancy probabilities; inferences in this setting can be made robust in a similar fashion, assuming correct specification of the model for the covariate effects on the marginal probabilities. We need, of course, to check the adequacy of such models. Intensity-based life history analyses, however, are more difficult to make robust since the validity of the likelihoods and score equations hinge critically on the correct specification of dependence on the process history. Model assessment, therefore, plays a crucial role in analysis and the interpretation of results. With delayed entry and right censoring affecting the information available for some transition intensities, certain kinds of model violations can be difficult to detect. It is sometimes helpful to consider the potential consequences of particular kinds of model misspecification. This can help in understanding the extent to which inferences may be sensitive to model violations. We comment briefly here on methods for determining the limiting behaviour of estimators from parametric and semiparametric models under possible misspecification.

Given a sample of n independent identical processes labeled $i = 1, \ldots, n$ and an estimating function $U(\theta) = \sum_{i=1}^{n} U_i(\theta)$, subject to mild regularity conditions the solution $\widehat{\theta}$ to $U(\theta) = 0$ satisfies $\sqrt{n}(\widehat{\theta} - \theta^{\dagger}) \sim N(0, \mathcal{A}^{-1}(\theta^{\dagger})\mathcal{B}(\theta^{\dagger})[\mathcal{A}^{-1}(\theta^{\dagger})]')$, where $\theta^{\dagger}$ is the solution to $E\{U_i(\theta)\} = 0$, $\mathcal{A}(\theta) = E\{-\partial U_i(\theta)/\partial\theta'\}$ and $\mathcal{B}(\theta) = E\{U_i(\theta)U_i'(\theta)\}$ with the expectations taken with respect to the true distribution (White, 1982). Struthers and Kalbfleisch (1986) studied issues of misspecification of the semiparametric Cox model and additional results on robustness were developed by Lin and Wei (1989). A detailed review of misspecification and robustness for proportional hazards and proportional intensity methods is given by O'Quigley and Xu (2014). Here we briefly consider related issues in the analysis of multistate data under a working Markov model, beginning with the Nelson-Aalen estimate.

The Nelson-Aalen estimate $\widehat{\Lambda}_{kl}(t)$ in (3.13) is based on the solution $d\widehat{\Lambda}_{kl}(u)$ to the estimating equation

$$\sum_{i=1}^{n} \bar{Y}_{ik}(u)\left\{dN_{ikl}(u) - d\Lambda_{kl}(u)\right\}, \quad u > 0, \tag{3.67}$$

and is a maximum likelihood estimate of the Markov cumulative transition intensity. Consider a general process, however, with intensity $\lambda_{kl}(t|\mathcal{H}_i(t^-))$ for individual i with history $\mathcal{H}_i(t) = \{Z_i(s), 0 \leq s \leq t\}$. Under completely independent censoring (i.e. $C_i \perp \{Z_i(s), s \geq 0\}$) the expectation of (3.67) can be taken first with respect to

$C_i \mid dN_{ik}(u), Y_{ik}(u^-), \mathcal{H}_i(u^-)$, which if $P(C_i \geq u) = G(u)$ gives

$$\sum_{i=1}^{n} G(u) Y_{ik}(u^-) \{dN_{ikl}(u) - d\Lambda_{kl}(u)\}, \quad u > 0.$$

Then taking the expectation with respect to $\{dN_{ikl}(u), Y_{ik}(u^-)\}$ and equating this to zero, we obtain $d\Lambda_{kl}^{\dagger}(u) = E\{dN_{ikl}(u) | Y_{ik}(u^-) = 1\}$, which is known as a marginal transition rate. The term *marginal* is used here since it represents an instantaneous risk of a $k \to l$ transition given occupancy of state k at time u^-, but it averages over all possible histories prior to u^-. Thus, the Nelson-Aalen estimate $\widehat{\Lambda}_{kl}(t)$ is in general consistent for $\Lambda_{kl}^{\dagger}(t) = \int_0^t d\Lambda_{kl}^{\dagger}(u)$, as discussed in Section 3.4.1. Under Markov models $\Lambda_{kl}^{\dagger}(t) = \Lambda_{kl}(t)$, but more generally $\Lambda_{kl}^{\dagger}(t)$ and the Nelson-Aalen estimate convey information on how marginal risk changes over time. As discussed in 3.4.1, Aalen-Johansen estimates may also be computed based on these marginal rates, and these yield consistent estimates for state occupancy probabilities under independent censoring even for non-Markov processes. When censoring is history dependent, the censoring process can be modeled to construct weights for inverse probability of censoring weighted estimating equations. We remark that occupancy probabilities have simple interpretations but marginal rate functions may not. In that case, we may wish to examine what rate functions result from specific types of non-Markov processes, either by direct calculation when possible, or simulation, as a guide to interpretation.

For the semiparametric setting, consider a working Markov model with $k \to l$ intensity of the form $\lambda_{kl}(t|\mathcal{H}_i(t^-)) = \lambda_{kl0}(t) \exp(x_i' \beta_{kl})$. Recall from Section 3.3.1 that the partial score equation for the regression coefficient β_{kl} is

$$U_{kl}(\beta_{kl}) = \sum_{i=1}^{n} U_{ikl}(\beta_{kl}) = \sum_{i=1}^{n} \int_0^{\infty} \bar{Y}_{ik}(u) \left\{ x_i - \frac{S_k^{(1)}(u; \beta_{kl})}{S_k^{(0)}(u; \beta_{kl})} \right\} dN_{ikl}(u) = 0, \quad (3.68)$$

where $S_k^{(r)}(u; \beta_{kl}) = \sum_{i=1}^{n} \bar{Y}_{ik}(u) x_i^{\otimes r} \exp(x_i' \beta_{kl})$. The estimate $\widehat{\beta}_{kl}$ obtained from (3.68) is consistent for $\beta_{kl}^{\dagger}$, the solution to

$$\int_0^{\infty} E\left[\bar{Y}_{ik}(u) \left\{ x_i - \frac{s_k^{(1)}(u; \beta_{kl})}{s_k^{(0)}(u; \beta_{kl})} \right\} dN_{ikl}(u) \right] = 0, \quad (3.69)$$

where $s_k^{(r)}(u; \beta_{kl}) = E\{S_k^{(r)}(u; \beta_{kl})\}$ and the expectations are taken with respect to the distribution of $\{Y_i(s), \bar{N}_i(s), s \geq 0; Z_i(0), X_i\}$. Then asymptotically,

$$\sqrt{n}(\widehat{\beta}_{kl} - \beta_{kl}^{\dagger}) \sim N(0, \mathcal{A}^{-1}(\beta_{kl}^{\dagger}) \mathcal{B}(\beta_{kl}^{\dagger}) [\mathcal{A}^{-1}(\beta_{kl}^{\dagger})]')$$

where $\mathcal{A}(\beta) = E\{-\partial U_{ikl}(\beta_{kl})/\partial \beta_{kl}\}$ and $\mathcal{B}(\beta_{kl}) = E\{U_{ikl}(\beta_{kl}) U_{ikl}'(\beta_{kl})\}$. If the specification of the intensity $\lambda_{kl}(t|\mathcal{H}_i(t^-))$ is incorrect, the limiting value $\beta_{kl}^{\dagger}$ is typically going to be a complex function of the correct intensities for the multistate process, the covariate distribution and, notably, the censoring distribution. Robustness is, therefore, difficult to achieve, and the consequences of misspecification are complex and multifaceted. These results are, however, useful for investigating the effects of particular kinds of model misspecification.

These methods can also be particularly useful in settings where there is confidence in the specification of the multistate model but concerns about the possible impact of dependent censoring. Here, various types of dependence of the censoring process on the event history can be specified, and the impact on the limiting distribution can be investigated. Often in such cases marginal features such as state occupancy probabilities are more seriously affected than regression coefficients. Use of inverse probability of censoring weights can address these biases, subject to correct specification of the model for the censoring intensity.

3.6   Design Issues

Consider the design of a prospective study in which individuals are to be followed over a fixed interval $(0, C^A]$, but individuals may drop out at some earlier time $C^R < C^A$; let $C = \min(C^R, C^A)$ denote the random period of observation for an individual. When exact transition times are known over $(0, C]$, the factors affecting the information provided from a prospective study are the sample size n, the planned duration of follow-up (C^A) and the distribution of the dropout time C^R. In addition, in planned studies we may be able to select individuals differentially according to baseline covariate values or their life history state at the time of selection. For example, in the Diabetes Control and Complications Trial (DCCT) described in Section 1.2.2, two groups of individuals were selected: (i) a "Primary Prevention" cohort, for which members had to have no retinopathy and Type I diabetes duration of $1-5$ years, and (ii) a "Secondary Intervention" cohort, for which individuals had only mild/moderate retinopathy and diabetes duration of $1-15$ years. A final factor that affects precision of estimation is of course the magnitude of the transition intensities, since these govern the occurrence of state transitions.

If time-dependent covariates are of interest, then a decision about how often to measure them is also needed. This affects our ability to examine their effect on the multistate process and while frequent measurement is desirable, it is often infeasible.

We do not attempt to give detailed guidelines, since study design is context specific. By way of practical guidance, we suggest identifying certain target parameters or functions of the process and then, within time and budgetary constraints for the study, examine the precision of relevant estimates. This requires the specification of a working model and associated parameter values for the multistate and censoring processes. Even for the special case of time-homogeneous Markov models, direct calculation of asymptotic variance via (3.14) or (3.16) is possible only in certain cases. More generally, we can simulate individual sample paths in order to assess the variability of estimators. This is necessary even for models where expressions for asymptotic variances are available (e.g. see (3.32)), since exact calculation of terms in the expressions is typically intractable.

Example 3.6.1:   Time-Homogeneous Markov Models

For a time-homogeneous Markov model with no covariates, (3.8) provides the asymptotic variance $\mathrm{var}(\widehat{\lambda}_{kl}) = n_{kl}^{-1} \lambda_{kl}^2$ for estimated transition intensities, where

n_{kl} is the observed total number of $k \to l$ transitions in the study cohort, a realization of $N_{kl} = \sum_{i=1}^{n} N_{ikl}(C_i)$. For a cohort of size n with $n_j = \sum_{i=1}^{n} Y_{ij}(0)$ individuals in state j at $t = 0$, consider for simplicity the case where the follow-up is equal to C^A for all individuals in the sample. Suppressing the dependence of $N_{kl}(C^A)$ on C^A, for a general time-homogeneous multistate process

$$E(N_{kl}) = n \int_0^{C^A} P_k(u)\,\lambda_{kl}(u)\,du = \lambda_{kl} \sum_{j=1}^{K} n_j \psi_{jk} \tag{3.70}$$

where

$$\psi_{jk} = \psi_{jk}(C^A) = \int_0^{C^A} P_{jk}(0,u)\,du,$$

is the expected time in state k over $(0, C^A]$ given occupancy of state j at $t = 0$. The terms in (3.70) are readily computed using (2.24).

For a 2-state model with $\lambda_{12} > 0$ and $\lambda_{21} > 0$, we obtain

$$E(N_{12}) = [n_1 \psi_{11} + n_2 \psi_{21}] \cdot \lambda_{12}$$

and

$$E(N_{21}) = [n_1 \psi_{12} + n_2 \psi_{22}] \cdot \lambda_{21}.$$

It is apparent that by selecting n_1 and n_2 in a particular way, we can change the precision of $\widehat{\lambda}_{12}$ and $\widehat{\lambda}_{21}$. The values of λ_{12} and λ_{21} must be specified for the determination of $\mathrm{var}(\widehat{\lambda}_{12})$ and $\mathrm{var}(\widehat{\lambda}_{21})$, and it is often of interest to assess how sensitive the asymptotic variances are to the specification. We note here that $\psi_{k.} = \psi_{k1} + \psi_{k2} = C^A$, for $k = 1, 2$.

For illustration, suppose $\lambda_{12} = 0.04$ and $\lambda_{21} = 1$ corresponding to a situation where individuals periodically spend short periods in state 2 and the rest of the time in state 1. Let $\alpha = \lambda_{12} + \lambda_{21}$ and $\pi = \lambda_{12}/\alpha$. Then (see Problem 1.2)

$$P(0,t) = \begin{pmatrix} 1 - \pi(1 - e^{-\alpha t}) & \pi(1 - e^{-\alpha t}) \\ (1-\pi)(1 - e^{-\alpha t}) & \pi + (1-\pi)e^{-\alpha t} \end{pmatrix}.$$

We consider three designs defined by different choices of (n_1, n_2): proportional selection whereby $(96, 4)$; balanced selection $(50, 50)$; and selection restricted to persons in state 2 $(0, 100)$. Table 3.7 shows values for $E(N_{12})$ and $E(N_{21})$ for the three choices for (n_1, n_2). We have asymptotic variances $\mathrm{var}(\widehat{\lambda}_{12}) = \lambda_{12}^2/E(N_{12})$ and $\mathrm{var}(\widehat{\lambda}_{21}) = \lambda_{21}^2/E(N_{21})$ based on the expected information matrix, and we see that by selecting individuals who are in state 2 $((n_1, n_2) = (0, 100))$, we substantially increase the precision for $\widehat{\lambda}_{21}$ while only slightly lowering the precision for $\widehat{\lambda}_{12}$.

3.7   Bibliographic Notes

Maximum likelihood estimation and inference methodology for parametric models is in principle straightforward. Andersen et al. (1993, Chapter 6 and Section 7.6) discuss regularity conditions and asymptotic results. Models for which transition intensities take certain common parametric forms can be fitted using survival

Table 3.7: Values of $E(N_{12})$ and $E(N_{21})$ for choices of C^A and $(n_1, n_2)^\dagger$.

| | Selection for Initial States (n_1, n_2) | | | | | |
| | (96, 4) | | (50, 50) | | (0, 100) | |
C^A	$E(N_{12})$	$E(N_{21})$	$E(N_{12})$	$E(N_{21})$	$E(N_{12})$	$E(N_{21})$
5	19.22	19.38	17.47	63.36	15.55	111.18
20	76.92	77.07	75.15	121.30	73.22	169.38
50	192.30	192.46	190.53	236.69	188.61	284.76

$\dagger$ The dependence on C^A is suppressed in the column headings.

analysis software; Kalbfleisch and Prentice (2002), Lawless (2003) and Klein and Moeschberger (2003) survey survival analysis methods. Early papers on multistate models focused on parametric estimation (e.g. Fix and Neyman, 1951; Sverdrup, 1965; Temkin, 1978); a good deal of early work was in actuarial science and demography (e.g. Hoem, 1971; Hoem, 1977).

Nonparametric estimation began with Altshuler (1970) and Nelson (1972), who introduced Nelson-Aalen estimation of cumulative hazard functions; Aalen (1975, 1978) extended this to Markov multistate processes and provided a rigorous development of asymptotics based on martingale theory. Aalen and Johansen (1978) and Fleming (1978a,b) gave the Aalen-Johansen estimator and associated asymptotic properties for Markov transition intensities; Aalen and Johansen stressed the representation involving product integration. Andersen et al. (1993, Chapter 4) describe these and other developments in great detail. Nonparametric estimation for semi-Markov models was developed in papers by Lagakos et al. (1978), Gill (1980) and Matthews (1984); see Andersen et al. (1993, Chapter 10) for a detailed survey.

Semiparametric modeling began with Andersen and Gill (1982), who extended the use of the Cox model (Cox, 1972) to more general counting processes. These multiplicative models are now widely used with Markov processes; see Andersen et al. (1993, Chapter 7) and for a less detailed but useful treatment, Aalen et al. (2008, Chapter 4). Therneau and Grambsch (2000) illustrate the use of Cox model software with multistate models. Semiparametric multiplicative intensity models (also known as relative risk models) are also used with semi-Markov processes. Here the development of asymptotic properties requires a different approach; see Dabrowska et al. (1994) and Lawless et al. (2001). Earlier papers on semi-Markov regression models were considered for the special case of recurrent or serial events by Gail et al. (1980) and Prentice et al. (1981); see also Cook and Lawless (2007, Chapters 4 and 5). Lawless and Fong (1999) and Cook and Lawless (2007, Chapter 4) consider issues related to delayed entry or left truncation. Additive intensity-based semiparametric regression models were introduced by Aalen (1980, 1989) and further developed by Lin and Ying (1994) and McKeague and Sasieni (1994). Andersen et al. (1993, Chapter 7) and Aalen et al. (2008, Chapter 4) describe basic methodology; a more comprehensive treatment is given by Martinussen and Scheike

(2006), who also consider combined multiplicative-additive intensity models and provide the R `timereg` package for analysis.

The Aalen-Johansen estimator was noted to provide robust estimation of state occupancy probabilities by Aalen et al. (2001) and Datta and Satten (2001), provided that censoring is independent of the multistate process. Datta and Satten (2002) introduced inverse probability of censoring (IPCW) weighting (Robins and Rotnitzky, 1992) to achieve robustness when censoring is state dependent; Glidden (2002) provided robust variance estimates. Gunnes et al. (2007) compare IPCW and Aalen-Johansen estimators with respect to bias and precision. Estimation for the illness-death process hass received considerable attention; see Vakulenko-Lagun et al. (2017) for recent results. Cook et al. (2009) consider applications involving recurrent and terminal events. Model assessment based on martingale residuals parallels their use in survival analysis; see Fleming and Harrington (1991), Andersen et al. (1993), Therneau and Grambsch (2000) and Kalbfleisch and Prentice (2002). Gandy and Jensen (2005a,b) consider such residuals for additive models. For recurrent events, both martingale and Cox-Snell or exponential residuals are described by Cook and Lawless (2007, Section 3.7.3) and the discussion extends to the assessment of intensity functions for multistate models. Goodness-of-fit tests based on model expansion and score statistics were developed by Hjort (1990), Grambsch and Therneau (1994) and Pena (1998) for event history models, and can be applied to multistate models; de Stavola (1988) is an early example. Grønnesby and Borgan (1996) consider tests of semiparametric survival models with expanded models of the form in Section 3.5.2; score statistics are of the type in (3.61) and involve sum of observed and expected martingale residuals. See also May and Hosmer (2004). Titman and Sharples (2010b) discuss methods based on comparisons of empirical (nonparametric) and parametric estimates of occupancy or transition probabilities; they emphasize intermittently observed data, but the approaches can also be used with complete data. Predictive model assessment is discussed in general by Gneiting et al. (2007) and van Houwelingen and Putter (2012, Chapter 3). Cortese et al. (2013) give an illustration for competing risks models. Additional discussion and references on prediction are given in Chapter 8.

3.8   Problems

Problem 3.1   Consider a 2-state time-homogeneous Markov process in which $\lambda_{12} = \exp(\theta_1)$ and $\lambda_{21} = \exp(\theta_2)$ denote the transition rates out of states 1 (healthy) and 2 (diseased), respectively. Individuals recruited to a study are initially observed to be in state 1 with probability π, and in state 2 with probability $1 - \pi$. The intention is to observe individuals over the interval $[0, C^A]$, but subjects may withdraw from the study so we suppose there is an independent exponential censoring time with rate ρ. Let n_{ik} denote the number of transitions out of state k made by individual i and $w_{ik}^{(r)}$ denote their rth sojourn time in state k, where $r = 1, 2, \ldots, n_{ik}$, $k = 1, 2$ and $W_{ik} = \sum_{r=1}^{n_{ik}} w_{ik}^{(r)}$, $i = 1, 2, \ldots, n$.

(a) Show that the Fisher information for θ_k is

$$\mathcal{I}(\theta_k) = \lambda_k \, E(W_k),$$

where $W_k = \sum_{i=1}^{n} W_{ik}$ and $E(W_k)$ denote the expected total time spent in state k over $[0, C^A]$, aggregating over all individuals and accommodating possible loss to follow-up.

(b) Show that

$$E(W_{ik}) = \pi \cdot G_{1k}(C^A) + (1 - \pi) \cdot G_{2k}(C^A),$$

where

$$G_{12}(s) = \frac{\lambda_{12}}{(\lambda_{12} + \lambda_{21})} \left[\frac{(1 - e^{-\rho s})}{\rho} + \frac{(1 - e^{-(\rho + \lambda_{12} + \lambda_{21})s})}{(\rho + \lambda_{12} + \lambda_{21})} \right]$$

$$G_{21}(s) = \frac{\lambda_{21}}{(\lambda_{12} + \lambda_{21})} \left[\frac{(1 - e^{-\rho s})}{\rho} + \frac{(1 - e^{-(\rho + \lambda_{12} + \lambda_{21})s})}{(\rho + \lambda_{12} + \lambda_{21})} \right]$$

and $G_{ll}(s) = (1 - e^{-\rho s})/\rho - G_{lk}(s)$, where $k, l = 1, 2$ and $k \neq l$.

(c) Suppose interest lies in planning a study to compare the $1 \to 2$ transition intensities between two groups of individuals. Discuss how the Fisher information can be used to determine the sample size necessary to meet power objectives in tests of hypotheses regarding these transition intensities.

(d) Suppose the process is in equilibrium and individuals are randomly sampled so that $\pi = \lambda_{21}/(\lambda_{12} + \lambda_{21})$. Give the revised Fisher information matrix reflecting the added information about $(\lambda_{12}, \lambda_{21})'$ from the initial distribution.

(Sections 3.1, 3.6)

Problem 3.2   Problem 3.1 dealt with the expected information for parameters of a time-homogeneous 2-state process to be observed over $[0, C^A]$. Now suppose that the transitions between states cannot be observed but rather the state is only known at R visits scheduled at periodic inspection times $a_r = r \cdot C^A/R$, $r = 0, 1, \ldots, R$.

(a) Derive the Fisher information matrix for $(\lambda_{12}, \lambda_{21})'$ under this observation process.

(b) If individuals are subject to early random withdrawal according to an exponential withdrawal rate, what is the revised Fisher information matrix?

(c) If there is a cost D_1 associated with the recruitment of an additional individual to a study and a cost D_2 associated with increasing R by one unit, find an expression that maximizes the precision of the λ_{12} estimator while minimizing the cost.

(Sections 3.1, 3.6)

Problem 3.3 The irreversible illness-death model is commonly used to characterize risk of tumour development and to assess the lethality of tumours in animal carcinogenicity experiments. State 1 represents a tumour-free state; state 2 the state of being alive with a tumour; and state 3, death. In such experiments, animals may be randomized to different doses of cancer-causing agents and followed over the interval $[0, C^A]$. Transitions into state 3 are usually observed subject to right censoring and transitions into state 2 are unobserved, but autopsy yields information on whether each animal that died had developed the tumour. If X denotes the dose of a carcinogen, interest lies in β_{12} in $\lambda_{12}(t|x) = \lambda_{12}\exp(\beta_{12}x)$ as a measure of the cancer-causing effect of the agent, and β_{23} in $\lambda_{23}(t|x) = \lambda_{13}\exp(\beta_{23})$ as a measure of the lethality of the tumour caused by the agent. We assume that $\lambda_{13}(t|x) = \lambda_{13}$, so that there is no effect of the carcinogen on death without tumour, and that there is no effect of the carcinogen on death following tumour onset.

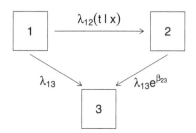

(a) Derive the form of the likelihood under the basic observation scheme where individuals are observed over $(0, \min(T_3, C^A)]$.

(b) Sacrifices can enhance information about the tumour-causing effect of the agent. Derive the likelihood for the case in which any animals alive at time C^A are sacrificed (killed), and an autopsy is carried out to determine whether they were in state 1 or state 2 at the time of sacrifice. Describe the increase in information about β_{12} and β_{23} over the study plan in (a).

(Sections 3.1, 3.6)

Problem 3.4 Obtain the estimator (3.5) for the Fisher information by showing that for $I_{kl}(\theta_{kl})$ given in (3.4), we have

$$E\{I_{kl}(\theta_{kl})\} = E\left\{\sum_{i=1}^{m}\int_0^\infty \bar{Y}_{ik}(u)\left\{\frac{\partial \log \lambda_{ikl}(u)}{\partial \theta_{kl}}\frac{\partial \log \lambda_{ikl}(u)}{\partial \theta'_{kl}}\right\}dN_{ikl}(u)\right\},$$

where for convenience we write $\lambda_{ikl}(u)$ for $\lambda_{ikl}(u; \theta_{kl})$. Use this to obtain (3.8) and the result immediately following it.

(Section 3.1)

Problem 3.5 Show that (3.27) is the profile maximum likelihood estimate of $\Lambda_{kl0}(t)$ when β_{kl} is known, by maximizing the likelihood function (2.20) with $\Lambda_{kl0}(t)$ restricted to be a step function with jumps only at times t for which a $k \to l$ transition was observed (that is, for which $d\bar{N}_{ikl}(t) = 1$ for one or more of $i = 1, \ldots, n$).

(Section 3.3)

Problem 3.6 Consider the following two methods for testing the adequacy of a time-homogeneous Markov model:

(a) Formulate a piecewise-constant model with transition intensities $\lambda_{kl}(t) = \lambda_{klr}$ for $b_{r-1} \leq t < b_r$ for a set of times $0 = b_0 < b_1 < \cdots < b_R$. Obtain both likelihood ratio and score tests of the hypothesis H_0: $\lambda_{klr} = \lambda_{kl}$ for $r = 1, \ldots, R$.

(b) Formulate a model with $\lambda_{kl}(t) = \lambda_{kl} \exp(\beta_{kl} g_{kl}(t))$, where the $g_{kl}(t)$ are specified functions. Derive the score test of H_0: $\beta_{kl} = 0$.

(c) Comment on the similarity of (a) and (b) in the case where we let R become large and the $b_r - b_{r-1}$ become small for (a).

(Section 3.5)

Problem 3.7

(a) Derive the expression (3.65) for the expected Brier score, using the fact that Y_t is binomial with $P(Y_t = 1 | X = x) = F(t|x)$.

(b) Explain why the first term in (3.65) is the minimum possible value for EBS(t). What are its maximum and minimum values? In what circumstances would EBS(t) be small?

(Section 3.5.3; see also Section 8.2)

Problem 3.8 Consider the illness-death model and a specification for the $2 \to 3$ transition intensity as

$$\lambda_{23}(t \mid \mathcal{H}(t^-)) = \lambda_{230}(t) \exp(\beta_{23} B(t)),$$

where $B(t) = t - t_2$ is the time since entry to state 2. Examine the score statistic for testing the hypothesis $\beta_{23} = 0$, based on the Cox partial likelihood function. Consider how the statistic and an associated variance estimate can be obtained either from the coxph function in R or by separate calculation.

(Sections 3.3, 3.5)

Problem 3.9   RECORD Trial Group (2005) report on a randomized clinical trial designed to evaluate the effect of oral calcium on the prevention of secondary fractures in elderly individuals who had experienced a low-trauma fracture in the previous 10 years. Consider the design of a similar study with a planned follow-up of $C^A = 5$ years. An illness-death model is adopted with state 1 representing alive and free of the secondary fracture; state 2 is entered when a secondary fracture occurs; and state 3 is entered upon death.

The incidence of the secondary fracture is expected to be about 15% during the course of follow-up in the placebo arm, and the effect of calcium is expressed as

$$\lambda_{12}(t \mid x) = \lambda_{120}(t)\exp(\beta_{12}x).$$

It is not expected that calcium would affect mortality before or after the secondary fracture, and so we adopt a Markov model with mortality rates $\lambda_{13}(t) = \lambda_{23}(t)$.

For simplicity suppose $\lambda_{120}(t) = \lambda_{12}$ and $\lambda_{13}(t) = \lambda_{23}(t) = \lambda_3$ and the goal is to test H_0: $\beta_{12} = 0$ versus H_A: $\beta_{12} \neq 0$, where $\beta_{12A} = \log 0.90$ is the effect of interest. Suppose the age at recruitment is uniformly distributed over 70−80 years of age, and the mortality rate is expected to be 35% over the course of the study.

(a) Derive the sample size requirement to ensure the study will have 80% power to detect the effect of interest with only administrative censoring at 5 years following randomization.

(b) Derive the revised sample size to accommodate study withdrawal assumed to occur uniformly over the course of follow-up so that $P(T_2 < \min(T_3, C^R, C^A)) = 0.60$.

(c) Confirm the calculations in (a) and (b) by conducting a simulation study.

(Section 3.6)

Problem 3.10   The design of longitudinal studies of aging or health involves issues similar to those in Problem 3.9. Suppose, for example, that we are interested in the age-specific incidence of dementia and its effect on mortality. Consider an illness-death model where state 1 is "alive without dementia", state 2 is "alive with dementia" and state 3 is dead. Suppose that X represents genetic or other risk factors for dementia.

(a) Discuss how you would assess the power of tests of association for X and dementia in a study for which n individuals are recruited with ages 60−65 years, and followed for up to 20 years. Consider what types of models for the intensity functions might be used.

(b) How could the adequacy of models be assessed using the study data?

(c) Suppose some individuals selected for the study were already in state 2. How could they be used when analyzing the effect of dementia on the death intensity?

(Sections 3.1, 3.5, 3.6)

Chapter 4

Additional Applications of Multistate Models

4.1 Competing Risks Analysis

4.1.1 Model Features and Intensity-Based Analysis

The competing risks phenomenon introduced in Example 1.3.2 arises in many settings and warrants special attention. The classic setting motivating much early work involves the analyses of death from particular causes and estimation of associated covariate effects, but the general framework can be used to characterize reasons for breakdowns of machines, reasons for discharge from hospital, and so on. The competing risks setting is also integral to the modeling of multistate processes in which a transition out of some state can be to two or more other states.

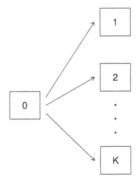

Figure 4.1: A multistate diagram for a competing risks problem with K causes of failure.

In the failure time setting, it is typically assumed that individuals can fail from at most one cause and that the particular cause is observable. With the multistate diagram in Figure 4.1 and restriction to fixed covariates, the event history conditioned upon in the intensities is simply the fact that failure has not yet occurred. The transition intensities are, therefore, often referred to as transition rates or cause-specific hazards in this context. For convenience, we number the states slightly differently in Figure 4.1, labeling the initial state 0 and the absorbing states $1, 2, \ldots, K$ to represent the K distinct causes of failure.

We first consider the case with no covariates and let $\Lambda_k(t) = \int_0^t \lambda_k(u) \, du$ denote the cumulative $0 \to k$ transition rate (or cause-specific hazard) for failure due to cause k. Since states $1, \ldots, K$ are each absorbing, the $(K+1) \times (K+1)$ matrix of cumulative transition intensities has zeros in all entries except for those in the first row and takes the form

$$Q(u) = \begin{bmatrix} -\lambda.(u) & \lambda_1(u) & \cdots & \lambda_K(u) \\ 0 & 0 & \cdots & 0 \\ \vdots & \vdots & \ddots & \vdots \\ 0 & 0 & \cdots & 0 \end{bmatrix},$$

where $\lambda.(u) = \sum_{k=1}^{K} \lambda_k(u)$. By (2.24) and product integration, we can compute the transition probability matrix as

$$P(0,t) = \prod_{(0,t]} \{I + Q(u)\,du\},$$

which we write as

$$P(0,t) = \begin{bmatrix} P_{00}(t) & P_{01}(t) & \cdots & P_{0K}(t) \\ 0 & 1 & \cdots & 0 \\ \vdots & \vdots & \ddots & \vdots \\ 0 & 0 & \cdots & 1 \end{bmatrix},$$

where $P_{00}(t) = P(T > t)$ is the probability of remaining in the initial state until time t; this is the overall survival function. The entry $P_{0k}(t)$ is the cumulative probability of failure from cause k by time t and when viewed as a function of time is referred to as the cumulative incidence function for type k events; we sometimes denote this by $P_k(t)$ or $F_k(t)$. The function $f_k(t) = dP_k(t)/dt$ is the sub-density function for failure due to cause k; it is improper since individuals can only fail from one cause and hence $\lim_{t \uparrow \infty} P_k(t) < 1$.

The data for such processes can be represented equivalently in terms of the failure time T and cause of failure ε, the multistate process $\{Z(s), s \geq 0\}$, where $Z(t)$ represents the state occupied at time t, or counting processes. We let

$$N_k(t) = I(T \leq t, \varepsilon = k)$$

indicate that a failure due to cause k occurred by time t, $\Delta N_k(t) = N_k(t + \Delta t^-) - N_k(t^-)$ and $dN_k(t) = \lim_{\Delta t \downarrow 0} \Delta N_k(t)$, $k = 1, \ldots, K$. With a sample of n individuals observed subject to right censoring, the likelihood (2.9) reduces to

$$L \propto \prod_{k=1}^{K} L_k, \tag{4.1}$$

where

$$\log L_k = \sum_{i=1}^{n} \int_0^{\infty} \bar{Y}_i(u) \{\log d\Lambda_k(u) \cdot dN_{ik}(u) - d\Lambda_k(u)\}, \tag{4.2}$$

with $\bar{Y}_i(u) = Y_i(u)Y_{i0}(u^-)$, $Y_i(u) = I(u \leq C_i)$ and $Y_{i0}(u) = I(Z_i(u) = 0)$, so that $Y_{i0}(u^-)$ indicates that individual i is at risk of a transition out of state 0 at time u, $i = 1, \ldots, n$. Notice that (4.2) has the same form as a log-likelihood for right-censored survival data where the effective censoring time for failure due to cause k

is the minimum of the actual right censoring time and the failure time due to any cause other than k.

If the different transition rates are functionally independent, parametric maximum likelihood estimates are easily obtained by assuming a parametric form for $\Lambda_k(t)$ and maximizing (4.2) for each $k = 1,\ldots,K$. For nonparametric estimation, we consider $d\Lambda_k(u)$ as the "parameter" of interest, differentiate (4.2) by this, and set the result equal to zero to obtain estimating equations

$$\sum_{i=1}^{n} \bar{Y}_i(u)\left\{dN_{ik}(u) - d\Lambda_k(u)\right\} = 0\,,$$

and solving this yields $d\widehat{\Lambda}_k(u) = d\bar{N}_{\cdot k}(u)/\bar{Y}_{\cdot}(u)$, where $d\bar{N}_{\cdot k}(u) = \sum_{i=1}^{n} \bar{Y}_i(u)\,dN_{ik}(u)$ and $\bar{Y}_{\cdot}(u) = \sum_{i=1}^{n} \bar{Y}_i(u)$. The Nelson-Aalen estimates of the cumulative transition intensities (cause-specific hazards) are

$$\widehat{\Lambda}_k(t) = \int_0^t d\widehat{\Lambda}_k(u)\,,$$

$k = 1,\ldots,K$. We define the $(K+1) \times (K+1)$ matrix of estimated transition intensities $\widehat{Q}(u)du$ as having top row

$$[-d\widehat{\Lambda}_{\cdot}(u),\ d\widehat{\Lambda}_1(u),\ d\widehat{\Lambda}_2(u)\,,\ \ldots,\ d\widehat{\Lambda}_K(u)],$$

where $d\widehat{\Lambda}_{\cdot}(u) = \sum_{k=1}^{K} d\widehat{\Lambda}_k(u)$, with all other entries zero. Then following (3.17), product integration yields

$$\widehat{P}(0,t) = \widehat{P}(t) = \prod_{(0,t]} \left\{I + \widehat{Q}(u)\,du\right\}.$$

Nonparametric estimates of the cumulative incidence functions are given as $\widehat{P}_k(t) = \widehat{P}_{0k}(t)$, $k = 1,\ldots,K$. These estimates may be written more explicitly as

$$\widehat{P}_k(t) = \int_0^t \widehat{P}_{00}(u^-)\,d\widehat{\Lambda}_k(u)\,, \qquad (4.3)$$

where $\widehat{P}_{00}(t)$ is the nonparametric estimate of the survivor function obtained from $\widehat{P}(0,t)$. Alternative estimates can be obtained if we use a Kaplan-Meier or other estimate of $P_{00}(u^-)$ in (4.3).

With fixed covariates, transition intensities are denoted $\lambda_k(t|x_i)$, where x_i is a $p \times 1$ covariate vector. We may then write a log-likelihood contribution analogous to (4.2) as

$$\log L_k = \sum_{i=1}^{n} \int_0^{\infty} \bar{Y}_i(u)\left\{\log d\Lambda_k(u\mid x_i)\cdot dN_{ik}(u) - d\Lambda_k(u\mid x_i)\right\}. \qquad (4.4)$$

Parametric analysis can again be carried out based on the methods described in Section 3.1, but nonparametric and semiparametric inference are often preferred in this setting. With multiplicative models of the form

$$d\Lambda_k(u\mid x) = d\Lambda_{k0}(u)\exp\left(x'\beta_k\right)\,,$$

we differentiate (4.4) with respect to the "parameters" $d\Lambda_{k0}(\cdot)$ and β_k to obtain the estimating equations

$$\sum_{i=1}^{n} \bar{Y}_i(u) \{dN_{ik}(u) - d\Lambda_k(u \mid x_i)\} = 0, \quad u > 0 \tag{4.5a}$$

$$\sum_{i=1}^{n} \int_0^\infty \bar{Y}_i(u) \{dN_{ik}(u) - d\Lambda_k(u \mid x_i)\} x_i = 0, \tag{4.5b}$$

where the left-hand side of (4.5b) is a $p \times 1$ function.

Solving (4.5a) for $d\Lambda_{k0}(u)$ gives the profile likelihood estimate

$$d\tilde{\Lambda}_{k0}(u; \beta_k) = \frac{d\bar{N}_{.k}(u)}{S^{(0)}(u; \beta_k)},$$

where $S^{(0)}(u; \beta_k) = \sum_{i=1}^{n} \bar{Y}_i(u) \exp(x_i'\beta_k)$, and substituting this into (4.5b) yields a $p \times 1$ estimating equation for β_k,

$$U_k(\beta_k) = \sum_{i=1}^{n} \int_0^\infty \bar{Y}_i(u) \left\{ x_i - \frac{S^{(1)}(u; \beta_k)}{S^{(0)}(u; \beta_k)} \right\} dN_{ik}(u) = 0, \tag{4.6}$$

where $S^{(1)}(u; \beta_k) = \sum_{i=1}^{n} \bar{Y}_i(u) x_i \exp(x_i'\beta_k)$. If $\hat{\beta}_k$ denotes the solution to (4.6), then $d\hat{\Lambda}_{k0}(u) = d\tilde{\Lambda}_{k0}(u; \hat{\beta}_k)$, $u > 0$.

If the covariate vectors are common across all transition intensities, we can carry out product integration to obtain the transition probability matrix conditional on $X = x$ as

$$P(0, t \mid x) = P(t \mid x) = \prod_{(0,t]} \{I + Q(u \mid x) du\}.$$

The explicit expression for the cumulative incidence function for failures due to cause k conditional on x is then

$$P_k(t \mid x) = P(Z(t) = k \mid Z(0) = 0, x) = \int_0^t P_{00}(u^- \mid x) d\Lambda_k(u \mid x). \tag{4.7}$$

As in (4.3), this is estimated by replacing unknown quantities by their corresponding estimates. It is apparent that (4.7) is a complex function of the cause-specific hazards and associated cause-specific covariate effects. This has led to the development of methodology for direct regression modeling of functions $P_k(t|x)$; we describe this next.

4.1.2 Methods Based on Cumulative Incidence Functions

Fine and Gray (1999) and Fine (2001) proposed formulating covariate effects directly on the state occupancy probabilities, or cumulative incidence functions. In the most general setting, transformation models are specified based on a monotone differentiable function $g(\cdot)$ on $(0,1)$ by writing

$$g(P_k(t \mid x)) = \alpha_k(t) + x'\beta_k, \tag{4.8}$$

where $\alpha_k(t)$ is a strictly monotonic function determining the probability of failure due to cause k by time t if $x = 0$, and β_k is a vector of regression coefficients. Since $P_{0k}(t|x)$ approaches a limit less than one as t increases, $\alpha_k(t)$ must approach a finite limit. In some settings, one may want to allow β_k to vary as a function of time, and this can be accommodated. We focus here on the case of constant coefficients β_k as in (4.8); they provide single summary statistics for each covariate effect. If x_i and x_j are $p \times 1$ covariate vectors associated with individuals i and j, respectively, then β_k is interpreted via

$$g(P_k(t \mid x_i)) - g(P_k(t \mid x_j)) = (x_i' - x_j')\beta_k, \tag{4.9}$$

so the rth element of β_k reflects the effect of a one unit change in the corresponding entry of x, when all other covariates are held fixed. The use of $g(u) = \log(-\log(1 - u))$ or $\log(u/(1-u))$ is common; the former transformation led Fine and Gray (1999) to characterize this model as a proportional hazards model on the sub-distribution function for failure due to cause k, and in the latter case the components of β_k are log odds ratios. The value of such models is mainly descriptive as the regression coefficients do not have simple interpretations in terms of process dynamics.

Various approaches have been suggested for estimation of the β_k and functions $\alpha_k(t)$. Fine and Gray (1999) gave a method for the case in which $g(u) = \log(-\log(1 - u))$ based on a weighted version of a partial likelihood as developed for the Cox model, and Fine (1999) considered a related rank-based method for general transformation models. A second approach (Fine et al., 2004; Scheike et al., 2008) is based on estimating functions for the binomial random variables $N_{ik}(t)$, and a third approach (Andersen et al., 2003) uses a similar method based on pseudo-values for the $N_{ik}(t)$. We begin with a description of the Fine and Gray (1999) method, which is implemented in the R `cmprsk` package, and then outline the other approaches.

Fine and Gray (1999) noted that $g(u) = \log(-\log(1 - u))$ gives the model

$$P_k(t \mid x) = 1 - \exp\{-\Gamma_k(t)\exp(x'\beta_k)\},$$

where $\Gamma_k(t) = \exp(\alpha_k(t))$. The similar structure to a proportional hazards model suggests the use of a Cox partial likelihood based on the time T_k of entry to state k, defining $T_k = \infty$ in cases where failure is due to any other cause. In this case, the indicator $Y_k^{\ddagger}(t) = I(T_k \geq t) = 1 - Y_k(t)$ is one provided an individual has not failed from cause k prior to t. If there is no censoring, the fact that

$$E\{Y_k^{\ddagger}(t)[dN_k(t) - \exp(x'\beta_k)\,d\Gamma_k(t)]\} = 0 \tag{4.10}$$

means that the quantity inside the expectation could be used as a basis for estimation of β_k and $\Gamma_k(t)$. The inclusion of individuals in the risk set beyond a time of failure due to a different cause $l \neq k$ ensures that the data are treated in a way that is consistent with the marginal interpretation of a cumulative incidence function.

When data are subject to right censoring, modifications are of course required. There are two types of individuals for whom the marginal "at risk" indicator $Y_k^{\ddagger}(t) = I(T_k \geq t)$ is known at time $t > 0$. First there are those who are uncensored and have yet to fail. These individuals satisfy the condition $t < \min(T, C)$ and so for them it is known that $Y_k^{\ddagger}(t) = 1$. Second, there are those who were observed to fail prior to t, and so for whom $T < \min(C, t)$. Among such individuals $Y_k^{\ddagger}(t) = 1$ if the observed failure is due to cause $l \neq k$, since we retain such individuals in the risk set beyond T_l and take $T_k = \infty$. If the observed failure is due to cause k, however, then it is known that $Y_k^{\ddagger}(t) = 0$. The key point is that when constructing the estimating function, we must restrict attention to individuals for whom $Y_k^{\ddagger}(t)$ is known, and such individuals satisfy either $t \leq \min(T, C)$ or $T < \min(C, t)$; when taken together these conditions simplify to $C \geq \min(T, t)$. Written another way, the marginal at-risk indicator $Y_k^{\ddagger}(t)$ is known provided $\bar{C}(t) = I(C \geq \min(T, t)) = 1$.

Individuals satisfying this condition constitute a biased sample since the condition involves T, so some adjustment is necessary to ensure consistent estimation of the parameters of interest. Under the assumption that the right censoring time $C = \min(C^R, C^A)$ is independent of (T, ε) given a vector x^*, which may include x and other covariates,

$$E\left\{ \frac{\bar{C}(t) Y_k^{\ddagger}(t)}{P(C^R \geq \min(T, t) \mid x^*)} \left[dN_k(t) - \exp(x'\beta_k)\, d\Gamma_k(t) \right] \right\} = 0.$$

This can be seen by first taking the conditional expectation of $\bar{C}(t)$ (or equivalently C) given (T, ε) and $X^* = x^*$, which yields (4.10).

If we re-introduce the subscript i indexing individuals, let $C_i = \min(C_i^R, C_i^A)$ and let $\bar{C}_i(t) = I(C_i \geq \min(T_i, t))$, we can consider the following partial pseudo-score function for β_k:

$$U_k^{\mathrm{w}}(\beta_k) = \sum_{i=1}^{n} \int_0^{\infty} \frac{\bar{C}_i(u) Y_{ik}^{\ddagger}(u)}{G_i^c(\min(T_i, u)^-)} \left\{ x_i - \frac{S^{(1,w)}(u; \beta_k)}{S^{(0,w)}(u; \beta_k)} \right\} dN_{ik}(u), \qquad (4.11)$$

where $Y_{ik}^{\ddagger}(u) = I(T_{ik} \geq u)$ and

$$S^{(r,w)}(u; \beta_k) = \sum_{i=1}^{n} w_{ik}(u)\, x_i^{\otimes r} \exp\left(x_i' \beta_k \right), \qquad r = 0, 1, 2, \qquad (4.12)$$

with $w_{ik}(u) = \bar{C}_i(u) Y_{ik}^{\ddagger}(u)/G_i^c(\min(T_i, u)^-)$ and $G_i^c(u) = P(C_i > u | x_i^*)$ the survival distribution for the net censoring time C_i. Note that this differs from the $G_i^c(u)$ function of Section 3.4.2, where the function there was related to the random censoring time C_i^R only.

We must assume that $G_i^c(u)$ is the true distribution for the censoring time C_i in (4.11) to ensure (4.11) is an unbiased estimating function. In practice, of course, $G_i^c(u)$ is unknown, and we seek a consistent estimator $\widehat{G}_i^c(u)$. Administrative censoring times can often be treated as fixed, and we model only random censoring. Sepcification of a model for $C_i = \min(C_i^R, C_i^A)$ can be more challenging when administrative censoring times vary across individuals and are associated with covariates; in that case we may need to model both administrative and random censoring.

As in other settings involving inverse probability of censoring weights (e.g. Section 3.4.2), these weights effectively create pseudo-risk sets that are representative of the risk sets that would be obtained had they not been chosen based on the censoring times (i.e. had we not needed to restrict the sample at time t to individuals satisfying the condition $\bar{C}_i(t) = 1$). Fine and Gray (1999) show for the estimator $\hat{\beta}_k$ defined as the solution to $U_k(\beta_k) = 0$, that $\sqrt{n}(\hat{\beta}_k - \beta_k)$ is approximately normal with mean zero and estimated covariance matrix $\hat{A}^{-1}\hat{B}\hat{A}^{-1}$ where $\hat{B} = \widehat{\text{asvar}}(n^{-1/2}U_k(\beta_k))$ and $\hat{A}$ is given by

$$\frac{1}{n}\sum_{i=1}^{n}\int_0^\infty w_{ik}(u)\left\{\frac{S^{(2,w)}(u;\hat{\beta}_k)}{S^{(0,w)}(u;\hat{\beta}_k)} - \left[\frac{S^{(1,w)}(u;\hat{\beta}_k)}{S^{(0,w)}(u;\hat{\beta}_k)}\right]\left[\frac{S^{(1,w)}(u;\hat{\beta}_k)}{S^{(0,w)}(u;\hat{\beta}_k)}\right]'\right\}dN_{ik}(u),$$

with $S^{(r,w)}(u;\beta_k)$ for $r = 0,1,2$ defined by (4.12).

In the original presentation of this approach, Fine and Gray (1999) did not consider a covariate-dependent censoring process but rather proposed a stabilized weight function in which $\bar{C}_i(u)$ is replaced with $I(C_i \geq \min(T_i,u)) \cdot G^c(u^-)$ giving the weight $w_{ik}(u) = \bar{C}_i(u) Y_{ik}^\ddagger(u) G^c(u^-)/G_i^c(\min(T_i,u)^-)$, where $G^c(u) = P(C > u)$ is the marginal survival distribution for the random censoring time. The introduction of $G^c(u)$ in the numerator yields less variable weights, which can improve efficiency of estimation. The crr function in the cmprsk library implements this by default, where $G^c(u)$ is replaced with the Kaplan-Meier estimate. While a general regression model for censoring is not implemented in crr, one can estimate $G_i^c(u) = P(C_i > u|x_i^*)$ through stratification based on mutually exclusive groups of individuals defined by discrete covariate vectors.

4.1.3   Methods Based on Direct Binomial Regression

An alternative approach to regression modeling in this context is based on binary indicators of whether the state k of interest is occupied at a particular time or set of times (Fine et al., 2004; Scheike et al., 2008; Grøn and Gerds, 2014). This approach to analysis can be applied to arbitary multistate models and to states other than absorbing states, but we consider its application to a competing risks problem here. In this case, an absorbing state is occupied at a particular instant if it has been previously entered. Let $g(u)$ be any monotonic function as in (4.8) and $h(u) = g^{-1}(u)$, so that $P_k(t|x) = h(\alpha_k(t) + x'\beta_k)$.

If we plan an analysis at a particular instant in time t, it is again necesary to recognize that because of censoring we may not know the status with respect to entry to state k for all individuals. As before, this is known if $C_i \geq \min(T_i,t)$, so under the same censoring assumption as in Section 4.1.2 (i.e. $C \perp (T,\varepsilon) \mid X^* = x^*$) we note that for any given $t > 0$, if $\bar{C}_i(t) = I(C_i \geq \min(T_i,t))$, then

$$E\{\bar{C}_i(t) N_{ik}(t)/G_i^c(\min(T_i,t)^-) \mid T_i,\varepsilon,x_i\} = I(C_i^A \geq t) \cdot N_{ik}(t).$$

Use of the pseudo-response $\bar{C}_i(t) N_{ik}(t)/G_i^c(\min(T_i,t)^-)$ amounts to reweighting the responses from the individuals who can be definitively classified with respect to the entry to state k, so that when taken in total, these individuals' responses represent the distrbution of responses in the uncensored sample.

Note that a fundamental difference between the Fine and Gray approach of Section 4.1.2 and this approach is that the former involves weighting estimating functions, and the latter involves weighting the responses. If $D^*(t, x_i)$ is a vector of fixed functions of the same dimension as $(\alpha_k(t), \beta_k')'$, the fact that

$$E\left\{D^*(t, x_i) I(C_i^A \geq t) \left[\frac{\bar{C}_i(t) N_{ik}(t)}{G_i^c(\min(T_i, t)^-)} - P_k(t \mid x_i)\right]\right\} = 0, \qquad (4.13)$$

means that consistent estimates of $\alpha_k(t)$ and β_k are obtained by solving

$$\sum_{i=1}^n D^*(t, x_i) I(C_i^A \geq t) \left[\frac{\bar{C}_i(t) N_{ik}(t)}{\widehat{G}_i^c(\min(T_i, t)^-)} - P_k(t \mid x_i)\right] = 0. \qquad (4.14)$$

Based on the fact that $N_{ik}(t)$ is binomial with $E\{N_{ik}(t)|x_i\} = P_k(t|x_i)$, the $D^*(t, x_i)$ terms typically adopted in generalized linear regression are a sensible choice:

$$\left(\begin{array}{c} \partial P_k(t \mid x_i)/\partial \alpha_k(t) \\ \partial P_k(t \mid x_i)/\partial \beta_k \end{array}\right) \frac{1}{P_k(t \mid x_i)(1 - P_k(t \mid x_i))}.$$

When estimation is to be conducted at several times $s_1 < s_2 < \cdots < s_R$, we can use (4.14) for each of these times giving separate estimates $\widehat{\alpha}_k(s_r)$, $\widehat{\beta}_k(s_r)$ for $r = 1, \ldots, R$. Alternatively, if the model (4.8) with fixed regression coefficients $\beta_k(t) = \beta_k$ is considered plausible, we can proceed as follows.

We first define the vectors $\tilde{N}_{ik} = (\tilde{N}_{ik1}, \ldots, \tilde{N}_{ikR})'$, where $\tilde{N}_{ikr} = \bar{C}_i(s_r) N_{ik}(s_r) / \widehat{G}_i^c(s_{ir}^-)$ with $\bar{C}_i(s_r) = I(C_i \geq \min(T_i, s_r))$ and $s_{ir} = \min(T_i, s_r)$. We let $\alpha_k(s_r) = \alpha_{kr}$, $r = 1, 2, \ldots, R$, $\alpha_k = (\alpha_{k1}, \ldots, \alpha_{kR})'$, write $\gamma_k = (\alpha_k', \beta_k')'$, and let $\dim(\gamma_k) = q = R + p$ where $\dim(x) = p$. With $P_{kr}(x_i) = P_k(s_r|x_i)$ and $P_k(x_i) = (P_{k1}(x_i), \ldots, P_{kR}(x_i))'$ an $R \times 1$ vector, let

$$D_i(\gamma_1) = \frac{\partial P_k(x_i)}{\partial \gamma_k'} = \left[\begin{array}{cccccc} A_{i1} & 0 & \cdots & \cdots & 0 & A_{i1}x_i' \\ 0 & A_{i2} & \ddots & & \vdots & A_{i2}x_i' \\ \vdots & \ddots & \ddots & \ddots & \vdots & \vdots \\ \vdots & 0 & \ddots & \ddots & 0 & \vdots \\ 0 & \cdots & \cdots & 0 & A_{iR} & A_{iR}x_i' \end{array}\right]$$

be a $R \times q$ matrix, where $A_{ir} = \partial P_k(s_r|x_i)/\partial \alpha_{kr}$. The generalized estimating equation for γ_k is then, for (4.14),

$$U^w(\gamma_k) = \sum_{i=1}^n U_i(\gamma_k) = 0 \qquad (4.15)$$

where

$$U_i^w(\gamma_k) = D_i'(\gamma_k) W_i(\gamma_k) (\tilde{N}_{ik} - P_k(x_i))$$

with

$$W_i(\gamma_k) = \text{diag}\left(\{P_{kr}(x_i)[1 - P_{kr}(x_i)]\}^{-1}, \ r = 1, \ldots, R\right),$$

an $R \times R$ working independence covariance matrix.

When attention is focused on the model with the fixed β_k, it is sensible to choose times s_r, $r = 1, \ldots, R$ that cover the range of times over which failures of type k tend to occur. The fact that the $N_{ik}(s_r)$ are correlated for $r = 1, \ldots, R$ suggests that R need not be very large since the incremental value of additional highly correlated responses may be modest. In the similar framework of pseudo-value estimation discussed in the next section, Andersen and Perme (2010) recommend $R \leq 10$.

Variance estimates for $\widehat{\alpha}_k(s_r)$ or $\widehat{\beta}_k$ are most easily obtained using the nonparametric bootstrap in which samples of n individuals are obtained by sampling with replacement from the study sample. We also note that smoothing can be applied to the estimates $\widehat{\alpha}_k(s_r)$ or $\widehat{\beta}_k(s_r)$ if desired.

Version 1.8.9 of the `timereg` package can fit much more general models than (4.8) by replacing the right-hand side with

$$v' A_k(t) + (\operatorname{diag}(t^p) x)' \beta_k \tag{4.16}$$

where v is a vector defining the covariates on which one wishes to stratify, and $A_k(t)$ is now a vector of functions such that $v' A_k(t)$ is strictly montonic for all v. This generalization, therefore, allows one to model the effect of some covariates nonparametrically by accommodating a separate function. The introduction of the t^p term enables one to parametrically model the effect of covariates that have time-varying effects for a given choice of $g(\cdot)$. Of course, if $p = 0$ then the covariate effects are assumed to be constant. The function `const` in the model statement for the `timereg` package contains the covariates with fixed coefficients β_k.

4.1.4 Models for State Occupancy Based on Pseudo-Values

The pseudo-value method was proposed by Andersen et al. (2003) as a way to model and estimate covariate effects on a state occupancy probability at a particular time or set of times. Like the direct binomial regression approach, this framework is quite general in that it can be used for non-progressive multistate processes, but application to the competing risks setting is very natural.

If t_0 denotes a particular time of interest and state k is the state of interest, the response at a specific time t is $N_{ik}(t) = I(Z_i(t_0) = k)$. The objective, as in the preceding section, is to fit regression models relating $P_k(t_0)$ to covariates x; as before, transformation models (4.8) are considered.

With a sample of n independent individuals and in the absence of censoring, we can estimate $P_k(t_0)$ in a model without covariates as

$$\tilde{P}_k(t_0) = \frac{1}{n} \sum_{i=1}^{n} I(Z_i(t_0) = k) = \frac{1}{n} \sum_{i=1}^{n} N_{ik}(t_0). \tag{4.17}$$

If $\tilde{P}_k^{-i}(t_0)$ denotes the estimate of $P_k(t_0)$ based on the sample excluding individual i, then a key observation from (4.17) is that

$$N_{ik}(t_0) = n \cdot \tilde{P}_k(t_0) - (n-1) \cdot \tilde{P}_k^{-i}(t_0). \tag{4.18}$$

When data are censored then $Z_i(t_0)$, and hence $N_{ik}(t_0)$ may be unknown, so (4.17) may not be available. The primary difference between the pseudo-value approach and the preceding approaches is that instead of using weighted estimating functions or weighted responses, we use pseudo-values $\tilde{N}_{ik}(t_0)$ as responses. This is achieved by noting that $P_k(t_0)$ can be estimated even with censored data and (4.18) can therefore be exploited to generate a pseudo-value. Specifically, let $\tilde{P}_k(t_0)$ be the Aalen-Johansen or some other estimate that is consistent and robust under the assumption of independent censoring; inverse probability weighting can be used to render a consistent estimate if censoring is not independent (see Section 3.4.2). The pseudo-observation for individual i is then defined by analogy with (4.18) as

$$\tilde{N}_{ik}(t_0) = n \cdot \tilde{P}_k(t_0) - (n-1) \cdot \tilde{P}_k^{-i}(t_0), \qquad (4.19)$$

where $\tilde{P}_k^{-i}(t_0)$ is the estimate obtained by excluding data from individual i. Note that while $N_{ik}(t_0)$ is a binary variable, $\tilde{N}_{ik}(t_0)$ obtained from (4.19) is not, but $E\{\tilde{N}_{ik}(t_0)\}$ is approximately equal to $P_k(t_0)$.

Regression with pseudo-observations is based on generalized linear models and the general strategy used in Sections 4.1.2 and 4.1.3. Specifically, we consider a monotone differentiable link function $g(\cdot)$ that maps the interval $(0,1)$ onto the real line, and set

$$g(P_k(t_0)) = \alpha_k(t_0) + x'\beta_k$$

as in (4.8). Software for fitting generalized linear models can be used to estimate $\alpha_k(t_0)$ and β_k, by treating $\tilde{N}_{ik}(t_0)$ as the response for individual i.

The specification of a single time-point of interest t_0 is often subjective and does not exploit the longitudinal nature of the data. As discussed in Section 4.1.3, it is common to consider a sequence of time-points $s_1 < s_2 < \cdots < s_R$ and an $R \times 1$ vector of pseudo-responses $\tilde{N}_{ik} = (\tilde{N}_{ik}(s_1), \ldots, \tilde{N}_{ik}(s_R))'$, where $\tilde{N}_{ik}(s_r)$ is the pseudo-value for individual i at s_r, $r = 1, \ldots, R$. We consider again the case where we constrain the effects of covariates to be common across time-points, as described in the preceding section; this is analogous to what is often done with longitudinal analyses based on generalized estimating equations where one often allows visit-specific intercepts but assumes common covariate effects. As before, we let x be a $p \times 1$ covariate vector and $\gamma_k = (\alpha_k', \beta_k')'$ be a vector of parameters where $\alpha_k = (\alpha_{k1}, \ldots, \alpha_{kR})'$ with $\alpha_{kr} = \alpha_k(s_r)$ and $\beta_k = (\beta_{k1}, \ldots, \beta_{kp})'$. We then let $P_k(x_i) = (P_{k1}(x_i), \ldots, P_{kR}(x_i))'$, where $P_{kr}(x_i) = P_k(s_r | x_i)$, and define the generalized linear model by

$$g(P_{kr}(x_i)) = \alpha_{kr} + x_i'\beta_k$$

for $r = 1, \ldots, R$. We may then specify a contribution to a set of $q = R + p$ estimating functions for individual i of the form

$$U_i(\gamma_k) = D_i'(\gamma_k) V_i^{-1}(\gamma_k) \left(\tilde{N}_{ik} - P_k(x_i) \right) \qquad (4.20)$$

where $D_i(\gamma_k) = \partial P_k(x_i)/\partial\gamma_k'$ and V_i is an $R \times R$ working covariance matrix; typically a working independence assumption is adopted. We then solve $U(\gamma_k) = \sum_{i=1}^{n} U_i(\gamma_k) = 0$ for $\hat{\gamma}_k$ and note that a consistent variance estimate is

$$\widehat{\mathrm{asvar}}(\sqrt{n}(\hat{\gamma}_k - \gamma_k)) = \hat{A}^{-1}(\hat{\gamma}_k) \, \hat{B}(\hat{\gamma}_k) \left[\hat{A}^{-1}(\hat{\gamma}_k) \right]' \qquad (4.21)$$

where

$$\hat{A}(\hat{\gamma}_k) = \frac{1}{n} \sum_{i=1}^{n} D_i'(\gamma_k) V_i^{-1}(\gamma_k) D_i(\gamma_k) \Big|_{\gamma_k = \hat{\gamma}_k}$$

and

$$\widehat{B}\left(\widehat{\gamma}_k\right) = \frac{1}{n} \sum_{i=1}^{n} U_i(\gamma_k) U_i'(\gamma_k)\Big|_{\gamma_k = \widehat{\gamma}_k}.$$

It is worth remarking that the general approach to inference based on pseudo-values is applicable to a much wider range of problems than the one considered here. Any other objects of inference that can be estimated with censored data can be handled this way as well, such as quantiles of survival time distributions, restricted mean survival times, and so on. We return to this in Section 4.3.

Models directed at covariate effects on the cumulative probability of a particular event in a competing, or semi-competing, risk setting warrant careful consideration when assessing the effect of randomized interventions. Use of models for treatment effects on the cumulative incidence function for death from a particular cause provide an inadequate reflection of the effect of treatment, as those not dying from that cause may be alive or may have died from another cause. An important supplementary analysis is therefore one directed at the cumulative incidence of death from another cause, or overall mortality. A similar point can be made for the analysis of non-fatal events in palliative trials. Analyses of treatment effects on the incidence of non-fatal events need to be supplemented with analyses of complementary outcomes. We return to this in Section 8.4.

4.1.5   A Competing Risks Analysis of Shunts in Hydrocephalus

4.1.5.1   Preliminary Descriptive Analysis

Here we consider the analysis of times to failure and causes of failure for shunts inserted in 839 children with hydrocephalus (Tuli et al., 2000). The shunts are intended to drain excess cerebospinal fluid from the cranium and thus prevent complications. The medical condition leading to the need for a shunt is classified into one of eight categories: a congenital abnormality (the reference condition), an intraventricular hemorrhage (et1), complications arising from meningitis (et2), aqueductal stenosis (et3), tumour (et4), trauma (et5), myelomeningocele (et6) or other causes (et7). Additional variables of interest include an indicator of whether the shunt drained fluid to the peritoneal cavity (s.type = 1), whether the shunt was inserted at the time of a concurrent surgery (o.surg = 1) and the age of the patient at the time of shunt insertion; age was classified into one of three categories with one reflecting surgery on premature infants before full gestation (approximately 40 weeks from conception), denoted as age < 0 year (age0 = 1), one representing surgery between age $= 0$ and age $= 1$ year (age1 = 1), and the reference category reflecting surgery conducted in children 1 year of age or older.

Shunts can fail for one of three reasons, including the development of an obstruction (fstatus = 1), infection (fstatus = 2) or another cause (fstatus = 3); the mortality rate is non-negligible in this population and so in Figure 4.2 we include a fourth absorbing state for death (fstatus = 4) entered if an individual dies with a functioning shunt. If T_k is the time from shunt insertion (in days) to entry to state k, $T = \min(T_1, \ldots, T_4)$, and C is the censoring time, ftime is $\min(T, C)$. The variable fstatus = 0 for individuals who are alive with a functioning shunt at

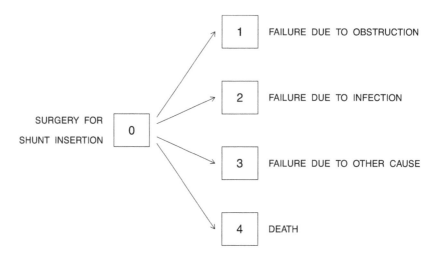

Figure 4.2: A multistate diagram for three causes of shunt failure in the presence of the competing risk of death.

the time of last contact. Following a shunt's failure, it is fully or partially replaced; the discussion here considers only the initial shunt for each child. The form of the dataframe is illustrated below.

```
> shunt
  id et1 et2 et3 et4 et5 et6 et7 s.type o.surg age0 age1 ftime fstatus
   1   0   0   0   1   0   0   0      1      0    0    0  1287       4
   2   0   0   0   1   0   0   0      1      0    0    0   457       4
   3   0   0   0   0   1   0   0      0      0    0    0    55       1
   4   0   1   0   0   0   0   0      1      0    0    0  1395       0
   5   0   0   0   0   0   1   0      0      0    0    0  3033       0
   6   0   0   0   1   0   0   0      1      0    0    0   391       4
   7   0   0   0   1   0   0   0      1      0    0    0  2980       0
   8   0   0   0   1   0   0   0      1      0    0    0  3771       0
   9   0   0   0   0   0   0   1      0      0    0    0    41       1
  10   0   0   0   0   0   0   1      1      0    0    0   879       4
  11   0   0   0   0   0   0   1      1      0    0    0    10       1
   :   :   :   :   :   :   :   :      :      :    :    :     :       :
```

The factorization of the likelihood in (4.1) and the form of the components of the log-likelihood (4.2) mean that we can estimate the transition intensities using software for survival analysis. The Nelson-Aalen estimates of the cumulative transition intensities for the four possible transitions in the absence of covariates are obtained using the **survfit** function with the specific command for the case of failures due to obstruction given as follows.

```
> library(survival)
> np1 <- survfit(Surv(ftime, fstatus == 1)~ 1, data=shunt, type="fh2")
> na1 <- data.frame(tt=np1$time, St=np1$surv, Ht=-log(np1$surv))
> na1
    tt       St       Ht
     0 0.997618 0.002385
```

```
   1 0.997618 0.002385
   2 0.994027 0.005991
   3 0.991619 0.008417
   4 0.987993 0.012080
   5 0.986780 0.013308
   :       :        :
3957 0.542041 0.612414
3979 0.542041 0.612414
```

The resulting estimates for each transition type are plotted in the left panel of Figure 4.3, and their shapes reveal some important features. The risk of failure due to infection appears to be negligible beyond the first 6 months following shunt insertion, and the risk of failure due to other causes persists over time but is quite low. The risk of failure due to obstruction, however, is considerably greater, and persists for the duration of follow-up; this is inferred from the steadily increasing cumulative intensity. Finally, we note that risk of death is greatest near the time of surgery, but it remains non-zero over the course of follow-up, reflecting the serious nature of the underlying conditions.

Estimates of the cumulative incidence functions given by (4.3) are computed using the cuminc function in the R cmprsk package. These are obtained as follows, with the estimates for each day obtained using the timepoints function; they are plotted in the right panel of Figure 4.3 for each cause of failure.

```
> library(cmprsk)
> fit <- cuminc(ftime=shunt$ftime, fstatus=shunt$fstatus)
> tt <- 1:(365*10)
> fitc <- timepoints(fit, tt)
> cif  <- data.frame(time=tt, obs=as.vector(fitc$est[1,]),
          inf=as.vector(fitc$est[2,]), oth=as.vector(fitc$est[3,]),
          died=as.vector(fitc$est[4,]))
> cif
  time      obs      inf      oth     died
     1 0.002384 0.001192 0.000000 0.003576
     2 0.005959 0.003576 0.000000 0.007151
     3 0.008343 0.005959 0.000000 0.008343
     4 0.011919 0.005959 0.001192 0.010727
     5 0.013111 0.007151 0.002384 0.011919
     :        :        :        :        :
```

```
> plot(0,0,xlim=c(0,10),ylim=c(0,0.4), xlab="YEARS  SINCE  SHUNT  INSERTION",
       ylab="CUMULATIVE  INCIDENCE  FUNCTION")
> lines(cif$time/365,cif$obs,type="s",lty=1)
> lines(cif$time/365,cif$inf,type="s",lty=2)
> lines(cif$time/365,cif$oth,type="s",lty=4)
> lines(cif$time/365,cif$died,type="s",lty=3)
> legend(0,0.4,c("OBSTRUCTION","INFECTION","OTHER CAUSES","DIED"),
         lty=c(1,2,4,3),bty="n")
```

The plots of the cumulative incidence functions convey a similar impression to the plots of the cumulative intensities in the left panel of Figure 4.3, but this is not necessarily always the case; the similarity arises here in part because the cumulative

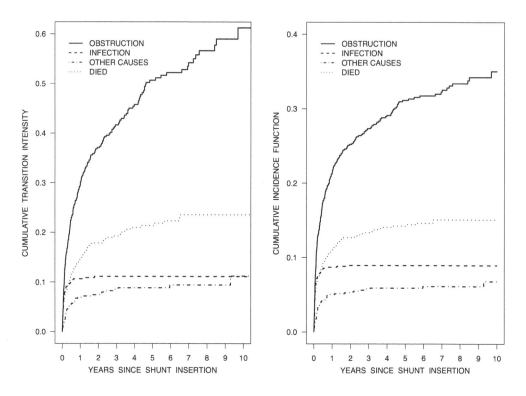

Figure 4.3: Plots of the estimated cumulative transition intensities $\Lambda_k(t)$, $k = 1, \ldots, 4$ (left panel) and cumulative incidence functions $P_k(t)$, $k = 1, \ldots, 4$ (right panel) for each cause of shunt failure and death.

intensities for many of the causes of failure are relatively flat beyond 1 or 2 years. Specifically, these estimates convey a steadily increasing cumulative probability of failure due to obstruction, with roughly 38% of the shunts failing due to this cause by 10 years post surgery. Risk of failure due to infection is greatest during the first year post surgery, and very few shunts fail due to infection after this period; failure due to other causes is likewise minimal beyond the second year. Approximately 15% of individuals die with a functioning initial shunt over the 10-year period post surgery. It is important to note that this does not represent total mortality over this period, since individuals whose initial shunts fail enter one of the other absorbing states; following such a transition they will typically receive another shunt and remain at risk of death.

4.1.5.2 Regression Analyses Based on Cause-Specific Hazards

Here we first fit several intensity-based Cox regression models; again we provide the code for the analyses of failures due to obstruction along with the associated output.

```
> cox1 <- coxph(Surv(ftime, fstatus == 1) ~ age0 + age1 + et1 + et2 + et3 + et4 +
             et5 + et6 + et7 + s.type + o.surg, data=shunt, method="breslow")
> summary(cox1)
```

```
n= 839, number of events= 263

          coef exp(coef) se(coef)       z Pr(>|z|)
age0    0.8856    2.4244   0.2675   3.310 0.000932
age1    0.6943    2.0023   0.2177   3.189 0.001425
et1     0.9888    2.6881   0.2744   3.604 0.000313
et2     0.5966    1.8159   0.3657   1.631 0.102830
et3     0.8778    2.4056   0.3196   2.746 0.006026
et4     0.5543    1.7407   0.3236   1.713 0.086694
et5     1.0901    2.9746   0.4829   2.257 0.023983
et6     0.8775    2.4048   0.2559   3.429 0.000606
et7     0.5880    1.8003   0.2777   2.117 0.034255
s.type -0.7068    0.4932   0.2033  -3.477 0.000507
o.surg  0.1824    1.2000   0.2225   0.819 0.412566
```

Table 4.1: Summary of estimates obtained from fitting cause-specific Cox models for failure due to obstruction, infection and other causes.

	OBSTRUCTION			INFECTION			OTHER		
	RR	95% CI	p	RR	95% CI	p	RR	95% CI	p
Age at Shunt Insertion			$0.002^\dagger$			$0.020^\dagger$			$0.244^\dagger$
>1 year	-	-	-	-	-	-	-	-	-
$0-1$ year	2.00	(1.31, 3.07)	0.001	2.65	(0.99, 7.07)	0.051	1.06	(0.45, 2.50)	0.892
<0 year	2.42	(1.44, 4.10)	<0.001	4.27	(1.44, 12.63)	0.009	2.24	(0.72, 6.91)	0.161
Condition			$0.007^\dagger$			$0.187^\dagger$			$0.028^\dagger$
Congenital	-	-	-	-	-	-	-	-	-
IV Hemorrhage	2.69	(1.57, 4.60)	<0.001	3.75	(1.07, 13.14)	0.039	0.61	(0.15, 2.41)	0.481
Meningitis	1.82	(0.89, 3.72)	0.103	3.35	(0.75, 15.00)	0.113	4.84	(1.58, 14.87)	0.006
AS	2.41	(1.29, 4.50)	0.006	3.55	(0.89, 14.24)	0.073	1.20	(0.29, 5.06)	0.802
Tumour	1.74	(0.92, 3.28)	0.087	1.92	(0.39, 9.46)	0.420	1.98	(0.59, 6.61)	0.268
Trauma	2.97	(1.15, 7.66)	0.024	8.81	(1.35, 57.60)	0.023	1.41	(0.15, 13.10)	0.763
MMC	2.40	(1.46, 3.97)	<0.001	3.84	(1.15, 12.81)	0.029	0.63	(0.19, 2.07)	0.450
Other	1.80	(1.04, 3.10)	0.034	3.42	(0.96, 12.18)	0.058	1.09	(0.34, 3.48)	0.890
VP Shunt	0.49	(0.33, 0.73)	<0.001	2.16	(0.52, 9.04)	0.291	1.38	(0.49, 3.92)	0.545
Concurrent surgery	1.20	(0.78, 1.86)	0.413	1.32	(0.61, 2.82)	0.481	2.30	(0.99, 5.35)	0.053

AS = Aqueductal stenosis; VP = ventriculoperitoneal.
† Likelihood ratio statistic for the overall effect.

The results for the model for shunt failure due to obstruction are summarized in the first set of columns in Table 4.1 and reveal that overall there is a significant effect of age at insertion (LRS = 12.62, d.f. = 2, $p = 0.002$). Specifically, there is a significantly increased risk of failure due to obstruction among individuals receiving the shunt before age = 0 (RR = 2.42; 95% CI: 1.44, 4.10; $p < 0.001$) and between ages = 0 and 1 year (RR = 2.00; 95% CI: 1.31, 3.07; $p = 0.001$) compared to individuals having surgery at 1 year of age or older. The condition leading to the need for a shunt is also significantly associated with the risk of failure due to obstruction (LRS = 19.29, d.f. = 7, $p = 0.007$). The largest hazard ratios and most significant increases in risk (relative to a congenital abnormality) are associated with intraventricular hemorrhage, aqueductal stenosis, trauma and MMC, but we observe that estimated hazard ratios relative to the congenital category exceed one for all other etiologies. Shunts that drain into the ventriculo-peritoneal cavity have significantly lower risk of failure due to obstruction (RR = 0.49; 95% CI: 0.33, 0.73; $p < 0.001$), but concurrent surgery at the time of shunt insertion does not show a significant change in risk.

The results of similar analyses for the other causes of failure are also given in Table 4.1. We note that age at first shunt insertion is significantly associated with failure due to infection, with younger children at elevated risk. Children receiving the shunt due to IV hemorrhage, trauma, MMC and other causes are at significantly greater risk of failure due to infection than those receiving it for congenital abnormalities, and estimated hazard ratios also exceed one for other etiologies. Children receiving their first shunt at the same time as another surgery appear to have an elevated risk of failure due to other causes.

4.1.5.3 Regression Based on Cumulative Incidence Functions

We next fit the Fine-Gray proportional sub-distribution hazards regression model of Section 4.1.2 using the competing risks regression function crr in the R cmprsk package. The function crr requires covariates we provided in the form of a matrix, which here is labeled covs.

```
> library(cmprsk)
> covs <- shunt[,c("age0","age1","et1","et2","et3","et4","et5","et6","et7",
                   "s.type","o.surg")]
> fg <- crr(ftime=shunt$ftime, fstatus=shunt$fstatus,
           cov1=covs, failcode=1, cencode=0)
```

The failcode argument indicates the value for the fstatus variable corresponding to the cause of failure of interest; here we focus on failure due to obstruction and so specify fstatus = 1. The specification of cencode = 0 indicates the fstatus variable corresponding to censored observations. As suggested in Section 4.1.2, it is possible to allow the censoring distribution to depend on covariates. This is achieved in crr by using Kaplan-Meier estimates of censoring distributions for mutually exclusive groups of individuals defined according to covariates and specifying the variable defining the groups with the cengroup argument. For simplicity, we first assume here a common censoring distribution and so omit this specification. Finally, as mentioned in Section 4.1.2, we note that (4.11) is specified here with $\bar{C}_i(u) = I(C_i \geq \min(T_i, u))\widehat{G}^c(u)$, which may be viewed as yielding a stabilized weight.

The default transformation in the ccr function is $g(u) = \log(-\log(1-u))$, which is adopted here. The estimated vectors of regression coefficients and associated statistics are as follows.

```
> summary(fg)

Competing Risks Regression
```

```
             coef exp(coef) se(coef)       z p-value
age0       0.5400    1.716   0.286  1.8901 0.05900
age1       0.5200    1.682   0.243  2.1405 0.03200
et1        0.9456    2.574   0.281  3.3710 0.00075
et2        0.4053    1.500   0.367  1.1042 0.27000
et3        0.7192    2.053   0.317  2.2698 0.02300
et4        0.2058    1.229   0.353  0.5832 0.56000
et5        0.7042    2.022   0.533  1.3221 0.19000
et6        0.8254    2.283   0.260  3.1743 0.00150
et7        0.4664    1.594   0.283  1.6452 0.10000
s.type    -0.7705    0.463   0.212 -3.6321 0.00028
o.surg    -0.0119    0.988   0.223 -0.0535 0.96000

Num. cases = 839
Pseudo Log-likelihood = -1676
Pseudo likelihood ratio test = 56  on 11 df
```

The results suggest an elevated risk of failure due to obstruction among children whose shunt was inserted at age < 0 $(\exp(\widehat{\beta}) = 1.68$, 95% CI: 1.04, 2.71; $p = 0.032)$ compared to those receiving it at 2 years of age or older. There was again evidence of a significant effect of the underlying condition leading to the need for a shunt $(p = 0.008)$ and a significantly lower risk of failure due to obstruction in shunts draining to the ventriculo-peritoneal cavity $(p < 0.001)$. The results are summarized in the columns labeled Fine-Gray in Table 4.2.

We next consider the analysis based on direct binomial regression (Section 4.1.3) of failure due to obstruction implemented using the comp.risk function in the timereg package in R. We first use set.seed(1000) to set the random number seed for the jackknife procedure adopted for variance estimation to ensure results can be replicated. The Event(time, fstatus) object is analogous to the Surv(time, status) object used in standard survival analysis, but here fstatus takes on integer values reflecting the cause of failure or censoring status. We do not consider stratification here, so the vector v_1 in (4.16) is simply the scalar 1. The function const specifies the covariates in x_1 in (4.16), and the default is to set $p = 0$ in (4.16) with the statement model = "prop" giving $g(u) = \log(-\log(1-u))$. The specification cause = 1 indicates that the model is based on failure due to obstruction where fstatus = 1. This yields the model

$$P_1(t \mid x_1) = 1 - \exp(-\exp(\alpha_1(t) + x_1'\beta_1)),$$

which is compatible with the preceding Fine-Gray model.

As discussed in Section 4.1, we may in principle have individual-specific censoring distributions with $G_i^c(u) = P(C_i^R > u|x_i)$. When cens.model = "KM" is specified, the Kaplan-Meier estimate is used to estimate a common censoring distribution $G_i^c(u) = P(C_i^R > u)$. We make this specification first to correspond most closely with the independent censoring assumption of the previous section. Unlike the Fine-Gray approach, which uses information from all times of failure due to the cause of interest, here we must specify the time-points at which we wish to use data for estimation. With time in days, we took $s_r = 730 \cdot r$, $r = 1, 2, \ldots, 5$ to correspond

Table 4.2: Summary of estimated regression coefficients based on the $g(u) = \log(-\log(1-u))$ transform of $P_1(t)$ for failure due to obstruction (cause 1).

	FINE-GRAY			DIRECT BINOMIAL[†]			PSEUDO-VALUES[‡]		
	$e^{\hat{\alpha}}$	95% CI	p	$e^{\hat{\alpha}}$	95% CI	p	$e^{\hat{\alpha}}$	95% CI	p
Age at									
Shunt Insertion									
> 1 year	-	-		-	-		-	-	
< 1 year	1.72	(0.98, 3.00)	0.059	1.61	(0.95, 2.74)	0.077	1.57	(0.92, 2.67)	0.099
0 − 1 year	1.68	(1.04, 2.71)	0.032	1.62	(1.05, 2.52)	0.031	1.62	(1.04, 2.52)	0.034
Condition									
Congenital	-	-		-	-		-	-	
IV Hemorrhage	2.57	(1.49, 4.46)	< 0.001	2.53	(1.44, 4.45)	0.001	2.79	(1.55, 5.01)	< 0.001
Meningitis	1.50	(0.73, 3.08)	0.270	1.48	(0.69, 3.17)	0.313	1.62	(0.74, 3.54)	0.229
AS	2.05	(1.10, 3.82)	0.023	2.04	(1.07, 3.88)	0.030	2.22	(1.15, 4.29)	0.018
Tumour	1.23	(0.62, 2.45)	0.560	1.26	(0.65, 2.46)	0.499	1.33	(0.66, 2.66)	0.427
Trauma	2.02	(0.71, 5.74)	0.186	1.93	(0.75, 4.95)	0.170	2.13	(0.79, 5.70)	0.133
MMC	2.28	(1.37, 3.80)	0.002	2.34	(1.39, 3.93)	0.001	2.46	(1.42, 4.25)	0.001
Other	1.59	(0.91, 2.78)	0.100	1.54	(0.87, 2.73)	0.138	1.68	(0.93, 3.05)	0.088
VP Shunt	0.46	(0.31, 0.70)	< 0.001	0.50	(0.33, 0.76)	0.001	0.52	(0.34, 0.80)	0.003
Concurrent Surgery	0.99	(0.64, 1.53)	0.957	0.95	(0.60, 1.50)	0.833	0.96	(0.60, 1.52)	0.859

AS = Aqueductal stenosis; VP = ventriculoperitoneal.
[†] Weights based on covariate independent censoring.
[‡] Standard errors are based on jackknife variance estimates.

to times roughly every 2 years in the **times** argument. Equation (4.14) can then be solved to obtain estimates, and then nonparametric bootstrap can be used for variance estimation. The code for this analysis follows and the results are displayed under the heading Direct Binomial in Table 4.2; these are in close agreement with those of the Fine-Gray approach.

```
> library(timereg)
> set.seed(1000)
> bin <- comp.risk(Event(time, fstatus) ~ const(age0) + const(age1) + const(et1) +
        const(et2) + const(et3) + const(et4) + const(et5) + const(et6) +
        const(et7) + const(s.type) + const(o.surg),
        data=shunt, model="prop", cause=1, cens.model="KM",
        times=365*c(2,4,6,8,10), n.sim=0)
> summary(bin)

  Competing risks Model
  No test for non-parametric terms
  Parametric terms :
              Coef.   SE Robust SE     z P-val
  const(age0) 0.478 0.270    0.270 1.771 0.077
  const(age1) 0.484 0.224    0.224 2.160 0.031
  const(et1)  0.928 0.288    0.288 3.227 0.001
```

```
const(et2)       0.392 0.388     0.388  1.009 0.313
const(et3)       0.713 0.328     0.328  2.172 0.030
const(et4)       0.231 0.341     0.341  0.677 0.499
const(et5)       0.658 0.480     0.480  1.371 0.170
const(et6)       0.848 0.265     0.265  3.199 0.001
const(et7)       0.433 0.292     0.292  1.484 0.138
const(s.type)   -0.691 0.210     0.210 -3.286 0.001
const(o.surg)   -0.049 0.233     0.233 -0.211 0.833
```

We next explore the possibility of covariate-dependent censoring and fit a Cox regression model for the censoring time involving the same covariates as those used in the response model. Here the status indicator in the function Surv corresponds to fstatus = 0, since this is the case when the censoring time is observed. We remark that in this dataset no early LTF times were recorded, so all censoring times C_i are administrative. However, individuals were enrolled in the database over a 10-year period and so the C_i vary widely. There were some calendar-time trends in the etiologies of persons enrolling, so it is of interest to consider the regression model below to account for this.

```
> fitC <- coxph(Surv(ftime, fstatus == 0) ~ age0 + age1 + et1 + et2 +
                et3 + et4 + et5 + et6 + et7 + s.type + o.surg,
                data=shunt, method="breslow")
> summary(fitC)

 n= 839, number of events= 329

           coef exp(coef) se(coef)     z Pr(>|z|)
age0    -0.0334    0.9672   0.2503 -0.13   0.8939
age1     0.0523    1.0536   0.1666  0.31   0.7538
et1      0.6474    1.9106   0.2318  2.79   0.0052
et2      0.6452    1.9064   0.3137  2.06   0.0397
et3      0.4690    1.5985   0.2636  1.78   0.0752
et4      0.3019    1.3524   0.2245  1.34   0.1787
et5      0.3607    1.4343   0.4176  0.86   0.3877
et6      0.2375    1.2680   0.1964  1.21   0.2266
et7      0.2589    1.2956   0.1971  1.31   0.1889
s.type   0.2309    1.2597   0.2119  1.09   0.2760
o.surg  -0.1890    0.8278   0.2756 -0.69   0.4927


Concordance= 0.558  (se = 0.019 )
Rsquare= 0.014   (max possible= 0.979 )
Likelihood ratio test= 12.2  on 11 df,   p=0.347
Wald test            = 12.3  on 11 df,   p=0.34
Score (logrank) test = 12.5  on 11 df,   p=0.328
```

A 2 d.f. Wald test of the effect of age yields $p = 0.875$, and the global (7 d.f.) Wald test of the effect of etiology yields $p = 0.150$. Nevertheless, there is some evidence from the particular contrasts of significantly different censoring intensities for different etiologies, so for completeness we repeat the direct binomial fit allowing for covariate-dependent censoring. If we specify cens.model = "cox" and omit the cens.formula argument, the comp.risk function will fit a Cox regression model with the covariates included in the response model to estimate the censoring distribution $G_i^c(u) = P(C_i^R > u|x_{ik})$. The relevant code and results are shown below.

```
> set.seed(1000)
> bin.cox <- comp.risk(Event(time, fstatus) ~ const(age0) + const(age1) + const(et1) +
              const(et2) + const(et3) + const(et4) + const(et5) + const(et6) +
              const(et7) + const(s.type) + const(o.surg),
              data=shunt, model="prop", cause=1, cens.model="cox",
              times=365*c(2,4,6,8,10), n.sim=0)
> summary(bin.cox)

  Competing risks Model
  No test for non-parametric terms
  Parametric terms :
                   Coef.    SE Robust SE      z P-val
  const(age0)      0.456 0.273    0.273  1.673 0.094
  const(age1)      0.508 0.224    0.224  2.266 0.023
  const(et1)       1.056 0.294    0.294  3.592 0.000
  const(et2)       0.464 0.389    0.389  1.192 0.233
  const(et3)       0.804 0.330    0.330  2.437 0.015
  const(et4)       0.296 0.338    0.338  0.874 0.382
  const(et5)       0.766 0.491    0.491  1.561 0.119
  const(et6)       0.898 0.266    0.266  3.378 0.001
  const(et7)       0.489 0.291    0.291  1.679 0.093
  const(s.type)   -0.664 0.213    0.213 -3.122 0.002
  const(o.surg)   -0.058 0.232    0.232 -0.248 0.804
```

These are different estimates of the covariate effects than those under the independent censoring assumption, but the estimates are quite similar and the conclusions regarding the significant factors do not change.

Finally, we consider the pseudo-values approach of Section 4.1.4 to assess the effects of covariates on failure due to obstruction. The R package **pseudo** is used for this analysis, along with the **geepack** package for the computation of robust standard errors based on (4.21). For comparison with the direct binomial regression, we use pseudo-values at the same times $s_r = 730 \cdot r$ days post surgery, $r = 1, 2, \ldots, 5$. The call to the **pseudoci** function computes the pseudo-values $\tilde{N}_{ik}(s_r)$ for the event that each failure state is occupied at the respective times in the **landmarks** vector. The letters **ci** in **pseudoci** reflect the use of the cumulative incidence function for computation of the pseudo-values. We focus here on the $\tilde{N}_{i1}(s)$ corresponding to failures due to obstruction.

```
> library(pseudo)
> library(geepack)
> landmarks <- 365*c(2,4,6,8,10)
> pfit <- pseudoci(time=shunt$ftime, event=shunt$fstatus, tmax=landmarks)
> attributes(pfit)
  $names
  [1] "time"   "cause"  "pseudo"

> attributes(pfit$pseudo)
  $names
  [1] "cause1" "cause2" "cause3" "cause4"
```

We create a dataframe containing the pseudo-values for failure due to obstruction and display the values at each of the specified times in the **landmarks** vector.

```
> shunt.new <- NULL
  for (j in 1:length(pfit$time)) {
     shunt.new <- rbind(shunt.new, cbind(shunt, pseudo=pfit$pseudo[[1]][,j],
                                         tpseudo=pfit$time[j]))
  }
> shunt.new <- shunt.new[order(shunt.new$id),]
> shunt.new
```

```
id et1 et2 et3 et4 et5 et6 et7 s.type o.surg age0 age1 ftime fstatus   pseudo tpseudo
 1   0   0   0   1   0   0   0      1       0    0    0 1287       4 -0.001376     730
 1   0   0   0   1   0   0   0      1       0    0    0 1287       4 -0.014754    1460
 1   0   0   0   1   0   0   0      1       0    0    0 1287       4 -0.028198    2190
 1   0   0   0   1   0   0   0      1       0    0    0 1287       4 -0.036690    2920
 1   0   0   0   1   0   0   0      1       0    0    0 1287       4 -0.045130    3650
 2   0   0   0   1   0   0   0      1       0    0    0  457       4 -0.000528     730
 2   0   0   0   1   0   0   0      1       0    0    0  457       4 -0.001199    1460
 2   0   0   0   1   0   0   0      1       0    0    0  457       4 -0.001650    2190
 2   0   0   0   1   0   0   0      1       0    0    0  457       4 -0.001935    2920
 2   0   0   0   1   0   0   0      1       0    0    0  457       4 -0.002218    3650
 :   :   :   :   :   :   :   :      :       :    :    :    :       :        :       :
```

The model fitting involves using $\tilde{N}_{i1}(t)$ at times $t = s_1, \ldots, s_r$ as outcomes, where $\tilde{N}_{i1}(t) = n\tilde{P}_1(t) - (n-1)\tilde{P}_1^{-i}(t)$. We let $w_{ir} = (1, I(r=2), \ldots, I(r=R))'$, which can be specified through the **factor(tpseudo)** specification, $\alpha_1 = (\alpha_{11}, \ldots, \alpha_{1R})'$, and x_i be the vector of covariates used in the preceding analyses. The generalized linear model takes the form $g(P_1(s_r)) = \alpha_{1r} + x_i'\beta_1 = v_{ir}'\gamma_1$, where $v_{ir} = (w_{ir}', x_i')'$ and again we use the clog-log link function $g(u) = \log(-\log(1-u))$. Under a working independence assumption for the covariance matrix, we obtain jackknife standard errors by setting the option **jack = TRUE** in the **geese** function.

```
> fit <- geese(pseudo ~ factor(tpseudo) + age0 + age1 + et1 + et2 + et3 +
             et4 + et5 + et6 + et7 + s.type + o.surg - 1,
             data=shunt.new, id=id, jack=TRUE, scale.fix=TRUE,
             family=gaussian, mean.link="cloglog", corstr="independence")

> est <- data.frame(mean=fit$beta, SD=sqrt(diag(fit$vbeta.ajs)))
> est$p <- 1 - pchisq((est$mean/est$SD)^2,1)
> est
                         mean       SD        p
factor(tpseudo)730  -1.610259 0.371743 0.000015
factor(tpseudo)1460 -1.441647 0.370017 0.000098
factor(tpseudo)2190 -1.337025 0.369899 0.000301
factor(tpseudo)2920 -1.270664 0.372596 0.000649
factor(tpseudo)3650 -1.215739 0.373908 0.001148
age0                 0.449399 0.272090 0.098604
age1                 0.480557 0.226212 0.033640
et1                  1.025043 0.299463 0.000619
et2                  0.481363 0.399882 0.228682
et3                  0.796255 0.336444 0.017949
et4                  0.282362 0.355783 0.427407
et5                  0.755215 0.502899 0.133169
et6                  0.898843 0.280192 0.001337
et7                  0.518510 0.303966 0.088042
```

```
s.type                 -0.658827 0.220859 0.002854
o.surg                 -0.041851 0.235456 0.858924
```

The results of the pseudo-value analysis are shown in the last set of columns of Table 4.2; they are in close agreement with results from the Fine-Gray and the direct binomial regression approach. There is significant evidence of elevated risk in children receiving shunts at an early age, and similar effects are seen for the condition leading to the need for the shunt, and the type of shunt is also significant.

It is of interest to compare the results in Table 4.2 with those in Table 4.1 for the cause-specific hazards analysis. Although the regression coefficients in the two analyses represent different things, we see that for many covariates, the results are qualitatively similar, with comparable p-values. For the age factor, however, there is a pronounced difference: age at shunt insertion is highly significant for the cause-specific hazard but much less so for the cumulative incidence function. The trauma and tumour etiologies also show contrasting p-values, once again with smaller p-values seen for the cause-specific hazard. Formula (4.3) shows that the effect of a cause-specific hazard function is modulated by the overall survivor function and, in particular, depends on the other cause-specific hazards. The cause-specific hazards analysis reflects shunt failure dynamics, whereas the cumulative incidence function analyses are directed at marginal features and mainly of descriptive value.

4.2   Alternative Methods for State Occupancy Probabilities

The methods in the preceding section based on binomial estimating functions and on pseudo-values can be used to estimate state occupancy probabilities in more general settings than competing risks. Here we consider these and other approaches more generally.

4.2.1   Estimation Based on State Entry Time Distributions

Another way to estimate occupancy probabilities is via distributions for entry and exit times for each state. Here we consider a general K-state process but assume there is a common state at the onset of the process. Let $T_k^{(r)}$ and $V_k^{(r)}$ denote the times of the rth entry and exit to state k, respectively. Then

$$P_k(t) = P(Z(t) = k) = \sum_{r=1}^{\infty} [P(T_k^{(r)} \leq t) - P(V_k^{(r)} \leq t)], \quad k = 1,\ldots,K$$

and by estimating the distribution functions of $T_k^{(r)}$ and $V_k^{(r)}$, say with Kaplan-Meier estimates, we obtain an estimate for $P_k(t)$. This approach is especially useful for nonparametric estimation in models where each state can be visited at most once. For example, with a progressive model with states $1,\ldots,K$ as in Figure 1.2(b), we have $T_1 = 0, T_2 = V_1,\ldots$ and

$$P_k(t) = P(T_k \leq t) - P(V_k \leq t) = P(T_k \leq t) - P(T_{k+1} \leq t) \qquad (4.22)$$

for $k = 1,\ldots,K$, with $T_{K+1} = V_K = \infty$. As another example, for the illness-death model in Figure 1.1 we have $P_1(t) = P(V_1 > t)$, $P_2(t) = P(T_2 \le t) - P(T_3 \le t)$ and $P_3(t) = P(T_3 \le t)$.

The estimators based on Kaplan-Meier estimation of state entry and exit time distributions are usually less efficient than Aalen-Johansen estimators discussed in Section 3.4.1. Moreover, the standard KM-based estimators require that censoring be independent of the multistate process in order to ensure that censoring is independent of the failure time variables $T_k^{(r)}$ and $V_k^{(r)}$; Problem 4.5 provides an illustration of bias that can arise when this is violated. If state-dependent censoring is present, inverse probability of censoring weighted KM estimation can be carried out (see Section 3.4.2) for estimation of $P(T_k^{(r)} \le t)$ and $P(V_k^{(r)} \le t)$. Variance estimation can be developed for certain models, but this is complicated and we prefer bootstrap methods, which are convenient and practical. Regression models for specific state entry or exit times can also be developed, but they do not lead to useful interpretation of covariate effects for occupancy probabilities. For this we refer to the methods in the next section.

4.2.2   Estimation Based on Binomial Data

Direct estimates of state occupancy probabilities are obtained from the estimating equations

$$\sum_{i=1}^{n} Y_i(t)[Y_{ik}(t) - P_k(t)] = 0, \quad k = 1,\ldots,K, \tag{4.23}$$

where $Y_i(t) = I(t \le C_i)$ and $Y_{ik}(t) = I(Z_i(t) = k)$ as before. The resulting estimate $\widehat{P}_k(t) = \bar{Y}_{\cdot k}(t)/Y_{\cdot}(t)$ is intuitive as it simply equals the proportion of persons with $C_i \ge t$ who are in state k at t. Here the process need not originate in the same state for all individuals. When this is the case, the estimates $\widehat{P}_k(t)$ obtained by solving (4.23) equal the Aalen-Johansen estimates when there are no censoring times prior to t (i.e. $Y_{\cdot}(t) = n$); see Problem 4.5. More generally, estimates based on (4.23) can be quite inefficient for larger values of t if $E\{Y_{\cdot}(t)\}$ is small. In addition, the estimates sum to one across states $k = 1,\ldots,K$, but in cases where a probability $P_k(t)$ is known to be monotonic, $\widehat{P}_k(t)$ may not be so for a specific dataset. Finally, by noting that the $Y_{ik}(t)$ $(k = 1,\ldots,K)$ are multinomial, conditional on $Y_{\cdot}(t)$, we see that the variance of $\widehat{P}_k(t)$ can be estimated as $\widehat{P}_k(t)[1 - \widehat{P}_k(t)]/Y_{\cdot}(t)$.

Again the validity of these estimates depends on censoring being completely independent of the multistate process so that $Y_i(t) \perp Y_{ik}(t)$; when this is violated $E(Y_{ik}(t)|Y_i(t) = 1) \ne P_k(t)$ so the estimating function in (4.23) is biased and therefore yields inconsistent estimators. Inverse probability weights can be introduced into (4.23) to give robustness to dependent censoring. The generalization of (4.23) is

$$\sum_{i=1}^{n} \frac{Y_i(t)}{G_i^c(t^-)} \{Y_{ik}(t) - P_k(t)\} = 0, \quad k = 1,\ldots,K \tag{4.24}$$

for a specific time t, where as before $G_i^c(t) = P(C_i^R > t|\mathcal{X}(t^-))$ is computed as in Section 3.4.2.

A new practical difficulty arises with the binomial approach when there is an absorbing state that precludes further observation. This was discussed in Section 2.2.2; what we may do is assume "observation" actually continues up to the administrative end of follow-up. We then set $Y_i(t) = 1$ in (4.24) when the process for individual i is still under observation or when $Z_i(t)$ is known because an absorbing state was entered at some time $s \leq t$. We correspondingly define the intensity for random censoring to be $\lambda^c(u|\mathcal{X}(u^-)) = 0$ when $\mathcal{X}_i(u^-)$ includes the information that $Z_i(u^-)$ is such an absorbing state. From (4.24) and provided we have an estimator $\widehat{G}_i^c(t)$ of $G_i^c(t)$, we can obtain the estimators

$$\widehat{P}_k(t) = \frac{\sum_{i=1}^n \widehat{w}_i(t) Y_{ik}(t)}{\sum_{i=1}^n \widehat{w}_i(t)}, \quad k = 1, \ldots, K \tag{4.25}$$

where $\widehat{w}_i(t) = Y_i(t)/\widehat{G}_i^c(t^-)$. Once again, we use bootstrap variance estimation. We note that when the model has one or more absorbing states, we can let T_i denote the time of entry to an absorbing state and then, $w_i(t)$ in (4.24) equals $Y_i(\min(T_i,t))/G_i^c(\min(T_i,t)^-)$, with a corresponding form for $\widehat{w}_i(t)$ in (4.25). A slightly different estimator was proposed by Scheike and Zhang (2007) with the denominator in (4.25) replaced by n.

Regression models can also be considered. For a given time t the state occupancy indicators $Y_{ik}(t)$ for $k = 1, \ldots, K$ are jointly multinomial and so one can consider joint models used for categorical data, such as the multivariate logistic model,

$$P_k(t \mid x) = \frac{\exp(\alpha_k(t) + x'\beta_k(t))}{\sum_{l=1}^K \exp(\alpha_l(t) + x'\beta_l(t))}, \quad k = 1, \ldots, K, \tag{4.26}$$

where for identifiability we set $\alpha_1(t) = \beta_1(t) = 0$. Although such models have some nice features, the effect of x on $P_k(t|x)$ depends on not just $\alpha_k(t)$ and $\beta_k(t)$ but also on the regression coefficients $\beta_l(t)$ for $l \neq k$. This has led to the proposed use of separate binomial models discussed in Section 4.1.4.

4.2.3   A Utility-Based Analysis of a Therapeutic Breast Cancer Clinical Trial

Multistate models are a powerful tool for cost-benefit analysis in situations where numerical values can be associated with time spent in specific states. In this example, we consider an application involving nonparametric estimation of state occupancy probabilities with a view to comparing quality-adjusted lifetime in a cancer trial.

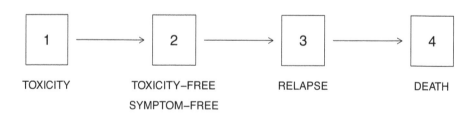

Figure 4.4: A multistate diagram for the disease course in a therapeutic breast cancer clinical trial.

Gelber et al. (1989) and Goldhirsch et al. (1989) report on a randomized trial designed to assess the quality-adjusted survival time over 84 months among breast cancer patients undergoing short- or long-duration chemotherapy. The response process can be represented by the 4-state model in Figure 4.4, where state 1 represents the situation in which individuals are experiencing toxicity due to the chemotherapy, state 2 represents the desirable state in which there is no toxicity and the individual is symptom free, state 3 represents relapse and state 4 represents death. The data are for 413 persons randomized to short-duration chemotherapy ($x = 0$) and 816 randomized to long-duration chemotherapy ($x = 1$). A portion of the dataframe giving the (possibly censored) transition times is given below, with the definitions of the variables given in Table 4.3.

```
> bc
id trt enum start    stop status from.state to.state stop2 status2  stop3 status3  stop4 status4
 1   1    1  0.00    7.00      1          1        2     7       1 114.84       0 114.84       0
 1   1    2  7.00  114.84      0          2        3     7       1 114.84       0 114.84       0
 2   0    1  0.00    3.00      1          1        2     3       1  72.76       1 113.82       0
 2   0    2  3.00   72.76      1          2        3     3       1  72.76       1 113.82       0
 2   0    3 72.76  113.82      0          3        4     3       1  72.76       1 113.82       0
 6   1    1  0.00    8.00      1          1        2     8       1 113.03       0 113.03       0
 6   1    2  8.00  113.03      0          2        3     8       1 113.03       0 113.03       0
 7   1    1  0.00    8.00      1          1        2     8       1  32.50       1  33.91       1
 7   1    2  8.00   32.50      1          2        3     8       1  32.50       1  33.91       1
 7   1    3 32.50   33.91      1          3        4     8       1  32.50       1  33.91       1
 :   :    :    :       :       :          :        :     :       :      :       :      :       :
```

Table 4.3: Variables and their meaning in multistate analysis of the therapeutic breast cancer clinical trial data.

Variable	Description
id	patient ID
trt	0 = short-duration chemotherapy; 1 = long-duration chemotherapy
enum	the line number
start	the time at the beginning of the period at risk
stop	the time at the end of the period at risk
status	indicator of a transition at stop time
from.state	the state occupied over the period at risk
to.state	the state entered at the end of the period at risk
stop2	time to toxicity-free or censored
status2	1 = toxicity-free; 0 = censored
stop3	time to relapsed or censored
states3	1 = relapsed; 0 = censored
stop4	time to death or censored
status4	1 = death; 0 = censored

The times T_k of entry to states $k = 2, 3, 4$ provide a useful comparison of the two treatment groups. Plots of $\widehat{F}_k(t|x)$, the Kaplan-Meier estimates of $P(T_k \leq t|x)$, $k = 2, 3, 4$ are given in Figure 4.5 for each group. We observe that the long-duration chemotherapy is naturally associated with longer times spent in state 1, but we see that it is also associated with longer times to relapse and death than the short-

duration therapy. The estimates are obtained by the following calls to the `survfit`
function.

```
> km2 <- survfit(Surv(stop2, status2) ~ 1, data=bc,
                 subset=((enum == 1) & (trt == 1)), type="kaplan-meier")

> km3 <- survfit(Surv(stop3, status3) ~ 1, data=bc,
                 subset=((enum == 1) & (trt == 1)), type="kaplan-meier")

> km4 <- survfit(Surv(stop4, status4) ~ 1, data=bc,
                 subset=((enum == 1) & (trt == 1)), type="kaplan-meier")
```

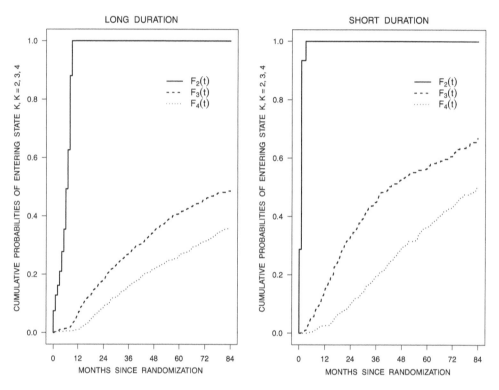

Figure 4.5: Kaplan-Meier estimates of $F_k(t) = P(T_k \leq t)$ for the long- and short-duration chemotherapy groups, $k = 2, 3, 4$.

We can also examine the average time spent in each of states 1, 2 and 3. Since the
time in a particular state k over $[0, t]$ can be expressed as $S_k(t) = \int_0^t I(Z(s) = k)\,ds$,
the expected time in state k over $[0, t]$ is

$$\psi_k(t) = E\{S_k(t)\} = \int_0^t (F_k(s) - F_{k+1}(s))\,ds.$$

Thus, $\psi_k(C^A)$ may be estimated by using occupancy probability estimates from Section 3.4 or 4.2, or just replacing $F_k(s)$ and $F_{k+1}(s)$ by their respective Kaplan-Meier
estimates, with the resulting estimate simply given by the area between $\widehat{F}_k(s)$ and
$\widehat{F}_{k+1}(s)$ in Figure 4.5 over $[0, C^A]$. Estimates of the restricted mean time spent in

each of the first 3 states up to $C^A = 84$ months are given separately for each treatment arm in Table 4.4 using the latter approach with standard errors obtained by the nonparametric bootstrap; estimates are almost identical when using the Aalen-Johansen estimates of the state occupancy probabilities. Note that the restricted mean sojourn time in a given state is similar to the restricted mean lifetime in survival analysis, which is defined as the average time in the "alive" state up to a specific time C^A.

Table 4.4: Restricted mean sojourn times (months) in transient states of Figure 4.4 by treatment group.

Duration	Toxicity			Toxicity-Free Symptom-Free			Relapse		
	TRUE	MEAN	SE[†]	TRUE	MEAN	SE[†]	TRUE	MEAN	SE[†]
Long	5.65	5.56	0.12	54.60	54.77	1.03	8.95	8.90	0.48
Short	0.52	0.51	0.04	47.79	47.75	1.54	15.89	15.86	0.97

[†] Standard errors are based on 500 bootstrap samples.

We see that long-duration chemotherapy is associated with longer times in both states 1 and 2, but shorter times in the relapse state 3. It is sometimes useful to synthesize these effects into a summary statement about the effect of prescribing the short- versus long-duration chemotherapy regimens. To this end a "quality-adjusted survival" analysis can be carried out by assigning utilities (quality scores) u_j to states $j = 1, 2, 3, 4$ in Figure 4.4. Assigning the utility $u_4 = 0$, we can then write the mean utility score at time t given $Z(0) = 1$ and $X = x$. The mean cumulative utility score over $[0, t]$ given $X = x$, denoted by $\mu(t|x)$, is then

$$\mu(t \mid x) = \int_0^t \sum_{j=1}^3 u_j P_j(s \mid x) \, ds = \sum_{j=1}^3 u_j \psi_j(t). \tag{4.27}$$

Estimates of $\mu(t|x)$ are obtained by using occupancy probability estimates for each treatment group. Glasziou et al. (1990) adopted utilities $u_1 = 0.1$, $u_2 = 0.5$ and $u_3 = 0.1$. Using the estimates $\widehat{P}_k(s) = \widehat{F}_k(s) - \widehat{F}_{k+1}(s)$ for $k = 1, 2, 3$ (with $\widehat{F}_1(s) = 1$ for all $s \geq 0$) discussed above, we obtain $\widehat{\mu}(84|x = 0) = 25.55$ (S.E. $= 0.73$) and $\widehat{\mu}(84|x = 1) = 28.76$ (S.E. $= 0.47$), where the standard error is again based on 500 bootstrap samples.

The estimates $\widehat{\mu}(t|x)$ are shown in Figure 4.6. They indicate a trend towards superior cumulative utility for the short-duration arm in the first three years after treatment, resulting in part from the shorter time spent in the toxicity state. The curves cross, however, due to the shorter times to relapse and higher mortality in the short-duration arm.

4.3 Analysis of State Sojourn Time Distributions

Grossman et al. (1998) discuss a community-based multicentre economic study of

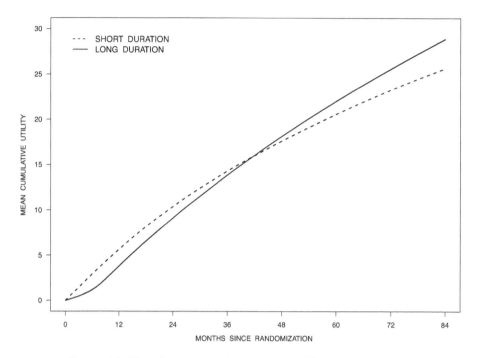

Figure 4.6: Plot of mean cumulative utility $\mu(t)$ by treatment strategy.

the cost-effectiveness of ciprofloxacin compared to standard antibiotic care in patients with chronic bronchitis with planned follow-up of one year. Individuals at least 18 years of age currently experiencing an exacerbation were eligible if they had at least three exacerbations in the previous year. Two hundred and twenty-two individuals were randomized with 115 allocated to receive ciprofloxacin. Data on the onset and resolution of all exacerbations were recorded with a view to computing the annualized number of days with symptoms of acute exacerbations (referred to as acute exacerbation of chronic bronchitis or AECB). Figure 4.7 displays the disease process for a sample of 10 individuals in each treatment arm. It is evident that there is significant variability in the frequency and duration of exacerbations between patients over the course of follow-up, even within treatment groups. We now consider an analysis of the total AECB symptom days over one year, with an emphasis on both treatment and risk factors for AECB.

We first consider the annual number of days with AECB symptoms, and focus on understanding the factors associated with a large number of symptom days, as well as the effect of treatment. Let $Z(s) = 1$ and 2 denote the AECB-free and AECB states, respectively, in Figure 4.8 and define $S_{i2}(t) = \int_0^t I(Z_i(s) = 2)\,ds$ as the total time spent in the exacerbation state over $[0,t]$ for individual i, $i = 1,\ldots,n$. If there is no early withdrawal from the study, then we observe $S_{i2}(C^A)$ for each individual. When individuals are randomly censored prior to C^A, then a simple *ad hoc* approach is to compute $S_{i2}(C_i)/C_i \times 365$ to obtain a prorated estimate of what would have been observed with complete follow-up. These could then be used in a regression analysis as the response with treatment and patient attributes as covariates.

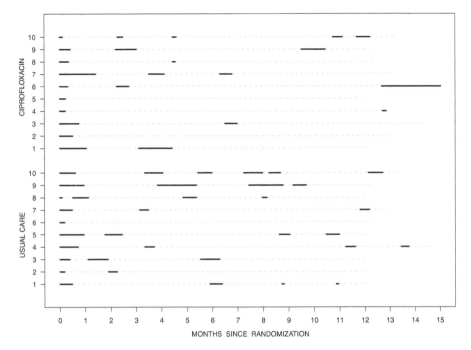

Figure 4.7: A display of the onset and resolution of acute exacerbations of chronic bronchitis for a sample of individuals for Grossman et al. (1998); solid lines represent periods with exacerbation, and dashed lines represent exacerbation-free periods.

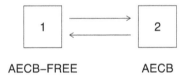

Figure 4.8: A 2-state diagram for the onset and resolution of acute exacerbations of chronic bronchitis.

An alternative approach is to obtain a robust estimate of the exacerbation state occupancy probability and note that since

$$\psi_2(t) = \int_0^t P(Z(s) = 2 \mid Z(0) = 2)\,ds,$$

is the expected number of days with symptoms over $[0, t]$, an estimate of $\psi_2(C^A)$ can be obtained by first estimating $P_{22}(0, t)$. This would then allow us to use the pseudo-value approach introduced in Section 4.1.4 to estimate the total number of symptom days over $[0, C^A]$ for individuals who were censored early. The approaches for estimation of transition probabilities discussed earlier include the Aalen-Johansen estimate (Section 3.4.1), estimates based on the marginal state entry and exit time distributions (Section 4.2.1), and the direct approach by binomial analysis (Section 4.2.2). For the Aalen-Johansen approach, robustness for non-Markov processes is achieved if censoring times are independent of the 2-state process; otherwise, we can use IPC weighting described in Section 3.4.2. An alternative would be to consider an expanded state space with exacerbation and exacerbation-free states defined according to the cumulative number of cycles through the states. This requires the construction of a large transition intensity matrix, so we instead adopt an approach based on the marginal entry time distributions. Following the approach outlined at the start of Section 4.2.1, we let $F_k^{(r)}(t) = P(T_k^{(r)} \leq t)$, where $T_k^{(r)}$ is the rth time state k is entered after $t = 0$. Let the maximum number of exacerbations beyond the first one, commencing over $[0, C^A]$, be denoted by R. We consider estimating $F_2^{(r)}(t)$, $r = 1, 2, \ldots, R$ and set $F_2^{(r)}(t) = 0$ for $r > R$. Likewise, we estimate $F_1^{(r)}(t)$, $r = 1, 2, \ldots, R+1$ to allow for resolution of the final exacerbation and set $F_1^{(r)}(t) = 0$ for $r > R+1$. Then we define

$$\widehat{\psi}_2 = \sum_{r=0}^{R} \int_0^{C^A} \left\{ [1 - \widehat{F}_1^{(1)}(s)] + \sum_{r=2}^{R+1} [\widehat{F}_2^{(r-1)}(s) - \widehat{F}_1^{(r)}(s)] \right\} ds$$

as an estimate; we suppress the dependence on C^A, since this is a common administrative censoring time. Following Andersen et al. (2003), let

$$\tilde{\psi}_{i2} = n \widehat{\psi}_2 - (n-1) \widehat{\psi}_2^{-i}$$

denote the pseudo-value for individual i, where $\widehat{\psi}_2^{-i}$ is the estimate obtained by excluding individual i from the data. We may then specify a generalized linear model where $g(E(\tilde{\psi}_{i2}|x_i)) = \theta_i = x_i'\beta$. We consider here a linear model with $g(u) = u$, so

$$\tilde{\psi}_{i2} = x_i'\beta + \epsilon_i$$

where ϵ_i is an error term, and we estimate β by least squares. With covariates gender (male = 1, female = 0), severity of disease (severe = 1, not severe = 0) and symptoms (number of AECB symptom days in the baseline exacerbation at randomization) as well as treatment (ciprofloxacin = 1, standard care = 0) and their effects have direct interpretations in terms of total symptom days.

Table 4.5: Estimates based on regression analysis of the expected number of AECB symptom days per year by the pseudo-value approach and *ad hoc* prorated approximation to the annual number of symptom days.

	PSEUDO-VALUES			APPROXIMATION		
	EST	95% CI	p	EST	95% CI	p
Treatment	-3.85	(-11.86, 4.16)	0.346	-3.75	(-21.37, 13.87)	0.677
Gender	10.02	(2.18, 17.86)	0.012	6.80	(-10.95, 24.56)	0.453
Severity	24.47	(9.91, 39.03)	< 0.001	12.25	(-15.00, 39.51)	0.378
Symptoms	0.31	(-0.11, 0.73)	0.148	0.16	(-1.13, 1.45)	0.812

Table 4.5 contains the regression estimates based on the pseudo-value method,

along with an estimate based on the *ad hoc* prorated approach. There is some difference between the two estimates for the gender and severity effects; the linear approximation method may incur bias when $\psi_2(t)$ is non-linear, so we prefer the pseudo-values estimates. The pseudo-values analysis shows that males tend to experience a higher number of days with exacerbation symptoms than females ($p = 0.012$) with an average of 10 additional days per year. Those with a severe form of chronic bronchitis have about 24 additional days per year of symptoms on average ($p < 0.001$) than those with less severe chronic bronchitis. Treatment and the number of symptom days at randomization for the baseline exacerbation are not significantly associated with the number of symptom days over the year. We remark that this example describes a situation where censoring of individuals might depend not just on the state occupied at a given time, but also on features of process history such as the cumulative time spent in the AECB state. There is no indication of such an effect here, but if there were we could use IPC weights described in Section 3.4.2, with suitably defined time-dependent covariates.

4.4 Bibliographic Notes

Competing risks has a long history, with many papers and a number of books. Some early papers are Cornfield (1957), Ederer et al. (1964), Altshuler (1970) and Hoel (1972); Kalbfleisch and Prentice (2002, Chapter 8), Lawless (2003, Chapter 9) and Crowder (2012) survey parametric, nonparametric and semiparametric competing risks methodology and provide many references. The edited volume by Klein et al. (2014) has several chapters on competing risks and the book by Beyersmann et al. (2012) provides illustrations based on R software. Prentice et al. (1978) emphasized the multistate formulation and the cause-specific intensities or hazard functions. Nonparametric Nelson-Aalen estimation of the cumulative intensities was introduced by Altshuler (1970) and studied further by Aalen (1976) and Fleming (1978b); estimators of cumulative incidence functions are also given. Direct regression modeling of cumulative incidence functions was introduced by Gray (1988) and Fine and Gray (1999), who developed estimation methods based on partial likelihoods and ranks. Bryant and Dignam (2004) consider alternative semiparametric estimators of cumulative incidence functions. Geskus (2011) extends nonparametric estimation and the Fine-Gray approach for cumulative incidence function regression to allow for left truncation of failure time. Binomial estimating functions were considered by Scheike and Zhang (2008, 2011) and Scheike et al. (2008); see also Gerds et al. (2012) and Grøn and Gerds (2014). Somewhat similar methods based on pseudo-values were considered by Klein and Andersen (2005). A review is given by Logan and Wang (2014); other recent papers on this approach include Graw et al. (2009) and Binder et al. (2014). Methods for simulating data from models satisfying the proportional sub-distribution hazards assumption are described in Beyersmann et al. (2012). Finally, some additional analysis of the cerebrospinal fluid shunt failure data is given by Tuli et al. (2000), Lawless et al. (2001) and Cook and Lawless (2007, Section 6.7).

Alternative methods of estimating state occupancy probabilities via entry and exit times to states were introduced by Pepe (1991), Pepe and Fleming (1991) and

Couper and Pepe (1997). Datta et al. (2000), Allignol et al. (2014) and others have considered this for the illness-death model. Direct nonparametric estimation based on binomial observations was considered by Pepe and Mori (1993). Cook et al. (2003) noted that these methods can be biased under state-dependent censoring and proposed inverse probability of censoring weighting (IPCW) adjustments. Scheike and Zhang (2007), Fiocco et al. (2008) and Grøn and Gerds (2014) consider direct binomial estimation, which conveniently handles regression models; Couper and Pepe (1997) considered a related method. Pseudo-value methods were considered by Andersen and Klein (2007), Andersen and Perme (2008, 2010), Fiocco et al. (2008) and Logan and Wang (2014). Some additional analysis of the breast cancer clinical trial data is given by Cook et al. (2003).

The estimation of state sojourn time distributions has mainly been studied for specific models, for example, progressive or alternating 2-state models (e.g. Lawless and Fong, 1999; Cook and Lawless, 2007, Sections 5.4, 6.5, 6.7). Mostajabi and Datta (2013) consider nonparametric estimation of entry, exit and sojourn time distributions; Satten and Datta (2002) consider waiting time distributions associated with multistate models. Vakulenko-Lagun et al. (2017) consider estimation of the joint distirbution of sojourn times in the illness-death model. The AECB data are discussed further by Cook and Lawless (2007, Section 6.7.2).

4.5   Problems

Problem 4.1   Consider a competing risks problem depicted in Figure 4.1 with $K = 2$ possible causes of failure and cause-specific hazards $d\Lambda_k(t|x) = d\Lambda_{k0}(t)\exp(x'\beta_k)$, where x is a vector of fixed covariates, $k = 1, 2$. Let $N_i(t) = (N_{i1}(t), N_{i2}(t))'$ denote the bivariate counting process for failures due to causes 1 and 2, C_i denote a right censoring time, $Y_i(t) = I(t \leq C_i)$, and $d\bar{N}_i(t) = Y_i(t)\,dN_i(t)$. The data that would normally be available for a sample of n independent individuals is $\{Y_i(u), \bar{N}_i(u), u \geq 0, x_i, i = 1, \ldots, n\}$. In some studies, however, it is not possible to determine the precise cause of failure for all individuals; let $R_i = 1$ if the cause of a failure is observable for individual i for all times $t > 0$ and $R_i = 0$ otherwise; suppose R_i is independent of $\{Y_i(u), d\bar{N}_i(u), u > 0\}$ given $X_i = x_i$. If $R_i = 0$, then all that is observed is $\{Y_i(u), d\bar{N}_{i\cdot}(u), u > 0, x_i\}$ where $d\bar{N}_{i\cdot}(u) = d\bar{N}_{i1}(u) + d\bar{N}_{i2}(u)$.

(a) Let $\pi_i = P(R_i = 1|x_i)$ and write the observed data likelihood for this setting.

(b) Write the complete data likelihood obtained by considering all causes as known and outline an expectation-maximization algorithm for estimation in a parametric setting.

(c) Consider a semiparametric setting with the additional assumption that $d\Lambda_{20}(t) = d\Lambda_{10}(t)\exp(\rho)$. Derive an expectation-maximization algorithm for this setting and give the form of the estimators at the kth maximization step in terms of the data and the estimate at the previous iteration.

(d) Many studies have an initial period of observation $[0, C^{IA}]$ during which data collection is more complete than it is during a second period of data collection $(C^{IA}, C^A]$. The reasons for this less-intensive phase are often related to budgetary constraints. Consider the same competing risks problem and model as in (b) but suppose

$$\pi_i(u) = P(R_i(u) = 1 \mid d\bar{N}_{i\cdot}(u) = 1, x_i)$$

where $\pi_i(u) = \pi_1$ for $u \leq C^{IA}$ and $\pi_i(u) = \pi_2 < \pi_1$ for $C^{IA} < u \leq C^A$. Discuss information about β_1 over $[C^{IA}, C^A]$ and the challenges in model checking as $\pi_2 \to 0$.

(Section 4.1; Goetghebeur and Ryan, 1995)

Problem 4.2   Exact determination of the cause of death can be challenging in some settings and different examiners may assign different causes. Consider a study in which all of n individuals are observed to die. Suppose there are K distinct causes of death, let D_i denote the recorded cause of death for individual i with true cause ε_i, assume

$$D_i \perp \{Y_i(u), \bar{N}_i(u), u > 0, X_i\} \mid \varepsilon_i$$

and let $\pi_{kj} = P(D_i = j|\varepsilon_i = k)$, for $j = 1, \ldots, K$; $k = 1, \ldots, K$.

(a) Suppose that the misclassification probabilities π_{kj} are known. Write the observed data likelihood and the "complete" data likelihood taking the ε_i as known.

(b) In practice, the π_{kj} are unknown, but auxiliary data may be available from a reliability exercise in which m raters ($m > 3$) classified the cause of death in a random sample of individuals in this study. These auxiliary data are then of the form $D_{i1}, \ldots, D_{im}$, where D_{ir} is the classification from rater r; assume $D_{i1}, \ldots, D_{im}$ are mutually independent given the true cause ε_i. Write the full likelihood incorporating this auxiliary data.

(c) How does the analysis change when individual i has administrative censoring time C_i^A, so that some persons are still alive at the end of follow-up?

(Section 4.1; Ebrahimi, 1996)

Problem 4.3   Suppose in the context of Problem 4.2 that auxiliary data instead arise from a validation study in which the cause of death is determined definitively by autopsy for a randomly chosen subsample of individuals who died. Write the corresponding likelihood and develop an expectation-maximization algorithm for estimation.

(Section 4.1; Lloyd-Jones et al., 1998)

Problem 4.4   Show that if none of the individuals $i = 1, \ldots, n$ is censored before time t, then the Aalen-Johansen estimator of $P_k(t)$ equals $\sum_{i=1}^{n} Y_{ik}(t)/n$, the observed fraction of individuals in state k at time t. Do this for the case where all individuals are in state 1 at $t = 0$, as well as the case where they may not be.

(Section 4.1)

Problem 4.5   Consider the simple empirical estimates

$$\widehat{P}_k(t) = \sum_{i=1}^{n} Y_i(t) Y_{ik}(t) \Big/ \sum_{i=1}^{n} Y_i(t), \quad k = 1, \ldots, K$$

of state occupancy probabilities.

(a) Show that if censoring times C_i are independent of the multistate processes $\{Z_i(t), t \geq 0\}$ for $i = 1, \ldots, n$, then $\widehat{P}_k(t)$ is unbiased.

(b) Consider a Markov illness-death model augmented to include a state for censoring as in the figure that follows, and suppose that the censoring intensities from states 1 and 2 are different, with $\lambda^c(t|Z(t^-) = 1) = \rho_1$ and $\lambda^c(t|Z(t^-) = 2) = \rho_2$. Prove that the estimator $\widehat{P}_1(t)$ above is biased unless $\rho_1 = \rho_2$.

(c) Prove also that the naive Kaplan-Meier estimator of $P(T_3 > t)$ is biased when $\rho_1 \neq \rho_2$, where T_3 is the entry time to state 3.

(Sections 4.1, 4.2)

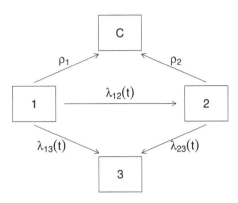

Problem 4.6 Consider a competing risks process with $K = 2$ causes of failure where, given a covariate x, the conditional cause-specific intensity functions are $\lambda_1(t|x) = \alpha_1 \exp(\beta_1 x)$ and $\lambda_2(t|x) = \alpha_2$.

(a) Obtain the cumulative incidence functions $P_1(t|x)$ and $P_2(t|x)$. Plot each of them for $x = 0$ and $x = 1$ in the case where $\alpha_1 = \alpha_2 = 1$ and $\beta_1 = \log 2$.

(b) Examine whether $P_1(t|x)$ is well approximated by models of the form (4.8) with $g(u) = \log(-\log(1-u))$ by plotting $\log(-\log(1 - P_1(t|x)))$ versus t for $x = 0$ and $x = 1$. Do the same for $P_2(t|x)$.

(c) What cautions would you suggest when comparing covariate effects in terms of the cause-specific intensities with effects based on transforms of the cumulative incidence functions?

(Section 4.1)

Problem 4.7 Thrombocytopenia is a medical condition in which affected individuals have very low platelet counts. Such patients routinely receive platelet transfusions, but there is interest in using platelets more conservatively to save resources and to reduce the exposure of thrombocytopenic patients to blood products from different donors. Heddle et al. (2009) report on a randomized trial designed to assess the impact of a lower dose of platelets upon platelet transfusion in the management of patients with this condition. Thrombocytopenic individuals were randomized to receive either the standard dose or a new lower dose of platelets upon the need for transfusion. Follow-up continued until recovery of platelet function or death.

The primary goal was to determine the impact of the low-dose transfusion strategy on the occurrence of clinically important bleeding. Individuals were examined daily with the bleeding status assessed using the World Health Organization 4-point bleeding scale; WHO Grade 2 or higher is considered clinically important, and we use this as the definition for state 2; $2 \to 1$ transitions occurred upon the resolution of a WHO Grade 2 or higher bleeding episode. Only two deaths are observed in this trial so we simply censor these individuals at this time here. The data are available in Section D.3.

(a) Construct suitable dataframes and obtain Aalen-Johansen estimates of the transition probability matrices for the multistate model depicted in Figure 4.9 separately for each group using the etm function.

(b) Using the result in (a), estimate the expected number of days of $\geq$ WHO Grade 2 bleeding over one month for an individual in the low-dose arm, and an individual in the standard dose arm.

(c) Using the results from (a) again, estimate the cumulative probability of platelet recovery for each treatment arm.

(d) Discuss strategies for formal comparisons between the two arms for testing treatment effects.

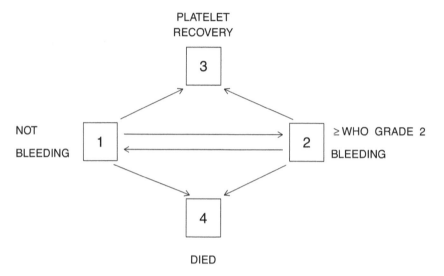

Figure 4.9: A 4-state model for the onset and resolution of clinically important bleeding along with platelet recovery and death.

Chapter 5

Studies with Intermittent Observation of Individuals

5.1 Introduction

In Chapter 1 we discussed settings in which subjects' states are known only at intermittent observation times. In such settings, the exact transition times, and in some cases, the number of transitions between successive observation times, are unknown. This type of data is common in cohort studies such as those described in Sections 1.2.3 and 1.2.4, where individuals are assessed and information is collected at periodic clinic visits.

The visit times may be fixed or random, and so the number and values can vary between individuals in general. We let $A_0 < A_1 < \cdots$ denote possibly random visit times for a particular individual over a period of time $[0, C^A]$ and let $a_0 < a_1 < \cdots < a_m$ denote their realized values. We assume for now that the states occupied and the values of external time-dependent covariates $X(t)$ are observed only at these times, so that the resulting data for such an individual are $\{a_r, Z(a_r), X(a_r), \ r = 1, \ldots, m\}$. Informally here, we let $\mathcal{H}^\circ(a_j) = \{a_r, Z(a_r), X(a_r), \ r = 0, 1, \ldots, j\}$ denote the history of the observed data up to the jth visit including the visit times. A more careful definition and discussion is given in Section 5.4. Then, conditional on $\mathcal{H}^\circ(a_0) = \{A_0 = a_0, Z(a_0), X(a_0)\}$, the probability of the observed data over the entire course of observation is

$$\prod_{j=1}^{m} P(A_j = a_j, Z(a_j), X(a_j) \mid \mathcal{H}^\circ(a_{j-1})). \tag{5.1}$$

In what follows, we outline the assumptions necessary to reduce (5.1) to a form on which inferences about the process $\{Z(s), s > 0\}$ can be made. A key one is conditional independence: the visit process is said to be *conditionally independent* if

$$A_j \perp \{Z(s), X(s), s > a_{j-1}\} \mid \mathcal{H}^\circ(a_{j-1}). \tag{5.2}$$

This means that conditional on $\mathcal{H}^\circ(a_{j-1})$, the states occupied and the covariate values *since the assessment at* a_{j-1} do not influence the time A_j of the next visit. This is in line with the definition of a sequential missing at random mechanism given by Hogan et al. (2004), but we prefer to characterize it as a *conditionally independent visit process* (CIVP), since it is less natural to conceptualize data as missing in this setting.

In disease registries of patients with relatively innocuous conditions, clinic visits may be set by physicians sequentially such that A_j is scheduled based on $\mathcal{H}°(a_{j-1})$, where $\mathcal{H}°(a_{j-1})$ contains all the pertinent information governing monitoring and care. If only visits scheduled this way are made, then the CIVP condition is satisfied. For more serious conditions, the CIVP assumption will often be implausible, since individuals naturally seek medical attention based on their health status and other related factors. Individuals whose condition worsens after a_{j-1}, for example, are more likely to schedule their next appointment with a physician sooner. This worsening condition may be reflected by transitions to more advanced disease states or changes in covariate values associated with the onset of new symptoms. This corresponds to what we term a *conditionally dependent visit process* (CDVP), since $\{Z(s), X(s), s > a_{j-1}\}$ and A_j are associated even conditional on the observed history $\mathcal{H}°(a_{j-1})$.

The validity of a CIVP assumption cannot be assessed solely from the observed data since $\{Z(s), X(s), s > a_{j-1}\}$ is unobserved until the next visit. Moreover, models that accommodate a CDVP assumption involve assumptions that cannot be verified on the basis of the observed data; sensitivity analyses are often recommended in such settings. We remark further on conditionally independent and dependent visit processes in Section 5.4, and until then proceed under the assumption of a CIVP.

We next discuss how (5.1) may be decomposed under the CIVP assumption to allow model fitting and inferences about the multistate process. The terms in (5.1) can be factored as

$$P(A_j = a_j \mid \mathcal{H}°(a_{j-1}))\ P(Z(a_j), X(a_j) \mid A_j = a_j, \mathcal{H}°(a_{j-1})).$$

The first term is typically considered noninformative as it will not involve parameters in the models for the $Z(t)$ and $X(t)$ processes, so we drop it and consider a factorization of the remaining partial likelihood contribution:

$$
\begin{aligned}
&P(Z(a_j), X(a_j) \mid A_j = a_j, \mathcal{H}°(a_{j-1})) \\
&= P(Z(a_j) \mid A_j = a_j, \mathcal{H}°(a_{j-1}))\ P(X(a_j) \mid A_j = a_j, Z(a_j), \mathcal{H}°(a_{j-1})).
\end{aligned}
\tag{5.3}
$$

Under the assumption of a non-informative covariate process, we may retain only the first term of the right-hand side of (5.3). The resulting partial likelihood for a sample of n independent individuals is then

$$L(\theta) = \prod_{i=1}^{n} \prod_{j=1}^{m_i} P(Z_i(a_{ij}) \mid A_{ij} = a_{ij}, \mathcal{H}_i°(a_{j-1})),
\tag{5.4}$$

where $a_{i0} < a_{i1} < \cdots < a_{im_i}$ are the observation times for individual i.

We let C_i^A denote the administrative censoring time for individual i, $i = 1, \ldots, n$. If random censoring for individual i is also possible due to early withdrawal at some time $C_i^R \leq C_i^A$, we typically make the conditional independence assumption that

$$C_i^R \perp \{Z_i(s), X_i(s), s > a_{i,j-1} \mid C_i^R > a_{i,j-1}, \mathcal{H}_i°(a_{j-1})\}.
\tag{5.5}$$

A censoring time may be unobserved because when a person becomes lost to follow-up there are no further visits, but under the CIVP assumption, (5.4) is still a valid partial likelihood and does not require C_i^R to be observed. An additional assumption that is made for a CIVP, however, is that the probabilities in (5.4) are the same as for the case where the a_{ij} are prespecified visit times, and can thus be calculated solely from the probabilistic specification for the $Z(t)$ process. This point is discussed further in Section 5.4. In such cases the visit process and withdrawal process governing C_i^R are sometimes called ignorable. We caution that the use of this term is context-specific and a particular observation process, which is "ignorable" for likelihood-based analyses, may require the specification and fitting of auxiliary models for the introduction of weights into marginal methods; we discuss this in Section 5.3.

We describe methods for fitting models and inference about Markov processes next.

5.2 Estimation and Analysis for Markov Models

5.2.1 Model Fitting

Computation of the terms in (5.4) is intractable for most types of processes, but much progress can be made for modulated Markov models. If $\mathcal{Z}_i^\circ(a_{ij}) = \{(a_{ir}, Z_i(a_{ir})), r = 1, \ldots, j\}$ and $\mathcal{X}_i^\circ(a_{ij}) = \{(a_{ir}, X_i(a_{ir})), r = 1, \ldots, j\}$, then under the CIVP assumption the terms in (5.4) equal $P(Z_i(a_{ij})|A_{ij} = a_{ij}, \mathcal{Z}_i^\circ(a_{i,j-1}), \mathcal{X}_i^\circ(a_{i,j-1}))$. For Markov models this reduces to $P(Z_i(a_{ij})|A_{ij} = a_{ij}, Z_i(a_{i,j-1}), \mathcal{X}_i^\circ(a_{i,j-1}))$, giving the partial likelihood function

$$L(\theta) = \prod_{i=1}^{n} \prod_{j=1}^{m_i} P(Z_i(a_{ij}) \mid A_{ij} = a_{ij}, Z_i(a_{i,j-1}), \mathcal{X}_i^\circ(a_{i,j-1})). \tag{5.6}$$

With fixed covariates X and multistate process history $\mathcal{H}(t) = \{Z(s), 0 < s \leq t; X\}$, parametric Markov models with transition intensities $\lambda_{kl}(t|\mathcal{H}(t^-)) = \lambda_{kl}(t|x; \theta_{kl})$ are relatively easy to fit. If θ contains all θ_{kl}, transition probabilities $P_{kl}(s, t|x; \theta)$ needed for (5.6) can be obtained by numerical approximation of product integrals (2.23) giving $P(s, t|x; \theta)$. Time-homogeneous models with transition intensities $\lambda_{kl}(x; \theta_{kl})$ are especially easy to handle, since if $Q(x; \theta)$ is the transition intensity matrix, by (2.24) the transition probability matrix is simply

$$P(s, t \mid x) = \exp\{(t - s) \cdot Q(x; \theta)\}, \quad s \leq t. \tag{5.7}$$

The likelihood for multistate processes under intermittent observation, unlike the likelihood (2.19) for right-censored data, does not factor into functionally independent components where each involves a different parameter vector. We give the log-likelihood function and information matrices based on it in Section 5.2.2, where we also discuss the information about parameters and the design of follow-up studies. The msm package in R (Jackson, 2011) provides software for fitting time-homogeneous models and models with piecewise-constant transition intensities. An illustration involving msm is given in Section 5.2.4.

For models with time-dependent covariates observed only at visit times, the likelihood (5.6) is usually constructed in one of two ways. The first is by assuming that $X_i(t)$ is fixed over the time interval $[a_{i,j-1}, a_{ij})$ at a value X_{ij} determined from $X_i(a_{i,j-1})$. Then the constant transition intensity matrix Q_{ij} applicable over $(a_{i,j-1}, a_{ij}]$ for individual i has (k,l) entry $\lambda_{kl}(x_{ij}; \theta_{kl})$; the transition probabilities $P_{kl}(a_{i,j-1}, a_{ij} | x_{ij})$ needed for the likelihood (5.6) can be obtained via the matrix exponential expression (5.7). This approach is practical in that it specifies models for which the observed process history up to time $a_{i,j-1}$ is used to model and predict the state occupied at time a_{ij}. When the covariate process is volatile or there is considerable variability in the times between assessments, this may be unappealing. The second main approach for dealing with time-dependent covariates in this setting is to specify a joint model for $Z(t)$ and $X(t)$. Models yielding tractable calculation of terms in (5.6) are rare, however, unless $X(t)$ takes on discrete values. When $X(t)$ is discrete, it is possible to create a categorical variable with, say, G distinct values so that one can model $\{Z(s), X(s), s > 0\}$ with a multistate model with states (k, g) for $k = 1, \ldots, K$ and $g = 1, \ldots, G$. If this model is assumed to be Markov, it can readily be fitted. We discuss joint models and issues involving time-dependent covariates in Section 8.3.

5.2.2   Parametric Information and Study Design

We assume here that the terms in the likelihood (5.6) may depend on fixed covariates or covariates $X_{i,j-1}$ that are based on the observed covariate history $\mathcal{X}^\circ(a_{i,j-1})$, giving the log-likelihood

$$\ell(\theta) = \sum_{i=1}^{n} \sum_{j=1}^{m_i} \log P(Z_i(a_{ij}) \mid Z_i(a_{i,j-1}), X_{i,j-1}; \theta).$$

We ignore notationally that the a_{ij} may be random, since under the CIVP assumptions they may be treated as prespecified values. For convenience, we also suppress the dependence on $X_{i,j-1}$ in the notation that follows and write $P_{kl}(a_{i,j-1}, a_{ij}; \theta)$ to represent $P(Z_i(a_{ij}) = l | Z_i(a_{i,j-1}) = k, X_{i,j-1}; \theta)$. With $Y_{ik}(t) = I(Z_i(t) = k)$, we rewrite $\ell(\theta)$ as

$$\ell(\theta) = \sum_{i=1}^{n} \sum_{j=1}^{m_i} \sum_{k=1}^{K} \sum_{l=1}^{K} Y_{ik}(a_{i,j-1}) Y_{il}(a_{ij}) \log P_{kl}(a_{i,j-1}, a_{ij}; \theta).$$

The observed information matrix $I(\theta) = -\partial^2 \ell(\theta)/\partial\theta\partial\theta'$ is then

$$I(\theta) = \sum_{i=1}^{n} \sum_{j=1}^{m_i} \sum_{k=1}^{K} \sum_{l=1}^{K} Y_{ik}(a_{i,j-1}) Y_{il}(a_{ij}) \left\{ -\frac{\partial^2 \log P_{kl}(a_{i,j-1}, a_{ij}; \theta)}{\partial\theta\partial\theta'} \right\}. \tag{5.8}$$

A Newton-Raphson algorithm can be used to obtain $\hat{\theta}$ and inferences can be based on the approximation $\hat{\theta} \sim N(\theta, I^{-1}(\hat{\theta}))$. Note that $-\partial^2 \log P_{kl}(s,t;\theta)/\partial\theta\partial\theta'$ can be written as

$$\frac{(\partial P_{kl}(s,t;\theta)/\partial\theta)(\partial P_{kl}(s,t;\theta)/\partial\theta')}{[P_{kl}(s,t;\theta)]^2} - \frac{\partial^2 P_{kl}(s,t;\theta)/\partial\theta\partial\theta'}{P_{kl}(s,t;\theta)}.$$

Kalbfleisch and Lawless (1985) propose an alternative algorithm for estimation, which avoids the need for second derivatives of the transition probabilities. This is achieved by replacing each (i,j) term in (5.8) with its conditional expectation given $Z_i(a_{i,j-1})$. Noting that

$$E\{Y_{il}(a_{ij}) \mid Y_{ik}(a_{i,j-1}) = 1\} = P_{kl}(a_{i,j-1}, a_{ij}; \theta),$$

we obtain

$$I_1(\theta) = \sum_{i=1}^{n} \sum_{j=1}^{m_i} \sum_{k=1}^{K} \sum_{l=1}^{K} \frac{Y_{ik}(a_{i,j-1})}{P_{kl}(a_{i,j-1}, a_{ij}; \theta)} \qquad (5.9)$$
$$\times \left(\frac{\partial P_{kl}(a_{i,j-1}, a_{ij}; \theta)}{\partial \theta} \right) \left(\frac{\partial P_{kl}(a_{i,j-1}, a_{ij}; \theta)}{\partial \theta'} \right).$$

Kalbfleisch and Lawless (1985) refer to the Newton-type algorithm for estimation based on (5.9) as a Fisher-scoring algorithm. The covariance matrix for $\hat{\theta}$ can likewise be estimated by $I_1^{-1}(\hat{\theta})$.

The precision with which model parameters are estimated depends on the gaps $\Delta a_{ij} = a_{ij} - a_{i,j-1}$ between successive observations of individuals, with some parameters more sensitive to the assessment times than others. To explore this we consider time-homogeneous Markov models without covariates, and assume that individuals have common visit times $0 = a_0 < a_1 < \cdots < a_m$. In this case we can write $P_{kl}(a_{j-1}, a_j; \theta)$ as $P_{kl}(\Delta a_j; \theta)$, where $\Delta a_j = a_j - a_{j-1}$, and by taking the expectation of $I_1(\theta)$ in (5.9), we get the Fisher information matrix with (u, v) component

$$\mathcal{I}_{uv}(\theta) = \sum_{j=1}^{m} \sum_{k=1}^{K} \sum_{l=1}^{K} \frac{E_k(a_{j-1}; \theta)}{P_{kl}(\Delta a_j; \theta)} \frac{\partial P_{kl}(\Delta a_j; \theta)}{\partial \theta_u} \frac{\partial P_{kl}(\Delta a_j; \theta)}{\partial \theta_v}, \qquad (5.10)$$

where $u, v = 1, \ldots, \dim(\theta)$ and

$$E_k(a_{j-1}; \theta) = \sum_{i=1}^{n} P(Z_i(a_{j-1}) = k \mid Z_i(0); \theta). \qquad (5.11)$$

Kalbfleisch and Lawless (1985) give expressions for the derivatives in (5.10), but a simple alternative is to use numerical differentiation based on approximations

$$\frac{\partial P_{kl}(w; \theta)}{\partial \theta_u} = \frac{P_{kl}(w; \theta + \delta_u) - P_{kl}(w; \theta - \delta_u)}{2\delta_u}$$

where δ_u is a vector of the same dimension as θ, with a small value δ_u in the position corresponding to θ_u, and zero elsewhere. The information matrix (5.10) is based on the assumption that individuals are each seen at all assessment times $a_1, \ldots, a_m$. If the length of follow-up varies between individuals because of staggered entry or losses to follow-up, then we may accommodate this by multiplying the ith term in (5.11) by the probability $\pi_i(a_{j-1})$ that the individual is observed at time a_{j-1}. This will, of course, result in a net loss of parametric information.

The asymptotic covariance matrix for $\sqrt{n}(\hat{\theta} - \theta)$ is the limit of $n\mathcal{I}(\theta)^{-1}$, where some initial distribution for $Y_i(0)$ is assumed. This can be used to study the effects of sample size n and the schedule of observation times on the precision of estimators, providing useful insights for planning longitudinal studies. One can also consider the impact of the initial distribution and use stratified sampling of individuals for inclusion in a study to ensure sufficient information is collected on all parameters.

Large gap times Δa_j can have a much more detrimental effect on the precision of estimators in multistate models featuring recurrent states (models for which certain states can be entered repeatedly) than in models comprised of only transient or absorbing states; we sometimes refer to the latter types of models as progressive. In progressive models, the loss of efficiency for a given transition intensity λ_{kl} relative to continuous observation is small when the gap times between visits are less than the average sojourn time in state k, and increases rather slowly as the Δa_j increase. For pairs of states (k,l) where both $\lambda_{kl} > 0$ and $\lambda_{lk} > 0$, the information on $\widehat{\lambda}_{kl}$ and $\widehat{\lambda}_{lk}$ drops rapidly as the times between visits increase, in part because there is no data on the number of transitions between these states. Transition probabilities, however, are more precisely estimable; Lawless and Nazeri Rad (2015) provide illustrations. Some insight into the loss of efficiency from intermittent observation can be obtained by considering a 2-state model with transition intensities λ_{12} and λ_{21}. The transition probability matrix $P(w)$ for this model is

$$P(w) = \begin{pmatrix} 1 - \pi\left(1 - e^{-\alpha w}\right) & \pi\left(1 - e^{-\alpha w}\right) \\ (1 - \pi)\left(1 - e^{-\alpha w}\right) & \pi + (1 - \pi)e^{-\alpha w} \end{pmatrix},$$

where $\alpha = \lambda_{12} + \lambda_{21}$ and $\pi = \lambda_{12}/(\lambda_{12} + \lambda_{21})$ is the equilibrium probability of being in state 2 (see Problems 1.2, 5.2). As w increases, the two rows of $P(w)$ approach the limiting distribution $(1 - \pi, \pi)$. It is apparent that for a study with large values of αw, there will be considerable information about π, but little information about α, λ_{12} or λ_{21}. Problem 5.3 considers this in more detail.

In models with covariates, the loss of efficiency for regression coefficients as the Δa_j increase is similar to that for transition intensities. The results described here have important ramifications for the planning of longitudinal studies. If the observation times a_j are far enough apart that multiple state changes are likely between successive observation times, then estimators of transition intensities, state duration distributions and covariate effects for recurrent states will be highly variable. The situation is better for transient states which can only be entered once. Estimates of transition probabilities and state occupancy probabilities are more precise. Thus, in a longitudinal cohort study on aging in which persons are seen every 3 or 4 years, it may be possible to obtain precise information about the onset and progression of cognitive impairment as a function of age. However, precise information on transient spells of disability may not be available unless it is possible to obtain accurate retrospective information on spells at the intermittent observation times. In the absence of this, information on the transition intensities between recurring states will be minimal, and associated estimators will be imprecise.

5.2.3   Model Checking

In Section 3.5.1 we discussed approaches to model assessment based on (i) comparison of parametric and nonparametric estimates, (ii) model expansion, (iii) examination of residuals or influence measures, and (iv) predictive assessment. When the states occupied by individuals are observed only intermittently, all approaches but model expansion are problematic except in special cases; nonparametric estimation of transition intensities and construction of residuals are not feasible. In some cases, we can obtain nonparametric estimates of state occupancy probabilities, or probability distributions for entry times that may be helpful provided observation times are ignorable. The case of non-ignorable observation times is discussed in Section 5.4. Nonparametric estimation of occupancy and transition probabilities is simple when individuals have common ignorable visit times $a_{ij} = a_j$, $j = 1, \ldots, m$. In this case, statistics like (3.58) can be used to assess parametric models, and comparison of parametric and nonparametric occupancy probability estimates at times $a_1, \ldots, a_m$ is straightforward. Problem 5.4 has a few results.

The types of model expansion discussed in Section 3.5.1 apply here as well, with the caveat that as models become more complex, they become harder to fit when processes are under intermittent observation. Piecewise-constant intensity alternatives to time-homogeneous models can be handled by the msm package, but Markov models with other forms of time dependence, or with non-Markov features, are more challenging to fit. Likelihood ratio tests are most convenient for formal comparison of nested models.

Two other points are important. The first is that longer times between observations make it more difficult to check the validity of any assumed models. Robust estimation of certain process features may be appealing in this case. Estimates are typically not found from intensity-based models for process dynamics but rather from models for marginal or partially conditional features; we consider this in Sections 5.3 and 5.4. A second point is that individuals may postpone or miss scheduled visits or, as mentioned earlier, may drop out of a study. We have assumed that a person's dropout time, or more generally the completeness of their data, is independent of their state beyond the last assessment time, conditional on their observed process history (see Section 5.1). This can more easily be violated when visits are far apart, and studies should aim to ensure that follow-up is as complete as possible.

5.2.4   Illustration: Progression of Diabetic Retinopathy

The Diabetes Control and Complications Trial (DCCT) was introduced in Section 1.2.2. The primary comparison of intensive insulin therapy (IT) and conventional therapy (CT) in the randomized trial was based on the progression of retinopathy in the two arms (Diabetes Control and Complications Trial Research Group, 1993, 1995). Here we will focus on the Primary Intervention (PI) Cohort, which consisted of individuals with no retinopathy at the time of randomization. Retinopathy was measured on the 23-point ordinal ETDRS scale at biannual visits. Box plots of the distribution of ETDRS scores by visit are given in Figure 5.1. There is a clear trend of an increasing mean in the ETDRS score over time with a greater increase in the conventional therapy arm. While such a plot conveys patterns in marginal features over time, it does not reflect the dynamics of the process, and the apparent trends may be influenced by the decreasing sample size and possible selection biases. The box plots also indicate substantial variation in the ETDRS scores at each visit. Figure 5.2 shows sample plots of ETDRS values for four subjects, who were seen every 6 months.

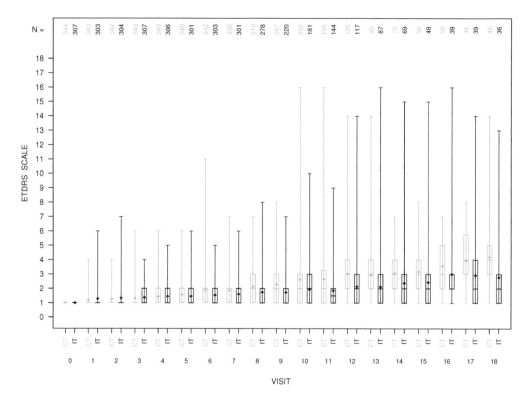

Figure 5.1: Box plots of the ETDRS scores over time for the conventional therapy (gray box) and intensive insulin therapy (black box). Lines show minimum, 25th, 50th and 75th percentiles along with the maximum scores; the diamond denotes the average value; the numbers N at the top of the plot are the total number of individuals on study at the respective visits. The baseline visit is at $t = 0$, the average time to the first visit is 0.7 years, and subsequent visits occur roughly every 6 months thereafter (i.e. 1.2, 1.7, 2.2, ... years).

To gain insight into the process dynamics, we consider a multistate analysis based on the two 5-state models in Figure 1.2. Each model has states 1: ETDRS = 1, 2: ETDRS = 2 or 3, 3: ETDRS = 4 to 6, 4: ETDRS = 7 to 9, and 5: ETDRS $\geq$ 10. Model M1 of Figure 1.2 allows transitions in both directions, whereas Model M2 is progressive. From Figure 5.2 it is apparent that only Model M1 is compatible with the data, since ETDRS scores can go either up or down on successive visits. Model M2 applies if an individual seen in a given state can never subsequently be seen in a lower state. The data may be modified to comply with this by replacing $Z_i(a_{ij})$ with $\max(Z_i(a_{i0}, Z_i(a_{i1}), \ldots, Z_i(a_{ij}))$, and the analysis under model M2 may then be viewed as modeling the highest degree of retinopathy recorded up to a given time. Another "operational" approach is to assume an individual has moved up one

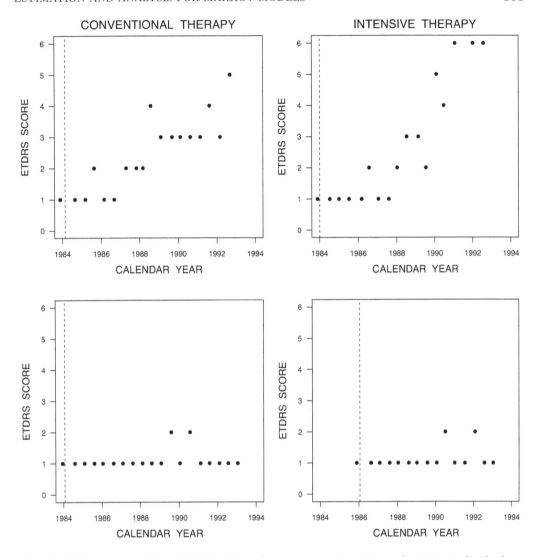

Figure 5.2: Profile plots of four individuals in the primary intervention cohort: two individuals on conventional therapy (left panels) and two individuals on intensive insulin therapy (right panels).

or more states only when they have been observed to be there for two consecutive visits; such a definition was used by the Diabetes Control and Complications Trial Research Group (1993), who took the time T from randomization to entry to state 3 as the response variable for treatment comparisons; the entry time to state 3 was then taken as the time of the second assessment at which they were known to be in state 3 or higher. This scenario is typical of situations where longitudinal measurements vary substantially over short time periods, though they may exhibit long-term trends. Given the variability and heterogeneity observed in the ETDRS patterns, no single best way to measure progression of retinopathy is apparent; next we fit models M1 and M2 in Figure 1.2 and comment on the interpretation of the results. In the dataset, individuals are recorded as moving from higher to lower states and so to fit M2, we ignore transition to lower states as suggested earlier and view the disease process in terms of the highest state of damage recorded up to a given time. In Section 6.4.2 we consider an alternative approach by modeling a latent progressive Markov process and accommodating misclassification of the true states of retinopathy; observed transitions to lower states are then assumed to arise due to misclassification of either the previous or current state.

We fit the two Markov models to each treatment group separately, first using time-homogeneous intensities, and then piecewise-constant transition intensities. This showed that accommodating time dependence is necessary; we settled on a two-piece model in which transition intensities are constant over 0–4 years from randomization and then constant but at different values thereafter. Models were fitted using the msm package in R, with M2 fitted to the derived "progressive" data. Progressive models are easily fitted, but reversible models may encounter problems, even though the intermittent observation times for individuals are only 6 months apart. Because there were few subjects observed in states 4 or 5 over the follow-up period, we constrain the intensities such that $\lambda_{34}(t) = \lambda_{45}(t)$ and $\lambda_{43}(t) = \lambda_{54}(t)$ for model M1; for model M2, we set $\lambda_{34}(t) = \lambda_{45}(t)$. Illustrative R code for model fitting with a reduced 3-state model discussed below is given in Section C.1.

Table 5.1 shows parameter estimates and 95% confidence intervals obtained as $\exp(\log \widehat{\lambda}_{kl} \pm 1.96 \ \text{s.e.}(\log \widehat{\lambda}_{kl}))$. Both models fit quite well as far as transitions among states are concerned; observed and expected half-yearly transition counts can be obtained as follows for the conventional therapy group. The observed counts for times a_{j-1}, a_j are given by

$$O^{(j)}_{kl} = \sum_{i=1}^{n} Y_i(a_j) Y_{ik}(a_{j-1}) Y_{il}(a_j), \quad k, l = 1, \ldots, 5 \tag{5.12}$$

where $a_0, a_1, a_2, \ldots$, are the common observation times. The nominal visit times in years from randomization are $a_0 = 0$, $a_1 = 0.7$, $a_2 = 1.2$, $a_3 = 1.7$ and so on. In fact, the observation times vary a little between and within individuals, and so we take $Z_i(a_j)$ to be the state occupied at the jth biannual visit. Subjects entered the trial over a period of years and so their lengths of follow-up vary, and thus also the total numbers of transitions observed for successive times (a_{j-1}, a_j). The corresponding expected transition counts are given by

$$E^{(j)}_{kl} = \sum_{i=1}^{n} Y_i(a_j) Y_{ik}(a_{j-1}) \widehat{P}(Z_i(a_j) = l \mid Z_i(a_{j-1}) = k)$$
$$= O^{(j)}_{k.} \widehat{P}(Z_i(a_j) = l \mid Z_i(a_{j-1}) = k), \tag{5.13}$$

where the estimated transition probabilities are the $\widehat{P}_{kl}(a_{j-1}, a_j)$ for the conventional therapy group based on the fitted model being assessed.

Table 5.1 shows that for the model M1 the upward intensities are smaller than the corresponding downward intensities in most cases. This could be a reflection of the measurement process (ETDRS scores are assigned by clinicians based on photographs of the eyes), as well as short-term fluctuations in the vascular condition of the eyes. After 4 years, $\widehat{\lambda}_{12} > \widehat{\lambda}_{21}$ in both treatment groups; this reflects the fact that retinopathy eventually worsens over time for many persons, and we note this

Table 5.1: Estimates of transition intensities and 95% confidence intervals for the reversible Markov model (M1) and the progressive Markov model (M2) fitted to data from the primary intervention cohort in the DCCT study ($m = 651$); the cut-point for the piecewise-constant intensities is 4 years.

Model	Period	Parameter	Conventional Therapy		Intensive Insulin Therapy	
			EST	95% CI[†]	EST	95% CI[†]
M1	$[0,4)$	λ_{12}	0.50	(0.45, 0.56)	0.55	(0.49, 0.63)
		λ_{21}	0.83	(0.71, 0.98)	1.08	(0.93, 1.26)
		λ_{23}	0.33	(0.25, 0.44)	0.23	(0.16, 0.34)
		λ_{32}	1.92	(1.37, 2.68)	2.61	(1.72, 3.94)
		$\lambda_{34} = \lambda_{45}^{\ddagger}$	0.10	(0.02, 0.50)	0.37	(0.02, 5.83)
		$\lambda_{43} = \lambda_{54}^{\ddagger}$	3.51	(0.82, 15.01)	8.80	(0.51, 152.27)
	$[4,\infty)$	λ_{12}	1.07	(0.90, 1.28)	0.51	(0.42, 0.63)
		λ_{21}	0.51	(0.40, 0.63)	0.41	(0.32, 0.51)
		λ_{23}	0.71	(0.59, 0.86)	0.38	(0.28, 0.52)
		λ_{32}	0.93	(0.73, 1.18)	1.92	(1.36, 2.71)
		$\lambda_{34} = \lambda_{45}^{\ddagger}$	0.19	(0.11, 0.34)	0.13	(0.05, 0.33)
		$\lambda_{43} = \lambda_{54}^{\ddagger}$	1.67	(0.85, 3.27)	0.42	(0.11, 1.61)
	log L		-2761.946		-2171.994	
M2	$[0,4)$	λ_{12}	0.33	(0.29, 0.37)	0.34	(0.30, 0.39)
		λ_{23}	0.14	(0.10, 0.17)	0.07	(0.05, 0.10)
		$\lambda_{34} = \lambda_{45}^{\ddagger}$	0.03	(0.01, 0.11)	0.04	(0.01, 0.14)
	$[4,\infty)$	λ_{12}	0.48	(0.34, 0.67)	0.32	(0.21, 0.47)
		λ_{23}	0.27	(0.22, 0.34)	0.08	(0.05, 0.12)
		$\lambda_{34} = \lambda_{45}^{\ddagger}$	0.06	(0.04, 0.10)	0.03	(0.01, 0.08)
	log L		-1388.908		-992.704	

[†] 95% CI computed as $\exp(\log \widehat{\lambda}_{kl}(t) \pm 1.96 \text{ s.e.}(\log \widehat{\lambda}_{kl}(t)))$.

[‡] Parameters constrained to be the same.

is more pronounced in the CT group. Progressive model M2 has higher estimated intensities for the CT group, except for λ_{12} and λ_{34} over $(0,4]$ years which are, however, roughly equal in the CT and IT groups.

As noted, few persons are observed in states 4 or 5 over the follow-up period, and intensities between states 3, 4 and 4, 5 are estimated imprecisely in model M1. In addition, examination of occupancy probabilities indicates the 2-piece intensity model is not very satisfactory beyond 4 years, but with the sparse data on higher transitions, fitting models with more pieces is problematic. We thus consider 3-state models in which states 3–5 in Figure 1.2 are combined, so that the revised state 3 now represents ETDRS scores ≥ 4; we refer to the resulting models as models M1B and M2B, respectively, with their multistate diagrams given in Figure C.1. These models provide a good assessment of treatment effects over the range of follow-up times; estimates for the probabilities of being in state 3 or higher that are based on 4-state and 5-state models agree closely with those for 3-state models. For the 3-state models we use three-piece intensities with different values for time intervals $[0,3)$, $[3,6)$ and $[6,\infty)$ years.

Table 5.2 shows estimates obtained by fitting models M1B and M2B to the CT and IT treatment groups. We see once again that for model M1B $\widehat{\lambda}_{12} < \widehat{\lambda}_{21}$ and $\widehat{\lambda}_{23} < \widehat{\lambda}_{32}$ over 0–3 years. This changes for $\widehat{\lambda}_{12}$ and $\widehat{\lambda}_{21}$ after 3 years and for $\widehat{\lambda}_{23}$ and $\widehat{\lambda}_{32}$ after 6 years in the CT group. This reflects the tendency for persons in both groups to experience some progression of retinopathy over time, with this occurring earlier for the CT group. Model M2B shows similar estimates for λ_{12} over $0-3$ years in the CT and IT groups, but thereafter the intensities are substantially higher for the CT group. Both models describe observed transitions well. Table 5.3 shows observed and expected transition counts for model M1B for the CT group. The only substantial discrepancy is over the interval $(2.7, 3.2)$ and $(3.2, 3.7)$ years for transitions from state 1. There is no clear explanation for this, but the fact that the discrepancies are in opposite directions for the two time intervals might reflect slight anomalies in the actual visit times versus the nominal times 2.7, 3.2, 3.7 or some small change in the way ETDRS scores are assigned. This discrepancy does not show up in the IT group.

We now address treatment comparison. Diabetes Control and Complications Trial Research Group (1993) compared the two groups in terms of T, the time of entry to state 3 under progressive model M2B with 3 states, though it also applies to the models with 4 or 5 states. For reversible models M1 and M1B the distribution of T_1, the time of first entry to state 3 can be obtained. With the 3-, 4-, or 5-state models, we find that the estimates of $F(t) = P(T \leq t)$ from model M2 agree well with the nonparametric Turnbull estimator (Turnbull, 1976) based on the interval-censored observation of the (presumed) first entry to state 3; see Figure 5.3 for the 3-state M2B. The estimates for $F_1(t) = P(T_1 \leq t)$ from model M1B, calculated by artificially making state 3 absorbing, are well above the M2B and Turnbull estimates. This reflects the fact that under model M1B the first entry to state 3 might go unobserved because an individual went back to state 2 before the next observation time. For these models a better measure of progression is given by prevalence or occupancy probabilities $P(Z(t) \geq 3)$. Figure 5.3 shows estimates of $P_3(t)$ for the 3-state piecewise homogeneous model M1B; estimates based on a 3-state progressive hidden Markov model (HMM) discussed in Section 6.4.2 are also shown. In this case, the estimates for M1B are substantially lower than those for the progressive models, reflecting the fact that for M1B a person may be in state 3 at one visit but return later to states 2 or 1. Empirical estimates of $P_3(t)$ for M1B are also shown in Figure 5.3, given by the fraction of subjects observed at a given visit time who are in state 3. The estimates from the fitted model M1B of Table 5.2 track the empirical estimates well.

Model M1B has the advantage of representing the actual measurement of retinopathy, though if there is enough information about the measurement process and short-term variations in the eyes, arguments can be made for concep-

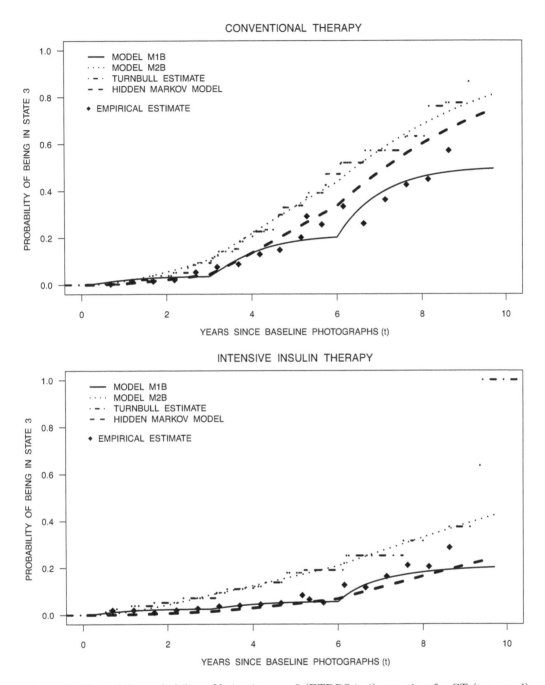

Figure 5.3: Plots of the probability of being in state 3 (ETDRS ≥ 4) over time for CT (top panel) and IT (bottom panel) using the estimates obtained from fitting 3-state nonhomogeneous models; M1B (reversible Markov model), M2B (progressive Markov model), and a hidden Markov model (HMM), along with the corresponding nonparametric Turnbull estimate for entry time to state 3 for M1B; empirical prevalence estimates are also represented.

Table 5.2: Estimates of transition intensities and 95% confidence intervals for the reversible 3-state Markov model (M1B) and the progressive 3-state Markov model (M2B) fitted to data from the primary intervention cohort in the DCCT study ($m = 651$); the cut-points for the piecewise-constant true intensity functions are at 3 and 6 years.

Model	Period	Parameter	Conventional Therapy		Intensive Insulin Therapy	
			EST	95% CI[†]	EST	95% CI[†]
M1B	$[0,3)$	λ_{12}	0.44	(0.38, 0.51)	0.52	(0.45, 0.61)
		λ_{21}	0.97	(0.80, 1.19)	1.20	(0.99, 1.44)
		λ_{23}	0.32	(0.21, 0.51)	0.25	(0.15, 0.42)
		λ_{32}	2.72	(1.60, 4.60)	2.98	(1.71, 5.19)
	$[3,6)$	λ_{12}	0.95	(0.82, 1.11)	0.61	(0.51, 0.72)
		λ_{21}	0.64	(0.53, 0.77)	0.63	(0.52, 0.77)
		λ_{23}	0.52	(0.43, 0.65)	0.26	(0.18, 0.38)
		λ_{32}	1.16	(0.89, 1.52)	2.13	(1.47, 3.08)
	$[6,\infty)$	λ_{12}	1.12	(0.77, 1.63)	0.45	(0.30, 0.68)
		λ_{21}	0.32	(0.20, 0.52)	0.27	(0.17, 0.44)
		λ_{23}	0.92	(0.68, 1.24)	0.51	(0.33, 0.77)
		λ_{32}	0.71	(0.49, 1.02)	1.17	(0.73, 1.87)
	log L		-2639.936		-2136.390	
M2B	$[0,3)$	λ_{12}	0.30	(0.26, 0.34)	0.34	(0.30, 0.39)
		λ_{23}	0.12	(0.09, 0.17)	0.08	(0.06, 0.12)
	$[3,6)$	λ_{12}	0.50	(0.39, 0.62)	0.33	(0.25, 0.44)
		λ_{23}	0.21	(0.17, 0.27)	0.07	(0.04, 0.10)
	$[6,\infty)$	λ_{12}	0.77	(0.33, 1.78)	0.45	(0.17, 1.20)
		λ_{23}	0.32	(0.20, 0.50)	0.09	(0.05, 0.18)
	log L		-1297.506		-960.261	

[†] 95% CI computed as $\exp(\log \widehat{\lambda}_{kl}(t) \pm 1.96 \text{ s.e.}(\log \widehat{\lambda}_{kl}(t)))$.

tual progressive models. Although models M1B and M2B give very different absolute measures of progression, Figure 5.3 clearly shows that the subjects in the IT group progress more slowly. Moreover, relative treatment measures such as $P_3(t|x = 1)/P_3(t|x = 0)$ agree fairly well for M1B and M2B, with values in the range 0.4−0.5 at most visit times. Box plots of the ETDRS scores at each visit also show a clear treatment difference; see Figure 5.1. It is clear from Figures 5.1 and 5.2 that scores vary widely across (and to some extent within) subjects. This suggests that models incorporating unobserved heterogeneity might be considered; they are discussed in Chapter 6. A difficulty in doing this for the multistate models, however, is that marginal prevalence estimates like $P_3(t)$ become more complicated. The estimates of marginal prevalence from the models used here are quite robust.

Table 5.3: Observed and expected transition counts for Markov model M1B: Conventional therapy group.

Interval*	State	$O^{(j)}_{kl}(a_{j-1}, a_j)$ Observed Counts $l=1$	2	3	$E^{(j)}_{kl}(a_{j-1}, a_j)$ Expected Counts $l=1$	2	3
(0.0, 0.7)	$k=1$	288	51	1	272.1	63.2	4.7
	2	0	0	0	0.0	0.0	0.0
	3	0	0	0	0.0	0.0	0.0
(0.7, 1.2)	$k=1$	248	36	3	241.1	43.4	2.6
	2	23	25	2	16.6	30.2	3.2
	3	1	0	0	0.2	0.5	0.3
(1.2, 1.7)	$k=1$	236	33	3	228.5	41.1	2.5
	2	24	37	1	20.6	37.4	4.0
	3	1	3	1	0.8	2.7	1.5
(1.7, 2.2)	$k=1$	214	42	4	218.4	39.3	2.3
	2	22	48	3	24.2	44.1	4.7
	3	1	4	0	0.8	2.7	1.5
(2.2, 2.7)	$k=1$	187	44	6	199.1	35.8	2.1
	2	27	60	7	31.2	56.8	6.1
	3	1	1	5	1.2	3.8	2.1
(2.7, 3.2)	$k=1$	142	68	5	162.7	48.1	4.2
	2	23	70	14	28.6	67.0	11.3
	3	1	10	7	2.1	8.5	7.4
(3.2, 3.7)	$k=1$	120	30	1	100.9	44.5	5.6
	2	34	85	15	26.7	86.7	20.7
	3	1	10	11	1.2	7.5	13.3
(3.7, 4.2)	$k=1$	81	42	8	87.6	38.6	4.9
	2	24	74	14	22.3	72.4	17.3
	3	1	9	12	1.2	7.5	13.3
(4.2, 4.7)	$k=1$	53	27	4	56.2	24.7	3.1
	2	18	66	12	19.1	62.1	14.8
	3	4	10	15	1.6	9.9	17.5
(4.7, 5.2)	$k=1$	34	20	2	37.4	16.5	2.1
	2	9	50	13	14.3	46.6	11.1
	3	0	9	16	1.4	8.6	15.1
(5.2, 5.7)	$k=1$	24	6	2	21.4	9.4	1.2
	2	16	35	12	12.5	40.7	9.7
	3	1	7	17	1.4	8.6	15.1
(5.7, 6.1)	$k=1$	18	11	3	22.3	8.5	1.2
	2	5	22	13	6.0	27.3	6.7
	3	1	5	16	0.7	6.0	15.4
(6.1, 6.6)	$k=1$	11	5	1	10.2	5.5	1.3
	2	5	23	5	3.1	20.2	9.7
	3	2	8	13	0.4	5.2	17.4
(6.6, 7.1)	$k=1$	9	4	0	7.8	4.2	1.0
	2	2	17	10	2.7	17.8	8.5
	3	0	3	11	0.2	3.2	10.6
(7.1, 7.6)	$k=1$	4	3	1	4.8	2.6	0.6
	2	1	15	6	2.0	13.5	6.5
	3	0	4	13	0.3	3.9	12.9
(7.6, 8.1)	$k=1$	3	2	0	3.0	1.6	0.4
	2	3	14	5	2.0	13.5	6.5
	3	1	3	15	0.3	4.3	14.4

* a_0, a_1, a_2, ... equal 0, 0.7, 1.2, ... in years.

Formal hypothesis tests for a treatment difference can be based on the multistate models, for example, using regression models $P_3(t|x)$ described in Sections 4.1 and 4.2. An alternative is to base a test on the time T to first entry to state 3 in model M1B or M2B. A third and perhaps preferable approach is to base a test on the empirical prevalence estimates $\tilde{P}_3(t|x)$ at some chosen time t; robust variance estimates are readily obtained. In particular, for either the IT ($x = 1$) or CT ($x = 0$) groups the empirical estimate at visit time t (where $t = 0.7, 1.2, 1.7, 2.2$ years, etc.) is $\tilde{P}_3(t) = Y_{.3}/n(t)$, where $n(t)$ is the number of subjects observed at time t and $Y_{.3}(t)$ is the number who are in state 3. The binomial variance estimate $\tilde{\text{var}}(\tilde{P}_3(t)) = \tilde{P}_3(t)(1 - \tilde{P}_3(t))/n(t)$ then applies, and we can test H_0: $P_3(t|x=0) = P_3(t|x=1)$ at a given time t. Tests assuming approximate normality of $\sqrt{n(t)}(\tilde{P}_3(t|x) - P_3(t|x))$ give p-values under 0.005 for times $t = 4.2$ to 6.2, and under 0.06 for all $t \geq 3.2$ years.

5.3   Nonparametric Estimation of State Occupancy Probabilities

Pure nonparametric estimation of cumulative transition intensity functions is intractable for most multistate models when processes are under intermittent observation. Two exceptions are competing risk models and illness-death models. However, the estimates are typically undefined over certain time intervals and are of limited usefulness. A more attractive approach in most cases is to use flexible parametric models to estimate intensities and cumulative intensity functions. Models with piecewise-constant intensity functions described in Section 3.1.1 can be fitted using the msm package and Joly et al. (2002) discuss how to fit flexible parametric models $\lambda_{kl}(t; \theta_{kl})$ using penalized likelihood. This approach can be implemented using the SmoothHazard package (Touraine et al., 2017). Both msm and SmoothHazard also handle regression models of the multiplicative intensity form.

Nonparametric estimation of process features such as state occupancy probabilities $P_k(t)$ and entry time distributions $F_k(t)$ is more feasible for states that can be entered just once. To estimate a distribution function $F_k(t)$, we can use the Turnbull (1976) estimate or other nonparametric estimates for interval-censored failure times, provided that interval-censored observation of the state entry time T_k is possible for all individuals. This is the case for progressive models and competing risk models, for example, but not for state 2 in the illness-death model in Figure 1.6, since for individuals observed to die following an assessment in state 1, it will not be known whether their path was through state 2. For competing risks models, Hudgens et al. (2001) showed that Turnbull estimator can be used for each of the cumulative incidence functions $F_k(t) = P_{0k}(t)$ in (4.7). Nonparametric estimates of distribution functions $F(t)$ from interval-censored data are provided by several R packages, including Icens, MLEcens, interval and gte.

For the illness-death model (1.6), we can readily estimate $F_3(t) = P(T_3 \leq t)$ and $F_1^*(t)$, the exit time distribution for state 1, which can in turn be used to estimate the state occupancy probability $P(Z(t) = 2|Z(0) = 1)$ as $\hat{P}_2(t) = \hat{F}_1^*(t) - \hat{F}_3(t)$. Frydman et al. (2013) consider the special case where it can be ascertained whether entry to state 3 was from state 1 or state 2, and give an estimator for $F_2(t) = P(T_2 \leq t)$, the (sub) distribution for time of entry to state 2. Estimates of distribution functions and occupancy probabilities are often undefined over certain time periods, and a simple adjustment is to use linear interpretation over such intervals. Confidence interval estimation is only feasible when the visit time processes are discrete, in which case nonparametric bootstrap sampling can be used as in Hudgens et al. (2001) and Frydman et al. (2013).

Another approach to estimating occupancy probabilities is to extend the method of Section 4.2.2 to the case of intermittent observation by using smoothing. The estimating function

$$U_k(t) = \sum_{i=1}^{n} \sum_{j=1}^{m_i} \frac{1}{b} \text{Ker}\left(\frac{a_{ij} - t}{b}\right) (Y_{ik}(a_{ij}) - P_k(t)), \qquad (5.14)$$

achieves this, where $\text{Ker}(x)$ is a kernel function, and $b > 0$ is a specified bandwidth. The rectangular kernel $\text{Ker}(s) = 0.5$, $-1 \leq s \leq 1$ is a simple choice though not necessarily optimal. The bandwidth is best chosen to be large enough to give at least $50-100$ visits within the window $(t - b, t + b)$ or else the estimate,

$$\tilde{P}_k(t) = \frac{\sum_{i=1}^{n} \sum_{j=1}^{m_i} \text{Ker}((a_{ij} - t)/b)\, Y_{ik}(a_{ij})}{\sum_{i=1}^{n} \sum_{j=1}^{m_i} \text{Ker}((a_{ij} - t)/b)}, \qquad (5.15)$$

obtained from setting $U_k(t) = 0$, will not be very precise. The estimates (5.15) are biased (see Problem 5.6), and the challenge is to choose b small enough to minimize bias but large enough to provide a precise estimate. In (5.15) the bandwidth b is shown as fixed but it could vary with t according to the frequency of visits at different times. Variance estimates and confidence intervals can be obtained via the bootstrap.

The msm package provides alternative estimates of $P_k(t)$ with its prevalence.msm function by imputing the state $Z_i(t)$ at a given time t for each individual that is still under follow-up, but not observed at time t. This is done by assigning as the imputed state $\tilde{Z}_i(t)$ the state occupied at the largest observation time a_{ij} that is $\leq t$; if C_i is a censoring time and $C_i < t$, then no such imputation is carried out for individual i. This usually gives less precise and, in many cases, more biased estimates than (5.15).

Finally, we remark that some occupancy probability functions $P_k(t)$ are monotonic, but that estimates such as $\tilde{P}_k(t)$ may not be. In that case it is simplest to restrict consideration to a fixed set of times $s_1, \ldots, s_R$ and then to monotonize $\tilde{P}_k(s_1), \ldots, \tilde{P}_k(s_R)$ if necessary by applying isotonic regression, as in Datta and Sundaram (2006) and Nazeri Rad and Lawless (2017).

An illustration of the estimators is deferred to the next section. It should be stressed that the estimators discussed here require that observation times be completely independent of the multistate process, a condition stronger than the CIVP condition needed for the likelihood analyses based on full specification of the multistate process in Section 5.1. In some models there are absorbing states that preclude further visits. This violates the independence condition. We discuss this complication in the next section, where we consider how to deal with state-dependent observation processes for marginal analyses.

5.4   Process-Dependent Observation Times

5.4.1   Further Remarks on Dependent Visit Processes

In Section 5.1 we introduced the concepts of conditionally independent and conditionally dependent visit processes for individuals in a clinical cohort or disease registry. We formalize these concepts here and provide a more detailed discussion of the issues. We begin by considering the setting where there are no covariates and interest lies only in modeling the multistate process under intermittent inspection. Let C^A denote an administrative censoring time and C^R a random time at which an individual may withdraw from the cohort. Then $C = \min(C^R, C^A)$ is the net censoring time and $Y(s) = I(s \leq C)$ indicates a person is on study at time s. We define $C(s) = I(C \leq s)$ and let $\{C(s), 0 < s\}$ denote the counting process for censoring.

As discussed in Section 5.1, the process by which individuals attend clinics and furnish information on their condition is often complex and influenced by an underlying pre-determined clinic schedule, appointments scheduled based on assessments made at prior clinic visits, and unplanned symptom-driven appointments. To characterize the visit process, we let $A(s)$ be a right-continuous process counting the number of assessments over $[0, s]$ and let $dA(s) = A(s) - A(s^-) = 1$ if an assessment is made at time s and $dA(s) = 0$ otherwise. We may then think in terms of random visit times $A_0 < A_1 < \cdots$, or equivalently in terms of the counting process $\{A(s), s \geq 0\}$ defining these times. Since visits can only be made for individuals still on study, withdrawal at C terminates the visit process. Thus, for $Z(s)$ to be observed, the individual must not have withdrawn from the study before s and must have a visit at time s.

To expand upon this, we consider a joint model for the inspection and multistate processes. We assume the random withdrawal time is conditionally independent of the response and visit process given the observed history. For ease of discussion and graphical representation, we consider a K-state process of interest in which transitions are only possible between adjacent states (i.e. $k \to k-1$ or $k \to k+1$ for $k = 2, \ldots, K-1$, with $1 \to 2$ and $K \to K-1$ also possible), but more general multistate processes can be handled in a similar fashion. Figure 5.4 contains a key portion of the state space diagram for the joint process where the state occupied for the process of interest is reflected by the column; the rows are defined by the cumulative number of assessments made after the baseline assessment at a_0 such that a downward transition is made upon the occurrence of a visit.

Let
$$\bar{\mathcal{H}}(t) = \{Y(u), A(u), Z(u), 0 \leq u \leq t\}$$

denote the complete history of all processes including the censoring process, the assessment process, and the multistate process over $[0, t]$. Of course, the multistate process is only under intermittent observation and so the multistate history $\mathcal{Z}(t) = \{Z(u), 0 \leq u \leq t\}$ is not fully observed. If $0 \leq a_0 < a_1 < \cdots < a_{A(s)}$ denote the realized assessment times over $[0, s]$, then at time t the *observed data history* is

$$\bar{\mathcal{H}}^\circ(t) = \{Y(u), A(u), 0 \leq u \leq t; (a_j, Z(a_j)), j = 1, \ldots, A(t)\}$$

and the observed history of the multistate process alone is $\mathcal{Z}^\circ(t) = \{(a_j, Z(a_j)), j =$

$0, 1, \ldots, A(t)\}$. The conditionally independent visit process condition has two parts, which we now discuss.

CIVP Condition 1: For a CIVP (under conditionally independent censoring), the intensity of the visit process must satisfy

$$\lim_{\Delta t \downarrow 0} \frac{P(\Delta A(t) = 1 \mid \bar{\mathcal{H}}(t^-))}{\Delta t} = \lim_{\Delta t \downarrow 0} \frac{P(\Delta A(t) = 1 \mid \bar{\mathcal{H}}^\circ(t^-))}{\Delta t} = Y(t) \lambda^a(t \mid \bar{\mathcal{H}}^\circ(t^-))$$

where $\Delta A(t) = A(t + \Delta t^-) - A(t^-)$.

CIVP Condition 1 implies that given the history of the visit process and $\mathcal{Z}^\circ(t^-)$, the visit intensity at time t does not depend further on $\mathcal{Z}(t^-)$; the visit process intensity can depend on the times of previous visits as well as the states occupied at these visits. The possible dependence on previously observed states accommodates situations in which future appointments may be based on an individual's state of health at a current assessment, or an observed trend in their health status over the recent past. The aspect of most practical importance here is that for a CIVP the visit intensity at time t cannot depend on the process of interest over $s > a_{A(t^-)}$. This would be violated if an individual sought medical attention because of, for example, declining health following their last visit at $a_{A(t^-)}$. As discussed in Section 5.1, this would often be the case in specialty clinics for individuals with chronic diseases. The model in Figure 5.4 can be specified so as to allow conditionally dependent visit processes, though they would not be estimable from the observed data considered here. We discuss this in Section 7.2.5. Finally, we note that the visit intensity is zero after the right censoring time C by definition.

Consider the construction of a likelihood function for an individual with m visits realized at $a_0 < a_1 < \cdots < a_m \geq C = \min(C^R, C^A)$. The partial likelihood based on the observed visit and multistate data is given by

$$\prod_{j=1}^m \left[\lambda^a(a_j \mid \bar{\mathcal{H}}^\circ(a_j^-)) \exp\left(-\int_{a_{j-1}}^{a_j} \lambda^a(u \mid \bar{\mathcal{H}}^\circ(u^-))\, du \right) P(Z(a_j) \mid a_j, dA(a_j) = 1, \bar{\mathcal{H}}^\circ(a_j^-)) \right].$$

It is apparent from (5.16) that if it were of interest, the conditionally independent visit process could be modeled and its parameters estimated based on observed data. If the visit process is noninformative, however (i.e. there are no shared parameters in the visit and multistate models), we may omit the terms involving $\lambda^a(u|\bar{\mathcal{H}}^\circ(u^-))$ and focus on the partial likelihood

$$\prod_{j=1}^m P(Z(a_j) \mid a_j, dA(a_j) = 1, \bar{\mathcal{H}}^\circ(a_j^-)). \tag{5.16}$$

To write these contributions in terms of the multistate model of interest, we require a further, more subtle, condition.

CIVP Condition 2: The conditional probabilities $P(Z(a_j)|a_j, dA(a_j) = 1, \bar{\mathcal{H}}°(a_j^-))$ satisfy

$$P(Z(a_j) \mid a_j, dA(a_j) = 1, \bar{\mathcal{H}}°(a_j^-)) = P(Z(a_j) \mid a_j, \mathcal{Z}°(a_j^-)), \qquad (5.17)$$

where on the right-hand side the probability is computed as if the visit times a_j, $j = 0, 1, 2, \ldots$ were fixed in advance.

CIVP Condition 2 ensures that the intensities governing the multistate process of interest are the same whether the process is under the particular observation scheme or not. Farewell et al. (2017) similarly make the distinction between realized visit times from a random visit process and fixed pre-specified visit times and refer to (5.17) as a "stability" condition. Under the additional assumption (5.17), we can write the partial likelihood (5.16) involving the parameters of the multistate model as

$$\prod_{j=1}^{m} P(Z(a_j) \mid a_j, \mathcal{Z}°(a_j^-)).$$

We note again that the CIVP assumptions cannot be checked solely using the observed data considered here. We discuss this further in Section 7.2.5.

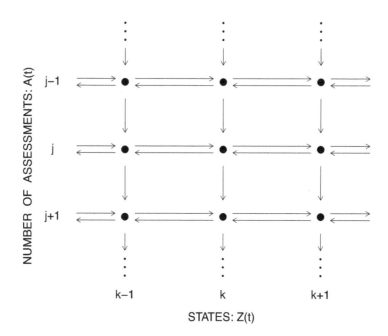

Figure 5.4: A state space diagram for joint consideration of a K-state process and an inspection process.

5.4.2   Marginal Features and Inverse Intensity of Visit Weighting

Under the CIVP condition, $Z_i(t)$ and $dA_i(t)$ are conditionally independent given the observed history $\bar{\mathcal{H}}_i°(t^-)$ of states, covariates and visits up to time t. For analyses based on full transition intensity functions and the likelihood function (5.4), we showed that inferences are valid under the CIVP condition. However, when estimating features such as $P_k(t)$ or $F_k(t)$ by the marginal methods discussed in Section 5.3, the same observation process may not be ignorable because $Z_i(t)$ and $dA_i(t)$ are neither marginally independent nor independent given only a fixed covariate X. Estimators based on functions such as (5.14) have undesirable properties in this setting. We consider this first for parametric estimation.

To be specific, consider a parametric model $P_k(t;\theta)$ for the occupancy probabilities of a specified state k; this is in the spirit of Sections 4.1 and 4.2 and does not necessarily correspond to a parametric transition intensity model. A marginal estimating function analogous to (5.14), but without the kernel function, may be specified with individual components

$$U_i(\theta) = \int_0^\infty Y_i(t)\left\{Y_{ik}(t) - P_k(t;\theta)\right\} g_i(t;\theta)\, dA_i(t), \tag{5.18}$$

where $Y_i(t) = I(t \leq C_i)$ and $Y_{ik}(t) = I(Z_i(t) = k)$ as before, and $g_i(t;\theta)$ is a specified vector of the same dimension as θ. Suppose for simplicity that given X, C_i is independent of the multistate and visit processes. For $E\{U_i(\theta)\}$ to equal 0, we require $E\{Y_{ik}(t)|dA_i(t) = 1\} = P_k(t;\theta)$, which is only satisfied if the visit process is completely independent of the multistate process. Under the CIVP assumption it will not be true, but a weighting adjustment to (5.18) can be made by modeling the visit process. We denote the visit process intensity by $\lambda_i^a(t|\bar{\mathcal{H}}_i^\circ(t^-))$ under the CIVP assumption. The estimating function

$$U_i^{\mathrm{w}}(\theta) = \int_0^\infty \lambda_i^a(t \mid \bar{\mathcal{H}}_i^\circ(t^-))^{-1} Y_i(t)\left\{Y_{ik}(t) - P_k(t;\theta)\right\} g_i(t;\theta)\, dA_i(t) \tag{5.19}$$

can be seen to be unbiased by taking the expectation of the integrand first with respect to $dA_i(t)$ given $(Y_i(t), Y_{ik}(t), \bar{\mathcal{H}}_i^\circ(t^-))$, and then with respect to $(Y_i(t), Y_{ik}(t), \bar{\mathcal{H}}_i^\circ(t^-))$. Of course, we do not know $\lambda_i^a(t|\bar{\mathcal{H}}_i^\circ(t^-))$ in general, and must assume that it can be estimated consistently through a family of models $\lambda_i^a(t|v_i(t))$, where $v_i(t)$ is a vector containing information in the observed history $\bar{\mathcal{H}}_i^\circ(t^-)$. The estimating function we use is then

$$U^{\mathrm{w}}(\theta) = \sum_{i=1}^n \sum_{j=1}^{m_i} \widehat{\lambda}_i^a(a_{ij} \mid v_i(a_{ij}))^{-1}\left\{Y_{ik}(a_{ij}) - P_k(a_{ij};\theta)\right\} g_i(a_{ij};\theta). \tag{5.20}$$

Estimating functions (5.19) and (5.20) are termed *inverse intensity of visit weighted (IIVW) estimating functions* (Lin et al., 2004). We discuss estimation of the visit process intensity in the next section.

The vector $g_i(t;\theta)$ in (5.20) must be specified. Since $Y_{ik}(t)$ is Binomial$(1, P_k(t;\theta))$ assuming the model is correct, we will use the vector

$$g_i(t;\theta) = \frac{\partial P_k(t;\theta)/\partial\theta}{P_k(t;\theta)(1 - P_k(t;\theta))}, \tag{5.21}$$

based on the form of the maximum likelihood score for binomial data. Fitting of regression models $P_k(t|x;\theta)$ for fixed X can proceed in the same way; see Section

4.1.4. It can be advantageous in that case to multiply the weights in (5.20) by a function of $X = x$, designed to render the weights less variable.

A further complication arises for processes where entry to a set $\mathcal{A}$ of absorbing states terminates an individual's follow-up. In this case the censoring time C_i is not independent of the multistate process. Moreover, inverse probability of censoring weights cannot be applied since $P(Y_i(t) = 1 | \bar{\mathcal{H}}_i^\circ(t^-)) = 0$ when $\bar{\mathcal{H}}_i^\circ(t^-)$ includes the information that $Z_i(t^-) \in \mathcal{A}$; this violates the positivity condition for censoring probabilities. In this case we assume, as described in Section 2.2.2, the random loss to follow-up cannot occur after absorption and the process is effectively observed continuously up to time C^A. We deal with this by thinking of the assessment process in discrete time and using the fact that a process cannot exit an absorbing state.

We first assume that visits may only occur at specified discrete times $0 = s_0 < s_1 < \cdots < s_R$ in this setting. We can often consider these discrete times of potential assessments as days or weeks, for example. We then define $dA_i^\dagger(s_r) = dA_i(s_r)$ when the state that was occupied at the most recent assessment prior to s_r is not in $\mathcal{A}$, and let $dA_i^\dagger(s_r) = 1$ otherwise. Thus, $\{A_i^\dagger(t), t > 0\}$ is a discrete-time counting process that counts the number of assessments made up to and including the assessment when the process was first observed to be in an absorbing state, and subsequently increments by one at each potential assessment time. Based on (5.19), we consider

$$U_i^{\mathrm{w}}(\theta) = \sum_{r=1}^R \lambda_i^a(s_r)^{-1} Y_i(s_r) \{Y_{ik}(s_r) - P_k(s_r; \theta)\} g_i(s_r; \theta) \, dA_i^\dagger(s_r), \qquad (5.22)$$

where

$$\lambda_i^a(s_r) = P(dA_i^\dagger(s_r) = 1 \mid \bar{\mathcal{H}}_i^\circ(s_r^-)),$$

is a probability and $Y_i(s_r) = I(s_r \leq C_i^\dagger)$, with $C_i^\dagger$ denoting C_i^R or a random censoring time observed prior to entry to $\mathcal{A}$. If the last observed state prior to s_r is not an absorbing state, then $\lambda_i^a(s_r) = \lambda_i^a(s_r | v_i(s_r))$ is the actual visit intensity (probability). If the process was observed to be in an absorbing state in $\mathcal{A}$ at some assessment before s_r, $dA_i^\dagger(s_r) = 1$ as noted above, and we specify $\lambda_i^a(s_r) = 1$ accordingly. In this case s_r represents a pseudo-visit and reflects the fact that we know $Z_i(s_r) = k$ for some $k \in \mathcal{A}$. With this adjustment to (5.19), we still have $E\{U_i^{\mathrm{w}}(\theta)\} = 0$, leading to consistent estimation of θ.

We can often use days as potential discrete visit times but if individuals' visits can be very far apart (e.g. 180 or 360 days), it may be better to use a coarser discretization in order to avoid small and highly variable $\lambda_i^a(s_r)$ values while avoiding the occurrence of more than one visit for an individual in a time interval. If visit times are actually continuous, then we define $Y_{ik}(s_r)$ as $Y_{ik}(a_{ij})$, where a_{ij} is the closest visit time to s_r. The effects of the trade-off between less variable weights and more accurate assignment of values $Y_{ik}(s_r)$ depends on the transition intensities in the multistate process. We consider this further in Sections 5.4.4 and 5.4.5.

5.4.3 Estimation of Visit Process Intensities

Visit processes can be modeled as recurrent event processes, and survival analysis software can be used to fit many models (Cook and Lawless, 2007). Pullenayegum and Lim (2016) discuss visit process modeling with an emphasis on multiplicative Markov models with or without random effects. Cox models with intensities $\lambda^a(t|v(t)) = \lambda_0^a(t) \exp(v'(t)\gamma)$ are often used, and Buzkova and Lumley (2008, 2009) have noted that the baseline intensity function $\lambda_0^a(t)$ can then be dropped from estimating functions (5.22). In many studies, the times between visits cluster around specific values in which case the visit intensity depends more strongly on the elapsed time since the last visit. A modulated renewal or semi-Markov process is then useful; a Cox model of this form has intensity $\lambda_i^a(t|\bar{\mathcal{H}}_i^\circ(t^-)) = \lambda_0^a(B_i(t)) \exp(v_i'(t)\gamma)$, where $B_i(t) = t - t_{A_i(t^-)}$. In this case, the baseline intensity function cannot be dropped from (5.22). An alternative approach is to retain the Markov model and to include functions of $B_i(t)$ in $V_i(t)$, but this would require a parametric specification of dependence on $B_i(t)$.

In the preceding section we introduced discrete-time visit processes in order to deal with situations where entry to an absorbing state terminates the follow-up process. Discrete time models for $P(dA_i(s_r) = 1|v_i(s_r))$ can be based on logistic regression or other binary response models. When the actual visit times a_{ij} are continuous, an alternative is to fit a continuous-time process and then to define $\lambda_i^a(s_r)$ based on this. For example, if $\bar{\mathcal{H}}_i^\circ(s_{r-1})$ includes the information that the last visit time was $a_{i,j-1}$ and the last state was non-absorbing, then we can use

$$
\begin{aligned}
&\lambda_i^a(s_r \mid \bar{\mathcal{H}}_i^\circ(s_{r-1})) \\
&= P(s_r - 0.5\Delta < a_{ij} \leq s_r + 0.5\Delta \mid \bar{\mathcal{H}}_i^\circ(s_{r-1})) \\
&= 1 - \exp\left\{ -\int_{s_r - 0.5\Delta}^{s_r + 0.5\Delta} \lambda_i^a(t \mid \bar{\mathcal{H}}_i^\circ(s_{r-1})) \, dt \right\} \\
&= 1 - \exp\left\{ -\left[\Lambda_0^a(s_r + 0.5\Delta - a_{i,j-1}) - \Lambda_0^a(s_r - 0.5\Delta - a_{i,j-1})\right] e^{v_i'(s_{r-1})\gamma} \right\}
\end{aligned}
$$

for a semi-Markov Cox model for the visit process.

5.4.4 Nonparametric Estimation of Occupancy Probabilities

Nonparametric estimation of an occupancy probability $P_k(t)$ along the lines of Section 5.4.2 is possible, but smoothing or grouping is needed when visit times are very irregular. We consider a discrete-time visit process as in the previous section and for convenience let $w_i(s_r) = Y_i(s_r) dA_i^\dagger(s_r)/\lambda_i^a(s_r)$. We assume interest lies in a particular state k, and let $\theta_r = P_k(s_r)$, $r = 1, \ldots, R$. Treating $\theta = (\theta_1, \ldots, \theta_R)'$ as the parameter vector in (5.22), we can take $g_i(s_l; \theta) = I(s_l = s_r)$ for a given r, in accordance with (5.21); this gives R different estimating functions

$$
U^{\mathrm{w}}(\theta_r) = \sum_{i=1}^n w_i(s_r) \{Y_{ik}(s_r) - \theta_r\}, \quad r = 1, \ldots, R, \tag{5.23}
$$

and setting $U^{\mathrm{w}}(\theta_r) = 0$ gives an estimate

$$
\tilde{\theta}_r = \tilde{P}_k(s_r) = \frac{\sum_{i=1}^n w_i(s_r) Y_{ik}(s_r)}{\sum_{i=1}^n w_i(s_r)}.
$$

As in Section 5.4.2, when visits occur in continuous time, we take $Y_{ik}(s_r) = 1$ only if the closest visit time a_{ij} to s_r has $Y_{ik}(a_{ij}) = 1$. The R function `prevalence.msm` in the `msm` package gives estimates for a similar approach basing $Y_{ik}(s_r)$ on $Y_{ik}(a_{ij})$ for the largest visit time $a_{ij} \le s_r$, and without the IIV weights $\lambda_i^a(s_r)^{-1}$.

For nonparametric estimation, there are conflicting considerations in the choice of R and $s_1, \dots, s_R$. Unless R is small enough (or $\Delta = s_r - s_{r-1}$ large enough) so that there are at least $50 - 100$ visits in $(s_{r-1}, s_r]$, the estimate $\tilde{\theta}_r$ is too imprecise to be of much use. However, when Δ is large there are more errors in assuming that $Y_{ik}(a_{ij}) = Y_{ik}(s_r)$ even for a_{ij} in $(s_r - 0.5\Delta, s_r + 0.5\Delta]$. A second issue arises for monotonic $P_k(s_r)$; the estimates $\tilde{\theta}_r = \tilde{P}_k(s_r)$ may not be monotonic. Estimates based on (5.23) can be adjusted to be monotonic, where necessary, by subjecting them to isotonic regression (e.g. Sun, 2006, p. 210); this is easily done using software such as the R function `isoreg`.

Nazeri Rad and Lawless (2017) propose an alternative method that uses smoothing. In settings where follow-up is not terminated when an individual enters an absorbing state, then kernel-weighted occupancy probability estimates take the form

$$\tilde{P}_k(s_r) = \frac{\sum_{i=1}^n \sum_{j=1}^{m_i} w_i(a_{ij}) \operatorname{Ker}((s_r - a_{ij})/b) \, Y_{ik}(a_{ij})}{\sum_{i=1}^n \sum_{j=1}^{m_i} w_i(a_{ij}) \operatorname{Ker}((s_r - a_{ij})/b)}, \tag{5.24}$$

where $w_i(a_{ij}) = \hat{\lambda}_i^a(a_{ij})^{-1}$, $\operatorname{Ker}(x)$ is a kernel function defined to be zero outside $[-1, 1]$ and $b > 0$ is a specified bandwidth. If $\operatorname{Ker}(x) = 0.5 \ (-1 \le x \le 1)$ and $b = 0.5\Delta$, then (5.24) is usually similar (but not identical) to the estimate obtained from (5.23) with $w_i(s_r)$ and $Y_{ik}(s_r)$ imputed using the nearest a_{ij}. Other kernel functions such as Gaussian or triangular densities give higher weight to observation times a_{ij} closer to s_r than those further away.

In cases where follow-up is terminated upon observed entry to a set of absorbing states, we make the alteration described in Section 5.4.2 and replace (5.24) with

$$\tilde{P}_k(s_r) = \frac{\sum_{i=1}^n \sum_{l=1}^R w_i(s_l) \operatorname{Ker}((s_r - s_l)/b) \, Y_{ik}(s_l)}{\sum_{i=1}^n \sum_{l=1}^R w_i(s_l) \operatorname{Ker}((s_r - s_l)/b)}, \tag{5.25}$$

where $w_i(s_l) = Y_i(s_l) \, dA_i^\dagger(s_l)/\lambda_i^a(s_l)$, with $\lambda_i^a(s_l) = dA_i^\dagger(s_l) = 1$ whenever entry to $\mathcal{A}$ has been observed prior to s_l. In this case there may be an advantage to using large R and taking b large enough so that there are at least $50-100$ visit times in $(s_l - b, s_l + b]$.

If $P_k(t)$ is monotonic, estimates that are monotonic can again be obtained by applying isotonic regression to the $\tilde{P}_k(s_r)$. Estimates $\tilde{P}_k(s_r)$ obtained via (5.24) or (5.25) sum to one over $k = 1, \dots, K$, but estimates with monotonic adjustment will not in general sum to one. One option to address this is to estimate $P_k(t)$ for $K - 1$ states $k \ne k_0$ only, and to let $\tilde{P}_{k_0}(t) = 1 - \sum_{k \ne k_0} \tilde{P}_k(t)$; it is best to choose state k_0 to be one for which $P_{k_0}(t)$ is non-monotonic.

Variance estimates for the $\tilde{P}_k(s_r)$ discussed here can be based on nonparametric bootstrap resampling. Asymptotic theory is difficult and would require continuous visit intensities that were bounded away from zero and a bandwidth sequence b_n that approached zero at a satisfactory rate as n increased. Limited simulation results (Nazeri Rad and Lawless, 2017) suggest that bootstrap variance estimation and confidence interval estimation based on normal approximation is adequate provided samples are sufficiently large. Visits must also be sufficiently numerous for nonparametric estimation to be effective. In many settings, parametric estimation of occupancy functions, based on flexible models such as splines, is preferable to nonparametric estimation.

5.4.5 Progression to Arthritis Mutilans

Section 1.2.3 introduced the University of Toronto Psoriatic Arthritis (PsA) Cohort, in which over 1200 persons with PsA have been followed for up to 40 years. The main indicators of disease progression include the number and severity of damaged joints, among the 64 joints in the body. We will consider here the number of severely damaged joints, defined as joints with grade 4 damage on the modified Steinbrocker scale (van der Heijde et al., 2005). When this number reaches five, a person is defined as having mutilans arthritis, and in order to model progression to this state we consider the multistate diagram with states 0 to 5 shown in Figure 5.5.

There is an interest in identifying and characterizing the effect of factors associated with progression, and we will consider several including two human leukocyte antigen (HLA) biomarkers (HLA-B27, HLA-C03) that have been found associated with disease progression in other studies. In addition, we will consider a person's age at onset of PsA, their sex, and three types of treatment: non-steroidal anti-inflammatory drugs (NSAIDs), disease-modifying anti-rheumatic drugs (DMARDs) and biologic therapies. The treatment covariates are time dependent and indicate whether the corresponding class of drugs has been prescribed as of time t since disease onset.

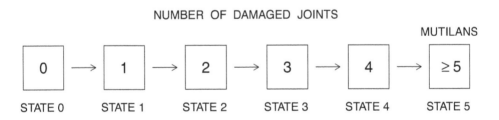

Figure 5.5: A progressive model for the number of severely damaged joints in the development of mutilans arthritis.

Joint damage was assessed at intermittent clinic visits, where x-rays are taken. Among the 1220 persons in the cohort, 937 had complete HLA measurements. In addition, we require that patients have known dates of birth and onset of PsA, that they have at least one x-ray assessment, and that the dates and results of the assessments (giving the number of severely damaged joints) be known. We base the following analyses on the 613 patients satisfying these criteria. The time origin is taken as the date of PsA onset. This date was typically before the date a person first attended the clinic and enrolled in the cohort, in about 20% of patients the gap between PsA onset and enrollment was over 10 years. The gaps between visits at which x-rays were taken are nominally about 2 years, but in fact these varied substantially both within and between patients, with some gaps of less than a year and some over 5 years. There is a concern that gaps between visits may be related to disease history, and we address this below.

We consider a main analysis based on the multistate model in Figure 5.5. In addition, we will estimate the distribution of the time T_5 to entry of the mutilans state 5 as a calibration check on the multistate model. The multistate analysis will not be affected if the time to the next visit depends on the state occupied at the current visit, but a simple failure time analysis of the time T_5 that does not condition on the observed history would give biased estimates. We remark that death is a competing risk. Here we simply treat it as a type of independent censoring, which leaves us with consistent estimates of transition intensities if the mortality intensity is the same regardless of which state k ($k = 0, 1, \ldots, 5$) a person is in. In this case, however, the proper interpretation of $\widehat{P}_k(t)$ is as an estimate of $P(Z(t) = k | Z(t) \in \{0, 1, \ldots, 5\})$. Lengths of follow-up vary widely according to when an individual's PsA onset occurred, and of the 613 patients, only 105 entered state 5 by the time of the last assessment. Analyses are conducted under the CIVP assumptions. We note that the initial x-ray visit times a_{i0} vary widely across individuals, and in some cases are 20 years or more after onset of PsA. It is a concern that the CIVP assumption may not hold for the a_{i0}, but it is impossible to check this adequately with the available data. Section 7.2.5 contains some discussion of non-CIVP models.

We initially fit multistate Markov regression models with multiplicative intensities as in (3.22) with $g(x; \beta) = \exp(x'\beta)$ and time-homogeneous baseline transition intensities. The covariates were age at PsA onset (in years), sex (female = 1, male = 0), binary HLA-B27 and HLA-C03 variables (positive = 1, negative = 0), and time-varying treatment variables indicating whether NSAIDs, for example, had ever been prescribed before time t. Analogous time-dependent indicators were defined for DMARDs and biologic therapy. We assume that regression coefficients $\beta_{k,k+1}$ are common for $k = 0, 1, \ldots, 4$. Model checks indicated that a time-homogeneous model is inadequate, and we were led to a model with piecewise-constant baseline intensities with separate constant intensities for the four time periods $(0, 6]$, $(6, 12]$, $(12, 18]$ and $(18, \infty)$; time is in years. Table 5.4 displays the estimates from the fitted model.

We observe that older age of onset is mildly associated with faster progression, and being positive for HLA-B27 is strongly associated with more rapid progression of severe joint damage. Among the treatments, taking biologics is associated with somewhat slower progression. We note also that the piecewise-constant baseline intensities decrease with time since onset. This may be due to patient heterogeneity related to unmeasured factors. As a crude check on this, we fitted mover-stayer models, described in Section 6.3, without covariates and compared them with four-piece Markov models without covariates. Because of the large times between x-ray visits, it was not possible to estimate the proportion of persons who are "movers" precisely, with proportions from about 0.6 up to 1.0 being highly plausible. Thus, unobserved heterogeneity in the form of persons who are either susceptible or non-susceptible to severe joint damage cannot be ruled out, but at the same time it is not needed to describe the patterns seen in the observed data.

Table 5.4: Estimates from the fitted 6-state Markov model with multiplicative intensities (Figure 5.5) based on data from the University of Toronto PsA Cohort, with piecewise-constant baseline intensities and covariate effects constant across transition types.

Covariate		EST	SE	RR	95% CI	p
Sex: Female vs. Male		0.157	0.088	1.17	(0.98, 1.39)	0.076
Age at Onset of PsA (per year)		0.007	0.004	1.01	(1.00, 1.01)	0.049
HLA B27: Yes vs. No		0.351	0.102	1.42	(1.16, 1.73)	< 0.001
HLA C3: Yes vs. No		-0.182	0.124	0.83	(0.65, 1.06)	0.143
Ever Received DMARDs: Yes vs. No		-0.123	0.125	0.88	(0.69, 1.13)	0.325
Ever Received NSAIDs: Yes vs. No		-0.090	0.124	0.91	(0.72, 1.17)	0.467
Ever Received Biologics: Yes vs. No		-0.375	0.175	0.69	(0.49, 0.97)	0.032

Baseline Transition Intensities	Interval	EST†	SE	$\widehat{\lambda}_{kl}(t)$	95% CI
$\log \lambda_{01}(t)$	$(0,6]$	-3.814	0.184	0.02	(0.02, 0.03)
	$(6,12]$	-3.967	0.230	0.02	(0.01, 0.03)
	$(12,18]$	-4.807	0.384	0.01	(< 0.01, 0.02)
	$(18,\infty)$	-4.319	0.251	0.01	(0.01, 0.02)
$\log \lambda_{12}(t)$	$(0,6]$	-0.895	0.253	0.41	(0.25, 0.67)
	$(6,12]$	-1.495	0.250	0.22	(0.14, 0.37)
	$(12,18]$	-1.947	0.322	0.14	(0.08, 0.27)
	$(18,\infty)$	-1.946	0.290	0.14	(0.08, 0.25)
$\log \lambda_{23}(t)$	$(0,6]$	-0.474	0.292	0.62	(0.35, 1.10)
	$(6,12]$	-1.461	0.267	0.23	(0.14, 0.39)
	$(12,18]$	-1.746	0.318	0.17	(0.09, 0.33)
	$(18,\infty)$	-1.606	0.295	0.20	(0.11, 0.36)
$\log \lambda_{34}(t)$	$(0,6]$	-0.135	0.347	0.87	(0.44, 1.72)
	$(6,12]$	-0.716	0.284	0.49	(0.28, 0.85)
	$(12,18]$	-1.435	0.349	0.24	(0.12, 0.47)
	$(18,\infty)$	-1.449	0.285	0.23	(0.13, 0.41)
$\log \lambda_{45}(t)$	$(0,6]$	-0.423	0.403	0.65	(0.30, 1.44)
	$(6,12]$	-0.746	0.292	0.47	(0.27, 0.84)
	$(12,18]$	-1.566	0.367	0.21	(0.10, 0.43)
	$(18,\infty)$	-1.780	0.271	0.17	(0.10, 0.29)

† Estimates in this column are for $\log \lambda_{kl}(t)$ and the adjacent column gives the corresponding standard error.

We now consider estimation of the entry time distribution for state 5, $F(t) = P(T_5 \leq t)$, using a failure time model. Comparison of such estimates with those from the multistate model provide a check on the latter. However, because times between x-ray visits may be disease related, the visit times may be non-ignorable for the simple failure time analysis. We therefore first fit models for the visit process; Table 5.5 shows the results for separate Cox models for (a) the time (from PsA onset) to the first visit a_{i0}, and (b) the times $\Delta a_{ij} = a_{ij} - a_{i,j-1}$, $j = 1, 2, \ldots$ between successive visits. We see that for (a), a person's sex and age at PsA onset are related to a_{i0}, and for (b) that age at onset and, to a mild extent, the number of damaged joints at the preceding visit, are associated with times between visits.

Table 5.5: Estimates from the fitted Cox model for the gaps between successive radiological assessments for patients in the University of Toronto PsA Cohort; follow-up censored at the minimum of date of death and July 15, 2013.

Covariate	EST	SE	RR	95% CI	p
Model (i): First Gap					
Sex: Female vs. Male	-0.206	0.083	0.81	(0.69, 0.96)	0.013
Age at Onset of PsA (per year)	0.029	0.003	1.03	(1.02, 1.04)	< 0.001
HLA B27: Yes vs. No	-0.174	0.106	0.84	(0.68, 1.03)	0.101
HLA C3: Yes vs. No	0.046	0.112	1.05	(0.84, 1.30)	0.679
Model (ii): Second and Subsequent Gaps					
Sex: Female vs. Male	-0.003	0.042	1.00	(0.92, 1.08)	0.948
Age at Onset of PsA (per year)	0.006	0.002	1.01	(1.00, 1.01)	0.001
HLA B27: Yes vs. No	-0.007	0.055	0.99	(0.89, 1.11)	0.905
HLA C3: Yes vs. No	0.007	0.055	1.01	(0.90, 1.12)	0.893
$Z(a_{j-1}) = 1$	0.058	0.096	1.06	(0.88, 1.28)	0.546
$Z(a_{j-1}) = 2$	0.128	0.106	1.14	(0.92, 1.40)	0.226
$Z(a_{j-1}) = 3$	0.079	0.132	1.08	(0.84, 1.40)	0.548
$Z(a_{j-1}) = 4$	0.154	0.126	1.17	(0.91, 1.49)	0.224
$Z(a_{j-1}) = 5$	-0.096	0.062	0.91	(0.80, 1.03)	0.122

In Figure 5.6, three estimates of $F(t)$ are plotted. One is a flexible piecewise-exponential failure time model with four pieces, and cut-points at 6, 12 and 18 years. The 95% pointwise confidence limits are also plotted. These estimates were obtained using the R msm package, since the failure time model is also a 2-state Markov model. The second estimate $\widehat{F}(t)$ is an IIV-weighted nonparametric estimate for times $t = 1, 2, \ldots, 30$ years. This was obtained from an expression similar to (5.25):

$$\widehat{F}(t) = \frac{\sum_{i=1}^{m} \sum_{j=1}^{m_i} \widehat{w}_{ij}(t) \, Y_{i5}(a_{ij})}{\sum_{i=1}^{m} \sum_{j=1}^{m_i} \widehat{w}_{ij}(t)}$$

where $\widehat{w}_{ij}(t) = I(t - 0.5 \leq a_{ij} \leq t + 0.5)/\widehat{P}_{ij}(t)$, with

$$\widehat{P}_{ij}(t) = \exp\left(-\widehat{\Lambda}_0(t - 0.5 - \min(a_{i,j-1}, t - 0.5)) e^{x'_{ij}\widehat{\beta}}\right) - \exp\left(-\widehat{\Lambda}_0(t + 0.5 - a_{i,j-1}) e^{x'_{ij}\widehat{\beta}}\right)$$

estimating the conditional probability of a visit in $(\max(a_{i,j-1}, t-0.5), t+0.5)$, based on the Cox models in Table 5.5. This estimate was then adjusted to be monotonic using the R isoreg function. The final estimate of $F(t)$ is that based on the 6-state Markov model, but excluding the time-dependent treatment covariates. This is obtained by averaging the multistate estimate using the empirical distribution of the baseline covariates as follows:

$$\widehat{F}(t) = \frac{1}{n} \sum_{i=1}^{n} \widehat{P}(Z_i(t) = 5 \mid x_i).$$

We remark that a Turnbull nonparametric estimate of $F(t)$ agreed well with the (unweighted) piecewise-exponential estimate in Figure 5.6. We elected to show the parametric estimate here since confidence limits can be obtained for it. We observe in Figure 5.6 that the piecewise-exponential estimate falls below the Markov process estimate for times over 12 years or so, but the latter is well within the 95% confidence limits for the piecewise-exponential model. The IIVW nonparametric estimate is imprecise but is seen to agree well with the Markov estimate.

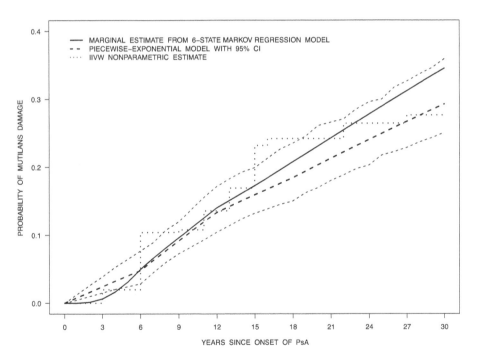

Figure 5.6: The estimated marginal distribution of the time to mutilans based on the 6-state Markov model and piecewise-exponential and nonparametric failure time distributions.

5.5 Intermittent Observation and Non-Markov Models

The terms in the likelihood function (5.4) can be difficult to calculate for non-Markov models that have many states or allow bi-directional transitions. For example, consider a reversible illness-death process with states 1, 2, 3 and transition intensity functions $\lambda_{kl}(t|\mathcal{H}(t^-))$ for $(k,l) = (1,2), (2,1), (1,3)$ and $(2,3)$, and suppose that we observe $Z(0) = 1$, $Z(a_1) = 2$, $Z(a_2) = 2$ at successive visit times $a_0 = 0, a_1, a_2$. Calculation of $P(Z(a_1) = 2, Z(a_2) = 2|Z(0) = 1)$ is complicated by the fact that the numbers of transitions over $(0, a_1]$ and $(a_1, a_2]$ are unknown: $1, 3, 5, \ldots$ transitions over $(0, a_1]$ and $0, 2, 4, \ldots$ transitions over $(a_1, a_2]$ are possible. If $t = 0$, and a_1 and a_2 are sufficiently close together that the probability of two or more transitions between visits is negligible, then we can use the approximation

$$P(Z(a_1) = 2, Z(a_2) = 2 \mid Z(0) = 1)$$

$$\simeq \int_0^{a_1} P(\text{1st transition at } t_2, \text{no other transitions by } a_2 \mid Z(0) = 1)\,dt_2$$

$$= \int_0^{a_1} \exp\left(-\int_0^{t_2^-} [\lambda_{12}(t) + \lambda_{13}(t)]\,dt\right) \lambda_{12}(t_2)$$

$$\times \exp\left(-\int_{t_2}^{a_2} [\lambda_{21}(t \mid t_2) + \lambda_{23}(t \mid t_2)]\,dt\right)\,dt_2,$$

where $\lambda_{12}(t)$ and $\lambda_{13}(t)$ are the intensities when $\mathcal{H}(t^-) = \{Z(u) = 1, 0 \le u < t\}$, and $\lambda_{12}(t|t_2)$ and $\lambda_{23}(t|t_2)$ are the intensities when $\mathcal{H}(t^-) = \{Z(u) = 1, 0 \le u < t_2; dN_{12}(t_2) = 1; Z(u) = 2, t_2 \le u < t\}$. While this is manageable, analogous calculations when there are more transitions between visits involve complicated multidimensional integrals.

Progressive or other unidirectional models are more tractable for parametric estimation, and for progressive models nonparametric estimation has also been considered for semi-Markov processes (e.g. Satten and Sternberg, 1999) An approach for semi-Markov models that is somewhat tractable is to assume phase-type sojourn time distributions for each transient state (Titman and Sharples, 2010b). This is achieved by associating sets of latent (unobservable) states with each of the observable states $1, \ldots, K$ and then assuming the latent process is a time-homogeneous Markov process. This facilitates the calculation of transition probabilities for the latent process, but probabilities of observable sample paths consist of sums over the latent process that can involve large numbers of terms. In addition, there is the issue of how many latent states to use; estimability problems arise if there are more than two or three latent states per observable state.

Non-Markov models also arise when random effects are included in a Markov model. For example, suppose that given a vector of random effects V, the process $\{Z(t), t > 0\}$ is Markov with transition intensities $g_{kl}(v)\lambda_{kl}(t)$, where the $g_{kl}(v)$ are specified functions. The marginal process then has intensity functions

$$\lambda_{kl}(t \mid \mathcal{H}(t^-)) = E_V\{g_{kl}(V) \mid \mathcal{H}(t^-)\}\lambda_{kl}(t).$$

Models with $\lambda_{kl}(t) = \lambda_{kl}$ are easily fitted, provided V is of low dimension since the likelihood function (5.4) for a single individual can be expressed as

$$\prod_{j=1}^{R} P(Z(a_j) \mid \bar{\mathcal{H}}_i^\circ(a_{j-1})) = E_V\left\{\prod_{j=1}^{R} P(Z(a_j) \mid Z(a_{j-1}), V) \mid \bar{\mathcal{H}}_i^\circ(a_{j-1})\right\}.$$

The terms $P(Z(a_j)|Z(a_{j-1}), V)$ in the right-hand side can be calculated via (5.7) for a given value of V, with numerical integration used to calculate the expectation. An illustration is deferred to Chapter 6, which describes models with random effects.

5.6 Mixed Observation Schemes

In some situations, certain transition times are unobservable but the exact times of others may be observed, subject only to right censoring. For example, for an illness-death model (Figure 1.6), the exact time of entry to state 3 may be known but not the time of entry to state 2 and, in some cases, whether entry to state 2 occurred at all. This would be the case when state 3 represented actual death, with state 2 a disease state that was observable only at intermittent visits to a medical facility. In cancer clinical trials, for example, disease progression status is only determined at periodic inspection times but times of death are observable; see Section 5.6.3. Because of their importance, we will first discuss illness-death models and then more general models.

5.6.1 Illness-Death Models

Consider the model shown in Figure 1.6 and assume for simplicity that $Z(0) = 1$ and all individuals are observed from time $t = 0$. Random variables T_r $(r = 2,3)$ denote times of entry to states $r = 2,3$ and for now, we ignore covariates. We denote the transition intensities by $\lambda_{12}(t)$, $\lambda_{13}(t)$ and $\lambda_{23}(t|\mathcal{H}(t^-)) = \lambda_{23}(t|t_2)$, where t_2 is the realized time of entry to state 2. We assume that a generic individual is followed over the time period $0 \le t \le C$, where the end-of-follow-up time C is independent of the multistate process, and the same condition holds for intermittent visit times A_j. Finally, we assume that if entry to state 3 occurs over $[0,C]$, T_3 is observed precisely. Thus, individuals may or may not be recorded to have entered state 2, and the entry time to state 3 may be observed or right censored. The four types of data that may be obtained, written as

(a) $Z(a_1) = \cdots = Z(a_m) = 1, Z(C) \in \{1,2\}$
(b) $Z(a_1) = \cdots = Z(a_{r-1}) = 1; Z(a_r) = \cdots = Z(a_m) = Z(C) = 2$, for some $r \le m$
(c) $Z(a_1) = \cdots = Z(a_m) = 1, T_3 = t_3$
(d) $Z(a_1) = \cdots = Z(a_{r-1}) = 1; Z(a_r) = \cdots = Z(a_m) = 2, T_3 = t_3$, for some $r \le m$.

In scenario (a), the individual is not recorded as having left state 1, but no assessment is made at C so they are only known not to be in state 3 at that time. In scenario (c), they are known to have entered state 3 at t_3 $(a_m < t_3)$, but it is not known which state they were in at t_3^-. Because it is known that the individual in scenarios (b) and (d) was in state 2 at a_m, if no entry to state 3 occurs over $(a_m, C]$ they are known to be in state 2 at C (scenario (b)) and the state they were in at t_3^- is known to be state 2 in (d). Likelihood expressions can be given for the general case where $\lambda_{23}(t|t_2)$ depends on t_2 as follows:

(a) $P_{11}(0,C) + P_{11}(0,a_m)\int_{a_m}^{C} P_{11}(a_m,t_2^-)\lambda_{12}(t_2)\exp\left\{-\int_{t_2}^{C}\lambda_{23}(t\,|\,t_2)\,dt\right\}dt_2$

(b) $P_{11}(0,a_{r-1})\int_{a_{r-1}}^{a_r} P_{11}(a_{r-1},t_2^-)\lambda_{12}(t_2)\exp\left\{-\int_{t_2}^{C}\lambda_{23}(t\,|\,t_2)\,dt\right\}dt_2$

(c) $P_{11}(0,t_3^-)\lambda_{13}(t_3) + \int_{a_m}^{t_3^-} P_{11}(0,t_2^-)\lambda_{12}(t_2)\exp\left\{-\int_{t_2}^{t_3^-}\lambda_{23}(t\,|\,t_2)\,dt\right\}\lambda_{23}(t_3\,|\,t_2)\,dt_2$

(d) $P_{11}(0,a_{r-1})\int_{a_{r-1}}^{a_r} P_{11}(a_{r-1},t_2^-)\lambda_{12}(t_2)\exp\left\{-\int_{t_2}^{t_3^-}\lambda_{23}(t\,|\,t_2)\,dt\right\}\lambda_{23}(t_3\,|\,t_2)\,dt_2.$

For a Markov process, $\lambda_{23}(t|t_2) = \lambda_{23}(t)$ and the expressions above simplify to

(a) $P_{11}(0,C) + P_{11}(0,a_m)\,P_{12}(a_m,C)$,
(b) $P_{11}(0,a_{r-1})\,P_{12}(a_{r-1},a_r)\,P_{22}(a_r,C)$,
(c) $P_{11}(0,a_m)\{P_{11}(a_m,t_3^-)\lambda_{13}(t_3) + P_{12}(a_m,t_3^-)\lambda_{23}(t_3)\}$,
(d) $P_{11}(0,a_{r-1})\,P_{12}(a_{r-1},a_r)\,P_{22}(a_r,t_3^-)\lambda_{23}(t_3)$,

where $P_{kl}(s,t) = P(Z(t) = l | Z(s) = k)$ for $t \geq s$. The msm function will fit Markov models with piecewise-constant intensities to this kind of data and SmoothHazard function will handle other parametric models.

In some situations, it may be possible to determine (e.g. by autopsy) whether a transition into state 3 was from state 1 or state 2. In that case, the expressions for outcome (c) are modified so that only the outcome $Z(t_3^-) = 1$ or $Z(t_3^-) = 2$ is included depending on the state that was occupied immediately prior to death.

5.6.2   General Models

Under the CIVP assumptions, it is possible to write the likelihood function for a general multistate process in which exact entry times to states in some set $\mathcal{E}$ are observable. Often $\mathcal{E}$ would include only absorbing states, but this is not essential. To deal with this situation, at visit time a_{j-1} with observed process history $\bar{\mathcal{H}}^{\circ}(a_{j-1})$, we consider four random variables: A_j = time of next "ordinary" visit; T_{Ej} = time of next entry to a state in $\mathcal{E}$; $\varepsilon_j = e_j$ if state e_j in $\mathcal{E}$ is entered at time T_{Ej}; $\delta_j = I(A_j < T_{Ej})$. Letting $a_j = \min(A_j, T_{Ej})$, we can write the partial likelihood function for a sequence of m observation times as

$$\prod_{j=1}^{m} P(Z(a_j)\,|\,\bar{\mathcal{H}}^{\circ}(a_{j-1}))^{\delta_j}\,P(T_{Ej} = a_j, \varepsilon_j = e_j\,|\,\bar{\mathcal{H}}^{\circ}(a_{j-1}))^{1-\delta_j}. \tag{5.26}$$

In this expression a_j is either an ordinary visit time or a "visit time" triggered by entry to a state in $\mathcal{E}$. The validity of (5.26) relies on the fact that A_j and $\{Z(t), t > a_{j-1}\}$ are conditionally independent, given $\bar{\mathcal{H}}^{\circ}(a_{j-1})$, as assumed for the derivation of (5.4).

The expressions in (5.26) are readily computed for Markov models. For example, consider the model in Figure 5.7 and suppose that exact times of entry to states

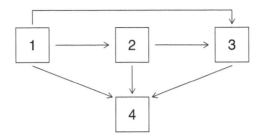

Figure 5.7: A 4-state model of cardiovascular risk levels and outcomes.

3 and 4 are observable. In a study of coronary heart disease, states 3 and 4 might represent a severe non-fatal event (e.g. heart attack, stroke) and death, with states 1 and 2 representing medium- and high-risk health status.

Suppose for illustration that $Z(a_{j-1}) = 2$. Then potential terms in the likelihood (5.26) are

$$P(Z(a_j) = 2 \mid \bar{\mathcal{H}}^\circ(a_{j-1})) = \exp\left\{ - \int_{a_{j-1}}^{a_j} [\lambda_{23}(t) + \lambda_{24}(t)]\, dt \right\}$$

$$P(T_{Ej} = a_j, \varepsilon_j = r \mid \bar{\mathcal{H}}^\circ(a_{j-1})) = P(Z(a_j^-) = 2 \mid \bar{\mathcal{H}}^\circ(a_{j-1})) \lambda_{2r}(a_j), \quad r = 3, 4.$$

If $Z(a_{j-1}) = 1$, outcomes are similarly easy to obtain. For example,

$$P(T_{Ej} = a_j, \varepsilon_j = 4 \mid \bar{\mathcal{H}}^\circ(a_{j-1})) = \sum_{l=1}^{2} P(Z(a_j^-) = l \mid \bar{\mathcal{H}}^\circ(a_{j-1})) \lambda_{l4}(a_j).$$

Covariates x can be included in parametric Markov intensity function specifications $\lambda_{kl}(t|x; \theta_{kl})$ without much difficulty. The msm package can deal with time-homogeneous multiplicative models, where $\lambda_{kl}(t|x; \theta_{kl}) = \lambda_{kl} \exp(x'\beta_{kl})$ and also with piecewise-constant baseline intensity functions $\lambda_{kl}(t)$.

When covariate effects for a marginal feature such as time of entry to some state k, or the occupancy probability $P_k(t|x)$ are of interest, direct models as in Chapter 4 and Section 5.4 can be considered. The weighted binomial estimating function methods can be applied provided we adopt a discrete-time visit process and we associate an entry time $T_{Ej} = a_j$ with the most recent potential visit time s_{r-1}. In that case, we define $dA^\dagger(s_r) = 1$ and

$$P(dA^\dagger(s_r) = 1 \mid \bar{\mathcal{H}}^\circ(s_{r-1}), T_{Ej} = s_{r-1}) = 1,$$

where E_j is an absorbing state that precludes further visits, and they remain equal to one for times s_{r+1}, s_{r+2}, ... up to the end of follow-up. If entry to E_j does not preclude further visits, we model $P\{dA^\dagger(s_r)|\bar{\mathcal{H}}^\circ(s_{r-1})\}$ as usual.

5.6.3 Progression-Free Survival in Cancer Trials

5.6.3.1 Progression-Free Survival Times and the Illness-Death Model

Phase III randomized clinical trials in cancer often aim to study the effect of a therapeutic intervention on the time to disease progression and the survival time.

The composite endpoint of progression-free survival, defined as the time to the first of progression or death, is frequently adopted as a basis for treatment comparisons. Challenges arise for the analysis of progression and progression-free survival times, because progression status is only determined at periodic assessment times. As a result, progression times, if progression is known to have occurred, are interval censored. Moreover, the ultimate progression status of individuals who have not progressed by the time of their last assessment is unknown because they may experience progression between then and time of death or censoring. The composite progression-free survival time T is therefore subject to a hybrid censoring scheme involving interval censoring for progression and right censoring for death.

In this section, we highlight the usefulness of the illness-death model (see Sections 1.2.1 and 1.6.1) for the joint consideration of progression and death and point out limitations of analyses of progression-free survival times. We also show that the illness-death model can be used to address the complex observation and censoring scheme for the events of interest. We first discuss the challenges in more detail.

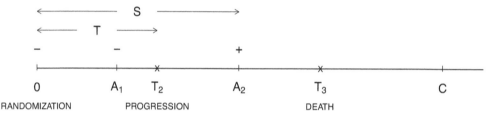

(a) A timeline diagram for events in which progression is detected.

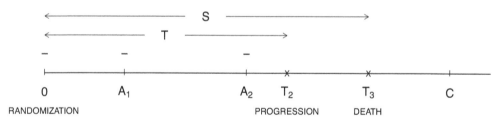

(b) A timeline diagram for events in which progression is not detected due to death.

Figure 5.8: Timeline diagrams for progression (T_1), death (T_2), assessments (A_1 and A_2) and censoring (C).

Figure 5.8(a) shows a timeline diagram for a hypothetical individual where T_2 and T_3 are the times of progression and death, respectively; A_k, $k = 1, 2, \ldots$ denote the assessment times; and C is a right censoring time. In this scenario, progression is first detected at A_2, so it is clear that $T = \min(T_2, T_3)$ lies in the interval $(A_1, A_2]$. The most common strategy for dealing with this kind of data in cancer clinical trials is to use right-endpoint imputation, whereby the surrogate $S = A_2$ is used in lieu of the actual progression-free survival time T, and standard methods for survival analysis are adopted. Figure 5.8(b) illustrates the complication that arises when death occurs without prior evidence of progression; here it is unknown whether or not progression occurred between the last negative assessment at A_2 and the time of death. The convention in this case is to assume individuals have not progressed and, hence, to use $S = T_3$ as the progression-free survival time.

To illustrate the bias from such analyses of progression-free survival, we simulate data from a clinical trial in which individuals are randomized with equal probability to one of two treatment arms ($X = 1$ for the experimental arm and $X = 0$ for the control arm). Given X, we next simulate data from a time-homogeneous illness-death process for a single sample of $m = 2000$ individuals; the transition intensities satisfy (i) $P(T_2 < T_3 | X = 0) = \lambda_{12}/(\lambda_{12} + \lambda_{13}) = 0.6$ or 0.8 so that 60% or 80% of individuals with $X = 0$ are expected to experience progression, (ii) $\lambda_{23}/\lambda_{13} = 1.5$ so that the mortality rate increases by 50% following progression and (iii) $\pi_A = P(C^A < T | X = 0) = 0.20$ so 20% of individuals are expected to be progression free and alive at the administrative censoring time C^A. We set $\beta_{12} = \beta_{13} = \beta = \log 0.4$ so there is a large common effect of treatment in delaying progression and on progression-free death, and $\beta_{23} = 0$ to represent the setting where there is no effect of treatment on mortality following progression. The random right censoring time C_i^R is simulated by an exponential distribution with the rate set to achieve 40% net censoring for the composite failure time T (i.e. $P(T > C) = 0.40$ where $C = \min(C^R, C^A)$).

Without loss of generality, we set $C^A = 1$ with $K = 4$ assessments scheduled at times $a_k = k/4$, $k = 1, \ldots 4$, but we add a bit of noise $\epsilon_k \sim N(0, \sigma_e^2)$ around these target assessment times to reflect variation in actual assessment times. We consider $\sigma_e = 1/24$ and $1/80$ to represent two modest levels of variation, whereby the chance that two nominally consecutive assessments are realized in the reverse order is small. A partial listing of the dataframe is shown below, with times recording the times of visits and states indicating the state occupied at the corresponding time. It is not apparent from the dataframe alone, but when states == 3 the times variable is an actual time of death; we will see shortly that this is specified in the call to the msm function. The variable stime is the surrogate progression-free survival time and sstatus is the indicator of whether this time is the failure time T or censoring time. The last column contains the treatment indicator. For convenience, stime is shown in each row for a given individual.

```
> simdata
  id enum      times states      stime sstatus x
   1    1 0.0000000      1 0.7416094       1 0
   1    2 0.2587894      1 0.7416094       1 0
   1    3 0.4726939      1 0.7416094       1 0
   1    4 0.7416094      3 0.7416094       1 0
   3    1 0.0000000      1 0.6112530       1 0
   3    2 0.2588814      1 0.6112530       1 0
   3    3 0.6112530      2 0.6112530       1 0
   3    4 0.6988743      2 0.6112530       1 0
   3    5 0.8030370    999 0.6112530       1 0
   5    1 0.0000000      1 0.2583610       1 0
   5    2 0.2306187      1 0.2583610       1 0
   5    3 0.2583610      3 0.2583610       1 0
   :    :          :      :          :     : :
```

In Figure 5.9, the true progression-free survival distributions (solid lines) for each treatment group are plotted for each of the four scenarios concerning probabilities of

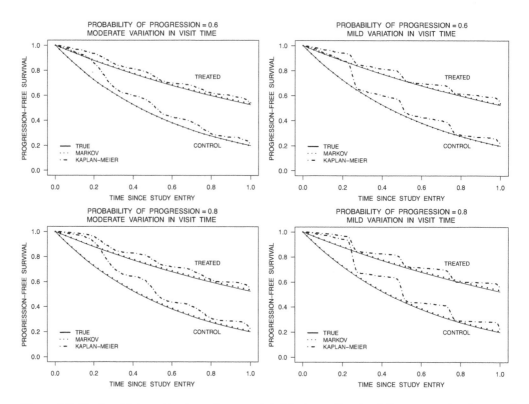

Figure 5.9: Kaplan-Meier estimate and parametric Markov estimate of the progression-free survival probability based on simulated data; 60% (top row) and 80% (bottom row) probability of progression in control arm, 40% net censoring by the random withdrawal time, $\sigma_e = 1/24$ (left side) and $\sigma_e = 1/80$ (right side), and $n = 2000$.

progression and models for the assessment time process. Also plotted is the Kaplan-Meier (KM) estimate based on the surrogate progression-free survival time for the simulated dataset of $n = 2000$ individuals. The KM estimates are obtained as follows:

```
> km0 <- survfit(Surv(stime, sstatus) ~ 1,
    data=simdata, subset=((enum == 1) & (x == 0)), type="kaplan-meier")

> km1 <- survfit(Surv(stime, sstatus) ~ 1,
    data=simdata, subset=((enum == 1) & (x == 1)), type="kaplan-meier")
```

Stepwise drops in the KM estimates based on the surrogate progression-free survival times from right-endpoint imputation are exhibited in many plots from cancer clinical trials and are evident here. This pattern is more pronounced with the smaller degree of variation around the fixed assessment times. The KM curves also sit well above the true progression-free survival curves. The estimate based on the correct Markov illness-death model fitted using the msm package (dotted line) closely tracks the true progression-free survival curve. The code for separate fits for the two treatment groups is given below, where the specification deathexact = 3 indicates that the times in the corresponding lines of the dataframe are actual death times.

```
> qmat <- matrix(0, nrow=3, ncol=3)
> qmat[1,c(2,3)] <- c(0.1,0.1); qmat[2,3] <- 0.1

> mfit0 <- msm(states ~ times, subject=id, data=simdata[simdata$x == 0,],
              qmatrix = qmat, deathexact=3, censor=999, censor.states=c(1,2),
              center=FALSE, opt.method="optim")

> mfit1 <- msm(states ~ times, subject=id, data=simdata[simdata$x == 1,],
              qmatrix = qmat, deathexact=3, censor=999, censor.states=c(1,2),
              center=FALSE, opt.method="optim")
```

5.6.3.2 Bone-Progression-Free Survival in Lung Cancer

Here we consider data from a trial of individuals with lung cancer metastatic to bone (Rosen et al., 2003). The goal of this trial, like the metastatic breast cancer trial discussed in Section 1.6.1, Section 3.2.2 and Section 3.3.2, is to evaluate the effect of a bisphosphonate compound called zoledronic acid on the incidence of skeletal complications. The secondary outcome considered here is progression of bone disease defined as the development of a new skeletal metastasis or the growth of a pre-existing lesion. Progression of bone disease may occur at any time but is detected only when study participants undergo bone scans, which occur roughly every 3 months during follow-up.

Table 5.6: Crude summary of outcome data on bone progression and death by treatment arm in Rosen et al. (2003).

			Died	
Treatment	n	Progressed	Post-Progression	Progression-Free
8 mg Zoledronic acid	129	46	38	70
Placebo	119	45	32	70

The study involves 371 individuals who had at least one follow-up bone scan and were randomized to one of three treatments: a monthly infusion of 4 mg of zoledronic acid, 8 mg of zoledronic acid, or a placebo infusion. For illustration, we restrict attention to the 8 mg zoledronic acid ($n = 129$) and placebo arms ($n = 119$). Individuals were to be followed for up to 13 months. Of the 129 individuals receiving zoledronic acid, 46 were documented to have experienced progression in bone disease through a positive bone scan, and among these 38 were subsequently observed to die (a pattern consistent with that of Figure 5.8(a)); 70 individuals were known to have died without a prior recorded progression in bone disease. Of the 119 individuals randomized to the placebo arm, 45 were documented to have experienced progression in bone disease through a positive bone scan, and among these 32 were subsequently observed to die; 70 individuals were known to have died without a prior recorded progression in bone disease (see Table 5.6).

The types of data represented in Figure 5.8 arise in this trial, so the remarks in Section 5.6.3.1 are relevant. A few lines of a dataframe follow in which **enum** records lines within individuals and the **times** variable records either the times of the assessments for progression, the time of death or the right censoring time. The variable states records the state occupied at each assessment time or the death state; the value 999 is used to represent a censoring state that can be entered from state 1 or 2. The variable **stime** is the surrogate progression-free survival time used in naive analyses, and **sstatus** is the associated status variable. The treatment indicator is **trt**, which is 1 for those receiving 8 mg of zoledronic acid and 0 otherwise.

```
> lung
  id enum times states stime sstatus trt
   1    1     0      1    81       1   1
   1    2     5      1    81       1   1
   1    3    81      2    81       1   1
   1    4   241      3    81       1   1
   2    1     0      1    82       1   0
   2    2    82      3    82       1   0
   5    1     0      1    78       1   0
   5    2    78      2    78       1   0
   5    3   638    999    78       1   0
  73    1     0      1    78       0   1
  73    2    78    999    78       0   1
   :    :     :      :     :       :   :
```

The following code enables us to fit the multistate model under the time-homogeneous assumption with common and separate treatment effects. We first initialize the values for the baseline transition rates.

```
> qmat <- matrix(0, nrow=3, ncol=3)
> qmat[1,c(2,3)] <- c(0.1,0.1); qmat[2,3] <- 0.1

> qmat.ini <- crudeinits.msm(states ~ times, subject=id, qmatrix=qmat,
                             data=lung, censor=999, censor.states=c(1,2))
> qmat.ini
            [,1]       [,2]      [,3]
  [1,] -0.007916  0.003118  0.004797
  [2,]  0.000000 -0.005595  0.005595
  [3,]  0.000000  0.000000  0.000000
```

In the call to the **msm** function, we specify a single covariate indicating treatment and first adopt the model with the constraint that $\beta_{12} = \beta_{13}$ leaving β_{23} free to vary; this is achieved with the option **constraint=list(trt=c(1,1,2))**. The **deathexact=3** specification indicates that lines in the dataframe with **states=3** contain the exact entry time to this state in the **times** column; these are the times of death. The **censor=999** specification indicates that when this appears in the **states** column, the value in the **times** variable is a right censoring time. The **censor.states=c(1,2)** specification is important in this setting since it indicates that individuals could be in states 1 or 2 at the time of censoring. The object **mfitc** contains the results from this analysis with the "c" to remind us that this is obtained with the constraint on the regression coefficients.

```
> mfitc <- msm(states ~ times, subject=id, data=lung, qmatrix=qmat.ini,
```

```
                   covariates=~trt, constraint=list(trt=c(1,1,2)),
                   deathexact=3, censor=999, censor.states=c(1,2),
                   center=FALSE, opt.method="optim")
> mfitc
  Maximum likelihood estimates
  Baselines are with covariates set to 0

  Transition intensities with hazard ratios for each covariate
                    Baseline                         trt
  State 1 - State 1 -0.027176 (-0.035758,-0.020654)
  State 1 - State 2  0.026677 ( 0.019995, 0.035593) 0.8116 (0.5563,1.184)
  State 1 - State 3  0.000499 ( 0.000035, 0.007114) 0.8116 (0.5563,1.184)
  State 2 - State 2 -0.005268 (-0.006478,-0.004285)
  State 2 - State 3  0.005268 ( 0.004285, 0.006478) 0.9253 (0.6970,1.228)

  -2 * log-likelihood:  2818.6501
```

The estimates reported in the **Baseline** column are of the log baseline intensities with the first estimate (for State 1 − State 1) the negative sum of log baseline intensities for $1 \to 2$ and $1 \to 3$ transitions. For rows corresponding to transitions, these point estimates and confidence intervals can be exponentiated for inferences about the intensities themselves. The entries under the **trt** variable are the relative rates of transitions for individuals receiving zoledronic acid versus placebo; the constraint $\beta_{12} = \beta_{13}$ which was specified results in common estimates in the second and third rows of the output. The multistate analysis with the common treatment effect for the $1 \to 2$ and $1 \to 3$ transitions gives a point estimate of a 19% reduction in the cause-specific hazard for progression and death without progression, but this is not a statistically significant effect (RR = 0.81; 95% CI: 0.56, 1.18; $p = 0.278$).

A second call to the **msm** function relaxes the constraint $\beta_{12} = \beta_{13}$ and therefore gives separate estimates; we use the estimates from the previous fit as initial values when fitting the more general model.

```
> mfitd <- msm(states ~ times, subject=id, data=lung,
      qmatrix=mfitc$Qmatrices$baseline, covariates=~trt,
      covinits=list(trt=c(mfitc$Qmatrices$trt[1,2], mfitc$Qmatrices$trt[1,3],
                    mfitc$Qmatrices$trt[2,3])),
      deathexact=3, censor=999, censor.states=c(1,2),
      center=FALSE, opt.method="optim")
> mfitd
  Maximum likelihood estimates
  Baselines are with covariates set to 0

  Transition intensities with hazard ratios for each covariate
                    Baseline                            trt
  State 1 - State 1 -0.0271997 (-3.612e-02,-0.020480)
  State 1 - State 2  0.0267195 ( 1.938e-02, 0.036839) 0.8244 (0.5340094,   1.273)
  State 1 - State 3  0.0004802 ( 4.418e-06, 0.052181) 0.5388 (0.0004098,708.570)
  State 2 - State 2 -0.0052726 (-6.533e-03,-0.004255)
  State 2 - State 3  0.0052726 ( 4.255e-03, 0.006533) 0.9307 (0.6921947,   1.251)

  -2 * log-likelihood:  2818.6089
```

When separate treatment effects are modeled, the estimated treatment effect on bone progression is comparable at RR = 0.82 (95% CI: 0.53, 1.37; $p = 0.384$), but there is essentially no information on the effect of treatment on the $1 \to 3$ transition intensity with the 95% confidence interval being (0.00, 708.57). A likelihood ratio test of H_0: $\beta_{12} = \beta_{13}$ versus H_A: $\beta_{13} \neq \beta_{13}$ gives a test statistic of 0.041, so p-value is $P(\chi_1^2 > 0.041) = 0.840$. Finally, in Table 5.7, we show the results from the 3-state model along with the results of fitting a Cox model for T using the imputed times S described in the preceding section. The Cox model results indicate a mildly significant increase in progression-free survival for the zoledronic acid arm. Interpretation is problematic, however, given the bias issues discussed earlier.

Table 5.7: Estimates of treatment effect (β) obtained from Cox regression based on surrogate progression-free survival times and an analysis based on a multistate Markov model (8 mg vs. placebo).

Cox Model		EST	SE	RR	95% CI	p
Prog-Free Survival		-0.252	0.133	0.78	(0.61, 1.00)	0.049
3-State Model		EST	SE	RR	95% CI	p
Common Effects		-0.209	0.193	0.81	(0.56, 1.18)	0.278
Separate Effects	Progression	-0.193	0.222	0.82	(0.53, 1.27)	0.384
	Prog-Free Survival	-0.618	3.664	0.54	(0.00, 708.57)	0.866

5.7 Bibliographic Notes

Intermittently observed life history processes produce what is sometimes referred to as panel data; Kalbfleisch and Lawless (1985) review this area. Conditionally independent visit process (CIVP) assumptions have been discussed by Grüger et al. (1991), Cook and Lawless (2007, Sections 7.1, 7.4), Cook and Lawless (2014) and others; Farewell et al. (2017) provide a general discussion of intermittently observed longitudinal responses. Tom and Farewell (2011) and Cook and Lawless (2014, Section 3.6) discuss intermittent observation of time-dependent covariates, for which there is an extensive literature in survival analysis (e.g. Andersen and Listøl, 2003; Sparling et al., 2006). Estimation and inference procedures for Markov models were discussed by Kalbfleisch and Lawless (1985) and earlier authors. Titman and Sharples (2010a) give a thorough review and include remarks on model checking, while Jackson (2011) describes the msm package in R, which implements a wide range of methods for Markov models with piecewise-constant intensities. Titman (2011) considers nonhomogeneous models. Efficiency of estimation and the design of panel studies for Markov processes are discussed by Cook (2000), Hwang and Brookmeyer (2003), Mehtala et al. (2015) and Lawless and Nazeri Rad (2015). Nonparametric estimation of cumulative intensity functions has been considered mainly for the illness-death model; see Frydman (1992, 1995), Frydman and Szarek

(2009) and Frydman et al. (2013). Joly et al. (2002) consider flexible parametric intensities and estimation via penalized likelihood. For estimation of state entry time distributions, nonparametric failure time methodology can be used; see Turnbull (1976) and Gentleman and Vandal (2002). Hudgens et al. (2001) and Frydman and Liu (2013) consider competing risks. Hudgens et al. (2014) consider parametric estimation of cumulative incidence functions.

Estimation of marginal state occupancy probabilities under process-related CIVP schemes is discussed by Nazeri Rad and Lawless (2017), using the IIVW estimating function methodology introduced by Lin et al. (2004) and developed further for longitudinal generalized linear models by Buzkova and Lumley (2007, 2008, 2009), Buzkova (2010), Pullenayegum and Feldman (2013) and Pullenayegum and Lim (2016). Zhu et al. (2017) and references therein consider the estimation of failure time distributions, which can be of interest for waiting times in multistate models. Estimation for non-Markov models has mainly been restricted to progressive semi-Markov models; see Satten and Sternberg (1999), Sternberg and Satten (1999), Kang and Lagakos (2007), Griffin and Lagakos (2010), Titman and Sharples (2010b), Yang and Nair (2011) and Lange and Minin (2013). Mixed observation schemes have been considered mainly for the illness-death model (e.g. Frydman and Szarek, 2009; Boruvka and Cook, 2016), but this is accommodated for Markov models by the msm package. A somewhat special type of mixed observation scheme arises for illness-death processes in the context of survival-sacrifice experiments; see, for example, Kalbfleisch et al. (1983), McKnight and Crowley (1984), Dewanji and Kalbfleisch (1986), Lindsey and Ryan (1993) and Problems 3.4, 5.10 and 5.11.

5.8   Problems

Problem 5.1   A homogeneous discrete-time Markov chain $\{Z(t), t = 0, 1, 2, \ldots\}$ is called *embeddable* if the $K \times K$ transition probability matrix P can be expressed as $P = \exp(Q)$, where Q is a $K \times K$ transition intensity matrix; that is, $Q = (\lambda_{kl})$ where $\lambda_{kl} \geq 0$ for $k \neq l$ and $\lambda_{kk} = -\sum_{l \neq k} \lambda_{kl}$. Under this condition the discrete time process can be generated from a homogeneous continuous-time Markov process observed at discrete time-points $t = 0, 1, 2, \ldots$. If $K = 2$, show that P is embeddable if and only if $p_{11} + p_{22} > 1$. Easily verified necessary and sufficient conditions for embeddability are not known for $K > 2$.

(Section 5.2; Kalbfleisch and Lawless, 1985)

Problem 5.2   Nagelkerke et al. (1990) report on a study of 84 children between the ages of 11 and 18 months in a village in the Kiambu district of Kenya. Stool samples from these children were examined weekly for the presence of *Giardia lamblia*, a protocol parasite affecting the lining in the small intestine. Samples were classified as positive or negative for the parasite, which we interpret as infected or non-infected. The data are given in Section D.6.

Let $\{Z_i(s), s \geq 0\}$ denote a continuous-time 2-state process where $Z_i(s) = 2$ if child i is infected at time s and $Z_i(s) = 1$ otherwise. Suppose the intensity for $1 \to 2$ ($2 \to 1$) transitions is λ_1 (λ_2); see Figure 5.10. Let $a_{i0} = 0$ denote the first assessment time for child i and a_{ir}, $r = 1, \ldots, m_i$ where m_i is the number of follow-up assessments, $i = 1, \ldots, n$. The assessment times are at weekly intervals, so assume that the CIVP conditions are satisfied.

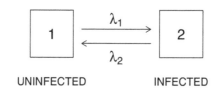

UNINFECTED   INFECTED

Figure 5.10: An alternating 2-state process for recurrent infections.

(a) Show that (see also Problem 1.2)

$$P(Z_i(t) = 2 \mid Z_i(s) = 1) = \frac{\lambda_1}{\lambda_1 + \lambda_2} (1 - \exp(-(\lambda_1 + \lambda_2)(t - s)))$$

and

$$P(Z_i(t) = 1 \mid Z_i(s) = 2) = \frac{\lambda_2}{\lambda_1 + \lambda_2} (1 - \exp(-(\lambda_1 + \lambda_2)(t - s))).$$

(b) Conditional on $Z_i(a_{i0})$, the likelihood given in (5.6) is

$$L(\lambda_1, \lambda_2) = \prod_{r=1}^{m_i} P(Z_i(a_{ir}) \mid a_{ir}, Z_i(a_{i,r-1})). \tag{5.27}$$

If the process is in equilibrium, however, the probability of the initial state being infected is $P(Z_i(a_{i0}) = 1) = \pi = \lambda_2/(\lambda_1 + \lambda_2)$. In this case, we can use an augmented likelihood

$$L_{AUG}(\lambda_1, \lambda_2) = P(Z_i(a_{i0})) L(\lambda_1, \lambda_2). \qquad (5.28)$$

If $m_i = 1$ for all $i = 1, \ldots, n$ in a sample of independent individuals assess the relative efficiency of estimators for λ_k, $k = 1, 2$, $\alpha = \lambda_1 + \lambda_2$, and π based on (5.27) and (5.28).

How does the relative efficiency change if $m_i = 2$ for all $i = 1, \ldots, n$?

(Section 5.2)

Problem 5.3   Consider a sample of n individuals with $P(Z_i(a_0) = 2) = \pi$ in the setting of Problem 5.2. Suppose each is seen at m times $w, 2w, \ldots, mw$. Obtain the Fisher information matrix (5.10) for the transformed parameters π and $\alpha = \lambda_1 + \lambda_2$, and consider what happens when w becomes large.

(Section 5.2)

Problem 5.4   Consider the scenario in Problem 5.3 involving the model in Figure 5.10 and observation of n individuals at equi-spaced times $0, w, 2w, \ldots, mw$, but where all individuals satisfy $Z(0) = 1$.

(a) Determine the form of the goodness-of-fit statistic (3.58) in this case.
(b) If $\widehat{p}_{11} + \widehat{p}_{12} > 1$, where $p_{kk} = P(Z(rw) = k | Z((r-1)w) = k)$ for $k = 1, 2$, show using the result in Problem 5.1 that $\widehat{p}_{11} = n_{11}/n_1.$ and $\widehat{p}_{22} = n_{22}/n_2.$, where for $k, l = 1, 2$,

$$n_{kl} = \sum_{i=1}^{n} \sum_{r=1}^{m} I(Z_i(rw) = l \mid Z_i((r-1)w) = k). \qquad (5.29)$$

(c) What is the limiting distribution of the statistics (3.58) as $n \to \infty$, with m fixed, in this case?

(Section 5.2)

Problem 5.5   Reversible models with intensities of semi-Markov form are difficult to fit when observation is intermittent. Consider a 3-state process with non-zero intensities $\lambda_{12}(t|\mathcal{H}(t^-)) = \lambda_{12}$, $\lambda_{21}(t|\mathcal{H}(t^-)) = h_{21}(B(t))$ and $\lambda_{23}(t|\mathcal{H}(t^-)) = h_{23}(B(t))$, where $B(t)$ is the time since the most recent entry to state 2. Suppose that an individual has $Z(0) = 1$ and is observed in states Z_1 and Z_2 at visit times a_1 and a_2, which are independent of the multistate process. Assuming that no more than two transitions occur between successive observation times, write the likelihood function for each of the seven possible outcomes (Z_1, Z_2).

(Section 5.5)

Problem 5.6   Consider the estimating function (5.14) for a given value of t. Find the functional value $P_k^*(t)$ for which $E\{U_k(t)\} = 0$, when the observation times $a_{ij} = a_j$ are the same for all individuals. Thus, show that $\widehat{P}_k(t)$ obtained from solving $U_k(t) = 0$ is biased.

(Section 5.3)

Problem 5.7   Consider the setting with intermittent observation of a continuous-time multistate process.

(a) Show that the estimating function (5.19) remains unbiased if a function $w_i^\circ(t)$ of t alone is included in the integrand.

(b) Weight stabilization commonly uses an estimate of the function $w_i^\circ(t) = \lambda_0(t|x_{i1}) = E\{dA_i(t|x_{i1})\}/dt$, where x_{i1} are fixed covariates used in $v_i(t)$, when modeling $\lambda_i^a(t|v_i(t))$. Explain intuitively why stabilized weights $\lambda_0(t|x_{i1})/\lambda_i^a(t|v_i(t))$ may be less variable than the unstabilized weights $\lambda_i^a(t|v_i(t))^{-1}$.

(Section 5.4)

Problem 5.8   Consider a failure process $\{Z(s), s \geq 0\}$ which begins in state 1 at $t = 0$ and let $T = T_2$ denote the $1 \to 2$ transition time governed by transition intensity function $\lambda(t)$. Suppose this process is intermittently observed over $[0, C^A]$ at assessment times governed by a counting process $\{A(s), s \geq 0\}$. Let $Y(t) = I(t \leq C^A)$ and $\bar{\mathcal{H}}(t) = \{Y(u), A(u), Z(u), 0 < u \leq t\}$. The resulting data may be represented by $\{(a_j, Z(a_j)), j = 0, 1, \ldots, m\}$, where $m = A(C^A)$.

(a) Let $\lambda^a(t|\mathcal{H}(t^-))$ denote the visit process intensity and suppose

$$\lim_{\Delta t \downarrow 0} \frac{P(\Delta A(t) = 1 \mid \bar{\mathcal{H}}(t^-))}{\Delta t} = Y(t)\,\lambda_0^a(t)\,\exp(\gamma\,Z(t_{A(t^-)})). \qquad (5.30)$$

Construct the likelihood in (5.16) based on this model for the inspection process.

(b) When failure times are interval-censored by such visit processes, the only visit times often reported are those satisfying

$$a^L = \max_{a_j:Z(a_j)=1} \{a_j\} \quad \text{and} \quad a^R = \min_{a_j:Z(a_j)=2} \{a_j\}$$

where $a^R = \infty$ if $Z(a_m) = 1$. Can the parameters of the visit process be estimated in this case?

(c) Show that under a CIVP

$$\prod_{j=1}^{m} P(Z(a_j) \mid a_j, \mathcal{Z}^\circ(a_j)) = P(a^L < T < a^R \mid A^L = a^L, A^R = a^R)$$

so that the likelihood reduces to the usual form for interval-censored failure time data.

(Sections 5.1, 5.4)

Problem 5.9 Consider the setting of Problem 5.8 with an intensity

$$\lim_{\Delta t \downarrow 0} \frac{P(\Delta A(t) = 1 \mid \bar{\mathcal{H}}(t^-))}{\Delta t} = Y(t) \lambda_0^a(t) \exp(\gamma Z(t^-)) \qquad (5.31)$$

for the visit process.

(a) Carry out a simulation study with $C^A = 1$, where $\lambda_{12}(t) = \lambda = -\log 0.80$ so that $P(T < 1) = 0.80$. Set $\lambda_0^a(t) = \lambda_0^a = 4$ and simulate a sample of $n = 500$ individuals with $\gamma = \log 0.5$. Maximize the misspecified partial likelihood

$$L(\lambda) = \prod_{i=1}^{n} \prod_{j=1}^{m} P(Z_i(a_{ij}) \mid a_{ij}, \mathcal{Z}_i^\circ(a_{ij}))$$

using software for parametric (exponential) modeling of interval-censored data and record the MLE. Repeat this 100 times and record the average value.

(b) Repeat the simulation study in (a) with $\gamma = \log 0.75, 0, \log 1.25$ and $\log 2$ and comment on the relationship between the empirical bias in λ and the value of γ.

(Section 5.4)

Problem 5.10 Animal tumorigenicity experiments to assess the carcinogenic effects of pesticides and other suspected carcinogens were described in Problem 3.3. Consider an experiment in which n mice are randomized to receive a potentially toxic compound or a placebo control; the illness-death model in Figure 7.5 can be used to describe the onset of tumours and death. Times of death are observed but may be administratively censored. The exact times that tumours develop are unknown, and it is customary to (i) autopsy animals who die to see whether a tumour is present, and (ii) to sample animals that are alive at one or more times during the course of follow-up, to sacrifice them, and to determine their tumour status. Lindsey and Ryan (1993) consider a Markov model for which $\lambda_{12}(t|x) = \lambda_{120} \exp(\beta x)$, $\lambda_{13}(t|x) = \lambda_{130} \exp(\delta x)$ and $\lambda_{23}(t|x) = \lambda_{130} \exp(\theta + \gamma x)$, with $x = I$(potentially toxic compound received), and $\lambda_{120}(t)$ and $\lambda_{130}(t)$ having piecewise-constant forms with cut-points at b_k, $k = 1, \ldots, K$.

(a) Consider a study with follow-up planned over $(0, C^A)$ and sacrifices scheduled at times a_k, $k = 1, \ldots, K$ with $a_K = C^A$. Let n_k denote the number of mice sacrificed at time a_k, assuming that they are randomly chosen from mice who are alive at that time. Write the likelihood for the resulting data.

(b) Discuss how an expectation-maximization (EM) algorithm can be used for estimation.

(c) Suppose that natural deaths are "observed" only at times a_k, so the exact time of death is interval censored. Write the revised likelihood and consider what parameters are most impacted by this loss of information.

(Section 5.6; Lindsey and Ryan, 1993)

Problem 5.11   Consider a tumorigenicity study on the effect of a known carcinogen 2-AAF on the development of tumours in female mice (Lindsey and Ryan, 1993). Mice could be observed to die with no tumour (DNT) or die with a tumour (DWT). Among the rats sacrificed during follow-up, they could be found to have a tumour (sacrificed with tumour present; SWT) or not (sacrificed with no tumour; SNT). Data pertaining to bladder and liver tumours are shown in Table 5.8 where the months with zero entries in the SNT and SWT columns are months where deaths were observed but no sacrifices were carried out for the piecewise-constant intensities. As in Problem 5.12 (c), the times of natural deaths are interval censored. Maximize the likelihood in (c) of Problem 5.10 using cut-points at 12 and 18 months for the piecewise-constant intensities. Assess the effect of exposure to 2-AAF versus control (no exposure) on the incidence of bladder and liver tumours.

(Section 5.6; Lindsey and Ryan, 1993)

Problem 5.12   Large cohort studies such as the Canadian Longitudinal Study on Aging (Raina et al., 2009) create a platform for research on risk factors for the development and progression of chronic diseases. Consider a study in which individuals 60−70 years of age are to be recruited and followed for 18 years, with detailed cognitive assessments planned at study entry and every 3 years thereafter. An illness-death model with state 1 representing intact cognition, state 2 representing dementia and state 3 death is used to help plan the study. It is expected that the mortality rate will be 60% over the 18 years from recruitment. Death times are observable subject only to right censoring, but dementia status is observed only at the 3-year cognitive assessments. If a goal is to assess the relationship of sex ($X = 1$ for female; $X = 0$ for male) to the incidence of dementia, consider a model with intensities

$$\lambda_{kl}(t \mid x) = \lambda_{kl0}(t) \exp(\beta_{kl}x)$$

where for $(kl) = (12)$, (13) and (23), $\lambda_{kl0}(t)$ has a piecewise-constant form with cut-points at 65, 70, 75, 80, and 85 years of age.

(a) Write the likelihood based on the data that will be acquired for an individual recruited at age a_0 who is in state 1.

(b) Assume that the age at recruitment is uniformly distributed over the interval $(60, 70)$ years of age. Discuss the kind of external information that is required to determine the power of a study with $n = 30,000$ individuals and an equal number of males and females recruited.

(c) Recruitment of individuals and each follow-up assessment incur costs. Let D_1 denote the cost of recruiting an individual to the cohort, and let D_2 denote the cost of each assessment. Derive the expected cost of this study assuming there is no random censoring due to study withdrawal before 18 years or death.

(Section 5.6; Raina et al., 2009)

Table 5.8: Summary data on the outcomes for 284 mice exposed to a carcinogen 2-AAF (150 parts per million (ppm)) and 387 exposed to a control, and assessed for the presence of tumours in the bladder and liver (Lindsey and Ryan, 1993).

Control

Month	Bladder DNT	DWT	SNT	SWT	Liver DNT	DWT	SNT	SWT
4	0	1	0	0	1	0	0	0
5	1	1	0	0	1	1	0	0
7	1	1	0	0	2	0	0	0
8	1	0	0	0	1	0	0	0
10	0	0	23	0	0	0	23	0
13	0	0	24	0	0	0	24	0
14	2	0	23	0	2	0	23	0
15	2	0	0	0	1	1	0	0
16	1	0	20	0	1	0	20	0
17	1	1	47	0	2	0	47	0
18	3	0	42	0	3	0	42	0
19	3	0	134	1	3	0	134	1
20	1	0	0	0	1	0	0	0
22	6	0	0	0	6	0	0	0
23	3	0	0	0	3	0	0	0
24	5	1	0	0	5	1	0	0
25	5	2	0	0	5	2	0	0
26	4	2	0	0	5	1	0	0
27	3	2	0	0	3	0	0	0
28	7	2	0	0	7	2	0	0
29	3	1	0	0	4	0	0	0
30	4	0	0	0	4	0	0	0
31	4	0	0	0	4	0	0	0
32	1	0	0	0	1	0	0	0

High Dose (150 ppm)

Month	Bladder DNT	DWT	SNT	SWT	Liver DNT	DWT	SNT	SWT
1	3	0	0	0	2	1	0	0
2	0	1	0	0	0	1	0	0
3	2	0	0	0	2	0	0	0
5	0	3	0	0	1	2	0	0
8	0	1	0	0	1	0	0	0
9	1	0	0	0	1	0	0	0
10	0	0	22	0	0	0	22	0
12	1	1	0	0	2	0	0	0
13	0	1	19	4	0	1	23	0
14	2	2	19	4	4	0	23	0
15	0	2	0	0	2	0	0	0
16	0	3	12	9	1	2	20	1
17	2	3	12	9	3	2	21	0
18	0	1	11	9	1	0	20	0
19	1	1	64	38	2	0	97	5
20	1	3	0	0	2	2	0	0
21	1	1	0	0	2	0	0	0
22	0	2	0	0	2	0	0	0
23	0	1	0	0	1	0	0	0
24	0	1	0	0	1	0	0	0
25	0	3	0	0	2	1	0	0
27	0	3	0	0	1	2	0	0
28	0	1	0	0	1	0	0	0
29	0	1	0	0	1	0	0	0
31	0	1	0	0	0	0	0	0
32	0	1	0	0	0	1	0	0
33	0	1	0	0	0	1	0	0

† DNT: died with no tumour, DWT: died with tumour, SNT: sacrificed with no tumour, SWT: sacrificed with tumour.
Counts represent number of events since the previous assessment time.

Problem 5.13   Consider a reversible illness-death model with transition intensities λ_{12}, λ_{21}, λ_{13} and λ_{23}. Suppose we have data on a cohort of individuals by time intervals $(0,1]$, $(1,2]$ and $(2,3]$ representing year of follow-up, $r = 1, 2$ and 3. The data consist of counts for the total number of transition n_{klr} of each type $(l \neq k)$ and the total time at risk in each state S_{kr}, for $k = 1, 2$ and $r = 1, 2, 3$; see the table below:

			$r = 1$		$r = 2$		$r = 3$	
$k = 1$	$l = 1$	$n_{121} = 395$	$S_{11} = 980$	$n_{122} = 400$	$S_{12} = 800$	$n_{123} = 280$	$S_{13} = 700$
	$l = 2$	$n_{131} = 1$		$n_{132} = 1$		$n_{133} = 1$	
$k = 2$	$1 = 1$	$n_{211} = 1040$	$S_{21} = 2500$	$n_{212} = 500$	$S_{22} = 2600$	$n_{213} = 320$	$S_{23} = 2500$
	$l = 2$	$n_{231} = 2$		$n_{232} = 1$		$n_{233} = 3$	

(a) Fit a continuous-time Markov model with piecewise-constant intensities for each year. Test the hypothesis that the intensities are actually constant across the 3 years.

(b) Give confidence intervals for the expected durations of sojourns in states 1 and 2.

(Section 5.6)

Chapter 6

Heterogeneity and Dependence in Multistate Processes

Despite the careful collection and use of information on endogenous and exogenous factors affecting life history processes, models often fail to describe such processes in their full complexity. In many contexts, one can view this inadequacy as arising from unobserved or latent factors that explain the excessive variation relative to a posited model. The effects of such unobserved variables can often be incorporated into a model by introducing random effects. The basic ideas of this approach were introduced in Section 2.3.4. In this chapter we discuss the role of random effect models for dealing with heterogeneity in life history processes more extensively, along with their role in modeling dependence between different multistate processes. Alternative approaches for dealing with dependence between multistate processes are also discussed, as are other applications of models involving latent processes.

6.1 Accommodating Heterogeneity in Life History Processes

6.1.1 Frailty Models in Survival Analysis

There has been considerable attention given to studying the effect of unexplained variation in risk in hazard-based models for survival data. If T is a survival time and X denotes a $p \times 1$ vector of covariates, consider a proportional hazards model $h(t|x) = h_0(t) \exp(x'\beta)$, where $h_0(t)$ is a baseline hazard function. If there is variation in risk between individuals in the population of interest, beyond that explained by the effect of X, it is often conveniently accommodated by introducing a latent individual-specific variable, which is typically assumed to be independent of X.

The hazard conditional on (X, V) is then defined as

$$\lim_{\Delta t \downarrow 0} \frac{P(T < t + \Delta t \mid T \geq t, X = x, V = v)}{\Delta t} = h(t \mid x, v). \tag{6.1}$$

Most commonly v is taken to be non-negative and assumed to arise from a distribution $G(\cdot)$ with mean 1 and variance ϕ, and the effect is modeled multiplicatively as $h(t|x, v) = v\, h_0(t) \exp(x'\beta)$. The constraint $E(V) = 1$ is introduced for identifiability reasons and also maintains the interpretation of $h_0(t)$ as a baseline hazard function. We refer to V as a latent variable, because it is unobserved, and we use the term *random effect* because V is a random variable. In the survival context, the term *frailty* is often used since in that setting V is conceptualized as reflecting unmeasured factors characterizing the robustness of an individual's health. For individuals with a given covariate vector $X = x$, values $v < 1$ would be held by individuals having attributes associated with lower risk (hazard) of failure compared to the average person with $X = x$, whereas $v > 1$ is associated with higher than average risk.

It is also possible to consider random coefficients, whereby the effect of covariates varies between individuals. This is routinely done in the class of linear mixed models but is less common in the multistate setting. The conditional nature of the transition intensities and the selection criteria often used in life history studies (see Section 7.1) can make it challenging to fit and interpret such models. We therefore focus here on multiplicative random effects acting on baseline intensities to reflect heterogeneity or dependencies between processes.

The hazard in (6.1) is called a conditional (or subject-specific) hazard, and the regression coefficient is interpreted as reflecting a subject-specific effect of a change in X, since it is defined conditionally on V. Under a multiplicative random effect model, the survival function for an individual with particular values (x, v) is $P(T \geq t|x, v) = \exp(-v\, H(t|x))$, where $H(t|x) = H_0(t)\exp(x'\beta)$ and $H_0(t) = \int_0^t h_0(u)\, du$ is the cumulative baseline hazard. To characterize the survival distribution for the population of individuals with $X = x$, we must average over the possible realizations of $V = v$ to obtain

$$P(T \geq t \mid X = x) = E_{V|X}\{P(T \geq t \mid X = x, V)\} \qquad (6.2)$$
$$= \int_0^\infty \exp(-v\, H(t \mid x))\, dG(v; \phi).$$

Temporal trends in "instantaneous risk" for particular individuals are reflected by the baseline hazard $h_0(t)$, or equivalently the slope of $H_0(t)$. If interest lies in characterizing temporal trends in risk in the population, however, it is necessary to examine the marginal (population) hazard function. The marginal hazard is $h^m(t|x) = -\partial \log P(T \geq t|x)/\partial t$, which by (6.2) is

$$h^m(t \mid x) = E_{V|T \geq t, X}\{V\, h(t \mid x)\} = h(t \mid x) \int_0^\infty v\, dG(v \mid T \geq t, x) \qquad (6.3)$$

where $dG(v|T \geq t, x) = P(T \geq t|x, v)\, dG(v)/P(T \geq t|x)$; $H^m(t|x) = \int_0^t h^m(s|x)\, ds$.

By writing the marginal hazard as in (6.3), two important observations can be made. First, when we restrict attention to individuals who are alive at some time $t > 0$, the distribution of the random effect V depends on t and x. Specifically, as t increases the sub-population of individuals remaining event-free and thus at risk is comprised of a higher proportion of individuals with lower frailty, and so the marginal hazard for death at time t at the population level is smaller than $h(t|x)$. It may therefore be difficult to distinguish between the heterogeneity of the form reflected by (6.1), and an inadequacy in the specification of $h_0(t)\exp(x'\beta)$; this inadequacy may, for example, be of a parametric specification of $h_0(t)$, or the multiplicative form of the covariate effects. A second point is that the marginal hazard (6.3) does not satisfy the proportional hazards assumption. The simple relative risk interpretation of the effect of X is confined to the setting where V is fixed (i.e. the elements of β can be interpreted as hazard ratios conditional on $V = v$).

To explore these phenomena further, it is instructive to consider a simple example where X is a Bernoulli random variable. Suppose $h(t|x, v) = v\, h(t|x)$, where

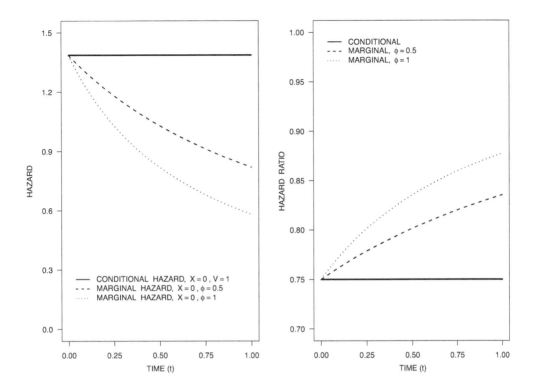

Figure 6.1: Plot of the conditional and marginal baseline hazard functions (left panel), and the conditional and marginal hazard ratio functions (right panel).

$h(t|x) = \lambda \exp(\beta x)$ is a constant (or exponential) hazard function. Since $h(t|x = 1, v)/h(t|x = 0, v) = \exp(\beta)$, β reflects the multiplicative effect of a one unit change in x for a given $V = v$. If V is gamma distributed with $E(V) = 1$ and $\mathrm{var}(V) = \phi$, then $P(T \geq t|x) = (1 + t\phi\lambda e^{\beta x})^{-\phi^{-1}}$ by (6.2). This gives $H^m(t|x) = \phi^{-1} \log(1 + t\phi\lambda e^{\beta x})$ and

$$h^m(t \mid x) = (\lambda \exp(\beta x))/(1 + t\phi\lambda \exp(\beta x)).$$

Note that $h^m(t|x) < h(t|x, v = 1)$ for $t > 0$ and that $h^m(t|x)$ is a decreasing function of t reflecting the phenomenon mentioned earlier. Moreover, the marginal hazard ratio for individuals with $X = 1$ versus $X = 0$ at time t is

$$\frac{h^m(t \mid X = 1)}{h^m(t \mid X = 0)} = \frac{1 + t\phi\lambda}{1 + t\phi\lambda e^{\beta}} \cdot e^{\beta}, \tag{6.4}$$

which depends on time whenever $\beta \neq 0$ and $\phi > 0$. If $\phi = 0$, then there is no heterogeneity, the model is exponential and the hazard ratio reduces to $\exp(\beta)$.

The left panel of Figure 6.1 displays the baseline hazard $h_0(t)$ in the conditional specification (6.1) when $\lambda = \log 4$ and $P(T > 1|x = 0, v = 1) = 0.25$. Also displayed are the marginal hazards (6.3) when $\phi = 0.5$ and $\phi = 1.0$. The decreasing marginal hazard induced by the increased proportion of lower-risk survivors is apparent, with the effect being greater for larger ϕ (i.e. the more heterogeneity there is in the population). The hazard ratio for the conditional model when $\exp(\beta) = 0.75$

and the marginal hazard ratios (6.4) when $\phi = 0.5$ and 1.0 are plotted in the right panel. The trend in the hazard ratio for the marginal model reflects the fact that the proportional hazards property does not hold unconditionally.

Hougaard (2000), Duchateau and Janssen (2008) and Wienke (2011) give comprehensive accounts of frailty modeling with censored data. There are several functions available for fitting survival models with random effects to accommodate heterogeneity or clustering in survival times. The `coxph` function in R fits Cox models with multiplicative random effects; illustrations are given in Therneau and Grambsch (2000). More elaborate random effect structures are accommodated with the `coxme` package for semiparametric Cox models (Therneau, 2012). General frailty distributions are accommodated in models with parametric baseline hazards with the `parfm` function (Munda et al., 2012). The suite of functions in the `frailtypack` package accommodates conditional Cox models with more complex dependence structures through shared, nested, joint and additive random effects (Rondeau et al., 2012).

6.1.2 A Progressive Multistate Model with Random Effects

Random effect models can play a useful role in the analyses of multistate processes for similar reasons to the survival setting, but there are issues warranting careful consideration. We will focus primarily on two specific types of processes. The first is a simple K-state progressive process $\{Z(s), s \geq 0\}$ with a state space depicted in Figure 6.2. Suppose that

$$\lim_{\Delta t \downarrow 0} \frac{P(Z(t + \Delta t^-) = k + 1 \mid Z(t^-) = k, \mathcal{H}(t^-), V = v)}{\Delta t} = \lambda(t \mid \mathcal{H}(t^-), v), \qquad (6.5)$$

where V is a scalar subject-specific random effect with mean one and variance ϕ, and $\mathcal{H}(t) = \{Z(s), 0 \leq s \leq t; X\}$ denotes the process history including an observed covariate X.

Figure 6.2: A progressive multistate process.

If each individual makes transitions according to a Markov process, then

$$\lambda(t \mid \mathcal{H}(t^-), v) = \lambda_k(t \mid x, v) \quad \text{if} \quad Z(t^-) = k.$$

A simple model with a multiplicative random effect, analogous to (6.1), takes $\lambda(t|\mathcal{H}(t^-), v) = v\,\lambda_k(t|x)$, with the random effect acting identically on all intensities. Thus, if $v > 1$ the rate at which transitions are made through the states is greater than the average given x; if $v < 1$ it is lower.

Given an individual is in state k at time t, then the observable process is governed by "marginal" intensities defined by

$$\lambda_k^m(t \mid \mathcal{H}(t^-)) = E\{V \lambda_k(t \mid x) \mid \mathcal{H}(t^-)\} = \lambda_k(t \mid x) E\{V \mid \mathcal{H}(t^-)\}, \qquad (6.6)$$

where $\mathcal{H}(t^-)$ includes the fact that $Y_k(t^-) = 1$. While analogous to (6.3), the evaluation of $E\{V|\mathcal{H}(t^-)\}$ is more complicated here because the event history, which in (6.3) is simply $T \geq t, X = x$, now involves more information. Specifically, if $t_0 = 0$ and $t_1 < \cdots < t_{k-1}$ denote the exit times from states 1 to $k-1$ ($k < K$), then to evaluate $E\{V|\mathcal{H}(t^-)\}$ in (6.6) we need to compute

$$\int_0^\infty v P(Z(u), 0 < u < t \mid x, v)\, dG(v) / P(Z(u), 0 < u < t \mid x)$$

where the denominator is

$$\prod_{j=1}^{k-1} \lambda_{j0}(t_j \mid x) \int_0^\infty v^{k-1} \exp\left(-v\left\{\sum_{j=1}^{k-1} \int_{t_{j-1}}^{t_j} \lambda_{j0}(u \mid x)\,du + \int_{t_{k-1}}^{t} \lambda_{k0}(u \mid x)\,du\right\}\right) dG(v),$$

and the numerator is of a similar form with v^{k-1} replaced with v^k. It is apparent that the intensity in (6.6) depends on the process history in a particular way, so the Markov property of the conditional model is not retained in the marginal model.

To explore this further, consider a 3-state Markov process given by Figure 6.2 with $K = 3$. Consider the case with no covariates and time-homogeneous baseline intensities $\lambda_k(t|\mathcal{H}(t^-), v) = v\lambda_k$, $k = 1, 2$. In this setting, if a subject is in state 2 at time t, their history is comprised only of the entry time to state 2 and the fact that they have not yet left state 2, so (6.6) becomes

$$\lambda_2^m(t \mid \mathcal{H}(t^-)) = \frac{Y_2(t^-)(1+\phi)\lambda_2}{1 + \phi(\lambda_1 t_1 + \lambda_2(t - t_1))} = \frac{Y_2(t^-)(1+\phi)\lambda_2}{1 + \phi[(\lambda_1 - \lambda_2)t_1 + \lambda_2 t]}.$$

Thus, while the process is Markov given v (i.e. for each individual), the intensity of a $2 \to 3$ transition at t depends on t_1 as well as the time since entry to state 2. In the multistate context, the random effect model therefore does not allow one to distinguish clearly between heterogeneity in a time-homogeneous process, and an intensity that is dependent on the process history.

Returning to the K-state model of Figure 6.2, if we let $K \to \infty$ then the model can be used to characterize a recurrent event process where the state reflects the cumulative number of events: $N(t) = Z(t) - 1$ counts the number of events (transitions) over the interval $[0, t]$. If we specify the intensity of the form

$$\lambda(t \mid \mathcal{H}(t^-), v) = v\rho(t \mid x) = v\rho_0(t)\exp(x'\beta),$$

the model corresponds to a mixed Poisson process with multiplicative covariate effects acting on a baseline rate function $\rho_0(t)$ conditional on $V = v$. In this case,

$$E\{N(t) \mid x\} = \mu(t \mid x) = \mu_0(t)\exp(x'\beta),$$

is called the cumulative mean function, where $\mu_0(t) = \int_0^t \rho_0(s)\,ds$. If x is a scalar covariate, then $\exp(\beta)$ represents the multiplicative effect of a one-unit increase of

x on the event rate, or equivalently, the effect on the expected number of events over a given interval of time. However, the covariate effect on the marginal event intensity is not multiplicative. If V is gamma distributed, for example, the marginal process is negative binomial, and the intensity (6.6) is of the form

$$\lambda^m(t \mid \mathcal{H}(t^-)) = \frac{1 + \phi N(t^-)}{1 + \phi \mu(t \mid X)} \rho(t \mid x).$$

In settings where there is some type of renewal upon the occurrence of each transition, semi-Markov processes are appealing; see Section 2.3.2. The focus in such settings turns to the analysis of sojourn times; if $T_1 = 0$, then $W_k = T_{k+1} - T_k$ denotes the sojourn time in state k, $k = 1, \ldots, K - 1$. If there is considerable heterogeneity in the sojourn times between individuals in a population, this can be accommodated through models (6.5) of the form

$$\lambda(t \mid Z(t^-) = k, \mathcal{H}(t^-), V = v) = h_k(B(t) \mid X, V = v) \tag{6.7}$$

where $B(t) = t - t_k$ is the time since entry to the current state, and $h_k(\cdot \mid x, v)$ is a hazard function for the sojourn time in state k given x and v. The full spectrum of modeling approaches available for the analysis of survival data may be considered for the analysis of sojourn times. Consider, for example, a sample of n individuals $i = 1, \ldots, n$ and let x_{ik} denote the covariate vector associated with the sojourn in state k with $x_i = (x'_{i1}, \ldots, x'_{i,k-1})'$. Location-scale models (Lawless, 2003) can be constructed of the form

$$\log W_{ik} = X'_{ik} \alpha_k + \log V_i + \sigma R_{ik}, \quad k = 1, \ldots, K - 1 \tag{6.8}$$

where it is assumed that X_i, V_i and R_{ik} are mutually independent, and R_{ik} has a standard error distribution in the location-scale family of models (e.g. extreme value, logistic, normal). In this case given $V_i = v_i$, covariate effects can be expressed as they are for accelerated failure time models. The introduction of the random effect in (6.8) induces a serial dependence between the gap times $W_{i1}, W_{i2}, \ldots, W_{i,k-1}$. Specifically,

$$\text{var}(\log W_{ij} \mid X_i) = \text{var}(\log V_i) + \sigma^2 \text{var}(R_{ij})$$

and

$$\text{cov}(\log W_{ij}, \log W_{ik} \mid X_i) = \text{var}(\log V_i).$$

A more general location-scale model can be obtained by letting

$$\log W_{ik} = X'_{ik} \alpha_k + R_{ik}, \quad k = 1, \ldots, K - 1, \tag{6.9}$$

and adopting a multivariate distribution for $R_i = (R_{i1}, \ldots, R_{i,K-1})'$, where marginally R_{ik} has an error distribution with dispersion parameter σ_k, $k = 1, \ldots, K - 1$. If each R_{ik} has a normal distribution, a multivariate normal distribution can be adopted for R_i with $\text{corr}(R_{ij}, R_{ik}) = \rho_{ik}$. More generally, copula models can be used to construct suitable joint distributions (see Section 6.2).

If the goal is to express covariate effects multiplicatively on a baseline hazard for a sojourn time distribution, we could have

$$h_k(B_i(t) \mid x_i, v_{ik}) = v_{ik} \, h_{k0}(B_i(t)) \exp(x_i'\beta_k) \tag{6.10}$$

where the V_{ik} have some frailty distribution, with a joint distribution for $V_i = (V_{i1}, \ldots, V_{i,k-1})$; copula models may play a useful role. The case where $v_{ik} = v_i$ corresponds to a shared random effect and hence stronger assumptions about the dependence structure.

6.1.3 Random Effect Models with Recurrent States

Continuous random effects can also be useful in other multistate processes, including ones where multiple sojourns may be realized in some states. The simplest example is the 2-state process depicted in Figure 6.3, which is useful in settings where there are alternating periods of disease activity and inactivity. Examples include recurrent outbreaks of symptoms in individuals infected with the herpes simplex virus, recurrent exacerbations in individuals with chronic bronchitis or recurrent hospitalizations for persons with psychiatric disorders. In such settings, there are often features that appear similar over time for a given individual. For example, some individuals with chronic bronchitis may tend to experience frequent exacerbations of a shorter than average duration, and other individuals may tend to have less frequent exacerbations but of longer than average durations when they occur. Available covariates may explain much of this variation but often there is insufficient information to adequately explain all of it. Random effects can play a useful role in both modeling the heterogeneity and accommodating the dependence in the repeated sojourn times.

Figure 6.3: A 2-state diagram for relapsing and remitting conditions.

We consider the formulation of a particular model for such processes. We let $\{Z_i(s), s \geq 0\}$ be a 2-state process depicted in Figure 6.3. If $X_i(t)$ denotes a vector of fixed and defined time-dependent covariates associated with individual i, then $\mathcal{H}_i(t) = \{Z_i(s), X_i(s), 0 \leq s \leq t\}$ denotes the history. We could then let (V_{i1}, V_{i2}) be a bivariate random effect, with

$$\lim_{\Delta t \downarrow 0} \frac{P(Z_i(t+\Delta t^-) = 3-k \mid Z_i(t^-) = k, \mathcal{H}_i(t^-), v_{ik})}{\Delta t} = v_{ik} \, \lambda_k(t \mid \mathcal{H}_i(t^-)), \tag{6.11}$$

$k = 1, 2$. This model features a general conditional intensity $\lambda_k(t|\mathcal{H}_i(t^-))$ and a multiplicative random effect that is unique to each individual and transition type.

In any given application, the specification of the conditional "baseline" intensities $\lambda_1(t|\mathcal{H}_i(t^-))$ and $\lambda_2(t|\mathcal{H}_i(t^-))$ should be based on the scientific context. In modeling the recurrent exacerbations in an observational study of patients with chronic bronchitis, for example, the time origin may be the date of disease diagnosis, in which case t represents the time since this diagnosis. The vector $X_i(t)$ may contain fixed covariates reflecting demographic factors such as age and sex, presence of comorbidities and information on smoking history. For the intensity governing the onset of symptoms, a Markov form is often reasonable because risk may change over time, for example, due to the cumulative effect of lung damage over the course of the chronic condition. A natural model is then

$$\lambda_1(t \mid \mathcal{H}_i(t^-)) = \lambda_1(t) \exp(x'_{i1}(t)\beta_1). \tag{6.12}$$

The covariates $X_{i1}(t)$ can include environmental or seasonal factors associated with exacerbation onset, and trends in baseline risk of exacerbations can be examined as a function of disease duration by plotting estimates of $\Lambda_1(t)$.

The intensity for the resolution of symptoms, $\lambda_2(t|\mathcal{H}_i(t^-))$, on the other hand, is often more reasonably based on a semi-Markov time scale since the duration of an exacerbation may depend primarily on the time since it began, and therapeutic interventions are often introduced when it begins or soon thereafter. The time since the onset of the underlying chronic condition and the age of individuals may also play an important role, of course, but these can be incorporated as elements of the covariate vector $X_{i2}(t)$ in a model such as

$$\lambda_2(t \mid \mathcal{H}_i(t^-)) = h_2(B_i(t)) \exp(x'_{i2}(t)\beta_2), \tag{6.13}$$

where $B_i(t)$ is the time since most recent entry to state 2. The function $h_2(\cdot)$ is a baseline hazard function for the sojourn time distribution for state 2 for an individual with $x_{i2}(t) = 0$.

The latent individual-specific terms $v_i = (v_{i1}, v_{i2})'$ in the intensities in (6.11) accommodate unexplained variation in the onset and resolution of symptomatic periods, and it is customary to assume they arise from a bivariate distribution $G(v_i)$. For identifiability and ease of interpretation, we typically constrain $E(V_{ik}) = 1$ and let $\text{var}(V_{ik}) = \phi_k$, $k = 1, 2$, and $\text{cov}(V_{i1}, V_{i2}) = \phi_{12}$. A bivariate log-normal distribution is often adopted in this setting, and when $\phi_{12} = 0$ the observed data likelihood can be written as a product of two marginal likelihoods, each based on log-normal frailty models. An alternative approach is to consider a more general class of bivariate distributions using copula functions to link any pair of marginal random effect distributions. We describe this next.

If U_{ik} is a uniform random variable on the range $[0, 1]$, $k = 1, 2$, then we let $\mathcal{C}(u_{i1}, u_{i2}; \phi_{12}) = P(U_{i1} \leq u_{i1}, U_{i2} \leq u_{i2}; \phi_{12})$ denote the bivariate cumulative distribution function of $U_i = (U_{i1}, U_{i2})$, where ϕ_{12} is a dependence or association parameter. Any such function $\mathcal{C}(\cdot, \cdot; \phi_{12})$ is called a copula, and there is a wide class of such functions (see Joe, 1997; Nelsen, 2006). In the Archimedean family of copulas, $\mathcal{C}(u_{i1}, u_{i2}; \phi_{12})$ can be written as

$$\mathcal{C}(u_{i1}, u_{i2}; \phi_{12}) = \mathcal{B}^{-1}(\mathcal{B}(u_{i1}; \phi_1) + \mathcal{B}(u_{i2}; \phi_2); \phi_{12}),$$

where $\mathcal{B}\colon [0,1] \to [0,\infty)$ is a continuous, strictly decreasing and convex generator function satisfying $\mathcal{B}(1;\phi_{12}) = 0$. Special copulas within the Archimedean family include the Clayton copula, which is widely used in survival analysis, with generator function $\mathcal{B}(s;\phi_{12}) = \phi_{12}^{-1}(s^{-\phi_{12}} - 1)$. The Frank copula has generator $\mathcal{B}(s;\phi_{12}) = -\log((\exp(-\phi_{12}s) - 1)/(\exp(-\phi_{12}) - 1))$, and the Gumbel copula has generator $\mathcal{B}(s;\phi_{12}) = (-\log s)^{\phi_{12}}$. If we let $G_k(v_{ik};\phi_k)$ denote the cumulative distribution function (c.d.f.) for V_{ik}, then the probability integral transform $U_{ik} = G_k(V_{ik};\phi_k)$ is a uniform random variable, $k = 1,2$. Thus, by letting $U_{ik} = G_k(V_{ik};\phi_k)$ in the copula, we may write $G(v_i;\phi) = \mathcal{C}(G_1(v_{i1};\phi_1), G_2(v_{i2};\phi_2);\phi_{12})$. Applying the probability integral transformation to the Clayton copula, for example, leads to the joint c.d.f. for V_i of the form

$$G(v_{i1}, v_{i2};\phi) = (G_1(v_{i1};\phi_1)^{-\phi_{12}} + G_2(v_{i2};\phi_2)^{-\phi_{12}})^{-1/\phi_{12}},$$

where $\phi = (\phi_1, \phi_2, \phi_{12})'$.

Association between V_{i1} and V_{i2} can be measured by Kendall's τ, a widely used dependence measure for non-negative random variables. Consider two draws from a bivariate distribution yielding $V_i = (V_{i1}, V_{i2})$ and $V_i' = (V_{i1}', V_{i2}')$. Kendall's τ is the probability that both elements of one vector are either larger or smaller than the corresponding elements of the other (concordance) minus the probability that one element is larger and one is smaller (discordance). We can express this as

$$\tau = P((V_{i1} - V_{i1}')(V_{i2} - V_{i2}') > 0) - P((V_{i1} - V_{i1}')(V_{i2} - V_{i2}') < 0).$$

For bivariate distributions in the Archimedean family, this can be expressed directly in terms of the generator function as

$$\tau = 1 + 4\int_0^1 \frac{\mathcal{B}(s;\phi_{12})}{\mathcal{B}'(s;\phi_{12})}\, ds.$$

The appeal of copula models for multivariate random effect distributions is that any marginal distributions can be adopted for V_{i1} and V_{i2} and linked with a copula function indexed by a functionally independent parameter characterizing the dependence between the random variables. A broad class of flexible bivariate distributions is therefore encompassed, including distributions with gamma margins; for the special case where $V_{i1} \perp V_{i2}$, these reduce to the familiar gamma frailty models. We next discuss how such models can be fitted in the context of a 2-state alternating process.

Consider a dataset in which individual i is observed over $[0, C_i]$, where $\mathcal{D}_{ik}$ is the set of their $k \to 3 - k$ transition times. The conditional likelihood given the random effect $v_i = (v_{i1}, v_{i2})'$, is proportional to the probability of the particular realization of the process given v_i, which by (2.9) is

$$\prod_{k=1}^2 \left\{ \prod_{t_r \in \mathcal{D}_{ik}} v_{ik}\, \lambda_k(t_r \mid \mathcal{H}_i(t_r^-)) \exp\left(-\int_0^\infty v_{ik}\, \bar{Y}_{ik}(s)\, \lambda_k(s \mid \mathcal{H}_i(s^-))\, ds\right) \right\}$$

with $\bar{Y}_{ik}(t) = Y_i(t)Y_{ik}(t^-)$, $Y_i(t) = I(t \le C_i)$ and $\bar{Y}_{ik}(t) = I(Z_i(t) = k)$. The observed data likelihood is given by the integral

$$\int_0^\infty \int_0^\infty \prod_{k=1}^2 \Big\{ \prod_{t_r \in \mathcal{D}_{ik}} v_{ik}\, \lambda_k(t_r \mid \mathcal{H}_i(t_r^-)) \tag{6.14}$$

$$\times\; \exp\!\Big(-\int_0^\infty v_{ik}\, \bar{Y}_{ik}(s)\, \lambda_k(s \mid \mathcal{H}_i(s^-))\, ds\Big) \Big\}\, dG(v_i).$$

Numerical integration can be used to compute (6.14) with estimation carried out using direct maximization of the observed data likelihood. Alternatively, an expectation-maximization algorithm can be adopted, which facilitates semiparametric modeling with unspecified baseline intensities.

6.1.4 Analysis of Exacerbations in Chronic Bronchitis

Chronic bronchitis is a type of chronic obstructive pulmonary disease in which affected individuals experience periodic inflammation of the bronchial tubes, which in turn leads to a build-up of mucus in the lungs causing difficulty in breathing, sneezing, tightness in the chest and coughing. Effective therapies open the airway passages, moderate or alleviate symptoms, and reduce inflammation.

Figure 6.4 displays profiles for a sample of 20 patients who took part in a health economic study designed to evaluate the cost-effectiveness of daily administration of ciprofloxacin versus standard care over one year of follow-up (Grossman et al., 1998). The lengths of the successive darker and lighter lines represent the duration of periods spent in each state. The 10 individuals in the top half of the plot were chosen from the ciprofloxacin arm and the 10 in the bottom half were from the standard care arm. There is evident heterogeneity in the frequency and duration of exacerbations, but it is unclear whether this is adequately explained by the available covariates, which we discuss below.

We adopt a conditional intensity for the onset of exacerbations given by (6.12). Because the treatment was assigned at study entry, here we take t as the time since randomization and use disease duration at the time of study entry as a covariate; we specify a piecewise-constant baseline rate $\lambda_1(t)$ with break-points every 60 days over a 360-day period. Covariates include treatment (ciprofloxacin vs. standard care), sex (female vs. male), severity of chronic bronchitis (severe vs. not severe), duration of symptoms of the exacerbation at the time of randomization (days), and disease duration at the time of recruitment (years). The conditional intensity for the resolution of exacerbations has the form of (6.13) with $B(t)$ the days since the onset of the exacerbation. As discussed in Section 6.1.3, this time scale is natural for the duration of exacerbations. The baseline hazard $h_2(s)$ is also taken to have a piecewise-constant form but with break-points every 10 days over a 50-day period. The same covariates used in the model for the onset of exacerbations were considered for this intensity.

In order to avoid complexities related to the initial conditions in the randomized study (individuals had to be suffering from an exacerbation at the time of contact to be recruited to the study), we omit the time from randomization to the resolution of the initial exacerbation; we discuss the issue of the initial conditions more fully in Section 7.1.5.

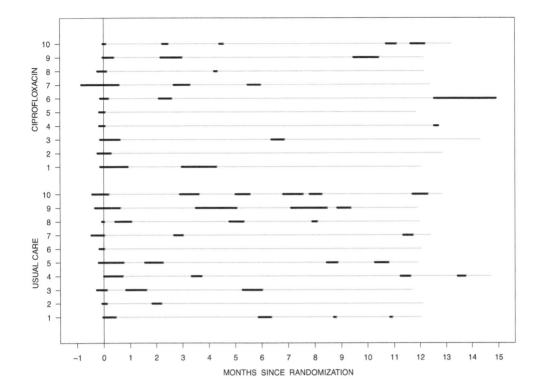

Figure 6.4: Sample profiles for 20 individuals from the study of recurrent exacerbations of patients with chronic bronchitis (Grossman et al., 1998); thicker segments represent periods of symptom exacerbation.

We adopt gamma distributions for the random effects V_1 and V_2 and define their joint distribution by a Gaussian copula. The log-likelihood based on (6.14) can then be computed using the `quad2d` function in the `pracma` package in R. The maximization was carried out using the all-purpose optimization function `nlm`, which also furnishes numerical second derivatives that can be used to obtain standard errors based on the observed information matrix. The results of fitting this model are given in the first column of Table 6.1.

The estimated Pearson correlation between the random effects from the Gaussian copula is 0.415 (95% CI: -0.412, 0.867), which is not statistically significantly different from zero. While this suggests a model with independent random effects can be fitted and interpreted, there is merit to retaining the more general structure because invoking the independence assumption puts one at risk of induced dependent censoring of gap times (Cook and Lawless, 2007, Section 4.4). The range of values in the 95% confidence interval for the correlation parameter in the Gaussian copula is quite large, and the fact that a statistically significant correlation was not declared does not imply an independence assumption is correct.

Based on the model with the Gaussian copula, we find mild evidence of women having a slightly higher risk of developing a new exacerbation ($\exp(0.239) = 1.27$, 95% CI: 0.98, 1.64; $p = 0.069$). People with a severe form of chronic bronchitis

Table 6.1: Results of fitting random effect models for the onset and resolution of exacerbations in individuals with chronic bronchitis with correlated gamma random effects under a Gaussian copula and under independence assumptions.

Parameter	Gaussian Copula Piecewise-Constant			Independent Random Effects Piecewise-Constant			Semiparametric		
	EST	SE	p	EST	SE	p	EST	SE	p
Conditionally Markov Intensity for Exacerbation-Free to Exacerbation Transitions									
Treatment	-0.043	0.130	0.740	-0.039	0.130	0.765	-0.035	0.130	0.787
Sex	0.239	0.132	0.069	0.235	0.132	0.075	0.231	0.132	0.080
Severity	0.547	0.180	0.002	0.546	0.180	0.003	0.548	0.180	0.002
Symptoms (Days)	-0.005	0.010	0.636	-0.005	0.010	0.640	-0.005	0.010	0.622
Bronchitis Duration (Yrs)	0.016	0.006	0.010	0.017	0.006	0.010	0.016	0.006	0.011
Conditionally Semi-Markov Intensity for Exacerbation to Exacerbation-Free Transitions									
Treatment	0.034	0.122	0.779	0.037	0.123	0.763	0.064	0.142	0.654
Sex	-0.073	0.125	0.557	-0.066	0.125	0.596	-0.060	0.143	0.677
Severity	-0.060	0.164	0.716	-0.075	0.166	0.653	-0.096	0.193	0.618
Symptoms (Days)	-0.010	0.009	0.280	-0.011	0.009	0.232	-0.014	0.010	0.171
Bronchitis Duration (Yrs)	-0.003	0.006	0.637	-0.003	0.006	0.614	-0.003	0.007	0.688
Parameters of Random Effect Distribution									
$\log \phi_1$	-1.204	0.301		-1.186	0.297		-1.195		
$\log \phi_2$	-2.197	0.656		-2.161	0.626		-1.429		
Correlation, ρ	0.415								
$\phi_{12} = \log((1+\rho)/(1-\rho))$	0.883	0.897	0.325						

have a much higher rate of developing an exacerbation ($\exp(0.547) = 1.73$, 95% CI: 1.21, 2.46; $p = 0.002$), and there is a slight but statistically significant increase in risk of exacerbations for each additional year since onset of chronic bronchitis ($\exp(0.016) = 1.02$, 95% CI: 1.00, 1.03; $p = 0.010$). The use of ciprofloxacin and the number of days from the onset of the baseline exacerbation to randomization did not have an effect on the risk of exacerbations. None of the covariates considered had an effect on the duration of exacerbations.

There was strong evidence of residual heterogeneity in the rate of exacerbations and the duration of exacerbations. The maximum likelihood estimate of ϕ_1 is $\widehat{\phi}_1 = 0.300$ with a 95% confidence interval $(0.166, 0.541)$, and for the duration of the exacerbations we find $\widehat{\phi}_2 = 0.111$ with a 95% confidence interval $(0.031, 0.402)$.

For illustration, we also fit the independence model, using both the piecewise-constant specifications and semiparametric Cox models, which can be fitted using the `coxph` function in R/S-PLUS with the `frailty` option; the results are given in the second and third columns of Table 6.1, where it can be seen that the findings from the two models are in broad agreement. Figure 6.5 displays the estimates of the cumulative baseline Markov rate $\Lambda_1(t)$ versus t in (6.12) and $H_2(s)$ versus s in (6.13) based on the piecewise-constant and semiparametric models under the

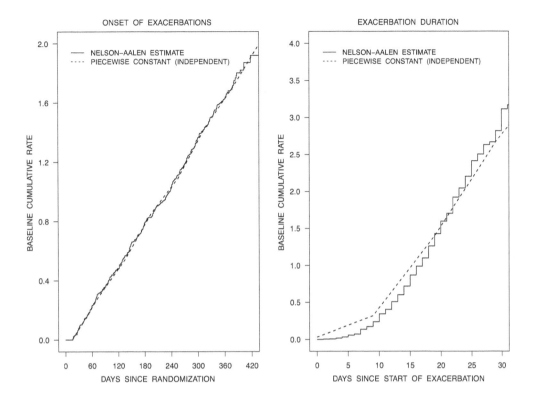

Figure 6.5: Estimated cumulative baseline rates and hazards from fits to the chronic bronchitis data using piecewise-constant and semiparametric intensities with independent gamma random effects.

independence assumption for V_1 and V_2. The two sets of estimates track each other well, suggesting the piecewise-constant specification is reasonable.

6.2 Modeling Correlated Multistate Processes

In studies involving paired or related organ systems, the nature of the dependencies between two or more processes may be of interest. In studies of glaucoma, age-related mascular degeneration (AMD) or other ocular diseases, interest may lie in modeling the declining visual acuity in left and right eyes. In diabetes studies, interest may lie in joint modeling of retinopathy and nephropathy over time; each condition can be modeled by a progressive multistate process, and as they are both driven by poor blood glucose control and the consequent circulatory impairment, some dependence between the two pathologies may be anticipated.

Let $\{Z_1(t), 0 \leq t\}$ and $\{Z_2(t), 0 \leq t\}$ denote two processes of interest. When they arise from paired organs such as eyes in the AMD setting, it may be natural to view the processes as clustered and to constrain the marginal processes to be the same. When the two processes are for different types of organs affected by a common underlying pathophysiology (e.g. poor control of blood sugar), dependence modeling may still be of interest but the marginal processes may have different state spaces and intensities. We assume for convenience that the processes have the same number of states and that interest lies in jointly modeling them to characterize their dependence. Figure 6.6 displays a simple illustrative setting with two 3-state progressive processes that we use for the purpose of discussion in what follows.

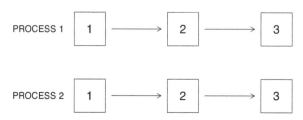

Figure 6.6: State space diagram for two parallel 3-state progressive processes.

Dependence models can be formulated using shared or correlated random effects in a similar spirit to the approaches of Section 6.1, via intensity-based models, or in certain contexts using copula models accommodating dependencies in some aspects (e.g. absorption times) of the processes. Valid joint inferences about parameters for the two processes can also be made by adopting a working independence assumption, fitting the two processes separately, and using robust variance estimates. We discuss each of these approaches in the sections that follow.

6.2.1 Dependence Models Based on Random Effects

We consider the processes as in Figure 6.6, assuming that there are only fixed covariates. Omitting subscripts for individuals, we let $\mathcal{H}_r(t) = \{Z_r(s), 0 \leq s \leq t; X\}$ denote the history for process r and $Y_{rk}(s) = I(Z_r(s) = k)$, $r = 1, 2$, and let

$$\lim_{\Delta t \downarrow 0} \frac{P(Z_r(t + \Delta t^-) = k+1 \mid Z_r(t^-) = k, \mathcal{H}_r(t^-), v_r)}{\Delta t} = \lambda_{rk}(t \mid \mathcal{H}_r(t^-), v_r) \quad (6.15)$$

denote the conditional intensity of a $k \to k+1$ transition for process r given the random effect v_r where $v_r = (v_{r1}, v_{r2})'$. A conditional Markov transition intensity will have the form

$$\lambda_{rk}(t \mid \mathcal{H}_r(t^-), v_r) = v_{rk} \lambda_{rk}(t \mid x), \quad k = 1, 2, \quad r = 1, 2. \quad (6.16)$$

If $v_{r1} = v_{r2}$, then the same random effects modulate the $1 \to 2$ and $2 \to 3$ baseline transition intensities, and the conditional model of Section 6.1.2 is retrieved for each marginal process. As mentioned in Section 6.1.2, such random effect models do not enable one to distinguish clearly between heterogeneity and a departure from the Markov property (i.e. a history dependence). The form of the conditional intensity must be presumed correct if the variance parameter of the random effect distribution is to be interpreted as a measure of between individual heterogeneity in the transition intensities. Subject to this assumption, this model can be useful for accommodating apparent variation in rates of passage through the various states.

When $V_{r1} \neq V_{r2}$, we assume $E(V_{rk}) = 1$ that V_{rk} has finite variance, and allow a dependence between V_{r1} and V_{r2}; we let $G_r(v_r)$ denote the bivariate cumulative distribution function. When using copula models (see Section 6.1.3) to construct the bivariate distributions, the random variables V_{r1} and V_{r2} could have different marginal distributions. Here if $\text{cov}(V_{r1}, V_{r2}) = 0$, then $\text{var}(V_{r1})$ and $\text{var}(V_{r2})$ reflect heterogeneity in the $1 \to 2$ and $2 \to 3$ transition intensities. If $\text{cov}(V_{r1}, V_{r2}) > 0$, then a particular form of history dependence is implied, with shorter sojourns in the initial state associated with shorter sojourns in the intermediate state; unlike the shared random effect model, negative associations are accommodated in this formulation since the covariance can be less than zero.

To consider two processes simultaneously in this framework, we must specify a four-dimensional random effect model for $V = (V_1', V_2')'$. Again the marginal univariate random effect distributions may each come from a different parametric family, but this flexibility is not typically exploited in practice.

Specification of the full random effect distribution enables one to write the observed data likelihood contribution of a single individual as

$$\int \left(\prod_{r=1}^{2} \left\{ \prod_{k=1}^{2} \left[\prod_{s \in \mathcal{D}_{rk}} v_{rk} \lambda_{rk}(s \mid x) \right] \exp\left(-\int_0^\infty v_{rk} Y_{rk}(u^-) \lambda_{rk}(u \mid x) \, du \right) \right\} \right) dG(v), \quad (6.17)$$

as was done in (6.14) for a single process.

While this random effect approach to dependence modeling has some appeal due to its connection with frailty models, computationally it can be challenging to evaluate and maximize the likelihoods when general multivariate random effect distributions are specified. Moreover, the ability to model and clearly describe the nature of the dependencies between processes hinges on the presence of heterogeneity for the marginal processes. Next we discuss an alternative approach that does not have this limitation.

6.2.2 Intensity-Based Models for Local Dependence

Another framework for jointly modeling multiple processes involves specifying event intensities dependent on the joint history for all processes. This idea was introduced in Section 2.3.5, but we consider it in more detail here.

For convenience, we consider a K_1-state process $\{Z_1(s), s \geq 0\}$ and a K_2-state process $\{Z_2(s), s \geq 0\}$ with histories $\mathcal{H}_r(t) = \{Z_r(s), 0 \leq s \leq t\}$, $r = 1, 2$, respectively; generalizations to more than two processes are straightforward in principle, although dependence concepts are more involved. A joint process can be conceived with states defined by the pair of states occupied for the individual processes. If $Z(t) = (Z_1(t), Z_2(t))$, the state space for $\{Z(s), s \geq 0\}$ is comprised of at most $K_1 \times K_2$ states, because some combinations of states may be impossible. Here the association between the two processes is modeled through the dependence on the joint history $\mathcal{H}(t) = \{Z(s), 0 \leq s \leq t\}$. The intensity function for a $k_r \to l$ transition for process r given $Z(t^-) = (k_1, k_2)$ is then

$$\lim_{\Delta t \downarrow 0} \frac{P(Z_r(t + \Delta t^-) = l \mid Z(t^-) = (k_1, k_2), \mathcal{H}(t^-))}{\Delta t} = \lambda_{k_1 k_2, l}^{(r)}(t \mid \mathcal{H}(t^-)), \quad (6.18)$$

$r = 1, 2$.

Process 1 is said to be *locally independent* of process 2 if

$$\lambda^{(1)}_{k_1 k_2, l}(t \mid \mathcal{H}(t^-)) = \lambda^{(1)}_{k_1 k'_2, l}(t \mid \mathcal{H}_1(t^-)), \quad \text{for } k'_2 = 1, \ldots, K_2.$$

This states that given $\mathcal{H}_1(t^-)$ the instantaneous risk of a $k_1 \to l$ transition for process 1 does not depend on the current state or history of process 2; local independence of process 2 to process 1 is similarly defined. Unlike the dependence models based on random effects (Section 6.2.1) or copula functions (Section 6.2.3), local dependence can be asymmetric in that one process can be locally dependent on another, but not vice versa.

Interest often lies in modeling local dependence when it is present. To facilitate this discussion, we now consider joint modeling of the two 3-state processes of Figure 6.6. Figure 6.8 in Section 6.2.4 shows a 9-state diagram illustrating how the two 3-state processes can be considered jointly in this framework. The fact that only one transition can occur at each instant in time is reflected in this figure by the presence of only vertical or horizontal arrows. Here the notation for the intensities in (6.18) can be simplified, since they are zero unless $l = k_r + 1$, so we write

$$\lim_{\Delta t \downarrow 0} \frac{P(Z_1(t + \Delta t^-) = k_1 + 1 \mid Z(t^-) = (k_1, k_2), \mathcal{H}(t^-))}{\Delta t} = \lambda^{(1)}_{k_1 k_2}(t \mid \mathcal{H}(t^-)),$$

$k_1 = 1, 2$, and

$$\lim_{\Delta t \downarrow 0} \frac{P(Z_2(t + \Delta t^-) = k_2 + 1 \mid Z(t^-) = (k_1, k_2), \mathcal{H}(t^-))}{\Delta t} = \lambda^{(2)}_{k_1 k_2}(t \mid \mathcal{H}(t^-)),$$

$k_2 = 1, 2$. The nature of the dependence between the processes can be inferred by comparing estimates of the transition intensity functions under suitable specifications. A parsimonious characterization of local dependence is obtained by specifying multiplicative regression models in which time-dependent covariates represent the state, or more generally the history, of the complementary process. For example, recall that $Y_{rk}(t) = I(Z_r(t) = k)$, and let $Y_r(t) = (Y_{r2}(t), Y_{r3}(t))'$ and $\beta^{(1)}_{k_1} = (\beta^{(1)}_{k_1 2}, \beta^{(1)}_{k_1 3})'$, $k_1 = 1, 2$. Local state dependence of process 1 on process 2 can be modeled via

$$\lambda^{(1)}_{k_1 k_2}(t \mid \mathcal{H}(t^-)) = \lambda^{(1)}_{k_1}(t \mid \mathcal{H}_1(t^-)) \exp(Y'_2(t^-) \beta^{(1)}_{k_1}) \tag{6.19}$$

where $\lambda^{(1)}_{k_1}(t \mid \mathcal{H}_1(t^-))$ is a baseline intensity of a $k_1 \to k_1 + 1$ transition for process 1, applicable when process 2 is in state 1, $k_1 = 1, 2$. The multiplicative term in (6.19) modulates this baseline intensity according to the state occupied by process 2 at t^-. The intensities for process 2 may likewise be defined as

$$\lambda^{(2)}_{k_1 k_2}(t \mid \mathcal{H}(t^-)) = \lambda^{(2)}_{k_2}(t \mid \mathcal{H}_2(t^-)) \exp(Y'_1(t^-) \beta^{(2)}_{k_2}) \tag{6.20}$$

where $\beta^{(2)}_{k_2} = (\beta^{(2)}_{k_2 2}, \beta^{(2)}_{k_2 3})'$, $k_2 = 1, 2$.

We let $\beta^{(1)} = (\beta^{(1)}_{12}, \beta^{(1)}_{13}, \beta^{(1)}_{22}, \beta^{(1)}_{23})'$ and $\beta^{(2)} = (\beta^{(2)}_{12}, \beta^{(2)}_{13}, \beta^{(2)}_{22}, \beta^{(2)}_{23})'$ represent the eight regression coefficients characterizing the dependence. As mentioned earlier this formulation accommodates an asymmetric dependence structure. It may be, for example, that the elements of $\beta^{(1)}$ are positive so that when process 2 is in higher states, the intensity for transitions to more advanced states is greater for process 1, but the elements of $\beta^{(2)}$ are zero or negative so that advanced states of process 1 do not alter, or perhaps reduce, the risk of progression for process 2. Other appealing features include the fact that dependencies are characterized by relative risks, and that these models are easily fitted. This approach, however, does not give a direct interpretation of covariate effects on features of the marginal processes. That is, if interest lies in assessing covariate effects on the two processes separately and dependence modeling is of secondary interest, then while we can introduce a covariate effect for X into (6.19) by writing

$$\lambda_k^{(1)}(t \mid \mathcal{H}_1(t^-)) \exp(Y_2'(t^-)\beta_k^{(1)} + \gamma x),$$

the coefficient γ is interpreted as the effect of a one unit increase in x on the instantaneous risk of a $k \to k+1$ transition for process 1 *conditionally on the state occupied for process 2*. We illustrate this point in the application of Section 6.2.4. In the next section we consider an approach to dependence modeling for a particular situation in which covariate effects retain their interpretation in joint models constructed using copula functions.

6.2.3 Dependence Models Retaining Simple Marginal Properties

We discussed the use of copula models for constructing flexible multivariate random effect distributions in Section 6.1.3. Copula models can also be applied for modeling dependencies in two or more processes directly when a set of sojourn or state entry times is of interest. This approach has advantages over the random effect or intensity-based approach to joint modeling if the marginal processes have a simple (e.g. Markov) structure and interest lies in expressing covariate effects on the intensities for a marginal process, as would be done if that process were solely of interest. In contrast, random effect models specify covariate effects conditionally on the random effects, and joint intensity-based models (see Section 6.2.2) specify covariate effects conditionally on the joint history; neither of these approaches yield estimates of covariate effects with a simple marginal interpretation. More generally, the simple models one can use for the marginal processes are typically incompatible with the joint models in the random effect and intensity-based frameworks for joint modeling.

Here we consider two settings where multistate processes can be jointly modeled using copula function. We focus again here on two 3-state processes as depicted in Figure 6.6 and consider the case where each process has the Markov property. Let T_{rk} be the entry time for state k in process r and $\mathcal{F}_r(t|x) = P(T_{r3} > t|X = x)$ denote the survival distribution for T_{r3}, $r = 1, 2$. Note that $\mathcal{F}_r(t|x) = P_{r11}(t|x) + P_{r12}(t|x)$, where $P_{rkl}(t|x) = P(Z_r(t) = l|Z_r(0) = k, X = x)$. In terms of the intensity functions, this can alternatively be written as

$$e^{-\Lambda_{r12}(t|x)} + \int_0^t \lambda_{r12}(s \mid x) \exp\left(-\Lambda_{r12}(s \mid x) - \Lambda_{r23}(s, t \mid x)\right) ds$$

where $\Lambda_{rkl}(s,t|x) = \int_s^t \lambda_{rkl}(u|x)\,du$. A copula function $C(u_1, u_2; \psi)$ may be used to accommodate an association between the absorption times T_{13} and T_{23} while allowing the model to retain the Markov property for the marginal processes. More generally, we can do this for the time of entry to any state in a process. However, it is important to note that the models for different pairs of states will not be compatible with any single model for the full bivariate process. This approach is therefore most appealing when entry to a specific state is of special interest.

In semi-Markov models, the analysis is based on the sojourn time in states, and copula models naturally lend themselves to dependence modeling among sojourn times. One may choose to model the dependence across processes in Figure 6.6 that may each be semi-Markov. For example, the joint distribution for the sojourn times (W_{1k}, W_{2k}) in state k may be constructed using a copula model. A more comprehensive joint model for $(W_{11}, W_{21}, W_{12}, W_{22})$ could also be considered. In the semi-Markov setting, the use of copula functions enables one to express covariate effects on sojourn time distributions using standard methods for survival analysis (e.g. proportional hazards models, or location-scale models as described in (6.9) of Section 6.1.2). In more complex multistate models, the number of sojourn times may be high, and if particular states may be entered repeatedly, then the dimension of the multivariate response may become quite large; the use of copula models can be less appealing in such cases.

In some contexts, interest may lie in simultaneous inferences regarding two or more processes, but not in the association between processes. In the DCCT study, for example, the effectiveness of the intensive glucose control program in delaying progression of nephropathy and retinopathy was of interest. In this case, a working independence assumption can furnish parameter estimates of effects on the marginal processes, and robust sandwich-type variances estimates can be used to ensure valid simultaneous inferences regarding two or more processes. We discuss this first in a more general context by considering composite likelihood functions as a basis for inference.

Composite likelihoods are based on components of a full likelihood, each of which would be valid likelihoods with some corresponding data (Varin et al., 2011). Let D_r represent a component of the data and $L_r(\psi) \propto P(D_r; \psi)$ a component likelihood indexed by a parameter ψ. A composite likelihood is the product of the components

$$CL(\psi) = \prod_{r=1}^{R} L_r(\psi). \tag{6.21}$$

Since each $L_r(\psi)$ is a likelihood, the composite likelihood will under regularity conditions provide a consistent estimator $\widehat{\psi}$. In particular, under mild regularity conditions, the component score functions $U_r(\psi) = \partial \log L_r(\psi)/\partial \psi$ satisfy $E\{U_r(\psi)\} = 0$, and the composite score

$$U(\psi) = \partial \log CL(\psi)/\partial \psi = \sum_{r=1}^{R} U_r(\psi)$$

satisfies $E\{U(\psi)\} = 0$. This approach is very flexible, and we note that individual components $L_r(\psi)$ may depend on only a subset of the parameter vector. For example, in the working independence framework, the rth component is associated with the rth marginal process and involves only its parameters.

With a sample of n independent individuals, let

$$CL(\psi) = \prod_{i=1}^{n} CL_i(\psi)$$

denote the overall composite likelihood where $CL_i(\psi) = \prod_{r=1}^{R} L_{ir}(\psi)$ with $L_{ir}(\psi)$ the contribution based on data D_{ir} from individual i, $r = 1, \ldots, R$, $i = 1, \ldots, n$. A consistent estimator $\widehat{\psi}$ is obtained by solving $U(\psi) = \sum_{i=1}^{n} U_i(\psi) = 0$, where $U_i(\psi) = \partial \log CL_i(\psi)/\partial \psi$. By the general theory of estimating functions (e.g. White, 1982),

$$\sqrt{n}(\widehat{\psi} - \psi) \to_D N(0, \mathcal{A}^{-1}(\psi)\mathcal{B}(\psi)\mathcal{A}^{-1}(\psi)), \qquad (6.22)$$

as $n \to \infty$, where

$$\mathcal{A}(\psi) = E\{-\partial U(\psi)/\partial \psi'\} \qquad (6.23)$$

and

$$\mathcal{B}(\psi) = E\{U(\psi)U'(\psi)\}. \qquad (6.24)$$

In the analysis of a particular dataset, the covariance matrix for $\widehat{\psi}$ is estimated by replacing the expectations in (6.23) and (6.24) with their empirical counterparts and evaluating the functions at $\psi = \widehat{\psi}$. Specifically, $\widehat{\text{cov}}(\sqrt{n}(\widehat{\psi} - \psi)) = \widehat{A}^{-1}(\widehat{\psi})\,\widehat{B}(\widehat{\psi})\,\widehat{A}^{-1}(\widehat{\psi})$, where

$$\widehat{A}(\widehat{\psi}) = -n^{-1}\sum_{i=1}^{n} \partial U_i(\psi)/\partial \psi'$$

and

$$\widehat{B}(\widehat{\psi}) = n^{-1}\sum_{i=1}^{n} U_i(\psi)U_i'(\psi)\Big|_{\psi=\widehat{\psi}}.$$

A natural question is how to select data $\{D_1, \ldots, D_R\}$ to construct a composite likelihood in a given setting. The general guideline is that the parts of the full likelihood that are kept in the composite likelihood should be informative, easily computed and contain parameters of interest; in contrast, the parts omitted are usually hard to formulate or evaluate, not very informative or pose a significant computational burden. With two associated multistate processes, one may, for example, adopt a working independence assumption for the two processes and use a composite likelihood based on the two marginal processes. This provides estimates for parameters specifying the individual processes, but association between them is not modeled. When marginal semi-Markov analyses are based on the sojourn times in successive states, it is important to realize that a working independence assumption cannot be adopted within individuals. This is because real dependence between successive gap times induces a dependent censoring mechanism, even with only administrative censoring at C acting on the process. For example, in a progressive model, the censoring time for W_k is $C - T_{k-1}$, which for $k \geq 2$ is not independent of W_k.

6.2.4   The Development of Axial Involvement in Psoriatic Arthritis

For illustration of methodology in this section, we consider analysis of the onset and progression of damage in the left and right sacroiliac (SI) joints among patients in the University of Toronto Psoriatic Arthritis Cohort who had normal SI joints at clinic entry. Involvement of these joints signifies the onset of axial disease, a term used to describe back involvement. The extent of damage was assessed based on applying the New York Criteria (Bennet and Wood, 1968) to radiographic images obtained at visits that are scheduled biannually. This method assigns a score of 0 for a normal joint, 1 if the presence of damage is equivocal, 2 if it is abnormal due to erosions of the bone surface or sclerosis, 3 if it is unequivocally abnormal, and 4 if there is evidence of ankylosis (abnormal stiffening and immobility due to bone fusion). In a 3-state process, we defined state 1 as having a New York Criteria (NYC) score of 0 or 1, state 2 as having a NYC score of 2, and state 3 for a NYC score of 3 or 4. Figure 6.6 shows a pair of 3-state processes that can represent the left and right joints. The analysis here is for 538 patients who had two or more radiological examinations.

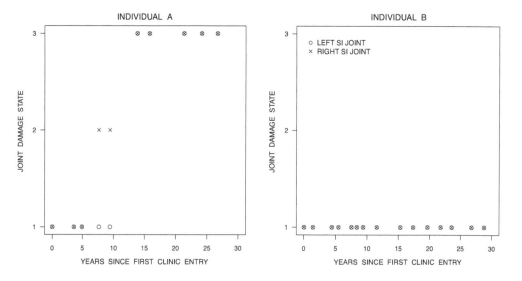

Figure 6.7: Plots of damage state for left (circle) and right (cross) sacroiliac joints for two individuals for the University of Toronto Psoriatic Arthritis Cohort.

Figure 6.7 contains a plot of data from two individuals with roughly 30 years of follow-up. Upon entry to the clinic, both SI joints were in state 1 for both individuals. For individual A (left panel) both the left and right SI joints were observed to enter state 3, but for individual B (right panel) no damage developed in either joint over the 30-year period. We first consider dependence modeling based on multivariate random effects models as described in Section 6.2.1. While any continuous random effect distribution could be employed, we adopt a four-dimensional log-normal distribution for $V = (V_1', V_2')'$ where $V_r = (V_{r1}, V_{r2})'$ with $r = 1, 2$ corresponding to the left and right sides, respectively; we set $E(V) = 1$. If $U_{rk} = \log V_{rk}$, $k = 1, 2$, then we let $U_r = (U_{r1}, U_{r2})'$, $r = 1, 2$, and $U = (U_1', U_2')' = (U_{11}, U_{12}, U_{21}, U_{22})'$. We write

$\text{cov}(U) = \Sigma$ in the form

$$\text{cov}(U) = \begin{bmatrix} \Phi_1 & \Psi \\ \Psi & \Phi_2 \end{bmatrix},$$

with

$$\Phi_r = \begin{bmatrix} \phi_{r1} & \phi_{r12} \\ \phi_{r21} & \phi_{r2} \end{bmatrix}, \quad r = 1, 2, \quad \text{and} \quad \Psi = \begin{bmatrix} \psi_{11} & \psi_{12} \\ \psi_{21} & \psi_{22} \end{bmatrix}.$$

The two processes are independent if the 2×2 submatrix $\Psi = 0$, but not otherwise. A simplified model with $\psi_{11} = \psi_{12} = \psi_{21} = \psi_{22} = 0$ is also fitted in the analysis that follows.

The likelihood contribution for an individual with observation times $a_0 < a_1 < \cdots < a_m$ is of the form

$$\int \left\{ \prod_{r=1}^{2} \prod_{j=1}^{m} P(Z_r(a_j) \mid Z_r(a_{j-1}), v_r, x) \right\} dG(v), \tag{6.25}$$

where $v = (v_1', v_2')'$ with $v_r = (v_{r1}, v_{r2})'$, $r = 1, 2$, and we omit the a_j in the conditioning argument. This uses the fact that the left and right SI processes are conditionally independent given the random effect vector $V = v$. The four-dimensional integration can be carried out using Gaussian quadrature as we do here. The overall likelihood is then obtained by taking the product across individuals of such terms. Table 6.2 contains the maximum likelihood estimates with time-homogeneous baseline transition intensities; the covariance matrix parameters are for the random effect distribution for U under the "Full Dependence Model". The nlm function in R was used for the optimization; the Hessian matrix is obtained by finite differencing of the observed data log-likelihood using this function by specifying hessian=TRUE, and the resulting observed information matrix furnished the standard errors. The baseline intensities for transitions from state 1 to state 2 corresponding to the onset of early signs of damage are comparable for the left and right joints, and the estimated baseline intensities are likewise similar for $2 \to 3$ transitions on the two sides. There is very strong evidence of a need to accommodate heterogeneity in the time-homogeneous transition intensities of all types, which persists in models fitted with piecewise-constant intensities (not shown here). For the process governing the left SI joint, the correlation between U_{11} and U_{12} is estimated to be

$$\widehat{\rho}_1 = \widehat{\phi}_{112} / \sqrt{\widehat{\phi}_{11} \widehat{\phi}_{12}} = 0.272,$$

which is not significantly different from zero (H_0: $\phi_{112} = 0$ versus H_A: $\phi_{112} \neq 0$ gives $p = 0.293$). For the right SI joint, the corresponding estimate is

$$\widehat{\rho}_2 = \widehat{\phi}_{212} / \sqrt{\widehat{\phi}_{21} \widehat{\phi}_{22}} = 0.158,$$

which again is not significantly different than zero ($p = 0.516$). While reduced models could be fitted with the constraints $\phi_{121} = \phi_{212} = 0$, we retain the general model in what follows.

A global assessment of between-process dependence can be carried out by a

Table 6.2: Estimates obtained with full and partial dependence models fitted to the left and right SI joints.

	Full Dependence Model				Partial Dependence Model ($\Psi = 0$)			
	Left ($r=1$)		Right ($r=2$)		Left ($r=1$)		Right ($r=2$)	
	EST	95% CI	EST	95% CI	EST	95% CI	EST	95% CI
λ_{r1}	0.222	(0.084, 0.587)	0.300	(0.115, 0.782)	0.058	(0.033, 0.100)	0.106	(0.053, 0.211)
λ_{r2}	0.260	(0.036, 1.869)	0.221	(0.031, 1.558)	0.253	(0.025, 2.551)	0.297	(0.023, 3.851)
ϕ_{r1}	6.229	(3.986, 8.472)	6.154	(4.015, 8.293)	1.971	(0.320, 3.622)	3.137	(1.247, 5.026)
ϕ_{r2}	6.102	(1.507, 10.696)	5.690	(1.584, 9.796)	4.168	(-0.474, 8.810)	3.458	(-0.629, 7.546)
ϕ_{r12}	1.677	(-1.450, 4.805)	0.935	(-1.887, 3.758)	0.116	(-2.190, 2.422)	-1.216	(-3.915, 1.482)
ψ_{11}	6.182	(2.996, 9.369)						
ψ_{22}	5.880	(-0.079, 11.839)			-	-	-	-
ψ_{12}	1.254	(-1.560, 4.068)			-	-	-	-
ψ_{21}	1.349	(-1.665, 4.364)			-	-	-	-
$\log L$		-965.107				-1097.806		

4 degree of freedom (d.f.) likelihood ratio test of the full model on the left side of Table 6.2 with the reduced model with $\Psi = 0$ on the right side of Table 6.2. Estimates for the reduced model were obtained by maximizing the likelihood under the constraint $\Psi = 0$. The likelihood ratio statistic $-2 \times (-1097.806 - (-965.107)) = 265.399$ gives a p-value of effectively zero ($P(\chi_4^2 > 265.399) < 0.001$), and so there is strong evidence of a dependence in the transition times for the development of damage in the left and right sides. The estimates of the correlation between the random effects are $\widehat{\text{corr}}(U_{11}, U_{21}) = 0.999$, $\widehat{\text{corr}}(U_{12}, U_{22}) = 0.211$, $\widehat{\text{corr}}(U_{21}, U_{12}) = 0.220$ and $\widehat{\text{corr}}(U_{21}, U_{22}) = 0.998$.

The correlation between the random effects for the same type of transitions across processes are extremely high, and the magnitude of the variance estimates are quite comparable. We therefore fit a shared random effect model with $U_1^* = U_{11} = U_{21}$ and $U_2^* = U_{12} = U_{22}$. The conditional intensity in this model is given by (6.17) upon replacing U_{rk}, $r = 1, 2$, with a common term U_k^*, $k = 1, 2$; the marginal likelihood has the same form as (6.25), but after replacing $v_r = (v_{r1}, v_{r2})$, $r = 1, 2$ with a common $v^* = (v_1^*, v_2^*)$ for $r = 1, 2$, it only requires integration over two dimensions. The results give estimates of the baseline intensities $\widehat{\lambda}_{11} = 0.220$ (95% CI: 0.098, 0.493), and $\widehat{\lambda}_{12} = 0.263$ (95% CI: 0.055, 1.250) for the left SI joint, and $\widehat{\lambda}_{21} = 0.302$ (95% CI: 0.133, 0.688) and $\widehat{\lambda}_{22} = 0.210$ (95% CI: 0.045, 0.982) for the right. These estimates are in close alignment with those from the full dependence model in Table 6.2, but the confidence intervals are quite a bit narrower here. We note that under either model, confidence intervals for the transition intensities are quite wide, reflecting the intermittent observation times and the heterogeneity of individuals. If we let $\text{var}(U_k^*) = \phi_k$, $k = 1, 2$, we obtain $\widehat{\phi}_1 = 6.181$ (95% CI: 4.401, 8.680) and $\widehat{\phi}_2 = 5.854$ (95% CI: 4.090, 8.379), and an estimated correlation $\widehat{\text{corr}}(U_1^*, U_2^*) = 0.218$. Thus, while the model with the four separate random effects is considerably more flexible, because certain correlations among the random effects between processes is so high and the corresponding variance estimates are comparable, inferences about the processes are similar in the simpler model with shared random effects. This model with the two-dimensional random effect is also easier to fit; we comment on the fit of a model with piecewise-constant baseline intensities shortly.

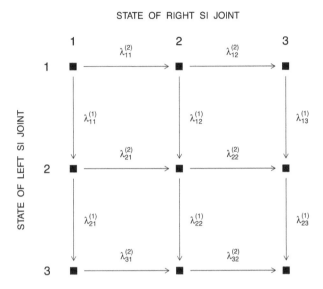

Figure 6.8: State space diagram for joint process for damage in left (1) and right (2) sacroiliac joints among patients with psoriatic arthritis using the notation of (6.18). For each side, state 1 represents no damage, state 2 moderate damage and state 3 clinically important damage.

As described in Section 6.2.2, we may alternatively consider intensity-based analyses based on local dependence concepts. Figure 6.8 shows the state space diagram for a joint time-homogeneous Markov model for the left and right SI joints. We can estimate the 12 transition intensities using the methods of Section 5.2. The estimates of the log intensities and their associated standard errors are given in Table 6.3 along with estimates of the transition intensities and their 95% confidence intervals. With $Z(t) = (Z_1(t), Z_2(t))$ denoting the joint process for the left ($r = 1$) and right ($r = 2$) SI joints, we may then obtain, for example,

$$P(Z_1(t) = k \mid Z(0) = (1,1)) = \sum_{l=1}^{3} P(Z(t) = (k,l) \mid Z(0) = (1,1))$$

as the marginal transition probability that the left joint is in state k, t years after disease onset. Figure 6.9 shows estimates for state 3 damage for the left and right SI joints based on the joint model and computed as described above, along with the probabilities $P(Z_r(t) = 3 | Z_r(0) = 1)$ estimated by fitting the 3-state models separately for the left and right processes, and nonparametric estimates based simply on the time to entry into state 3 (Turnbull, 1976). There is good agreement between these three estimates. The estimate of progression to state 3 damage is slightly higher in the right SI joint than in the left one. We also plot the estimated probability of state 3 damage based on a shared two-dimensional random effect model with piecewise-constant baseline intensities having cut-points at 4 and 8 years after disease onset; there is quite good agreement between this estimate and the others with slight deviations after 15 years from disease onset.

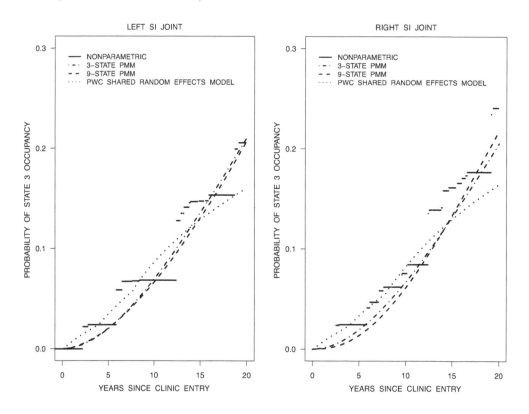

Figure 6.9: Plot of the cumulative probability of state 3 damage for the left and right SI joints from separate nonparametric estimation of time to state 3 entry for left and right sides, separate parametric Markov models (PMMs) with time-homogeneous intensities, a joint 9-state parametric Markov model with time-homogeneous intensities and a model with shared random effects and piecewise-constant intensities with cut-points at 4 and 8 years from disease onset.

One may also be interested in joint probabilities such as

$$P(Z_1(t) \geq k, Z_2(t) \geq k \mid Z_1(0) = 1, Z_2(0) = 1).$$

A nonparametric estimate of this can be obtained using Turnbull's algorithm for interval-censored data by taking the right endpoint as the first assessment time an individual was known to have entered the set of states $(k,l), k \geq 2, l \geq 2$, and the left endpoint of the censoring interval would be the time of the prior assessment. Figure 6.10 shows the nonparametric estimates for $k = 2$ (left panel) and $k = 3$ (right panel), along with joint probabilities based on the 9-state Markov model, the four-dimensional random effect model with time-homogeneous baseline intensities, and the two-dimensional shared random effect model with time-homogeneous and piecewise-constant baseline intensities with cut-points at 4 and 8 years from disease onset. When estimating the joint probability of grade 3 damage for both the left and right SI joints, the estimates based on the random effect models track the nonparametric estimates slightly better with the 9-state Markov model giving more conservative probabilities early during the course of the disease. When estimating the cumulative probability of at least grade 2 damage, a similar relationship between the estimates is observed; the model with the two-dimensional shared random effect and piecewise-constant baseline rates yields the estimate with the best agreement with the nonparametric estimate.

Table 6.3: Estimated transition intensities from joint 9-state model and associated estimated measures of local dependence between left and right SI joints.

Side	Transition	State of Other Side	EST	SE	λ	95% CI		EST	95% CI	p
Left	$1 \to 2$	1	-3.857	0.146	0.02	(0.02, 0.03)		-	-	-
$(r=1)$		2	-1.150	0.190	0.32	(0.22, 0.46)	$e^{\beta_{12}^{(1)}}$	14.99	(8.74, 25.69)	<0.001
		3	0.289	0.395	1.33	(0.62, 2.90)	$e^{\beta_{13}^{(1)}}$	63.17	(27.67, 144.21)	<0.001
	$2 \to 3$	1	-3.503	0.351	0.03	(0.02, 0.06)		-	-	-
		2	-3.792	0.443	0.02	(0.01, 0.05)	$e^{\beta_{22}^{(1)}}$	0.75	(0.24, 2.34)	0.618
		3	-1.450	0.260	0.23	(0.14, 0.39)	$e^{\beta_{23}^{(1)}}$	7.79	(3.18, 19.09)	<0.001
Right	$1 \to 2$	1	-3.723	0.133	0.02	(0.02, 0.03)		-	-	-
$(r=2)$		2	-2.350	0.248	0.10	(0.06, 0.16)	$e^{\beta_{12}^{(2)}}$	3.95	(2.09, 7.44)	<0.001
		3	-1.438	0.467	0.24	(0.10, 0.59)	$e^{\beta_{13}^{(2)}}$	9.83	(3.80, 25.39)	<0.001
	$2 \to 3$	1	-2.301	0.294	0.10	(0.06, 0.18)		-	-	-
		2	-3.324	0.323	0.04	(0.02, 0.07)	$e^{\beta_{22}^{(2)}}$	0.36	(0.15, 0.88)	0.025
		3	-2.072	0.633	0.13	(0.04, 0.44)	$e^{\beta_{23}^{(2)}}$	1.26	(0.30, 5.25)	0.754

The 9-state model enables one to characterize the dependence between the processes, as discussed in Section 6.2.2 using (6.19) and (6.20). Relative risks and 95% confidence intervals obtained by fitting (6.19) and (6.20) are also given in Table 6.3. From these it is evident that progression of damage in one side has a highly significant effect on progression intensities in the other side. For example, individuals with moderate damage (state 2) in the right SI joint have a highly significant 15-fold higher risk of moderate damage developing in the left SI joint. Likewise, individuals with a left SI joint in state 2 have a highly significant four-fold increase in the risk of developing moderate damage in the right SI joint. These associations, while expressed differently in terms of relative risks, are in broad alignment with the inferences from the random effects analyses where the estimate of $\widehat{\psi}_{11} = 6.189$ (95% CI: 2.980, 9.397) reflects a strong positive association in the transition times out of state 1 for the SI joints on the left and right sides. There is an insignificant increase in risk in the $2 \to 3$ intensity for the left side upon occurrence of moderate damage on the right. Those with clinically important damage on the right have a significant almost eight-fold increased risk of developing clinically important damage in the leftover persons with moderate damage on the right. Interestingly, there is evidence that any damage on the left side leads to higher transition rates on the right side for all transitions with the exception of state 3 damage for $2 \to 3$ transitions.

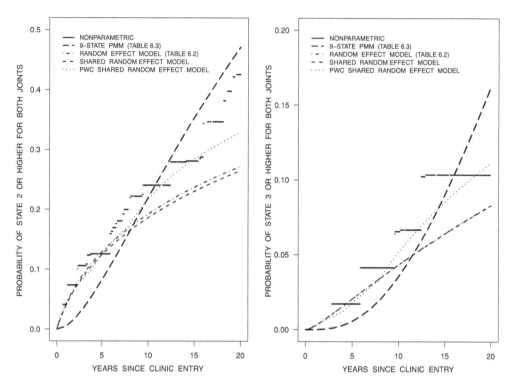

Figure 6.10: Plot of the cumulative probability of having at least state k damage for both the left and right SI joints from nonparametric estimation of the time to entry into states $(2, 2)$ in the joint state space, a joint 9-state Markov model with time-homogeneous intensities, a four-dimensional random effect model with time-homogeneous conditional intensities, a two-dimensional shared random effect model with time-homogeneous conditional intensities and a two-dimensional shared random effect model with piecewise-constant intensities having cut-points at 4 and 8 years.

6.3   Finite Mixture Models

6.3.1   Notation and Maximum Likelihood Estimation

Models accommodating heterogeneity through the use of continuous random effects involve very particular assumptions on the nature of heterogeneity and are hard to contemplate in non-progressive processes with many states. Sometimes discrete mixtures are preferable. In many contexts, for example, the proportion of individuals remaining in their initial state is much higher than would be expected from a given model based solely on observed factors. This type of heterogeneity can often be dealt with by allowing a class (or sub-population) of individuals to be at zero risk of transitions. A mixture model can be specified to reflect this in which a latent Bernoulli random variable indicates the class to which an individual belongs. Such models are called mover-stayer models (Goodman, 1961; Frydman, 1984) since individuals may be characterized as either "movers" if they make transitions through the multistate process, or "stayers" if they do not.

A generalization of the mover-stayer model is to allow $G > 2$ classes of individuals labeled $1, \ldots, G$, where the multistate processes for individuals in the same class have common features. There is often a scientific rationale for the accommodation of distinct classes of individuals. It may be thought, for example, that there are different sub-types of a disease that has traditionally been viewed as a homogeneous disease. Alternatively, the distinct classes may be used as a basis for studying genetic determinants of the disease course. In other settings, finite mixture models are motivated by apparent clusters of life history paths that share common features. In cohort studies, for example, some individuals experience rapid disease progression, some progress at much slower rates and some may not progress at all, even after extensive follow-up. Of course, in this setting it is best to consider the need for finite mixture models after assessing the fit of simpler models. Estimating the most appropriate number of classes is challenging, and we focus our discussion on the case where G is specified.

We let W be a latent random variable indicating the sub-population to which a particular individual belongs, with

$$P(W = g \mid X; \gamma), \quad g = 1, \ldots, G, \quad \text{where} \quad \sum_{g=1}^{G} P(W = g \mid X; \gamma) = 1.$$

With $\mathcal{H}(t) = \{Z(s), 0 \le s \le t, x\}$ we denote the transition intensities for individuals in class g as

$$\lambda_{kl}(t \mid \mathcal{H}(t^-), W = g) = \lambda_{kl}^{(g)}(t \mid \mathcal{H}(t^-); \theta_g), \quad k \ne l, \tag{6.26}$$

where θ_g is a set of parameters specifying the intensities, $g = 1, \ldots, G$, $\theta = (\theta_1', \ldots, \theta_G')'$ and $\psi = (\theta', \gamma')'$. In some classes, particular transition intensities may be zero, in which case the number of effective states is smaller.

Here we consider the problem of fitting a model comprised of a finite mixture of Markov processes when individuals are under intermittent observation. For an individual with assessment times $a_0 < a_1 < \cdots < a_m$, we denote the observed multistate data as $\mathcal{Z}^\circ(\infty) = \{(a_r, Z(a_r)), r = 0, 1, \ldots, m\}$ and let $\mathcal{H}^\circ(\infty) = \{(a_r, Z(a_r)), r = 0, 1, \ldots, m; x\}$. Suppressing the dependence on the visit times, and under the assumption that $Z(0) \perp W \mid X$, the observed data likelihood contribution is

$$\sum_{g=1}^{G} P(\mathcal{Z}^\circ(\infty) \mid Z(a_0), W = g, X; \theta_g) P(W = g \mid X; \gamma). \tag{6.27}$$

Since the process for each class is Markov,

$$P(\mathcal{Z}^\circ(\infty) \mid Z(a_0), W = g, X; \theta_g) = \prod_{j=1}^{m} P(Z(a_j) \mid Z(a_{j-1}), W = g, X; \theta_g),$$

where $P(Z(a_j) \mid Z(a_{j-1}), W = g, X; \theta_g)$ is of the form (5.6) but with the probability evaluated according to the Markov model for class g, $g = 1, \ldots, G$. The overall likelihood is obtained by taking a product of terms like (6.27) for a sample of n independent processes.

Optimization of the overall likelihood is particularly amenable to an expectation-maximization (EM) algorithm. To see this, consider a complete data likelihood contribution defined when W is observed, given by

$$L_C(\psi) = \prod_{g=1}^{G} \left[\prod_{j=1}^{m} P(Z(a_j) \mid Z(a_{j-1}), W = g, X; \theta_g) \, P(W = g \mid X; \gamma) \right]^{I(W=g)}.$$

If ψ^r is the estimate of ψ at the rth iteration, then let

$$\mathcal{Q}(\psi; \psi^r) = E\{\log L_C(\psi) \mid \mathcal{Z}^\circ(\infty); \psi^r\}$$

be the objective function to be maximized. If we let $p_{kl}(s, t \mid x; \theta_g) = P(Z(t) = l \mid Z(s) = k, x, W = g; \theta_g)$ be the transition probability function for process g and $n_{jkl} = I(Z(a_{j-1}) = k, Z(a_j) = l)$, then with weight $\omega_g^r = P(W = g \mid \mathcal{H}^\circ(\infty); \psi^r)$, we can write $\mathcal{Q}(\psi; \psi^r) = \mathcal{Q}_1(\theta; \psi^r) + \mathcal{Q}_2(\gamma; \psi^r)$, where

$$\mathcal{Q}_1(\theta; \psi^r) = \sum_{g=1}^{G} \sum_{j=1}^{m} \sum_{k=1}^{K} \sum_{l=1}^{K} \omega_g^r \cdot n_{jkl} \cdot \log p_{kl}(a_{j-1}, a_j \mid x; \theta_g), \qquad (6.28)$$

and

$$\mathcal{Q}_2(\gamma; \psi^r) = \sum_{g=1}^{G} \omega_g^r \cdot \log P(W = g \mid X; \gamma). \qquad (6.29)$$

The full expectation of the complete data log-likelihood is the sum of such terms over all individuals in a sample. Note that the G distinct contributions to (6.28) can be maximized separately provided the θ_g are functionally independent, and the $\omega_g^r \cdot n_{jkl}$ term can be treated as a pseudo-transition count. The Fisher-scoring algorithm of Kalbfleisch and Lawless (1985) discussed in Section 5.1 could be adopted. Because (6.29) has the form of a multinomial likelihood, it can be maximized using software for multinomial regression where the ω_g^r play the role of pseudo-responses. Following the maximization of $\mathcal{Q}(\psi; \psi^r)$, we obtain ψ^{r+1}, which can be used to obtain a new

$$\omega_g^{r+1} = P(W = g \mid \mathcal{H}^\circ(\infty); \psi^{r+1})$$

given by

$$\frac{\prod_{j=1}^{m} P(Z(a_j) \mid Z(a_{j-1}), W = g, X; \theta_g^{r+1}) \, P(W = g \mid X; \gamma^{r+1})}{\sum_{g=1}^{G} \prod_{j=1}^{m} P(Z(a_j) \mid Z(a_{j-1}), W = g, X; \theta_g^{r+1}) \, P(W = g \mid X; \gamma^{r+1})}.$$

The estimation procedure can proceed iteratively until the difference between ψ^r and ψ^{r+1} is below a desired tolerance.

Care is needed in fitting these finite mixture models. In Section 6.1.2 we remarked that it can be difficult to distinguish the need to accommodate heterogeneity versus more involved forms of history dependence. In the finite mixture model, setting the need to accommodate distinct sub-populations hinges critically on the adequacy (or inadequacy) of the specified forms of the conditional models. We give an illustrative example in the following section. Considerable thought must also go into how best to incorporate covariate effects. While in theory covariates may be used to model both class membership and the transition intensities, parameter estimation can be difficult when one or more covariates appear in both parts of the model. Many questions can often be addressed by modeling only class membership as a function of covariates.

6.3.2 Modeling Variation in Disease Activity in Lupus

The University of Toronto Lupus Clinic maintains a registry of 1823 patients diagnosed with systemic lupus erythematosus (SLE), a complex episodic autoimmune disease in which many organ systems can be affected over time. The multifaceted nature of the condition makes it difficult to characterize disease severity at a given time as well as to measure change over time. As a result, a composite systemic lupus erythematosus disease activity index (SLEDAI) has been developed, which provides a global measure of activity through a score on a 105-point scale, with high values corresponding to a more extreme level of activity (Gladman et al., 2002).

The researchers at the clinic believe there are three distinct sub-populations of patients exhibiting different patterns of disease activity over time. It is anticipated that one sub-population will experience highly variable relapsing and remitting disease activity, one will experience persistently active disease and another will exhibit a monophasic pattern characterized by high activity at the point of study recruitment and subsequently lower activity. The disease processes in each sub-population will be better understood through estimation of the parameters governing the process dynamics (transition intensities) as well as a multinomial model for the probability individuals are in each of the three sub-populations. For each sub-population, the disease process will be modeled using a multistate Markov model, well-suited for the intermittent observation scheme.

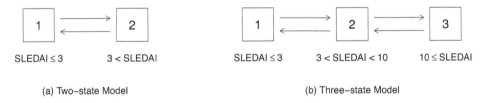

Figure 6.11: Two-state and 3-state models for finite mixture modeling of the lupus disease activity data.

For a first pass at modeling, we adopt a 2-state time-homogeneous Markov model conditional on W, with state 1 representing SLEDAI ≤ 3 corresponding to mild disease activity, and state 2 representing SLEDAI > 3 corresponding to more appreciable disease activity; see Figure 6.11(a). We fit a trinomial logistic regression model for class membership with a single covariate x_1 indicating that a patient had an age of onset at least 30 years. Letting $x = (1, x_1)'$ and using class 1 as the reference class, we specify

$$P(W = g \mid X = x; \gamma) = \frac{\exp(x'\gamma_g)}{1 + \exp(x'\gamma_2) + \exp(x'\gamma_3)}, \quad g = 1, 2, 3, \tag{6.30}$$

where $\gamma_1 = (0,0)'$, $\gamma_2 = (\gamma_{20}, \gamma_{21})'$, $\gamma_3 = (\gamma_{30}, \gamma_{31})'$ and $\gamma = (\gamma_2', \gamma_3')'$. In a second set of analyses, we keep (6.30) but model the disease activity in more detail using 3-state models for each class with states defined as SLEDAI ≤ 3 (state 1), $4 \leq$ SLEDAI $<$ 10 (state 2) and $10 \leq$ SLEDAI (state 3). The multistate diagrams are depicted in Figure 6.11(b).

Table 6.4: Parameter estimates from fitting finite mixture models to the lupus data on disease activity with $G = 3$ classes and 2-state (left) or 3-state (right) processes; Age indicates SLE diagnosis ≥ 30 years of age.

	2-State Processes			3-State Processes		
	Estimated Transition Intensities					
Class	Transition	EST	95% CI	Transition	EST	95% CI
1	$1 \rightarrow 2$	1.43	(1.27, 1.61)	$1 \rightarrow 2$	1.19	(1.06, 1.35)
	$2 \rightarrow 1$	0.43	(0.39, 0.47)	$2 \rightarrow 1$	0.59	(0.53, 0.65)
				$2 \rightarrow 3$	0.39	(0.32, 0.46)
				$3 \rightarrow 2$	1.42	(1.25, 1.61)
2	$1 \rightarrow 2$	2.03	(1.86, 2.22)	$1 \rightarrow 2$	2.38	(2.16, 2.64)
	$2 \rightarrow 1$	2.34	(2.14, 2.57)	$2 \rightarrow 1$	2.91	(2.51, 3.39)
				$2 \rightarrow 3$	1.89	(1.67, 2.15)
				$3 \rightarrow 2$	4.10	(3.66, 4.58)
3	$1 \rightarrow 2$	0.42	(0.38, 0.45)	$1 \rightarrow 2$	0.52	(0.47, 0.58)
	$2 \rightarrow 1$	1.91	(1.75, 2.10)	$2 \rightarrow 1$	2.79	(2.51, 3.10)
				$2 \rightarrow 3$	0.79	(0.66, 0.95)
				$3 \rightarrow 2$	4.44	(3.83, 5.14)

Estimates for Polychotomous Logistic Model for Class Membership (Reference: Class 1)

Class	Parameter	EST	95% CI	Transition	EST	95% CI
2	Intercept, γ_{20}	-0.624	(-0.862, -0.385)		-0.502	(-0.765, -0.239)
	Age, γ_{21}	0.496	(0.142, 0.850)		0.654	(0.301, 1.007)
3	Intercept, γ_{30}	-0.393	(-0.586, -0.199)		-0.308	(-0.521, -0.095)
	Age, γ_{31}	0.844	(0.570, 1.118)		0.993	(0.704, 1.281)

We consider data from 1767 individuals in the University of Toronto Lupus Clinic with an average of 10.5 years of follow-up. We use data from the time of entry to the clinic, which is the point at which the disease began to be managed by a specialist and detailed information on disease activity became available. We condition on $(Z(a_0), X)$ and assume $Z(a_0) \perp W \mid X_1$; in Chapter 7 we discuss an alternative approach that does not make this assumption.

For the 2-state mixture model 1300 (73.6%) individuals were in state 2 upon entry to the clinic. The starting values for the parameters in the transition intensities for the different classes were obtained by fitting a standard model with one class as in Section 5.2.1. The regression coefficients for x_i in the mixture model were initially set to 0, and the starting values for the intercepts were chosen to give equal probabilities for the three classes. The fitting was carried out using the EM algorithm of Section 6.3.1, and the standard errors were obtained by numerical differentiation using the `nlm` function in R applied to the observed data log-likelihood. Estimated log time-homogeneous intensities and parameters of the mixture components for the 2-state mixture model (Figure 6.11(a)) are displayed in the left column of Table 6.4. The estimates from the trinomial logistic regression model indicate that individuals with later onset of SLE have significantly higher odds of belonging to class 2 than class 1 (OR = 1.64; 95% CI: 1.15, 2.34; $p < 0.001$) and also a significantly higher odds of belonging to class 3 compared to class 1 (OR = 2.32; 95% CI: 1.76, 3.05; $p < 0.001$). The probabilities of class membership are reported in Table 6.5 for those with age of onset < 30 ($X_1 = 0$) and ≥ 30 ($X_1 = 1$) years of age. The transition intensity patterns in Table 6.4 vary qualitatively across the three classes. For each model, class $g = 1$ shows rapid progression from state 1 and a relatively low rate of return to state 1, class $g = 2$ is characterized by frequent back-and-forth transitions between states 1 and 2 and class $g = 3$ shows a tendency not to progress.

The results of fitting the 3-state finite mixture model are given in the right-hand columns of Table 6.4. Note that the labels of the classes are arbitrary and successive calls to an optimization algorithm may yield estimates corresponding to a different labeling. Choosing suitable starting values can mitigate the chance of this happening to some degree. The effects of age at diagnosis on class membership were broadly similar to the 2-state model with later onset associated with higher odds in class 2 versus 1 (OR = 2.70; 95% CI= 2.02, 3.60; $p < 0.001$) and higher odds in class 3 versus 1 (OR = 1.92; 95% CI= 1.35, 2.74; $p < 0.001$). The probabilities of class membership for the 3-state model are again shown in Table 6.5.

Estimates of the transition intensities for the three classes were aligned with those of the 2-state mixture model in the left columns of Table 6.4; we comment on these shortly. To better characterize the disease course in the three classes, it is helpful to compute additional summary statistics based on the fit. For each class, we compute the expected sojourn time in each state (with the proportion of time in the highest state representing high disease activity, a representation of the burden of disease) and the expected number of transitions out of each state (representing the volatility of the disease). For a time-homogeneous 2-state Markov model with intensities λ_{12} and λ_{21}, the mean sojourn time in state k is $\mu_k = \lambda_{k,3-k}^{-1}$ and the mean cumulative time spent in state k over the interval $(0, C^A]$ given they were initially in state k is $\int_0^{C^A} P_{kk}(u; \lambda)\, du$, where $\lambda = (\lambda_{12}, \lambda_{21})'$. The expected total number of transitions out of state k, given initial occupancy in state k, can likewise be expressed as

$$E\left\{ \int_0^{C^A} I(Z(u) = k)\lambda_{k,3-k}\, du \mid Z(0) = k \right\} = \int_0^{C^A} P_{kk}(u^-; \lambda)\, \lambda_{k,3-k}\, du.$$

For the 2-state model, it can be seen from the estimates in Table 6.5 that individuals in class 1 have relatively long average sojourns in the active disease state ($\simeq 2.3$

years) and spend the majority of their time in state 2. Individuals in class 2 have much shorter average sojourns in state 2 and spend roughly half of their time in the active disease state; the higher transition intensities reported in Table 6.4 also mean they tend to exhibit considerable volatility in their disease process, which is reflected by the much greater expected number of transitions between states. Individuals in class 3, in contrast, spend the majority of their time in the low disease activity state and have comparatively short sojourn times in state 2; they also make considerably fewer transitions on average compared to those in class 2. The three classes for the 3-state model have qualitatively similar features; see Table 6.5.

Table 6.5: Summary process features by class for a mixture of three time-homogeneous Markov processes.

| | | | | Model-Based Expectations[†] | | |
| | | | | Sojourn Time | Prop. Time | # Transitions |
Class (g)	X_1	$P(W = g\|X_1)$	State	in State	in State	from State
			Finite Mixture of 2-State Processes			
1	0	0.45	1	0.70	0.23	3.89
	1	0.29	2	2.34	0.77	3.35
2	0	0.24	1	0.49	0.54	11.10
	1	0.26	2	0.43	0.46	11.17
3	0	0.31	1	2.40	0.82	3.46
	1	0.46	2	0.52	0.18	4.10
			Finite Mixture of 3-State Processes			
1	0	0.43	1	0.84	0.33	3.89
	1	0.24	2	1.02	0.40	5.72
			3	0.70	0.27	2.96
2	0	0.26	1	0.42	0.48	11.20
	1	0.28	2	0.21	0.24	18.32
			3	0.24	0.28	7.87
3	0	0.31	1	1.93	0.79	4.29
	1	0.48	2	0.28	0.11	6.34
			3	0.23	0.09	2.35

[†] Expected sojourn time per visit, proportion of time in the state, and total number of transitions out of the state over a 10-year period, based on the fitted models.

6.4 Hidden Markov Models

6.4.1 Models and Estimation

States are often based on distinct conditions such that the state definitions are clear and there is no difficulty in determining what state an individual is in at any given time. There is typically no ambiguity, for example, as to whether an individual is alive or dead, or whether they have experienced a fracture or some other debilitating event. In other settings, the definitions of the states may be explicit but it can be difficult to determine whether individuals satisfy the criteria. For example, when states are defined based on radiological assessment, imperfect images and subjectivity in their interpretation can lead to disagreements among physicians as to the condition of an individual. This is the case in assessing the presence of diffuse bilateral infiltrates in the lung based on chest x-rays (Meade et al., 2000), a key component in the diagnosis of acute respiratory distress syndrome in the intensive care unit (Stewart et al., 1998). Analogous difficulties arise when assessing the precise state of damage in joints of patients with arthritis.

In other settings, states may be based on categories of a continuous marker, but the marker may be measured with error. Satten and Longini Jr (1996) consider the health of the immune system in individuals with HIV infection where states are defined based on intervals of CD4 cell counts. Measurement error in CD4 cell counts can result in misclassification of states. In this section we consider settings where states may be misclassified or otherwise imperfectly ascertained. We next define notation and formulate the likelihood for the case where the true underlying process is Markov.

As before, we let $Z(a_j)$ represent the true state occupied at time a_j, but let $W(a_j)$ denote the state recorded at a_j. With a perfect classification procedure $P(W(a_j) = Z(a_j)|x) = 1$, but more generally we let

$$\nu_{kh} = P(W(s) = k \mid Z(s) = h, x)$$

denote the conditional probability that an individual is recorded to be in state k given they are in state h at time s, with $\sum_{k=1}^{K} \nu_{kh} = 1$, for each $h = 1, 2, \ldots, K$. In this case, the process $\{Z(s), s \geq 0\}$ is latent (unobserved) or "hidden", and when it is assumed to be a Markov process the term *hidden Markov model* (HMM) is used. It should be noted that the observed process $\{W(s), s \geq 0\}$ is not Markov.

We consider the case where the processes are under an intermittent inspection process subject to right censoring, as in Chapter 5. As in Section 5.4, we let $\bar{\mathcal{H}}(t)$ denote the complete history of all processes, which here includes the history of the latent process, so

$$\bar{\mathcal{H}}(t) = \{Y(u), A(u), Z(u), W(u), 0 \leq u \leq t\},$$

suppressing the notation for fixed covariates. A history incorporating information on the states and the potentially misclassified states occupied at the assessment times is

$$\bar{\mathcal{H}}^*(t) = \{Y(u), A(u), 0 \leq u \leq t; (a_j, Z(a_j), W(a_j)), j = 0, 1, \ldots, A(t)\}.$$

Finally, we let $\bar{\mathcal{H}}^\circ(t) = \{Y(u), A(u), 0 < u \leq t; (a_j, W(a_j)), j = 0, 1, \ldots, A(t)\}$ denote the observed process history. Similar to what was done in (5.16), we consider the joint probability of true and misclassified states $\{(Z(a_j), W(a_j)), j = 0, 1, \ldots, m\}$ and the inspection times $0 = a_0 < a_1 < \cdots < a_m$ over $(0, C = \min(C^R, C^A))$ as

$$P(Z(a_0), W(a_0) \mid a_0) \times$$

$$\prod_{j=1}^{m} \left\{ \lambda^a(a_j \mid \bar{\mathcal{H}}^*(a_j^-)) \exp\left(-\int_{a_{j-1}}^{a_j} \lambda^a(u \mid \bar{\mathcal{H}}^*(u^-)) \, du \right) P(Z(a_j), W(a_j) \mid a_j, \bar{\mathcal{H}}^*(a_j^-)) \right\}.$$

To factor the likelihood and thereby omit terms involving the visit process intensity, we require $\lambda^a(t|\bar{\mathcal{H}}^*(t^-)) = \lambda^a(t|\bar{\mathcal{H}}^\circ(t^-))$, which allows for the visit intensity to depend on the number and times of past visits as well as the observed states.

Defining $\mathcal{Z}^\circ(a_j) = \{(a_r, Z(a_r)), r = 0, 1, \ldots j\}$ as before, we then make the following assumptions:

A.1 $P(Z(a_j) \mid a_j, \bar{\mathcal{H}}^*(a_j^-)) = P(Z(a_j) \mid a_j, \mathcal{Z}^\circ(a_{j-1}))$

A.2 $P(W(a_j) \mid Z(a_j), a_j, \bar{\mathcal{H}}^*(a_j^-)) = P(W(a_j) \mid Z(a_j))$.

Assumption A.1 means that the state occupied at assessment a_j is conditionally independent of the recorded states, given the history of the true states, covariates and past inspection times. Assumption A.2 states that the classification of the state occupied at each inspection time depends only on the current true state and possibly covariates but is conditionally independent of the observation process and prior recorded and true disease states. Both assumptions are reasonable in many settings.

Conditioning on the fixed covariates and the fact that the first assessment was at $a_0 = 0$, a complete data likelihood in which we consider the latent process as observed can be written as

$$L_C \propto P(W(a_0), Z(a_0) \mid X, a_0) \prod_{j=1}^{m} P(W(a_j), Z(a_j) \mid a_j, \bar{\mathcal{H}}^*(a_j^-)) \quad (6.31)$$

$$= P(W(a_0) \mid Z(a_0), X, a_0) P(Z(a_0) \mid X, a_0)$$

$$\times \prod_{j=1}^{m} P(W(a_j) \mid Z(a_j), a_j, \bar{\mathcal{H}}^*(a_j^-)) P(Z(a_j) \mid a_j, \bar{\mathcal{H}}^*(a_j^-)).$$

By Assumptions A.1 and A.2, (6.31) can be rewritten as

$$L_C \propto \left[P(W(a_0) \mid Z(a_0)) \prod_{j=1}^{m} P(W(a_j) \mid Z(a_j)) \right]$$

$$\times \left[P(Z(a_0) \mid a_0, X) \prod_{j=1}^{m} P(Z(a_j) \mid a_j, \mathcal{Z}^\circ(a_j^-), X) \right], \quad (6.32)$$

where the first term in square brackets relates to the misclassification process, and the second term in square brackets pertains to the latent process model.

The observed data likelihood can be obtained by summing (6.32) over all possible realization of $\{Z(a_j), j = 0, 1, \ldots, m\}$; this is done in Jackson and Sharples (2002), Jackson et al. (2003) and the msm package. When the number of states in the process is large or if there are a large number of assessment times, this summation may involve a large number of terms. An EM algorithm can alternatively be used since the complete data likelihood factors into two functionally independent parts. The E-step requires taking the expectation of $\log L_C$ with respect to the entire path $\{Z(a_j), j = 0, 1, \ldots, m\}$, since none of these values are observed. The advantage of the EM algorithm in any particular setting may lie in the availability of software or algorithms for the maximization step; the second term in (6.32) pertains to the latent process model and its parameters. The Kalman filter offers an alternative computationally convenient framework for estimation in this setting (e.g. Fahrmeir and Tutz, 2001, Chapter 8).

With limited sample sizes, estimation of the misclassification probabilities and the transition probabilities are confounded to some extent, and maximum likelihood estimation can be challenging. This is especially true when the times between visits are sufficiently large that multiple transitions might occur. An additional caveat is that model checking can be difficult.

6.4.2 A Hidden Markov Model for Retinopathy in the DCCT

In Section 5.2.4 we considered two Markov models for describing the progression of retinopathy as measured by a 23-point ordinal scale. For reasons discussed there, we considered two 3-state models. Model M1B accommodated improvement in retinopathy by allowing transitions to states representing less severe retinopathy, while model M2B was a progressive process in which transitions were modeled only if a state corresponding to a more advanced stage of the disease was recorded. The two models differ in that M1B models the *observed data*, which includes improvements in the recorded state of damage. The apparent improvement recorded may reflect (i) natural minor variation in the degree of retinopathy and (ii) measurement error in the assessment of the degree of retinopathy.

Table 6.6: Estimates of transition intensities and misclassification probabilities and associated 95% confidence intervals for a progressive 3-state hidden Markov model (HMM) fitted to data from the primary intervention cohort in the DCCT period ($n = 651$); the cut-points for the piecewise-constant true intensity functions are at 3 and 6 years.

		Conventional Therapy		Intensive Therapy	
Period	Transition Intensities	EST	95% CI[†]	EST	95% CI[†]
$[0,3)$	λ_{12}	0.19	(0.16, 0.23)	0.13	(0.10, 0.16)
	λ_{23}	0.07	(0.03, 0.13)	0.03	(0.01, 0.10)
$[3,6)$	λ_{12}	0.33	(0.26, 0.43)	0.14	(0.10, 0.20)
	λ_{23}	0.21	(0.15, 0.29)	0.04	(0.02, 0.11)
$[6,\infty)$	λ_{12}	0.34	(0.15, 0.77)	0.08	(0.02, 0.29)
	λ_{23}	0.35	(0.22, 0.58)	0.10	(0.03, 0.42)
	Misclassification Probabilities	EST	95% CI[†]	EST	95% CI[†]
	ν_{12}	0.06	(0.05, 0.081)	0.11	(0.09, 0.13)
	ν_{13}	< 0.01	(< 0.01, 0.01)	< 0.01	(< 0.01, 0.01)
	ν_{23}	0.05	(0.03, 0.07)	0.06	(0.04, 0.09)
	ν_{21}	0.18	(0.14, 0.23)	0.11	(0.08, 0.15)
	ν_{31}	< 0.01	(0.00, 0.69)	< 0.01	(< 0.01, 1.00)
	ν_{32}	0.28	(0.18, 0.40)	0.32	(0.09, 0.69)
log L		-2534.382		-2047.418	

[†] 95% CI computed as $\exp(\log \widehat{\lambda}_{kl} \pm 1.96\,\text{s.e.}(\log \widehat{\lambda}_{kl}))$.

If one views the process as purely progressive, then the transitions to states representing less severe retinopathy may be viewed as arising due to measurement error; we view the observed states as possibly misclassified versions of a progressive underlying condition. The results of fitting a hidden progressive 3-state Markov model to the same data used to fit models M1B in Section 5.2.4 is given in Table 6.6.

Model M2B of Section 5.2.4 and the HMM here are fitted using different data, but we observe that the transition intensity patterns in Tables 5.2 and 6.6 are qualitatively quite similar. The HMM intensities are smaller because the estimated misclassification error probabilities imply slower real rates of progression. Misclassification errors are especially frequent with regard to true state 2 being classified as state 3 (probabilities $\hat{\nu}_{32}$ about 0.30). We emphasize that the plots for the HMM in Figure 5.3 show probabilities $P_3^{\circ}(t) = P(W(t) = 3)$ for the observed process. Figure 6.12 is the same as Figure 5.3, except that $P_3(t) = P(Z(t) = 3)$ is shown for the HMM. In this case, the plot of $\hat{P}_3(t)$ for the IT group is close to that for model M1B; for the CT group $\hat{P}_3(t)$ agrees well with model M1B up to 4 years from onset, and gives higher values after 4 years. There is thus a fair degree of consistency between state 3 prevalences under model M1B and the HMM. The difference after 4 years in the CT group in part reflects the collapsing of all ETDRS scores of 4 and above into a single state. Four-state models where state 4 represents ETDRS ≥ 7 give values of $\hat{P}_3(t) + \hat{P}_4(t)$ for the CT group that track empirical and M1 estimates up to about 6 years from randomization.

6.5   Bibliographic Notes

The literature on heterogeneity with survival data has grown considerably over the past 40 years. Vaupel et al. (1979) consider the impact on the construction and interpretation of life tables in demography. Hougaard (1984) considered mixture models that describe the heterogeneity through random effects that act multiplicatively on the conditional hazard. Other early work includes Vaupel and Yashin (1985) and Hougaard (1986). Aalen et al. (2015b) provide a discussion of frailty and related biological factors in human disease. There are several books on frailty models with survival data including Duchateau and Janssen (2008) and Wienke (2011). The topic has also received detailed treatment in other books on survival and event history analysis, including Therneau and Grambsch (2000, Chapter 9) who discuss the various roles of frailty models, algorithms for fitting Cox models, and applications. Hougaard (2000) gives a comprehensive treatment including inference and computing (Chapter 8), the construction of joint models (Chapters 7 and 9), and multivariate frailty models (Chapter 10); see also Crowder (2012, Chapter 8). Yashin et al. (1995), Xue and Brookmeyer (1996) and others who have considered multivariate random effect distributions after pointing out the limitations of using univariate random effects to characterize dependence in multiple failure times. Van den Berg (2001) provides a comprehensive review of univariate and multivariate frailty models with an emphasis on econometric applications.

For the multistate setting, Aalen (1987) describes the construction of random effect models for a reversible illness-death process with time-homogeneous transition

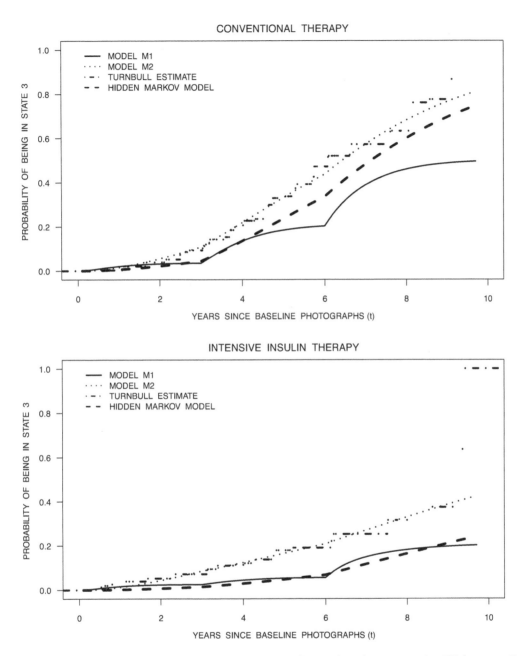

Figure 6.12: Plots of the probability of being in state 3 (ETDRS ≥ 4) over time for CT (top panel) and IT (bottom panel) using the estimates obtained from fitting nonhomogeneous models: M1B (reversible Markov model), M2B (progressive Markov model) and hidden Markov model based on a latent progressive Markov model, along with the corresponding nonparametric Turnbull estimator for entry time distribution.

intensities and separate multiplicative random effects for each intensity; intensity functions for the observable process and likelihoods are expressed in terms of multivariate Laplace transforms of the random effects. Satten (1999) considered a K-state progressive time-homogenous model with a common random effect and pointed out that the marginal likelihood could be obtained in closed form if the random effect had a Laplace transform, when the process was only under intermittent observation. Cook (1999) considered a model for a reversible 2-state conditionally Markov model with a random effect on the equilibrium probability as well as the total rate of transitions and gave the likelilhood when such processes were under intermittent observation. Mealli and Pudney (1999) discuss specification tests for random effects models. Cook et al. (2004) describe K-state progressive models with a discrete multivariate random effect distribution and a component accommodating a "stayer" sub-population. Sutradhar and Cook (2008) considered multivariate normal random effects. Random effect models for alternating 2-state processes have been developed for modeling episodic exacerbations in chronic diseases (Ng and Cook, 1997; Cook et al., 1999). Xu et al. (2010) consider illness-death models and Lange and Minin (2013) consider model fitting with intermittent observation. Putter and van Houwelingen (2015) consider the utility of frailty models when describing the association between sojourn times or transitions in conditionally Markov multistate processes. Relatively little software exists for fitting multistate models with random effects, but the `frailty` option in the `coxph` function deals with Cox models with independent multiplicative random effects V_{kl}, and special packages such as `coxme` (Therneau, 2012) and `frailtypack` (Król et al., 2017) fit certain models.

Aalen et al. (2008) discuss intensity-based models for examining dependencies between processes through "dynamic path analysis"; see also Cook and Lawless (2014). When interest lies in inferences about features of individual processes, copula models can be adopted to accommodate dependencies in entry or sojourn times across processes (Diao and Cook, 2014); more details are given in Diao (2013). Working independence assumptions can also be used. Lee and Kim (1998) develop robust sandwich-type variance estimates for settings involving multivariate Markov processes under intermittent observation; this approach is similar in spirit to that of Wei et al. (1989) for multivariate Cox regression. Such methods can be used for both clustered and multivariate multistate processes.

Cure rate models are motivated by settings in that there is an unidentified sub-population of individuals at zero risk of the event (Farewell, 1982, 1986). Aalen (1988) introduced a compound Poisson mixing distribution that accommodates a non-susceptible fraction of individuals along with continuous variation in risk between susceptible individuals; this was further developed in Aalen (1992). Recent developments include Taylor (1995), Peng et al. (1998) and Peng and Dear (2000) who developed an expectation-maximization (EM) algorithm for fitting Cox models with a cured fraction. Mover-stayer models represent the multistate generalization of cure-rate models in that the population is comprised of a sub-population of individuals who will pass through a multistate process and a sub-population of individuals who will remain in the initial state. Goodman (1961) developed methods for consistent parameter estimation, and Frydman (1984) developed maximum likelihood methods. Fuchs and Greenhouse (1988) developed an EM algorithm for estimation with cure rate models. Multistate models incorporating dynamic mover-stayer indicators include Heckman and Walker (1987), Yamaguchi (1994, 1998, 2003), Cook et al. (2002) and O'Keeffe et al. (2013).

The use of hidden Markov models (HMMs) has grown rapidly over the last 30 years with applications including environmental research, speech recognition and molecular biology. Jackson and Sharples (2002) and Jackson et al. (2003) consider their use for disease processes, and Titman and Sharples (2010a) describe analysis using the `msm` package.

6.6 Problems

Problem 6.1   Consider a pair of 3-state processes depicted in Figure 6.6 with intensities of the form (6.16), where $V_r = (V_{r1}, V_{r2})$ is a bivariate random effect for process r with $V_r \perp X$, $V_{r1} > 0$, $v_{r2} > 0$, $r = 1, 2$, and joint cumulative distribution functions $G(v_r)$.

(a) Show that

$$dG_r(v_r \mid \mathcal{H}_r(t^-), Z_r(t^-) = 2)$$
$$= \frac{v_{r1}\, \lambda_{r1}(t_{r2} \mid x)\, \exp\left(-v_{r1}\, \Lambda_{r1}(t_{r2} \mid x)\right) \exp\left(-v_{r2}\, \Lambda_{r2}(t_{r2}, t \mid x)\right) dG_r(v_r)}{\int v_{r1}\, \lambda_{r1}(t_{r2} \mid x)\, \exp\left(-v_{r1}\, \Lambda_{r1}(t_{r2} \mid x)\right) \exp\left(-v_{r2}\, \Lambda_{r2}(t_{r2}, t \mid x)\right) dG_r(v_r)}$$

where t_{r2} is the time of entry to state 2 for process r and $\Lambda_{rk}(s, t \mid x) = \int_s^t \lambda_{rk}(u \mid x) du$.

(b) Does the marginal model for process r retain the Markov property if $V_{r1} \perp V_{r2}$?

(c) Suppose the conditional intensities are semi-Markov with $\lambda_{rk}(t \mid \mathcal{H}_r(t^-), v_r) = v_{rk}\, h_{rk}(B(t) \mid x)$ where $h_{rk}(\cdot \mid x)$ is a hazard for the sojourn time in state k for process r and $B(t)$ is the time since the start of the process if $k = 1$ and the time since entry to state 2 if $k = 2$. Under what conditions do the marginal processes retain the semi-Markov property?

(d) Consider the calculation of $P(Z(a_1) = (1,1), Z(a_2) = (2,1))$ where $Z(0) = (1,1)$ and V_1, V_2 are not independent, and comment on the difficulties of applying this and other models with associated random effects.

(Sections 6.1, 6.2)

Problem 6.2   Consider an illness-death model with conditional Markov intensities

$$\lambda_{kl}(t \mid V = v) = v\, \lambda_{kl}(t), \quad (kl) = (12), (13), (23),$$

where V is a random variable having a gamma distribution with mean 1 and variance ϕ. Determine the marginal intensity functions. Examine $\lambda_{23}(t \mid t_2)$, where t_2 is the time of entry to state 2 and comment on the flexibility of this model and one with no random effect but a non-Markov parametric form for $\lambda_{23}(t \mid t_2)$.

(Section 6.1; Xu et al., 2010)

Problem 6.3   Consider the progressive multistate model of Problem 2.5. Suppose $\{Z_i(s), s > 0\}$ is the process for individual i, who makes transitions according to intensities of the form

$$\lim_{\Delta t \downarrow 0} \frac{P(Z_i(t + \Delta t^-) = k+1 \mid Z_i(t^-) = k, \mathcal{H}_i(t^-), V_i = v_i)}{\Delta t} = v_i\, \lambda_k,$$

where V_i has a distribution $G(\cdot)$ with mean 1 and variance ϕ. Suppose individuals are under intermittent observation, and let $0 = a_{i0} < a_{i1} < \cdots < a_{im_i}$ denote the observation times for individual i under a conditionally independent visit process in a sample of n independent individuals, $i = 1, \ldots, n$.

(a) Derive the form of the observed data likelihood in terms of the Laplace transform of V.

(b) Describe how to carry out a likelihood ratio test of the null hypothesis H_0: $\phi = 0$ versus H_A: $\phi > 0$.

(Section 6.1; Liang and Self, 1996)

Problem 6.4 Section 1.6.4 describes a study in which HIV-infected individuals have a co-infection involving the cytomegalovirus (CMV). Let T_{i1} and T_{i2} denote the times to viral shedding in the urine and blood for individual i, $i = 1, \ldots, n$. Let $0 = a_{i0} < a_{i1} < \cdots < a_{im_i}$ denote the unique times at which individual i provides a blood and/or urine sample to be tested for evidence of viral shedding. Let $\delta_{ij1} = 1$ if a urine sample is obtained at a_{ij} for individual i with $\delta_{ij1} = 0$ otherwise, and $\delta_{ij2} = 1$ if a blood sample is obtained at a_{ij} with $\delta_{ij2} = 0$ otherwise; $\delta_{ij} = (\delta_{ij1}, \delta_{ij2})$. If C_i is a censoring time and $Y_i(t) = I(t \leq C_i)$ the assessment process can be viewed as a marked point process and denoted by $\{Y_i(t), dA_i(t), \delta_i(t), t > 0\}$ where $\delta_i(t) = \delta_{ij}$ if $t = a_{ij}$ and is zero otherwise, $j = 1, \ldots$. Suppose interest lies in an intensity-based analysis using a Markov model with a state space depicted in Figure 1.8 and functionally independent intensities $\lambda_{kl}(t)$ with $(k, l) \in \{(1, 2), (1, 3), (2, 4), (3, 4)\}$. The observed process history is then $\mathcal{Z}_i^\dagger(t) = \{a_{ij}, \delta_{ij}, \mathcal{S}_i(a_{ij}), j = 0, 1, \ldots, A_i(t)\}$ where $\mathcal{S}_i(a_{ij})$ is the set of states that could be occupied at a_{ij} based on the available data.

(a) Derive conditions analogous to the CIVP conditions of Section 5.4, which justify the use of the observed data partial likelihood

$$\prod_{i=1}^n \left\{ \sum_{\mathcal{Z}_i^\circ(a_{im_i}) \in \mathcal{P}_i} \prod_{j=1}^{m_i} P(Z_i(a_{ij}) \mid a_{ij}, \delta_{ij}, \mathcal{Z}_i^\circ(a_{ij}^-)) \right\} \qquad (6.33)$$

where $\mathcal{P}_i$ is the set of possible paths for the process that are compatible with the observed data.

(b) Describe how this observed data likelihood could be maximized via an EM algorithm where the complete data likelihood is based on the scenario where $\delta_{ij} = (1, 1)$ for $j = 1, \ldots, m_i$. Describe how the maximization step could be carried out by adopting one of the algorithms of Section 5.2.2.

(c) Suppose instead an analysis was planned based on a working independence assumption between the times to viral shedding in the urine and blood. Consider a likelihood for interval-censored univariate failure time data (Lawless, 2003; Sun, 2006). Give conditions under which this standard partial likelihood is valid.

(Sections 5.4, 6.2)

Problem 6.5 The data for the study discussed in Section 1.6.4 and Problem 6.4 are described in Section D.4 and available at the website for this book.

(a) By maximizing the observed data likelihood in (6.33) or implementing the EM algorithm in Problem 6.4(b) fit the multistate model in Figure 1.8 to the data using piecewise-constant intensities with cut-points at 2 and 12 months.

(b) Plot the estimate of the marginal distribution function $F_1(t) = P(T_1 \leq t)$ based on $\widehat{P}_{12}(0,t) + \widehat{P}_{14}(0,t)$ from the fitted model and compare that to a marginal nonparametric estimate of $F_1(t)$ obtained using the `prodlim` function in the `prodlim` package in R. You could alternatively use the `KaplanMeier` in S-PLUS (TIBCO Spotfire S+® 8.2, 2010). Repeat this for the time to viral shedding in the blood.

(c) What do you conclude regarding the association between the times to viral shedding in the urine and blood based on the intensity-based model?

(Sections 5.4, 6.2)

Problem 6.6   Consider the study in Problem 6.4.

(a) Consider a frailty model where $T_{i1} \perp T_{i2} | V_i$ with V_i a scalar random effect with distribution function $G(v_i; \phi)$, $E(V_i) = 1$ and $\text{var}(V_i) = \phi$. Under a multiplicative shared random effect model

$$\lim_{\Delta t \downarrow 0} \frac{P(t \leq T_{ij} \leq t + \Delta t \mid t \leq T_{ij}, V_i = v_i)}{\Delta t} = v_i \, h_j(t),$$

$j = 1, 2$, let $h_j(t) = \alpha_{jk}$ if $t \in \mathcal{B}_{jk} = [b_{j,k-1}, b_{jk})$, where $0 = b_{j0} < b_{j1} < \cdots < b_{jK_j} = \infty$ define K_j pieces in a conditionally piecewise-constant hazard model. Write a complete data likelihood where the complete data includes known t_{i1}, t_{i2} and v_i. Describe the iterative steps of an EM algorithm to maximize the observed data likelihood.

(b) Explain how existing software for Poisson regression can be exploited at the maximization step.

(c) Interest lies in estimating the distribution of the time to the first evidence of viral shedding in blood or urine in the population, defined by $T_i = \min(T_{i1}, T_{i2})$ for individual i. Describe how you would estimate the median time to viral shedding in either source.

(Sections 5.4, 6.2; Jackson, 2011, Section 3.4)

Problem 6.7   Consider a study of individuals with age-related macular degeneration where visual acuity is graded on a K-point scale with scores representing progressively worsening vision (Age-Related Eye Disease Study Research Group, 2005). Let $\{Z_r(s), s \geq 0\}$ denote the K-state process with states labeled $k = 1, \ldots, K$ for the left $(r = 1)$ and right $(r = 2)$ eyes, respectively. Consider an inception cohort of individuals recruited at time $a_0 = 0$ with $Z_r(a_0) = 1$, $r = 1, 2$. We adopt a model such that visual acuity at time t for eye r is governed by

$$\lim_{\Delta t \downarrow 0} \frac{P(Z_r(t + \Delta t^-) = k + 1 \mid Z_r(t^-) = k, V = v)}{\Delta t} = v \lambda_k(t),$$

$k = 1, \ldots, K - 1$ where $V > 0$ is a gamma distributed random effect with $E(V) = 1$ and $\text{var}(V) = \phi$ accommodating heterogeneity in the rates of declining visual acuity.

(a) Describe how to make a prediction for $I(Z_r(t^\ddagger) = K)$ at some time $t^\ddagger$ after disease onset based on the marginal model for $\{Z_r(s), s \geq 0\}$.

(b) Suppose it was determined that $Z_r(t^\dagger) = k_r^\dagger$ $(1 \leq k_r^\dagger \leq K)$ at some intermediate time $t^\dagger$, $0 < t^\dagger < t^\ddagger$. How would you update your prediction in light of this information?

(c) How would you revise your prediction knowing $Z_r(t^\dagger) = k_r^\dagger$, $r = 1, 2$ at $t^\dagger$?

(d) Give an expression for the distribution of the time to total blindness defined as $T = \max(T_{1K}, T_{2K})$.

(Sections 6.1, 6.2)

Problem 6.8 Cancer patients undergoing surgery may have their tumour completely removed and then be at zero risk of recurrence. We consider individuals following cancer surgery as being in one of two classes: individuals where the entire tumour was removed are in class 1 and all other individuals are in class 2. It is generally unknown if the entire tumour has been removed following surgery, so we consider a latent class model where V indicates membership in class 2 and the conditional intensity for recurrence is $\lambda_{12}(t|\mathcal{H}(t^-), v) = v \lambda_{12}(t)$, with $\lambda_{13}(t|\mathcal{H}(t^-), v) = \lambda_{13}(t)$ and $\lambda_{23}(t|\mathcal{H}(t^-), v) = \lambda_{23}(t|t_2)$ denoting death intensities in the illness-death model shown below. In addition let $P(V = 1|X = x) = \text{expit}(x_i'\beta)$ where $\text{expit}(x) = \exp(x)/(1 + \exp(x))$, $x = (1, x_1, \dots, x_{p-1})'$ is a $p \times 1$ covariate vector, and β is an associated vector of regression coefficients.

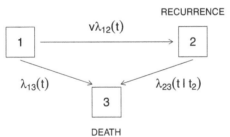

Let C_{ik} be the censoring time for T_{ik}, $k = 2, 3$, $i = 1, \dots, n$ and consider a study yielding data

$$\{\min(T_{ik}, C_{ik}), \delta_{ik} = I(T_{ik} \leq C_{ik}), k = 2, 3, X_i, i = 1, \dots, n\}.$$

Note that $C_{i2} = T_{i3}$, $\delta_{i2} = 0$ if $T_{i3} \leq C_{i3}$, $T_{i3} < T_{i2}$.

(a) Write the observed data likelihood.

(b) Describe how to fit models and consider ways to carry out inference for parametric analysis.

(c) Describe how you would fit a model with semiparametric specifications for the transition intensities.

(Section 6.3; Conlon et al., 2015)

Problem 6.9 Consider a progressive K-state process $\{Z(s), s \geq 0\}$ with states $1, 2, \ldots, K$ and $P(Z(0) = 1) = 1$. Let V denote an individual-specific random effect with distribution function $G(v)$, and

$$\lim_{\Delta t \downarrow 0} \frac{P(Z(t + \Delta t^-) = k+1 \mid Z(t^-) = k, \mathcal{H}(t^-), V = v)}{\Delta t} = v \lambda_k(t),$$

$k = 1, \ldots, K - 1$. Let $a_0 = 0$ and $\{A(s), s \geq 0\}$ denote an intermittent assessment process and $A_1, A_2, \ldots$ the assessment times with

$$\lim_{\Delta t \downarrow 0} \frac{P(\Delta A(t) = 1 \mid \bar{\mathcal{H}}^\circ(t^-), v)}{\Delta t} = v \lambda^a$$

where $\bar{\mathcal{H}}^\circ(t) = \{Y(u), A(u), 0 \leq u \leq t; (a_j, Z(a_j)), \ j = 0, 1, \ldots, A(t)\}$.

  (a) Write the likelihood given $V = v$ for the multistate data and the assessment times, and then derive the observed data likelihood when $\lambda_k(t) = \lambda_k$.
  (b) Derive the form of the visit process intensity:
$$\lim_{\Delta t \downarrow 0} \frac{P(\Delta A(t) = 1 \mid \bar{\mathcal{H}}^\circ(t^-))}{\Delta t}.$$

  (c) Discuss the utility of this model for dealing with a CDVP if the multistate process is Markov.

(Sections 5.4, 6.2.1)

Problem 6.10 Consider the data from Nagelkerke et al. (1990) discussed in Problem 5.2 and provided in Section D.6. The test for the presence of the parasite in the stool samples does not have perfect sensitivity, so the underlying process $\{Z_i(s), s > 0\}$ is not accurately observed. Let $\{Z^*(t), t > 0\}$ denote the observable process such that

$$P(Z^*(t) = 1 \mid Z(t) = 2) = \pi$$

is the false-negative rate where $0 \leq \pi \leq 1$; assume the specificity is perfect.

  (a) Write the likelihood function for the observed data under the assumption that the underlying process is as described in Problem 5.2 and that it is in equilibrium at the time of recruitment.
  (b) Maximize the likelihood in (5.28) under the assumption of perfect sensitivity and compare these with the estimates obtained by maximizing the likelihood in (a). Report the estimated false-negative rate of the test and use this to explain the differences seen in the two analyses.
  (c) Plot contours of the profile relative likelihoods
$$PRL(\lambda_1, \pi) = \frac{L(\lambda_1, \tilde{\lambda}_2(\lambda_1, \pi), \pi)}{L(\widehat{\lambda}_1, \widehat{\lambda}_2, \widehat{\pi})} \quad \text{and} \quad PRL(\lambda_2, \pi) = \frac{L(\tilde{\lambda}_1(\lambda_2, \pi), \lambda_2, \pi)}{L(\widehat{\lambda}_1, \widehat{\lambda}_2, \widehat{\pi})},$$

where $\tilde{\lambda}_k(\lambda_{3-k}, \pi)$ is the profile maximum likelihood estimate of λ_k, and $\widehat{\lambda}_1$, $\widehat{\lambda}_2$ and $\widehat{\pi}$ are the maximum likelihood estimates. Comment on the information regarding λ_k and π and the estimability issues for the associated parameters.

(Section 6.4; Nagelkerke et al., 1990; Rosychuk and Thompson, 2001)

Chapter 7

Process-Dependent Sampling Schemes

7.1 History- and State-Dependent Selection

7.1.1 Types of Selection Schemes and Likelihoods

Many studies of chronic diseases, or other processes, involve the selection of individuals whose observed process history satisfies particular conditions. Here we consider examples of such conditions along with methods of estimation that account for such selection criteria. We focus for now on progressive processes $\{Z_i(s), s \geq 0\}$ with state space depicted in Figure 7.1. State 0 represents the condition of being alive and disease-free, and state 1 is entered upon disease onset. States $2, \ldots, K-1$ represent worsening disease states that may be entered as the disease progresses, and state K represents death. If an individual does not develop the disease during their lifetime they make a $0 \to K$ transition, but individuals who develop the disease ultimately make a $k \to K$ transition for some value of k, $k = 1, \ldots, K-1$. We consider a population of individuals whose processes are independent.

Any study will involve sampling individuals from the population and collection of data. In what follows, the time scale is individual-specific and unless stated otherwise is assumed to be age, with $t = 0$ often corresponding to a person's birth. Let A_{i0} denote a possibly random time at which an individual i is contacted, their state $Z(a_{i0})$ is determined and they are considered for selection. The history of the process at $A_{i0} = a_{i0}$ is denoted by $\mathcal{H}_i(a_{i0}) = \{Z_i(s), 0 \leq s \leq a_{i0}\}$. Information on $\{Z_i(s), s > a_{i0}\}$ can be acquired through prospective follow-up. The mechanism by which individuals are identified and the times A_{i0} are generated, depends on the sampling scheme. In many studies, individuals are randomly selected from a population or some well-defined sub-population. The UK Whitehall II Study (Marmot et al., 1991), for example, considered the relationships between social class, morbidity and mortality among British civil servants. The Physicians' Health Study examined the effect of aspirin and beta-carotene on cardiovascular mortality in a sample of 22,071 male physicians in the United States, and the effect of estrogen therapy on cardiovascular disease among post-menopausal women was studied on 48,470 nurses in Stampfer et al. (1991). In studies directed at progression of disease, individuals are often identified through disease registries, administrative records, tertiary care facilities, and so on.

Prospective cohort studies aiming to estimate the incidence rate of disease or the effect of interventions on disease prevention would recruit individuals in state 0 at a_{i0} and follow them prospectively to record disease onset, represented by tran-

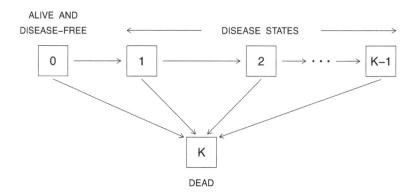

Figure 7.1: A multistate model for the onset and evolution of a progressive disease process and death.

sitions into state 1, and perhaps information on disease progression. Such designs admit straightforward analyses but are often prohibitively expensive if the disease incidence is low. Moreover, if the disease course is slow, unless samples are large and there is considerable follow-up, relatively little information will be obtained on disease progression. Incident cohort studies, in contrast, may aim to collect information on the disease process from the time of onset; in this case individuals may need to be in state 1 at time a_{i0} to be recruited for study. To ensure that the sample represents individuals with a recent onset, the selection condition may be further refined to $Z_i(a_{i0}) = 1$ and $a_{i0} - t_{i1} < D$, where D is a maximum permissable disease duration. Here and subsequently, T_{ik} and t_{ik} represent the time of entry to state k, $k = 1, \dots, K$. Some examples of selection criteria in particular studies are discussed next.

Example 7.1.1: The Diabetes Control and Complications Trial

The Diabetes Control and Complications Trial (DCCT) on Type 1 diabetes was introduced in Section 1.2.2. and some analyses related to the progression of retinopathy in affected persons were discussed in Sections 5.2.4 and 6.4.2. The time origin for a multistate model representing the retinopathy process could naturally be set as the onset time of diabetes, defined here as the time insulin therapy is initiated. In this setting state 0, rather than representing "alive and disease-free" as in Figure 7.1, represents the condition of being an insulin-dependent diabetic with no signs of retinopathy; states 1 to $K - 1$ represent worsening stages of retinopathy, and state K represents death.

The DCCT was comprised of two sub-studies. The primary prevention sub-study was designed to assess whether an intensive program for glycemic control reduced the risk of developing retinopathy, so it required there be no evidence of diabetic retinopathy at the time $A_{i0} = a_{i0}$ of the screening assessment. An added requirement that individuals were on continuous insulin therapy for at most 5 years meant the selection conditions were $Z_i(a_{i0}) = 0$ and $a_{i0} < 5$ years. The secondary intervention sub-study aimed to provide information on the progression of retinopathy once it had begun and subjects were required to have had insulin-dependent diabetes for less than 15 years ($A_{i0} < 15$), with very mild to moderate retinopathy.

Example 7.1.2:   The Centre for Prognosis Studies in Rheumatic Diseases

Psoriatic arthritis (PsA) is an autoimmune disease in which affected individuals suffer from psoriatic plaques on their skin and inflammation and pain in their joints, which can ultimately lead to joint destruction. The onset of psoriasis (Ps) typically precedes the development of joint involvement and so we might take t as age, state 0 as representing a healthy state, state 1 as entered upon the onset of psoriasis and state 2 as entered upon the development of PsA; state $K = 3$ represents death.

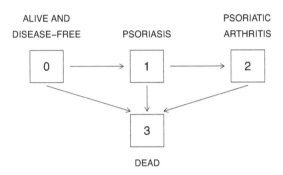

Figure 7.2: A 4-state model for the onset of psoriasis and psoriatic arthritis with the competing risk of death.

We introduced the University of Toronto Psoriatic Arthritis Clinic (UTPAC) registry in Section 1.2.3. This registry was launched in 1977 with a view to better understanding the disease course and progression rates of this complex disorder. A primary method of recruitment to this registry is through a population-based screening tool comprised of a 10-item questionnaire. Individuals completing this questionnaire and suspected of having psoriatic arthritis are invited to attend a clinic at the Centre for Prognosis Studies in Rheumatic Diseases for a more definitive diagnosis; those found to have the disease are invited to join the UTPAC. The sampling condition for this registry is therefore $Z_i(a_{i0}) = 2$, and upon entry to the UTPAC a detailed history is taken during which the values t_{i1} (the age of onset of psoriasis) and t_{i2} (the age of onset of psoriatic arthritis) are retrospectively obtained.

Interest also lies in characterizing the course of psoriasis, including risk factors for the development of PsA. Because the retrospective data on the $1 \rightarrow 2$ transition times for persons who have already progressed to PsA are not particularly informative about the incidence of PsA, a second registry was formed in 2008 at the University of Toronto Psoriasis Clinic (UTPC). Screened patients identified as having psoriasis (but not PsA) are recruited and upon entry to the registry they undergo a detailed clinical examination and provide samples for genetic testing. They are then followed prospectively according to a standardized protocol, and when they develop PsA this is noted. For the UTPC registry the selection condition is therefore $Z_i(a_{i0}) = 1$ with the onset time of PsA noted during follow-up of recruited individuals.

We consider now the general setting in which individuals are recruited for a study if they are in one of a set of states $\mathcal{S}$ at the time of the assessment, under the assumption that their selection does not otherwise depend on the process history. To simplify the notation, we drop the subscript i and consider a generic individual. Special cases include studies that simply require individuals to be alive at time a in which case $\mathcal{S} = \{0, 1, \ldots, K-1\}$, or studies aiming to recruit diseased individuals, in which case $\mathcal{S} = \{1, 2, \ldots, K-1\}$. Under the assumption that the contact time A_0 is independent of $\{Z(t), t > 0\}$, the probability of a particular sample path, conditional on $A_0 = a_0$ and the selection condition, is

$$\frac{\prod_{k=0}^{K-1} \prod_{l>k} \left\{ \prod_{t_j \in \mathcal{D}_{kl}} \lambda_{kl}(t_j \mid \mathcal{H}(t_j^-)) \exp\left(-\int_0^\infty Y_k(u^-) \lambda_{kl}(u \mid \mathcal{H}(u^-)) \, du\right) \right\}}{P(Z(a_0) \in \mathcal{S} \mid Z(0) = 0, A_0 = a_0)}, \quad (7.1)$$

where $\mathcal{D}_{kl}$ is the set of $k \to l$ transition times for $l > k$, $k = 0, 1, \ldots, K-1$.

Due to loss to follow-up at C, say, we may only observe $\{Z(s), s > 0\}$ over $(0, \min(T_K, C)]$. In this case a censored data analogue to (7.1) is used, where $\mathcal{D}_{kl}$ is the set of $k \to l$ transitions observed over $(0, \min(T_K, C)]$ and $Y_k(u^-)$ is replaced with $\bar{Y}_k(u) = Y(u) Y_k(u^-)$ where $Y(u) = I(u \leq C)$.

A key issue is whether data on the process history prior to A_0 are available. The likelihood (7.1) includes such data, and to consider this issue we decompose it as

$$L = L_1 \cdot L_2 \quad (7.2)$$

where

$$L_1 \propto P(Z(s), 0 < s \leq a_0 \mid Z(a_0) \in \mathcal{S}, A_0 = a_0, Z(0) = 0) \quad (7.3)$$

represents the contribution from the retrospective and current data at a_0, and

$$L_2 \propto P(Z(s), a_0 < s \leq C \mid Z(u), 0 \leq u \leq a_0, A_0 = a_0) \quad (7.4)$$

is the contribution based on the prospective data given the history up to and including time a_0. The term L_1 in (7.3) can be further factored as $L_1 = L_{1R} \cdot L_{1C}$, where

$$L_{1R} \propto P(Z(s), 0 < s < a_0 \mid Z(a_0), Z(a_0) \in \mathcal{S}, A_0 = a_0) \quad (7.5)$$

represents the contribution from the strictly retrospective data at $A_0 = a_0$, and

$$L_{1C} \propto P(Z(a_0) \mid Z(a_0) \in \mathcal{S}, A_0 = a_0) \quad (7.6)$$

represents the contribution of the cross-sectional data at $A_0 = a_0$. If detailed historical data up to time a_0 are available, then an analysis can, in principle, be based on L_1 alone, but there is often relatively little information on features of interest. Prospective data in L_2 are often more informative, but we note that depending on the model, details about $\{Z(s), 0 \leq s < a_0\}$ are still needed for conditioning. We often seek models for which coarsened data on the disease course until the time of recruitment are sufficient. One can also consider a likelihood based on $L_{1C} \cdot L_2$, but

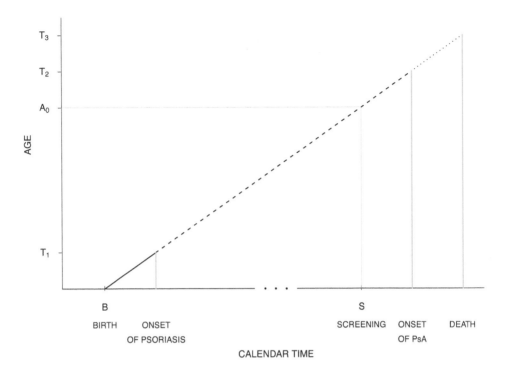

Figure 7.3: Lexis diagram depicting birth, other life history events and screening dates in calender time (horizontal axis) and ages at state transitions on the vertical axis.

computation of L_{1C} and L_2 is challenging for general processes if historical data are unavailable, coarsened, costly to retrieve, or unreliable.

Thus far, we have said little about how A_0 is generated. This depends on the population screening scheme which may sometimes be difficult to characterize. This then makes it challenging to determine the probabilities in (7.1)−(7.6). To examine this it is helpful to consider the special case of an epidemiological study involving screening a population of individuals and sampling from those screened, subject to selection conditions. For illustration, consider a process of screening the population for individuals with psoriasis for recruitment to the University of Toronto Psoriasis Clinic (UTPC), with reference to the multistate model in Figure 7.2. Figure 7.3 is a Lexis diagram (Keiding, 2011) in which the horizontal and vertical axes are calendar time and age, respectively. The date of birth of an individual is depicted at calendar time B on the horizontal axis. Using the convention that age and calendar time are in the same units, the 45° solid line emanating from the birth date represents the time they spend alive and disease-free (i.e. in state 0). This solid line may terminate upon death, or change to a dashed line upon the development of psoriasis. The dashed line segment would terminate upon death with psoriasis or, as in the case of Figure 7.3, may change to a dotted line upon the development of psoriatic arthritis. The calendar time of the subsequent death is also depicted at which point the dotted line terminates. The times of the transitions are depicted in calendar time on the horizontal axis and in terms of the individual's age on the vertical axis. If a screening scheme is carried out at calendar time S, the age of a person who is alive at time S is $A_0 = S - B$, so the distribution of A_0 is governed by the birth process in the population, the disease and survival process, and any trends in these processes over time. We note incidentally that in some contexts we might choose to let B denote the time of some initial event in a person's life rather than their date of birth, but for discussion we refer here to B as the birth date.

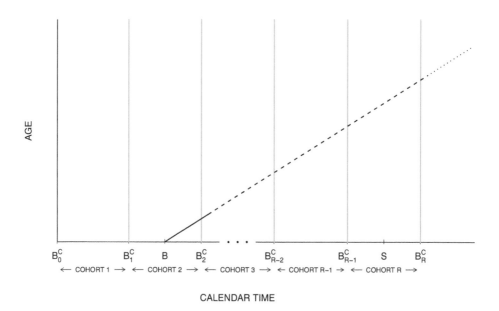

Figure 7.4: A Lexis diagram with R cohorts defined based on calendar time of birth.

If there are trends in the disease process over calendar time, we may define birth cohorts delineated by the calendar times $B_0^C, B_1^C, \ldots, B_R^C$ in Figure 7.4. Often cohorts of individuals are defined by 5- or 10-year intervals of calendar time, so we may set $B_0^C = 1950$, $B_1^C = 1960$, ..., $B_6^C = 2010$, $B_7^C = 2020$, for example. If $B \in [B_r^C, B_{r+1}^C)$ for an individual, they are said to be in cohort r, $r = 1, \ldots, R$. Because S is fixed, if $B \in [B_r^C, B_{r+1}^C)$ then $A_0 \in (S - B_{r+1}^C, S - B_r^C]$, so membership in cohort r can of course be defined in terms of the corresponding interval for A_0. If there is interest in examining trends in the disease process over birth cohorts, probability models for the multistate processes may be stratified by cohort, or differences in the disease processes across cohorts can be modeled so that trends can be summarized parametrically.

The age at recruitment is also the delayed entry time to the prospective part of the study. Independent delayed entry implies that in (7.4), for example,

$$P(Z(s), a_0 < s \le C \mid Z(u), 0 < u \le a_0, A_0 = a_0) = P(Z(s), a_0 < s \le C \mid Z(u), 0 < u \le a_0)$$

so that the relevant probability for the likelihood is the same as the probability we would use if A_0 was predetermined to be equal to a_0. There are settings in which there may be an association between the disease course and the age at recruitment, as when there is a birth cohort effect, and then the information $A_0 = a_0$ is related to the prospective disease process. Chaieb et al. (2006) give examples in the context of survival analysis; see also Keiding (1991). In this case, we can stratify on period of birth, as described above.

Finally, we note that the process by which individuals in a population are selected for screening at time S must be considered. The referral mechanism to the UTPC is complex, multifaceted, and not easy to characterize adequately. The discussion here is based on an idealization of the recruitment process that applies when individuals are first identified through random screening of a population. Such mechanisms are more representative of how individuals are recruited for prevalent cohort studies, which we discuss in Section 7.1.3, or sub-studies of larger prospective cohort studies.

7.1.2   Empirical Studies of Design Efficiencies for Markov Processes

Study design has several aspects, some of which were discussed in Sections 3.6 and 5.2.2. Here we present a small simulation study to illustrate differences in information from retrospective, cross-sectional, and prospective data on multistate processes sampled subject to different state-dependent selection conditions. We consider a Markov process, in which case $\lambda_{kl}(t|\mathcal{H}(t^-)) = \lambda_{kl}(t)$, computation of L_{1C} simply requires the transition probability matrix $P(s,t)$, and (7.4) reduces to

$$L_2 \propto \prod_{k=k_0}^{K-1} \prod_{l>k} \prod_{t_j \in \mathcal{D}_{kl}} \lambda_{kl}(t_j) \exp\left(-\int_0^\infty \bar{Y}_k(u)\,\lambda_{kl}(u)\,du\right), \qquad (7.7)$$

where here $\mathcal{D}_{kl}$ is the set of $k \to l$ transition times over $(a_0, C]$.

When individuals are only observed intermittently following recruitment, we usually denote the first assessment time by a_0 and let $a_1 < \cdots < a_m$ denote m follow-up assessment times. Suppressing the visit times, we can then consider the likelihood

$$L = L_1 \cdot \prod_{j=1}^m P(Z(a_j)\,|\,Z(a_{j-1})) \qquad (7.8)$$

if the transition times over $(0, a_0]$ are available, or alternatively $L_{1C} \cdot L_2$ if they are not. If no historical data are available, conditioning on $Z(a_0)$ yields

$$L_2 = \prod_{j=1}^m P(Z(a_j)\,|\,Z(a_{j-1}), Z(0)=0), \qquad (7.9)$$

where $Z(a_0) \in \mathcal{S}$ and jointly modeling $Z(a_0)$ and the prospective data lead to $L_{1C}L_2$.

Consider the time-homogeneous disease process depicted in Figure 7.1 with $K = 4$. Since this is a disease process affecting individuals, we think of t as age in years. For illustration we consider a time-homogeneous Markov model, so that $\lambda_{k-1,k}(t) = \lambda_k$ $(k=1,2,3)$ and $\lambda_{k4}(t) = \gamma_k$ $(k=0,1,2,3)$. We let the lifetime probability of disease be 0.15, which implies $\lambda_1/(\lambda_1 + \gamma_0) = 0.15$, and set the probability of progression (rather than death) among individuals in state 1 as 0.50 (i.e. $\lambda_2/(\lambda_2 + \gamma_1) = 0.50$). We assume individuals progress with a 10% higher intensity from $2 \to 3$ than they do from $1 \to 2$ by letting $\lambda_3 = r_1\lambda_2$ with $r_1 = 1.1$. We let $\gamma_1/\gamma_0 = r_2 = 1.05$, so that the mortality rate is 5% higher after disease onset, and assume $\gamma_j = r_3\gamma_{j-1}$, $j = 2, 3$ with mortality rates higher by a factor of $r_3 = 1.1$ following progression to a higher disease state. Finally, we set γ_0 so that $P(Z(100) = 4|Z(0) = 0) = 0.95$ (i.e. so that the probability an individual has died by $t = 100$ years of age is 0.95). These constraints lead to transition intensities $\lambda_1 = 0.005$, $\lambda_2 = 0.031$, $\lambda_3 = 0.034$, $\gamma_0 = 0.029$, $\gamma_1 = 0.030$, $\gamma_2 = 0.034$ and $\gamma_3 = 0.037$.

We simulate the calendar times of births in a population according to a time-homogeneous Poisson process and consider recruiting individuals for a study at an instant in calendar time according to one of two sampling schemes. In the first sampling scheme, given a person's age (time since birth) is $A_{i0} = a_{i0}$ at the calendar time of screening, we sample them only subject to the condition that they are alive at $A_{i0} = a_{i0}$, so $\mathcal{S} = \{0, 1, 2, 3\}$. In the second sampling scheme, we select only individuals who are alive and diseased, so $\mathcal{S} = \{1, 2, 3\}$. We consider both a continuous prospective observation scheme with administrative censoring 15 years after recruitment, and an intermittent observation scheme by which individuals are examined every 3 years for up to 15 years following recruitment. At these assessment times the disease state is recorded, but we assume that the times of death over the planned period of follow-up are observed precisely. In general, it will not be known which state was left upon death, so a partial likelihood like (5.26) given in Section 5.6.2 is used. We consider the frequency properties of maximum likelihood estimators exploiting retrospective, cross-sectional and prospective data $(L_1 \cdot L_2)$, cross-sectional and prospective data $(L_{1C} \cdot L_2)$, and simply prospective data (L_2).

Regarding the data simulation, we assume births occur in calendar time according to a stationary process and following births the life histories arise according to the multistate model. Since the birth process is stationary, however, we may first generate the life history process for a very large sample of individuals, generate $A_{i0} \sim \text{unif}(0, T_4^{\max})$ for each individual where $T_4^{\max} = \max_i \{T_{i4}\}$, and then apply the appropriate selection condition to obtain the sample of a desired size. The first selection condition with $\mathcal{S} = \{0, 1, 2, 3\}$ will always be satisfied under this approach, but individuals found to be in state 0 at a_{i0} will not be chosen according to the second sampling scheme with $\mathcal{S} = \{1, 2, 3\}$. In this case another A_{i0} may be generated until $Z(a_{i0}) \in \{1, 2, 3\}$. We repeat this until we obtain a sample of $n = 1000$ individuals, and generate 200 such samples. The empirical standard error of estimators for all three likelihoods for each selection and observation scheme are examined in order to give insight into the relative efficiency of the different procedures. The empirical biases are negligible in all cases as one would expect since these are all correctly specified likelihoods; we therefore do not report these.

The results in the top half of Table 7.1 correspond to the design with $\mathcal{S} = \{0, 1, 2, 3\}$. The percentages of individuals in states 0, 1, 2 and 3 at the time of sampling based on this design were 86.4%, 7.3%, 3.3% and 3.0%, respectively. Under the right-censored prospective observation scheme, the gain in efficiency from modeling the state occupied at the time of screening (i.e. from using $L_{1C}L_2$ vs. L_2) is apparent by the smaller empirical standard errors for most parameters. Incorporation of historical data on transition times through use of L_1L_2 further reduces the empirical standard error but to a lesser extent. The same general trend is apparent under an intermittent observation scheme with the results summarized in the right-hand columns of Table 7.1. There is only a slight loss of information from the intermittent versus continuous observation scheme; this often occurs for time-homogeneous processes (see Section 5.2.2).

Table 7.1: Empirical standard errors of estimators for a sample of $n = 1000$ individuals sampled from a population with requirement that $Z(a_0) \in \mathcal{S}$ under a right-censored or intermittent observation scheme; statistics computed for 200 simulated samples.

	Right-Censored			Intermittent Observation		
	$L_1 \cdot L_2$	$L_{1C} \cdot L_2$	L_2	$L_1 \cdot L_2$	$L_{1C} \cdot L_2$	L_2
Selection Condition: $Z(a) \in \{0,1,2,3\}$						
$\log \lambda_1$	0.081	0.084	0.110	0.083	0.086	0.119
$\log \lambda_2$	0.102	0.112	0.142	0.109	0.118	0.152
$\log \lambda_3$	0.130	0.140	0.187	0.139	0.149	0.209
$\log \gamma_0$	0.040	0.041	0.041	0.040	0.040	0.040
$\log \gamma_1$	0.119	0.129	0.141	0.129	0.144	0.154
$\log \gamma_2$	0.149	0.168	0.187	0.155	0.176	0.190
$\log \gamma_3$	0.119	0.135	0.178	0.123	0.139	0.184
Selection Condition: $Z(a) \in \{1,2,3\}$						
$\log \lambda_1$	0.012	0.027	-	0.012	0.028	-
$\log \lambda_2$	0.040	0.053	0.063	0.042	0.054	0.065
$\log \lambda_3$	0.055	0.060	0.075	0.059	0.065	0.080
$\log \gamma_0$	0.068	0.154	-	0.068	0.155	-
$\log \gamma_1$	0.057	0.063	0.063	0.059	0.066	0.066
$\log \gamma_2$	0.065	0.073	0.076	0.068	0.077	0.080
$\log \gamma_3$	0.053	0.064	0.066	0.054	0.065	0.067

The bottom half of Table 7.1 shows the results when the selection scheme requires $Z(a) \in \mathcal{S} = \{1,2,3\}$. The percentage of individuals in states 1, 2 and 3 at the time of sampling here were 53.5%, 24.3% and 22.1%, respectively. Based on $L_1 L_2$ more precise estimates of all estimable parameters are obtained from this selection scheme except for γ_0 since more individuals are in the disease states at the time of sampling. However, λ_1 and γ_0 are not estimable from this design if only prospective data are used. Moreover, since there is no information for these parameters in L_2, the information about them is the same for the right-censored and intermittent observation schemes. The relative efficiency of estimators across the two observation schemes (right censoring and intermittent observation) and three likelihoods otherwise parallels that in the top half of Table 7.1.

Table 7.2 contains results for an alternative sampling design that employs quota sampling. Here the $n = 1000$ individuals are sampled such that 500 are recruited in state 0 and 500 are recruited in the disease states (i.e. $Z(a_0) = 1, 2$ or 3). We otherwise assume that the processes and ages at screening are generated as before.

The percentage of individuals in states 0, 1, 2 and 3 at the recruitment time are now 50%, 26.7%, 12.2% and 11.1%, respectively. Based on L_2 the standard errors of estimators are smaller for the transition intensities among the disease states since the sample contains more individuals occupying these states. This reflects the potential utility of purposefully constructing samples for prospective follow-up. The price paid is a loss of information on disease incidence; see the larger standard error of 0.156 for $\log \hat{\lambda}_1$ for the maximum likelihood estimator based on the prospective right-censored data (i.e. L_2), compared to the value of 0.110 in Table 7.1.

This simulation study is simply to illustrate some qualitative design effects. The relative performance of the various approaches for sampling and observing individuals and analyzing the data depends in general on the transition intensities, which in most realistic cases are not time-homogeneous. More realistic disease processes featuring age-dependent Markov intensities and age-dependent selection criteria are routinely employed; see the discussion of the Canadian Longitudinal Study on Aging in Section 1.1. There may also be calendar-time trends in how the disease progresses with increased availability of effective interventions for disease management. Finally, we note that cohort studies sometimes employ complex sampling designs that may require the incorporation of survey weights in estimating functions. This is beyond our scope here, but see Section 7.2 for estimating equations involving other weights.

Table 7.2: Empirical standard errors of estimators for a sample of $n = 1000$ individuals obtained by quota sampling so as to obtain 500 individuals with $Z(a_0) = 0$ and 500 individuals with $Z(a_0) \in \{1, 2, 3\}$; statistics shown are based on 200 simulations.

	Right-Censored			Intermittent Observation		
	$L_1 \cdot L_2$	$L_{1C} \cdot L_2$	L_2	$L_1 \cdot L_2$	$L_{1C} \cdot L_2$	L_2
$\log \lambda_1$	0.154	0.154	0.156	0.164	0.164	0.165
$\log \lambda_2$	0.055	0.063	0.089	0.058	0.066	0.095
$\log \lambda_3$	0.076	0.080	0.106	0.080	0.085	0.113
$\log \gamma_0$	0.055	0.062	0.061	0.055	0.062	0.062
$\log \gamma_1$	0.069	0.081	0.081	0.072	0.086	0.085
$\log \gamma_2$	0.081	0.093	0.097	0.087	0.101	0.105
$\log \gamma_3$	0.072	0.076	0.093	0.074	0.078	0.095

7.1.3   Prevalent Cohort Sampling and Failure Times

There is an extensive literature on studies designed to assess the distribution of time between an initial event (termed *onset* here for convenience) and some subsequent event (termed *failure*). Familiar examples concern the occurrence of death following HIV infection (Hogg et al., 1994), or following a diagnosis of severe dementia (Brookmeyer et al., 2002; Fitzpatrick et al., 2005). Prevalent cohort studies are ones where individuals are required to have experienced onset prior to their time of selection. The relevant multistate process for an individual is the illness-death process shown in Figure 7.5, where we let T_k denote the age at entry to state k, and let $T = T_2 - T_1$ denote the time of interest. More generally, $T_2 \mid T_1 = t_1$ or $\lambda_1(t|t_1)$ is of interest. Understanding of the failure process following onset is best achieved through study of the failure intensity $\lambda_1(t|t_1)$ following onset at time (age) t_1. For convenience, we suppress any dependence on covariates in the notation.

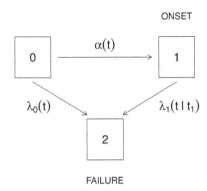

Figure 7.5: An illness-death model for onset and failure; time t represents age and t_1 is the onset time.

Studies operate in calendar time, and we continue to distinguish this from the age time scale for an individual. This is done in Figure 7.6, where we portray a situation in which individuals recruited have experienced onset over some period of calendar time (B_L, B_R) of width $\Delta B = B_R - B_L$ and are alive (i.e. have not yet failed) at B_R. We let U_0 and U_1 denote the calendar times of onset and failure, respectively, so $T = U_1 - U_0$. In studies of dementia, for example, one might define the target population as individuals diagnosed with dementia within the decade prior to a particular study selection date. In this case, we treat B_R as the calendar time of selection, at which point the individual depicted is A_0 years of age, and suppose that the study involves follow-up to calendar time B_E where E denotes the end of the study.

For an individual born at calendar time B, the study selection conditions are then $T_1 \in (A_L, A_0)$ and $T_2 > A_0$, where $A_L = B_L - B$ and $A_0 = B_R - B$. On the age scale we are in the general framework of Section 7.1.1, and we can consider two specific likelihood functions for an individual. The prospective likelihood based on data observed over (A_0, C) and assuming both T_1 and T_2 are observable, is

$$L_2 \propto P(\tilde{T}_2 = \tilde{t}_2, \delta_2 \mid T_1 = t_1 > a_L, T_2 > a_0) \tag{7.10}$$

which is

$$\lambda(\tilde{t}_2 \mid t_1)^{\delta_2} \exp\left(-\int_{a_0}^{\tilde{t}_2} \lambda(s \mid t_1)\, ds\right),$$

where $\tilde{T}_2 = \min(T_2, C)$, $\tilde{t}_2 = \min(t_2, C)$ and $\delta_2 = I(\tilde{t}_2 = t_2)$. The full likelihood, based on retrospective plus prospective data, is

$$L \propto P(T_1 = t_1, \tilde{T}_2 = \tilde{t}_2, \delta_2 \mid T_1 \in (a_L, a_0), T_2 > a_0)$$

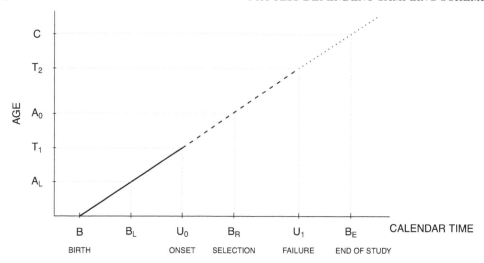

Figure 7.6: A Lexis diagram illustrating the relation between calendar time and individual-specific age scales for a study with prevalent cohort sampling.

which is given by

$$\frac{\alpha(t_1)\exp(-\int_{a_L}^{t_1}(\lambda_0(u)+\alpha(u))\,du)\,\lambda(\tilde{t}_2\mid t_1)^{\delta_2}\exp(-\int_{t_1}^{\tilde{t}_2}\lambda(u\mid t_1)\,du)}{\int_{a_L}^{a_0}\alpha(s)\exp(-\int_{a_L}^{s}(\lambda_0(u)+\alpha(u))\,du)\exp(-\int_{s}^{a_0}\lambda(u\mid s)\,du)\,ds}.$$

In some contexts a convenient approach is to focus on the process of initial events in a population, rather than at the individual level. This is the case, for example, when study individuals are sampled from a registry of persons who have initially experienced the event. The "initial events process" is a function of births in the population plus an individual-level model for onset like that in Figure 7.5. For discussion, we assume that the birth process is a Poisson process with intensity function $\rho(u)$, beginning at some calendar time origin that is far in the past. We let $f(t\mid u_0,t_1)$ denote the conditional distribution of time T from onset to failure for a person with onset at age t_1 and calendar time U_0. The case where T is independent of U_0 and T_1 has received much study, though it applies in situations where the intensity for the $1 \to 2$ transition in Figure 7.5 is semi-Markov and there is no trend in the disease process in calendar time. In settings where the onset event is the development of dementia, the semi-Markov assumption requires that the risk of death following the onset of dementia is independent of the age of onset, which is implausible. For the assumption of no trend over calendar time in the distribution for sojourn time T to be reasonable, it would be necessary for there to be a consistent approach to the diagnosis and management of the condition over time. If $\tilde{T}=\min(U_1-u_0,B_E-u_0)$ and $\delta=I(\tilde{T}=U_1-u_0)$, then in the general case the conditional likelihood L_2 becomes

$$L_2 \propto P(\tilde{T}=\tilde{t},\delta_2\mid U_0=u_0,t_1,U_0\in(B_L,B_R),U_1>B_R)$$
$$=\frac{f(\tilde{t}\mid t_1,u_0)^{\delta_2}\,\mathcal{F}(\tilde{t}\mid t_1,u_0)^{1-\delta_2}}{\mathcal{F}(B_R-u_0\mid t_1,u_0)}, \tag{7.11}$$

where $\mathcal{F}(s|t_1, u_0) = P(T > s \,|\, T_1 = t_1, U_0 = u_0)$. Note that (7.11) is the same as (7.10). The full likelihood exploits the information in u_0. Here $L = L_1 L_2$ is

$$L \propto P(U_0 = u_0, \tilde{T} = \tilde{t}, \delta_2 \,|\, U_0 \in (B_L, B_R), U_1 > B_R, T_1 = t_1) = L_1 L_2,$$

where $L_1 \propto P(U_0 = u_0, U_1 > B_R \,|\, U_0 \in (B_L, B_R), t_1)$ may be written as

$$L_1 \propto \frac{P(U_0 = u_0 \,|\, t_1)\mathcal{F}(B_R - u_0 \,|\, t_1, u_0)}{\int_{B_L}^{B_R} P(U_0 = u_0 \,|\, t_1)\mathcal{F}(B_R - u_0 \,|\, t_1, u_0)\, du_0}. \tag{7.12}$$

The need to consider age-specific initial event rates is important when these rates and survival time T depend on age. In that case L here and L following (7.10) are the same, and we condition on age at time B_R. Much of the methodological literature assumes that the event intensity in not age-dependent, nor is the survival time T. The case where the initial events follow a homogeneous Poisson process is especially well studied (Asgharian et al., 2014). Problem 7.1 considers this special case.

In some settings a complication arises because the exact onset time (U_0 or T_1) is unknown. Often the time T_1 is taken as the age an individual is diagnosed with a particular chronic disease. In the cases of dementia, diabetes, and many other conditions there can be an appreciable delay from the actual onset of the disease and the formal diagnosis. If it can at least be determined that $U_0 \in (B_L, B_R)$, then the conditioning event becomes only $\{U_0 \in (B_L, B_R), U_1 > B_R\}$. Similarly, for L_1 we use instead of (7.12) the probability $P(U_1 > B_R \,|\, U_0 \in (B_L, B_R))$.

Example 7.1.3: The Incidence of Psoriatic Arthritis in Psoriasis Patients

Here we consider the challenge of estimating the incidence of psoriatic arthritis in patients diagnosed with psoriasis using data from the University of Toronto Psoriasis Clinic (UTPC) mentioned in Example 7.1.2. The multistate diagram in Figure 7.2 characterizes the process of interest. For simplicity, we again consider a homogeneous population and let $\mathcal{H}(t) = \{Z(s), 0 \le s \le t\}$ denote the history of the disease process with general intensities $\lambda_{kl}(t|\mathcal{H}(t^-))$ governing $k \to l$ transitions at age t; the progressive nature of the process in Figure 7.2 means we can write $\lambda_{1l}(t|\mathcal{H}(t^-)) = \lambda_{1l}(t|t_1)$, $l = 2, 3$, and $\lambda_{23}(t|\mathcal{H}(t^-)) = \lambda_{23}(t|t_1, t_2)$. We consider the contribution of a particular individual to the UTPS where U_0 denotes the onset time of psorasis. As this is an adult psoriasis clinic with no restrictions on the age of onset of psoriasis, we can set $B_L \to -\infty$ so the selection conditions are $B < B_R - 18$ and $T_1 < A_0 < \min(T_2, T_3)$. Figure 7.7 portrays the selection of individuals for the UTPC and UTPAC datasets. If there is no trend in the disease process over calender time, the prospective likelihood can then be written as

$$L_2 \propto P(Z(s), a_0 < s \le \tilde{t}_2 \,|\, T_1 = t_1, A_0 = a_0, a_0 < \min(T_2, T_3))$$

$$\propto \lambda_{12}(\tilde{t}_2 \,|\, t_1)^{\delta_{12}} \lambda_{13}(\tilde{t}_2 \,|\, t_1)^{\delta_{13}} \exp\left(-\int_{a_0}^{\tilde{t}_2} (\lambda_{12}(s \,|\, t_1) + \lambda_{13}(s \,|\, t_1))\, ds\right)$$

$$\times \left\{\lambda_{23}(\tilde{t}_3 \,|\, t_1, t_2)^{\delta_{23}} \exp\left(-\int_{t_2}^{\tilde{t}_3} \lambda_{23}(s \,|\, t_1, t_2)\, ds\right)\right\}^{\delta_{12}}$$

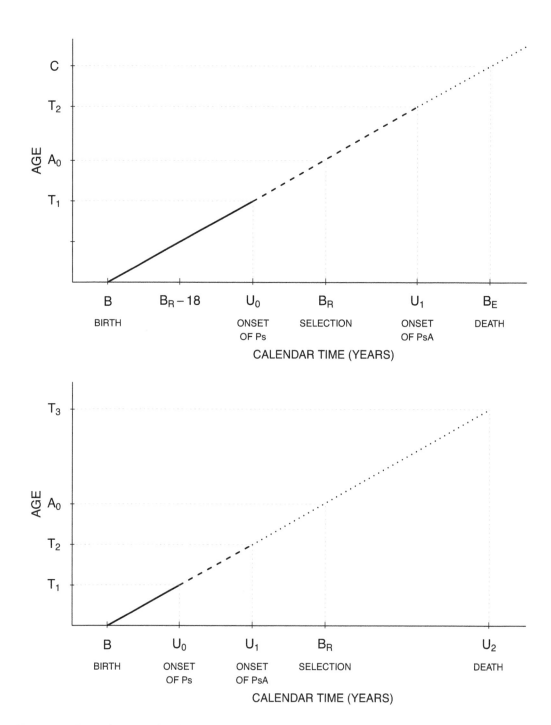

Figure 7.7: Lexis diagrams for the selection and follow-up of patients in the UTPC (top panel) and UTPAC (bottom panel).

where $\tilde{t}_2 = \min(t_2, t_3, C)$, $\tilde{t}_3 = \min(t_3, C)$ and $\delta_{kl} = I(k \to l$ transition observed). By writing the intensities in this way, it is clear that the intensity for psoriatic arthritis may depend on the age of onset of psoriasis, and the age of onset of psoriasis and psoriatic arthritis may alter the risk of death. The full likelihood incorporating the retrospective data is of the form

$$\frac{\lambda_{01}(t_1)\exp\left(-\int_{a_L}^{t_1}(\lambda_{01}(u)+\lambda_{03}(u))du\right)P(Z(u), a_0 < u < \tilde{t}_2 \mid T_1 = t_1, A_0 = a_0 < \min(T_2, T_3))}{\int_0^{a_0}\lambda_{01}(s)\exp\left(-\int_{a_L}^{s}(\lambda_{01}(u)+\lambda_{03}(u))du\right)P(Z(u), a_0 < u < \tilde{t}_2 \mid T_1 = s, A_0 = a_0 < \min(T_2, T_3))ds}$$

and it is immediately apparent that this cannot be maximized in the absence of external information on mortality rates among disease-free individuals, or without assumptions or external information about the relation between the mortality rates from different disease states (e.g. if the onset of psoriasis does not alter the risk of death so $\lambda_{13}(t|t_1) = \lambda_{03}(t)$.

If the incidence of PsA is low among individuals with psoriasis, there may be relatively little information on the intensity of death following onset of psoriasis. To learn more about this, we incorporate information from the University of Toronto Psoriatic Arthritis Clinic. As mentioned in Example 7.1.2 individuals were recruited if they were in state 2 of Figure 7.2 at the time of contact. The prospective likelihood contribution from such individuals is then proportional to

$$L_{2B} \propto P(Z(s), a_0 < s \le \tilde{t}_3 \mid T_1 = t_1, T_2 = t_2, A_0 = a_0 < T_3)$$

$$\propto \lambda_{23}(\tilde{t}_3 \mid t_1, t_2)^{\delta_{23}} \exp\left(-\int_{a_0}^{\tilde{t}_3} \lambda_{23}(u \mid t_1, t_2)\, du\right)$$

where $\tilde{t}_3 = \min(t_3, C)$ and $\delta_{23} = I(2 \to 3$ transition is observed). A prospective likelihood $L_2 \cdot L_{2B}$, exploiting information from both registries, will yield more efficient estimation about transition intensities out of states 1, 2 and 3.

Here we report on the results of modeling the prospective data from the UTPC and UTPAC by maximizing $L_2 \cdot L_{2B}$. We let $\lambda_{12}(t|t_1) = \lambda_{120}(t)\exp(\alpha \log(t_1/30))$ to allow the risk of PsA to depend on the age of onset of psoriasis, with the baseline intensity applicable to an individual with onset at 30 years of age. A piecewise-constant intensity was specified with cut-points at 50, 60 and 70 years of age. Similarly we let $\lambda_{13}(t|t_1) = \lambda_{130}(t)\exp(\beta \log(t_1/30))$ and let

$$\lambda_{23}(t \mid t_1, t_2) = \lambda_{230}(t)\exp(\gamma_1 \log(t_1/30) + \gamma_2 \log(t_2/40))$$

be the intensities for death among individuals with psoriasis and psoriatic arthritis, respectively. Only 3 out of 637 psoriasis patients were observed to die, while the times of death are known for 150 of the 1378 patients recruited to the psoriatic arthritis clinic; survival times were otherwise censored at their age as of December 31, 2016. It is expected that the mortality is comparable among psoriasis and psoriatic arthritis patients so for estimability we set $\lambda_{130}(t) = \lambda_{230}(t)$ and again adopt a piecewise-constant form with cut-points at 50, 60, 70 and 80 years of age.

Table 7.3 displays the estimates obtained from fitting the model. A Wald test of H_0: $\alpha = 0$ vs. H_A: $\alpha \neq 0$ gives $p = 0.151$ so there is no evidence of an association between the age of onset of psoriasis and the risk of psoriatic arthritis. A 3 degree of freedom likelihood ratio test of H_0: $\lambda_{120}(t) = \lambda_{120}$ yields $p = 0.675$ so a time-homogeneous baseline intensity for the onset of PsA could be specified, but we retain the model with the covariate and the piecewise-constant baseline intensity for generality. There is no evidence of an association between the risk of death and the age of onset of psoriasis among individuals with psoriasis ($p = 0.942$). Among patients with psoriatic arthritis, there is likewise no association between death and the age of onset for psoriasis or psoriatic arthritis.

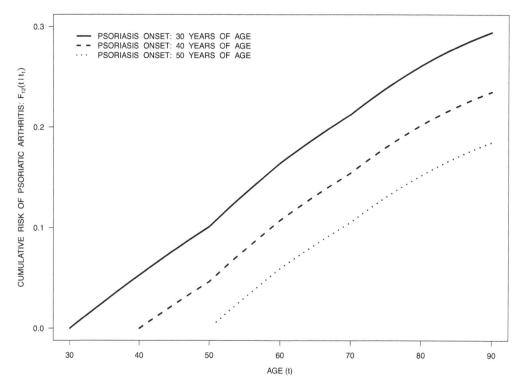

Figure 7.8: Plot of the cumulative risk of psoriatic arthritis by age and according to age of onset of psoriasis.

To get a sense of the risk of PsA among patients with psoriasis onset at different ages, we plot an estimate

$$F_{12}(t \mid t_1) = \int_{t_1}^{t} \exp\left(-\int_{t_1}^{s}(\lambda_{12}(u \mid t_1) + \lambda_{13}(u \mid t_1))\,du\right)\lambda_{12}(s \mid t_1)\,ds$$

in Figure 7.8 for $t_1 = 30$, 40 and 50. For an individual developing psoriasis at 30 years of age the estimated probability, they are diagnosed with psoriasis by age 90 is about 0.30; this is closer to 0.20 for those with psoriasis onset at age 50.

Table 7.3: Results of fitting a multistate model to data from the University of Toronto Psoriasis Clinic and the University of Toronto Psoratic Arthritis Clinic.

Transitions	Parameter	Interval	EST	SE	Exponentiated Values		
					EST ($\times 10^{-1}$)	95% CI	p
$1 \to 2$	$\log \lambda_{120}(t)$	$[0, 50)$	-4.915	0.244	0.07	(0.05, 0.12)	
		$[50, 60)$	-4.531	0.283	0.11	(0.06, 0.19)	
		$[60, 70)$	-4.646	0.354	0.10	(0.05, 0.19)	
		$[70, \infty)$	-4.407	0.502	0.12	(0.05, 0.33)	
	α		-0.192	0.134	8.25	(6.35, 10.73)	0.151
$1 \to 3$	$\log \lambda_{130}(t)$	$[0, 50)$	-6.225	0.245	0.02	(0.01, 0.03)	
		$[50, 60)$	-5.793	0.225	0.03	(0.02, 0.05)	
		$[60, 70)$	-4.850	0.157	0.08	(0.06, 0.11)	
		$[70, 80)$	-4.001	0.154	0.18	(0.14, 0.25)	
		$[80, \infty)$	-3.511	0.203	0.30	(0.20, 0.44)	
	β		-0.021	0.286	9.80	(5.59, 17.16)	0.942
$2 \to 3$	$\log \lambda_{230}(t)$	$[0, 50)$	-6.225	0.245	0.02	(0.01, 0.03)	
		$[50, 60)$	-5.793	0.225	0.03	(0.02, 0.05)	
		$[60, 70)$	-4.850	0.157	0.08	(0.06, 0.11)	
		$[70, 80)$	-4.001	0.154	0.18	(0.14, 0.25)	
		$[80, \infty)$	-3.511	0.203	0.30	(0.20, 0.44)	
	γ_1		0.314	0.202	13.69	(9.22, 20.32)	0.120
	γ_2		0.255	0.348	12.90	(6.53, 25.50)	0.464

† The piecewise-constant baseline intensities $\lambda_{130}(t)$ and $\lambda_{230}(t)$ are constrained to be the same because only three deaths were observed in the psoriasis cohort.

7.1.4 Design Based on Probabilistic State-Dependent Sampling

Here we consider a setting in which individuals in a population are screened and sampled for a study based on selection probabilities that may depend on the state they occupy. This is an extension of the quota sampling scheme represented in Table 7.2. We restrict attention to Markov models of the form shown in Figure 7.1 and assume here that historical data $\mathcal{H}_i(a_{i0})$ are not available, but what is known is that $Z_i(a_{i0}) = k_0$, $A_{i0} = a_{i0}$ and the covariate X_i. When individual i is screened, we consider a model used to select them (or not) based on the available data. If $\Delta_i = 1$ if individual i is sampled and $\Delta_i = 0$ otherwise, individual i sampled with probability $P(\Delta_i = 1 | Z_i(a_{i0}), X_i, A_{i0} = a_{i0}) \leq 1$, where previously this was taken to be 1 for all screened individuals satisfying $Z_i(a_{i0}) \in \mathcal{S}$.

The partial likelihood contribution for individual i for a Markov process can be written

$$L_i \propto L_{i1C} \cdot L_{i2}^{\Delta_i}, \tag{7.13}$$

where the information from the screening data is represented in the cross-sectional contribution $L_{i1C} = P(Z_i(a_{i0}) | A_{i0} = a_{i0}, Z_i(0) = 0, X_i)$ and individuals selected for prospective follow-up contribute through L_{i2}. If information about the state occupied at the time of screening is unavailable for unselected individuals, one can condition on $Z_i(a_{i0})$ and omit the term L_{i1C} in (7.13); the resulting loss of information can be appreciable if a large sample from the population is screened compared to the number of individuals sampled.

The partial log-likelihood resulting from such a sampling scheme and (7.13) is

$$\log L_{i1C}(\theta) + \Delta_i \log L_{i2}(\theta) \tag{7.14}$$

where θ represents parameters in the multistate model. This yields the observed information matrix

$$I_{i1}(\theta) + \Delta_i I_{i2}(\theta) \tag{7.15}$$

with

$$I_{i1}(\theta) = -\left[\frac{\partial^2 \log P(Z_i(a_{i0}) = k_0 \mid A_{i0} = a_{i0})}{\partial\theta\partial\theta'}\right]$$

the information from the cross-sectional data that is available for all screened individuals and

$$I_{i2}(\theta) = -\frac{\partial^2 \log L_{i2}}{\partial\theta\partial\theta'},$$

the observed information matrix based on the prospective data available for selected individuals.

With intermittent observation following recruitment, we may consider the likelihood analogous to (7.13) of the form

$$L_i \propto L_{i1C} \cdot \left\{ P(\Delta_i = 1 \mid Z_i(a_{i0}) = k_0, X_i, A_{i0} = a_{i0}) \prod_{j=1}^{m_i} P(Z_i(a_{ij}) \mid Z_i(a_{i,j-1}), X_i) \right\}^{\Delta_i} \tag{7.16}$$

where $a_{i1} < a_{i2} < \cdots < a_{im_i}$ are the m_i follow-up times for individual i. Here we presume no transition times are observed exactly and that only data at the assessment times are available. Relating this to (5.10) we can then write the observed information for a Markov process with state space as in Figure 7.1 as

$$-\sum_{k_0=1}^{K-1} Y_{ik_0}(a_{i0}) \frac{\partial^2 \log P(Z_i(a_{i0}) = k_0 \mid Z_i(a_{i0}) < K, X_i, A_{i0} = a_{i0})}{\partial\theta\partial\theta'} \tag{7.17}$$

$$-\Delta_i \sum_{j=1}^{m_i} \sum_{k=k_0}^{K-1} \sum_{l>k} Y_{ik}(a_{i,j-1}) Y_{il}(a_{ij}) \frac{\partial^2 \log P(Z_i(a_{ij}) = l \mid Z_i(a_{i,j-1}) = k, X_i)}{\partial\theta\partial\theta'}.$$

To discuss efficient sampling schemes we consider the expected information matrix computed with a given selection model. For illustration purposes, we assume no covariates X_i are involved. The first term obtained from (7.17) is then independent of X and the expectation conditional on A_{i0} is then averaged over A_{i0} to obtain $\mathcal{I}_1$.

The second term has expectation, conditional on $Z_i(a_{i0})$, (see (5.10))

$$-\sum_{k_0=1}^{K-1} P(Z_i(a_{i0}) = k_0 \mid Z_i(a_{i0}) < K, A_{i0} = a_{i0}) P(\Delta_i = 1 \mid Z_i(a_{i0}) = k_0, A_{i0} = a_{i0})$$

$$\times \sum_{j=1}^{m_i} \sum_{k=k_0}^{K-1} \sum_{l>k} \frac{E_k(a_{i,j-1})}{P_{kl}(a_{i,j-1},a_{ij})} \frac{\partial P_{kl}(a_{i,j-1},a_{ij})}{\partial \theta} \frac{\partial P_{kl}(a_{i,j-1},a_{ij})}{\partial \theta'}$$

where

$$E_k(a_{i,j-1}) = P(Z_i(a_{i,j-1}) = k \mid Z_i(a_{i0}), A_{i0} = a_{i0}),$$

and again the Fisher information matrix is obtained by then taking the expectation of this with respect to A_{i0} to obtain $\mathcal{I}_2$. The total expected information is then $\mathcal{I}(\theta) = \mathcal{I}_1(\theta) + \mathcal{I}_2(\theta)$.

The expected information and efficiency of $\hat{\theta}$ depend on design parameters δ that specify the selection probabilities $\pi_i(\delta) = P(\Delta_i = 1 | Z_i(a_{i0}), A_{i0} = a_{i0}; \delta)$. In practice one might wish to set δ to maximize the precision of estimators of particular features of the multistate process, such as covariate effects on transition intensities or sojourn time distributions. Relatively little work has been done on this aside from the case of failure time models, perhaps because of the complexity of many processes and the need to specify all parameters. Adaptive approaches are possible in which a preliminary convenience sample is collected, interim estimates of the process parameters are obtained and a selection model is specified to maximize the precision of a target estimator. Since several process features are typically of interest, however, exploration of efficiency is usually multidimensional. The pros and cons of alternative sampling plans can be investigated with the expressions here, or extensions of them when covariates are present. In this case, designs that increase efficiency for some parameter may decrease it for others.

7.1.5 Selection and Initial Conditions with Heterogeneous Processes

As discussed in Chapter 6, there is often more heterogeneity across processes than can be accounted for by the incorporation of known covariates. Random effects models are appealing in such settings. Consider the random effects model in (6.1) for the analysis of survival times where it is assumed that $X \perp V$. If we consider a cohort study in which individual i is sampled at age a_{i0} and followed prospectively to observe their lifetime, the left-truncation condition (i.e. $T_i > a_{i0}$) must be accounted for in the likelihood construction. Specifically since V_i is associated with $T_i|X_i$, $V_i \not\perp I(T_i > a_{i0}) \mid X_i$ and so the likelihood contribution for the prospectively observed survival data is

$$\int [v_i h(\tilde{t}_i \mid x_i)]^{\delta_i} \exp\left(-\int_{a_{i0}}^{\tilde{t}_i} v_i h(u \mid x_i) \, du\right) dG(v_i \mid T_i > a_{i0}, x_i)$$

where

$$dG(v_i \mid T_i > a_{i0}, x_i) = \frac{P(T_i > a_{i0} \mid x_i, v_i) \, dG(v_i)}{\int P(T_i > a_{i0} \mid x_i, v_i) \, dG(v_i)}.$$

Note that this conditional frailty distribution is analogous to the one used in (6.3) to derive the marginal hazard function directly in terms of the conditional hazard.

Now consider a heterogeneous multistate process with $k \to l$ intensity, given fixed covariates X_i and random effect $V_i = v_i$, of the form

$$\lambda_{ikl}(t \mid \mathcal{H}_i(t^-), v_i) = v_{ikl}\, \lambda_{ikl}(t \mid \mathcal{H}_i(t^-))$$

where V_{ik} is sub-vector of V_i with elements V_{ikl}, and $\mathcal{H}_i(t) = \{Z_i(s), 0 \leq s \leq t, x_i\}$; again we assume $X_i \perp V_i$. Suppose individual i is first observed at time a_{i0} and is to be followed prospectively from that time until a completely independent right censoring time C_i. The relevant distribution function for V_i for the modeling of the prospective data is $G(v_i \mid \mathcal{H}_i(a_{i0}))$, since v_i affects $\mathcal{H}_i(a_{i0})$. If complete information is available on $\mathcal{H}_i(a_{i0})$ then one can derive $G(v_i \mid \mathcal{H}_i(a_{i0}))$ and construct a valid likelihood for the prospective data as

$$\int \prod_{\substack{k=1}}^{K} \prod_{\substack{l=1 \\ l \neq k}}^{K} \prod_{t_r \in \mathcal{D}_{ikl}} v_{ikl}\, \lambda_{kl}(t_{ir} \mid \mathcal{H}_i(t_{ir}^-)) \tag{7.18}$$

$$\times\, \exp\left(-\int_0^\infty v_{ikl}\, \bar{Y}_{ik}(u^-)\, \lambda_{ikl}(u \mid \mathcal{H}_i(u^-))\, du\right) dG(v_i \mid \mathcal{H}_i(a_{i0}))$$

where $\mathcal{D}_{ikl}$ is the set of $k \to l$ transition times observed over $(a_{i0}, C_i]$. Very challenging complications arise when individuals are not observed from the start of the process or the history $\mathcal{H}_i(a_{i0})$ is incomplete for other reasons.

Individuals may also be sampled from the population of processes at some time $A_{i0} = a_{i0}$ and selected for study subject to conditions on $\mathcal{H}_i(a_{i0})$. This is acceptable and is naturally dealt with in the likelihood (7.18) provided the history $\mathcal{H}_i(a_{i0})$ is available.

If individuals are sampled for inclusion in a study based on $Z_i(a_{i0})$, but no information about the process history $\mathcal{H}_i(a_{i0})$ is available, it is necessary to evaluate

$$dG(v_i \mid Z_i(a_{i0}), a_{i0}, X_i) = \frac{P(Z_i(a_{i0}) \mid a_{i0}, v_i, X_i)\, dG(v_i)}{\int P(Z_i(a_{i0}) \mid a_{i0}, v_i, X_i)\, dG(v_i)}.$$

For most processes, this will be intractable; an exception is the conditionally Markov processes for which $\{Z_i(s), s > 0\} \mid v_i, x_i$ is Markov.

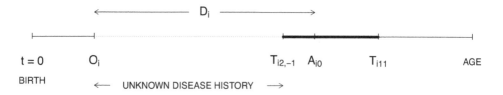

Figure 7.9: A timeline diagram of disease onset O_i and study entry time A_{i0}.

In Section 6.1.4 we considered the analysis of data from a trial of patients with chronic obstructive pulmonary disease (COPD) based on an alternating 2-state model discussed in Section 6.1.3. This model involved separate correlated random effects acting on the two transition intensities. The intensity for the onset of exacerbations had a Markov time scale with the time origin taken as the date of diagnosis of COPD, whereas a semi-Markov time scale was used for the intensity for the resolution of exacerbations. In Section 6.1.4 we remarked that to be eligible for inclusion in the study, individuals were required to be experiencing an exacerbation at the time of the initial contact. Figure 7.9 illustrates the situation. If O_i represents the age of onset of COPD for an individual and A_{i0} is their age at screening, then the disease duration at recruitment is $D_i = A_{i0} - O_i$. In addition, we let $T_{i2,-1}$ represent the onset time of the exacerbation being experienced at A_{i0}, and T_{i11} represent the time of transition back to the exacerbation-free state 1.

The onset time for this baseline exacerbation is available in this study, so a natural approach is to consider the duration of the initial exacerbation ($W_{i1} = T_{i11} - T_{i2,-1}$) conditional on it being longer than $a_{i0} - t_{i2,-1}$. This has conditional density

$$v_{i2}\, h_2(w_{i1} \mid x_{i2}(t_{i2,-1}))\, \exp\left(-v_{i2} \int_{a_{i0}-t_{i2,-1}}^{w_{i1}} h_2(s \mid x_{i2}(s))\, ds\right).$$

where $x_{i2}(t_{i2,-1})$ represents covariates known at time $t_{i2,-1}$. In this case, it would be appropriate to use the conditional distribution $dG(v_i \mid Z_i(a_{i0}) = 2, A_{i0} = a_{i0}, W_{i1} > a_{i0} - t_{i2,-1}, x_{i2}(t_{i2,-1}))$. This cannot be obtained in a simple form here, and one could instead adopt a working model with some type of dependence on O_i and $a_{i0} - t_{i2,-1}$.

7.1.6 Initial Conditions with a Finite Mixture Model

In Section 6.3.2 we presented an analysis of data from the University of Toronto Lupus Clinic based on finite mixture models with three classes. Those analyses were directed at modeling the disease course in the clinic and the state occupied at clinic entry was conditioned upon and assumed to be independent of the class occupied. The time from disease onset to clinic entry is highly variable, however, with some patients having been diagnosed more than 20 years before recruitment. This makes clinic entry an arbitrary and unnatural time origin for models directed at characterizing the full course of the disease process. For such an objective the time of disease diagnosis is a more sensible time origin and we consider this here. Suppressing the subscript i we consider a generic individual and let $Z(0)$ denote the state occupied at disease onset.

A challenge is that the state occupied at the time of disease onset ($Z(0)$) is unknown, except for individuals in an inception sub-cohort S_I for whom $a_0 = 0$. To address this we introduce the initial distribution $P(Z(0) \mid W, X)$ to model the latent state occupied at the time of disease onset for individuals not in the inception sub-cohort. In this approach the state occupied at the time of clinic entry can be informative about the transition intensities and class membership. While the initial distribution may differ between classes, in practice this may be difficult to estimate and imposing the constraint of a common initial distribution may be necessary. We retain the general formulation in the derivation that follows.

Consider a generic individual with assessments at $a_0 < a_1 < \cdots < a_m$, define $\mathcal{Z}^\circ(a_j) = \{(a_r, Z(a_r)), r = 0, 1, \ldots, j\}$ as before, and let $\mathcal{Z}^\circ(\infty) = \{(a_r, Z(a_r)), r = 0, 1, \ldots, m\}$ and $\mathcal{H}^\circ(\infty) = \{(a_r, Z(a_r)), r = 0, 1, \ldots, m; X\}$. To simplify the notation we suppress the visit times in some of the expressions that follow and treat them as fixed, which is allowable under the CIVP conditions. We now think of $(Z(0), W)$ as missing for individuals not in the inception sub-cohort and W as the only missing information for individuals in the inception sub-cohort. Under a finite mixture of Markov processes the contribution to a complete data likelihood from a generic individual is

$$P(\mathcal{Z}^\circ(\infty) \mid Z(0), W, X) \cdot P(Z(0) \mid W, X) \cdot P(W \mid X), \qquad (7.19)$$

where the first term $P(\mathcal{Z}^\circ(\infty) \mid Z(0), W, X)$ is given by

$$P(Z(a_0) \mid Z(0), W, X) \prod_{j=1}^{m} P(Z(a_j) \mid Z(a_{j-1}), W, X; \theta), \qquad (7.20)$$

by the Markov property of the class W model; we index this by θ as in Section 6.3.1. For an individual in the inception cohort, $a_0 = 0$ so the first time on the right-hand side of (7.20) is one. The second term in (7.19) involves the initial distribution $P(Z(0) \mid W, X)$, which is multinomial and indexed by η. The final term $P(W|X)$ is the multinomial model for class membership indexed by γ. Here we let $\psi = (\theta', \eta', \gamma')'$.

We again formulate this optimization problem via an EM algorithm and the rth maximization step involves maximizing

$$\mathcal{Q}(\psi; \psi^{(r-1)}) = \mathcal{Q}_1(\theta; \psi^{(r-1)}) + \mathcal{Q}_2(\eta; \psi^{(r-1)}) + \mathcal{Q}_3(\gamma; \psi^{(r-1)})$$

where

$$\mathcal{Q}_1(\theta; \psi^{(r-1)}) = E\{\log P(\mathcal{Z}^\circ(\infty), Z(a_0) \mid Z(0), W, X) \mid \mathcal{H}^\circ(\infty); \psi^{(r-1)}\}, \qquad (7.21\text{a})$$

$$\mathcal{Q}_2(\eta; \psi^{(r-1)}) = E\{\log P(Z(0) \mid W, X; \eta) \mid \mathcal{H}^\circ(\infty); \psi^{(r-1)}\}, \qquad (7.21\text{b})$$

$$\mathcal{Q}_3(\gamma; \psi^{(r-1)}) = E\{\log P(W \mid X; \gamma) \mid \mathcal{H}^\circ(\infty); \psi^{(r-1)}\}. \qquad (7.21\text{c})$$

For an individual in the inception cohort $\mathcal{S}_I$ the expectations are only with respect to W using the conditional distribution

$$P(W \mid \mathcal{H}^\circ(\infty)) = \frac{P(\mathcal{Z}^\circ(\infty), Z(0) \mid W, X)\, P(W \mid X)}{E_W\{P(\mathcal{Z}^\circ(\infty), Z(0) \mid W, X) \mid X\}}.$$

For individuals not in the inception cohort, the expectations for (7.21a) and (7.21b) are with respect to $(Z(0), W)$ and based on $P(Z(0), W \mid \mathcal{H}^\circ(\infty))$ given by

$$\frac{P(\mathcal{Z}^\circ(\infty), Z(a_0), Z(0) \mid W, X)\, P(W \mid X)}{E_{Z(0), W}\{P(\mathcal{Z}^\circ(\infty), Z(a_0) \mid Z(0), W = g, X) \mid X\}}.$$

Here we consider a revised analysis of the data from the lupus clinic considered in Section 6.3.2. There are 755 out of 1767 individuals in the inception cohort. We adopt a common initial distribution across the three classes and omit the covariate to give $P(Z(0) \mid W, X) = P(Z(0); \eta)$, where

$$P(Z(0) = k; \eta) = \frac{\exp(\eta_2 I(k = 2)) + \exp(\eta_3 I(k = 3))}{1 + \exp(\eta_2) + \exp(\eta_3)}$$

for the state occupied at the time of disease onset, where $\eta = (\eta_2, \eta_3)'$ and $k = 1, 2, 3$.

The results of the analysis described in this section are displayed in the last three columns of Table 7.4, while the results in the previous columns are those from Section 6.3.2, reproduced here for comparison. There is a modest difference in the estimates of the transition intensities for the different classes but generally broad agreement between the new and earlier results in the kinds of dynamics of the three classes. The estimates of η yield estimated probabilities for the initial distribution of $\widehat{P}(Z(0) = 1) = 0.19$, $\widehat{P}(Z(0) = 2) = 0.41$ and $\widehat{P}(Z(0) = 3) = 0.40$ for the model of this section. In terms of the class membership, we find that for those with disease onset before 30 years of age the proportions in classes 1, 2 and 3 are estimated to be 0.41, 0.25 and 0.34, respectively, while for those with later onset they are 0.23, 0.28 and 0.49, respectively. These can be compared to 0.43, 0.26 and 0.31 for those with early onset and 0.24, 0.28 and 0.48 for late onset from the earlier analyses. The close agreement arises in part because a high proportion of the patients were recruited to this clinic shortly after disease onset and diagnosis.

7.2 Outcome-Dependent Subsampling and Two-Phase Studies

7.2.1 Two-Phase Studies

In some settings, it is infeasible or costly to measure certain variables for everyone in a population or study group. Two-phase studies are ones in which values of difficult to measure variables are ascertained for only a subset of the full group. Suppose for discussion that a response variable Y and covariates W are known for all individuals $i = 1, \ldots, N$ in a "Phase 1" sample denoted $\{Y_i, W_i, i = 1, \ldots, N\}$. A subsample of these individuals is chosen to create a Phase 2 sample; we let R_i indicate whether individual i is selected ($R_i = 1$) or not ($R_i = 0$) for the subsample. The value of X_i is then measured for individuals in the Phase 2 sample. The distribution $f(y|x, w)$ of Y given X and W is of interest, and by judicious choice of selection probabilities

$$\pi_i = \pi(Y_i, W_i) = P(R_i = 1 \mid Y_i, W_i),$$

we can increase the efficiency of estimation. In the context of this book, Y is typically a multistate process denoted with our previous notation by $\{Z(s), s \geq 0\}$. For example, $Z_i(t)$ might represent levels of retinopathy in a cohort of persons with Type 1 diabetes as in Sections 5.2.4 and 7.1 and W_i (or $W_i(t)$) could represent covariates measured on everyone. Let $\mathcal{Z}_i(t) = \{Z_i(s), 0 \leq s \leq t\}$ and $\mathcal{W}_i(t) = \{W_i(s), 0 \leq s \leq t\}$ denote the histories of the respective processes. If we wished to select a subsample of individuals at some point for measurement of genomic variables X, then selection probabilities π_i could be based on the process and covariate histories $\mathcal{Z}_i(C_i)$, $\mathcal{W}_i(C_i)$ up to the current follow-up time C_i for each individual. If we were searching for factors associated with faster progression of retinopathy, then it might be advantageous to over-sample persons in higher states.

Table 7.4: Estimates obtained from fitting a finite mixture of 3-state processes for disease activity in lupus conditioning on the state at clinic entry as in Section 6.3.2 and by modeling the state at disease onset; Age indicates SLE diagnosis $\geq$ 30 years of age.

Class	Transition	From Clinic Entry[1]		From Disease Onset[2]	
		EST	95% CI	EST	95% CI
1	$1 \to 2$	1.19	(1.06, 1.35)	1.28	(1.13, 1.44)
	$2 \to 1$	0.59	(0.53, 0.65)	0.55	(0.50, 0.61)
	$2 \to 3$	0.39	(0.32, 0.46)	0.39	(0.34, 0.45)
	$3 \to 2$	1.42	(1.25, 1.61)	1.22	(1.09, 1.37)
2	$1 \to 2$	2.38	(2.16, 2.64)	2.45	(2.23, 2.69)
	$2 \to 1$	2.91	(2.51, 3.39)	2.69	(2.39, 3.04)
	$2 \to 3$	1.89	(1.67, 2.15)	2.02	(1.77, 2.31)
	$3 \to 2$	4.10	(3.66, 4.58)	3.83	(3.42, 4.28)
3	$1 \to 2$	0.52	(0.47, 0.58)	0.58	(0.52, 0.64)
	$2 \to 1$	2.79	(2.51, 3.10)	2.57	(2.35, 2.81)
	$2 \to 3$	0.79	(0.66, 0.95)	0.71	(0.61, 0.81)
	$3 \to 2$	4.44	(3.83, 5.14)	3.61	(3.13, 4.16)

Estimates for Common Initial Distribution (Reference: Initial State 1)

Initial State	Parameter	EST	95% CI	EST	95% CI
2	Intercept, η_2			0.754	(0.558, 0.949)
3	Intercept, η_3			0.722	(0.525, 0.919)

Estimates for Polychotomous Logistic Model for Class Membership (Reference: Class 1)

Class	Parameter	EST	95% CI	EST	95% CI
2	Intercept, γ_{20}	-0.502	(-0.765, -0.239)	-0.494	(-0.745, -0.243)
	Age, γ_{21}	0.654	(0.301, 1.007)	0.671	(0.316, 1.026)
3	Intercept, γ_{30}	-0.308	(-0.521, -0.095)	-0.187	(-0.386, 0.012)
	Age, γ_{31}	0.993	(0.704, 1.281)	0.918	(0.636, 1.201)

[1] Taking the time origin as clinic entry and conditioning on the state occupied at that time analyses of Section 6.3.2 with naive use of marginal multinomial model for $P(W = g)$.
[2] Modeling initial distribution at disease onset.

There has been limited work on two-phase sampling designs for the study of general multistate processes, but methods for time-to-event (or failure time) outcomes are very well developed. The best-known designs are case-cohort and nested case-control designs (e.g. Borgan and Samuelsen, 2014), where individuals in a cohort who have failed and those who have not failed by a specific time are selected with different probabilities. In particular, if T_i and C_i represent failure and censoring times, with $Y_i = \min(T_i, C_i)$ and $\Delta_i = I(T_i \leq C_i)$, then selection probabilities π_i may be dependent on these and a fixed covariate vector W_i, in which case we denote it by $\pi(Y_i, \Delta_i, W_i)$. In settings where failures are fairly rare, it is common to take $\pi(Y_i, \Delta_i = 1, W_i) = 1$ so that all individuals observed to fail are selected for Phase 2. There has been extensive development of methods based on weighted estimating functions for fitting Cox models (e.g. Borgan and Samuelsen, 2014) and other semiparametric models. We focus here on methods that could be used in the multistate setting. We start by describing maximum likelihood and weighted likelihood approaches in a general context and then in the next section discuss their application to multistate models.

First we return to the general model $f(y|x, w)$ for a response variable Y given covariate vectors X, W, where X is only observed for individuals selected for Phase 2. We assume that there is a parametric specification $f(y|x, w; \theta)$ and also let $g(x|w; \gamma)$ denote a model for X given W. Since $P(R_i = 1|Y_i, X_i, W_i) = P(R_i = 1|Y_i, W_i)$, the X-values for individuals with $R_i = 0$ are missing at random (MAR) in the terminology of Rubin (1976) and the likelihood function for (θ, γ) based on observed Phase 1 and Phase 2 data is proportional to

$$L(\theta, \gamma) = \prod_{i=1}^{N} \{f(y_i \mid x_i, w_i; \theta) \, g(x_i \mid w_i; \gamma)\}^{R_i} \, f_1(y_i \mid w_i; \theta, \gamma)^{1-R_i}, \qquad (7.22)$$

where

$$f_1(y \mid w; \theta, \gamma) = \int f(y \mid x, w; \theta) \, dG(x \mid w; \gamma). \qquad (7.23)$$

In (7.23), $G(x|w; \gamma)$ is the distribution function corresponding to the probability density or mass function $g(x|w; \gamma)$ for X given W. The Phase 2 selection probabilities π_i do not enter (7.22) explicitly, although they do affect the information about (θ, γ) and the asymptotic distributions of maximum likelihood estimators $(\widehat{\theta}, \widehat{\gamma})$ obtained by maximization of (7.22).

The maximum likelihood estimating functions from (7.22) and the expression (7.23) can be written in a form that facilitates the use of EM algorithms. Let $U_{i1}(Y_i, X_i, W_i; \theta) = \partial \log f(Y_i|X_i, W_i; \theta)/\partial\theta$ and $U_{i2}(X_i, W_i; \gamma) = \partial \log g(X_i|W_i; \gamma)/\partial\gamma$, where we write out the arguments explicitly to facilitate discussion regarding expectation. Then if $U_1(\theta, \gamma) = \partial \log L(\theta, \gamma)/\partial\theta$ and $U_2(\theta, \gamma) = \partial \log L(\theta, \gamma)/\partial\gamma$, it can be shown that in a sample of size N they can be written as

$$\sum_{i=1}^{N} \{R_i \, U_{i1}(Y_i, X_i, W_i; \theta) + (1 - R_i) \, E_X[U_{i1}(Y_i, X, W_i; \theta) \mid Y_i, W_i]\}$$

$$\sum_{i=1}^{N} \{R_i \, U_{i2}(X_i, W_i; \gamma) + (1 - R_i) \, E_X[U_{i2}(X, W_i; \gamma) \mid Y_i, W_i]\},$$

respectively. Zhao et al. (2009), Zeng and Lin (2014) and others consider EM algorithms for solving the associated estimating equations.

A disadvantage of (7.22) is the need to model the covariate distribution $g(x|w)$, which may be difficult in some settings. Methods that avoid this are available. One approach is through conditional likelihoods based on the distributions $P(Y_i|X_i, W_i, R_i = 1)$; in this case, only data from individuals in the Phase 2 subsample are used. A simpler alternative is provided by weighted likelihood estimating functions. We note that

$$U_1^{\mathrm{w}}(\theta) = \sum_{i=1}^{N} \frac{R_i}{\pi(Y_i, W_i)} U_{i1}(Y_i, X_i, W_i; \theta) \tag{7.24}$$

is an unbiased estimating function, since

$$E\left\{ E\left[\frac{R_i}{\pi(Y_i, W_i)} U_{i1}(Y_i, X_i, W_i; \theta) \mid Y_i, X_i, W_i \right] \right\} = E\{U_{i1}(Y_i, X_i, W_i; \theta)\} = 0.$$

This estimating function also uses data only from individuals in the Phase 2 sub-sample. Efficiency of estimation can be improved by replacing π_i in (7.24) with estimates $\hat{\pi}_i$ (e.g. Lawless et al., 1999; Lumley et al., 2011); estimation of π_i uses information from persons with $R_i = 0$. It is also possible to "augment" (7.24) by adding terms for individuals with $R_i = 0$ (see Rotnitzky, 2009). A constraint on the use of (7.24) is that we require $\pi(Y, W) > 0$ for all (Y, W); this is not needed for maximum likelihood.

7.2.2   Multistate Processes

Consider a cohort of individuals in which the data $\{\mathcal{Z}_i(C_i), \mathcal{W}_i(C_i)\}$, $i = 1, \ldots, N$ have been observed by some point in time, with $\{W(s), s \geq 0\}$ representing for now an external covariate process. We use this general notation to represent observed histories of the multistate and covariate process $\{W_i(s), s \geq 0\}$ at C_i, which covers both the case of right-censored data and intermitently observed data. There are many ways that Phase 2 selection probabilities could be defined, but here we focus on the following, which would be useful in many settings. First we form strata $S_{(1)}, \ldots, S_{(M)}$ based on $\mathcal{W}_i(C_i)$ and $\mathcal{Z}_i(C_i)$, and then define

$$\pi_i = P(R_i = 1 \mid \mathcal{Z}_i(C_i), \mathcal{W}_i(C_i), C_i) = P(R_i = 1 \mid \delta_i, C_i),$$

where $\delta_i \in \{1, \ldots, M\}$ specifies which stratum to which individual i belongs. In the special case of a failure time model with two states, this encompasses what is called stratified case-cohort sampling (e.g. Ding et al., 2017). In this more general setting, however, letting $\bar{\mathcal{H}}_i^{\circ}(C_i) = \{\mathcal{Z}_i(C_i), \mathcal{W}_i(C_i), C_i\}$ for simplicity, the likelihood function analogous to (7.22) is

$$L(\theta, \gamma) = \prod_{i=1}^{N} P(\bar{\mathcal{H}}_i^{\circ}(C_i) \mid C_i, X_i, \mathcal{W}_i(C_i), \bar{\mathcal{H}}_i^{\circ}(0))^{R_i} \tag{7.25}$$

$$\times \ P(\bar{\mathcal{H}}_i^{\circ}(C_i) \mid C_i, \mathcal{W}_i(C_i), \bar{\mathcal{H}}_i^{\circ}(0))^{1-R_i},$$

where the contribution for individuals with $R_i = 0$ does not condition on the unknown X_i; here $\bar{\mathcal{H}}_i^{\circ}(0) = \{Z_i(0), W_i(0)\}$. This likelihood is difficult to deal with if $\{W_i(s), s \geq 0\}$ is truly time-varying, since elaborate modeling would be needed to characterize $g(x|\mathcal{W}_i(C_i))$ for X_i given $\mathcal{W}_i(C_i)$. Even in the case where W_i is fixed, specification of $g(x|w_i)$ and subsequent calculation of the second terms (for $R_i = 0$) in (7.25) is complicated in most situations; we provide an illustration below.

In contrast, weighted likelihood estimation using unbiased estimating functions analogous to (7.24) is straightforward. In this framework, we use

$$U_1^{\mathrm{w}}(\theta) = \sum_{i=1}^{N} \frac{R_i}{\pi_i} U_{i1}(\theta), \tag{7.26}$$

where $U_{i1}(\theta)$ is written more explicitly as $U_{i1}(\bar{\mathcal{H}}_i^\circ(C_i) \mid C_i, X_i, \mathcal{W}_i(C_i), \bar{\mathcal{H}}_i^\circ(0); \theta)$ and given by

$$\frac{\partial \log P(\bar{\mathcal{H}}_i^\circ(C_i) \mid C_i, X_i, \mathcal{W}_i(C_i), \bar{\mathcal{H}}_i^\circ(0); \theta)}{\partial \theta}.$$

This simply requires that we use the usual likelihood estimation procedure on individuals with $R_i = 1$, but incorporating weights π_i^{-1}. Under mild conditions, solving $U_1^{\mathrm{w}}(\theta) = 0$ is equivalent to maximizing the weighted log-likelihood function. Variance estimation for estimators follows from standard theory for estimating functions. Even though the π_i in (7.26) are specified by design, efficiency gains for $\widehat{\theta}$ are obtained by replacing π_i with estimates $\widehat{\pi}_i$ (e.g. Lawless et al., 1999; Lumley et al., 2011). Phase 1 samples are often stratified as described above, in which case individuals are placed into one of M strata $S_{(1)}, \ldots, S_{(M)}$. A random Phase 2 sample of specified size n_j can then be selected from $S_{(j)}$, $j = 1, \ldots, M$. If there are N_j of the Phase 1 individuals in $S_{(j)}$, then the selection probability for an individual $i \in S_{(j)}$ is n_j/N_j. For a fixed sample size n_j in stratum j, the selection indicators R_i for persons in the same stratum are not independent; this complicates variance estimation slightly. For simplicity, we give variance estimates for the case of Bernoulli sampling, in which case the R_i are treated as mutually independent, with $\pi_i = \sum_{j=1}^{M} I(i \in S_{(j)}) p_j$ where p_j is the selection probability for $S_{(j)}$. In this case, efficiency is improved if we replace design probabilities p_j with "estimates" $\widehat{p}_j = n_j/N_j$, where n_j is the number selected from $S_{(j)}$.

The case of Bernoulli sampling can be described by a general formulation where $\pi_i(\alpha) = \pi(\bar{\mathcal{H}}_i^\circ(C_i); \alpha)$, with α a parameter vector specifying the selection probabilities. To estimate α we use the fact that the R_i are independent, with $P(R_i = 1 | \bar{\mathcal{H}}_i^\circ(C_i)) = \pi_i(\alpha)$. We estimate α by considering the following estimating function, based on $R_1, \ldots, R_N$:

$$U_2(\alpha) = \sum_{i=1}^{N} U_{i2}(\alpha) = \sum_{i=1}^{N} \frac{R_i - \pi_i(\alpha)}{\pi_i(\alpha)(1 - \pi_i(\alpha))} \frac{\partial \pi_i(\alpha)}{\partial \alpha}. \tag{7.27}$$

We estimate θ by first solving $U_2(\alpha) = 0$ to obtain $\widehat{\alpha}$, and then replace π_i by $\pi_i(\widehat{\alpha})$ in (7.26) to solve for $\widehat{\theta}$. This is equivalent to solving simultaneously the two estimating functions

$$U_1^{\mathrm{w}}(\theta, \alpha) = 0 \quad \text{and} \quad U_2(\alpha) = 0,$$

where $U_1^{\mathrm{w}}(\theta, \alpha)$ is (7.26) with π_i given by $\pi_i(\alpha)$. Writing

$$A = \begin{pmatrix} A_{11} & A_{12} \\ 0 & A_{22} \end{pmatrix} = \frac{1}{N} \begin{pmatrix} E(-\partial U_1^{\mathrm{w}}(\theta, \alpha)/\partial \theta') & E(-\partial U_1^{\mathrm{w}}(\theta, \alpha)/\partial \alpha') \\ 0 & E(-\partial U_2(\alpha)/\partial \alpha') \end{pmatrix}$$

and

$$B = \begin{pmatrix} B_{11} & B_{12} \\ B_{21} & B_{22} \end{pmatrix} = \frac{1}{N} \operatorname{var} \begin{pmatrix} U_1^{\mathrm{w}}(\theta, \alpha) \\ U_2(\alpha) \end{pmatrix},$$

we have the asymptotic distribution of $\sqrt{N}(\widehat{\theta} - \theta)$ with mean 0 and covariance matrix $\Sigma(\widehat{\theta})$ given by the upper left block of the matrix $A^{-1}B[A^{-1}]'$. Using the fact that $A_{22} = B_{22}$ and $A_{12} = B_{12}$, this gives

$$\Sigma(\widehat{\theta}) = A_{11}^{-1}[B_{11} - A_{12}A_{22}^{-1}A_{12}'][A_{11}^{-1}]'. \tag{7.28}$$

The equality $A_{22} = B_{22}$ follows from the fact that $U_2(\alpha)$ is a likelihood score function, and the equality $A_{12} = B_{12}$ follows by noting that the ith term of NB_{12} is

$$E\left\{ \frac{R_i U_{i1}(\theta)}{\pi_i(\alpha)} \frac{R_i - \pi_i(\alpha)}{\pi_i(\alpha)(1 - \pi_i(\alpha))} \frac{\partial \pi_i(\alpha)}{\partial \alpha'} \right\} = E\left\{ U_{i1}(\theta) \frac{\partial \log \pi_i(\alpha)}{\partial \alpha} \right\} \tag{7.29}$$

by virtue of the fact that $E\{R_i | \bar{\mathcal{H}}_i^\circ(C_i)\} = \pi_i(\alpha)$. Similarly, the ith term of NA_{12} is easily shown to equal (7.29). We can then estimate $\Sigma(\widehat{\theta})$ by replacing A_{11}, A_{12}, A_{22} and B_{11} with estimates

$$\widehat{A}_{11} = -\frac{1}{N} \sum_{i=1}^{N} \frac{R_i}{\widehat{\pi}_i} \left(\frac{\partial U_{i1}(\theta)}{\partial \theta'} \right) \Big|_{\widehat{\theta}},$$

$$\widehat{A}_{12} = -\frac{1}{N} \sum_{i=1}^{N} \frac{R_i}{\widehat{\pi}_i} U_{i1}(\widehat{\theta}) \left(\frac{\partial \log \pi_i(\alpha)}{\partial \alpha'} \right) \Big|_{\widehat{\alpha}},$$

$$\widehat{A}_{22} = -\frac{1}{N} \sum_{i=1}^{N} \left(\frac{\partial \log \pi_i(\alpha)}{\partial \alpha} \right) \left(\frac{\partial \log(1 - \pi_i(\alpha))}{\partial \alpha'} \right) \Big|_{\widehat{\alpha}},$$

$$\widehat{B}_{11} = \frac{1}{N} \sum_{i=1}^{N} \frac{R_i}{\widehat{\pi}_i^2} U_{i1}(\widehat{\theta}) U_{i1}'(\widehat{\theta}),$$

where for simplicity we write $\widehat{\pi}_i$ for $\pi_i(\widehat{\alpha})$.

Example 7.2.1:   A 3-State Progressive Model

To contrast the maximum likelihood and weighted likelihood approaches, we consider a very simple progressive Markov model with states 1, 2 and 3. We suppose that the multistate processes $\{Z_i(s), s \geq 0\}$ and a fixed covariate W_i are observed for all N individuals in a cohort, and that covariates X_i are to be measured only for a subset of n individuals. For simplicity, we assume that given the covariates X_i and W_i the multistate process is time-homogeneous with conditional intensity functions

$$\lambda_{12}(t \mid x, w) = \lambda_1 \exp(x'\beta_1 + w'\gamma_1)$$
$$\lambda_{23}(t \mid x, w) = \lambda_2 \exp(x'\beta_2 + w'\gamma_2)$$

We consider Phase 2 sampling plans where selection probabilities

$$\pi_i = P(R_i = 1 \mid \mathcal{Z}_i(C_i), C_i, W_i)$$

may depend on data observed up to specified follow-up times C_i, $i = 1, \ldots N$. For individual i, let $d_{i1} = N_{i12}(C_i)$ and $d_{i2} = N_{i23}(C_i)$ indicate whether a $1 \to 2$ and $2 \to 3$ transition occurred over $[0, C_i]$, and let T_{i1} and T_{i2} denote the observed time spent in states 1 and 2, respectively. For convenience suppose X_i and W_i are both one-dimensional. If X_i and W_i are both known ($R_i = 1$), the likelihood contribution is $L_i(\theta) = L_{i1}(\theta_1) \cdot L_{i2}(\theta_2)$, where $\theta_1 = (\lambda_1, \beta_1, \gamma_1)'$, $\theta_2 = (\lambda_2, \beta_2, \gamma_2)'$, and

$$L_{i1}(\theta_1) = \lambda_1^{d_{i1}} \, e^{(\beta_1 x_i + \gamma_1 w_i) \, d_{i1}} \, \exp\{-\lambda_1 T_{i1} e^{\beta_1 x_i + \gamma_1 w_i}\},$$
$$L_{i2}(\theta_2) = \lambda_2^{d_{i2}} \, e^{(\beta_2 x_i + \gamma_2 w_i) \, d_{i2}} \, \exp\{-\lambda_2 T_{i2} e^{\beta_2 x_i + \gamma_2 w_i}\}.$$

If only W_i is known, however (i.e. $R_i = 0$), the likelihood contribution is

$$L_i^*(\theta, \gamma) = \lambda_1^{d_{i1}} \lambda_2^{d_{i2}} \, e^{(\gamma_1 d_{i1} + \gamma_2 d_{i2}) w_i}$$
$$\times \int e^{(\beta_1 d_{i1} + \beta_2 d_{i2}) x} \, \exp\{-\lambda_1 T_{i1} e^{\beta_1 x + \gamma_1 w_i} - \lambda_2 T_{i2} e^{\beta_2 x + \gamma_2 w_i}\} \, dG(x \mid w_i)$$

where the parameter γ indexes $G(x|w) = G(x|w; \gamma)$. We see from $L_i^*(\theta, \gamma)$ the difficulty with maximum likelihood estimation. The contributions for individuals with $R_i = 0$ do not factor into separate pieces for θ_1 and θ_2, and integration over $G(x|w_i)$ is required. When X is discrete and especially if it is binary, generalized linear models are convenient, and summation over X is straightforward. The score functions for θ and γ can then be expressed in forms like those used in a EM algorithm as mentioned in Section 7.2.1. Continuous univariate X can also be handled, though both here and in the discrete case, model checks for $G(x|w)$ can be challenging.

The weighted likelihood score functions of the form (7.26), in contrast, are easily handled. Defining $\beta_{10} = \log \lambda_1$ and $\beta_{20} = \log \lambda_2$, we find that $U_1^{\mathrm{w}}(\theta) = ([U_{11}^{\mathrm{w}}(\theta_1)]', [U_{12}^{\mathrm{w}}(\theta_2)]')'$, where

$$U_{11}^{\mathrm{w}}(\theta_1) = \sum_{i=1}^{N} \frac{R_i}{\pi_i} \, (d_{i1} - T_{i1} \exp(\beta_{10} + \beta_1 x_i + \gamma_1 w_i)) \begin{pmatrix} 1 \\ x_i \\ w_i \end{pmatrix}$$

$$U_{12}^{\mathrm{w}}(\theta_2) = \sum_{i=1}^{N} \frac{R_i}{\pi_i} \, (d_{i2} - T_{i2} \exp(\beta_{20} + \beta_2 x_i + \gamma_2 w_i)) \begin{pmatrix} 1 \\ x_i \\ w_i \end{pmatrix}.$$

For given $\hat{\pi}_i$ the estimating equations $U_{11}^{\mathrm{w}}(\theta_1) = 0$ and $U_{12}^{\mathrm{w}}(\theta_2) = 0$ are easily solved, and covariance matrices for $(\hat{\theta}_1, \hat{\theta}_2)$ can be obtained from (7.28). In Section 7.2.4 we consider choices for sampling probabilities π_i.

7.2.3 Inference for Models with Semiparametric Multiplicative Intensities

Weighted pseudo-partial likelihood methods have been developed for case-cohort and other two-phase failure time studies. We show here how these methods can be applied for the semiparametric transition intensity function models discussed in Section 3.3 when the Phase 1 data are subject to right censoring. We will do this by first considering general parametric models of multiplicative form, with intensities

$$\lambda_{kl}(t \mid \mathcal{H}_i(t^-)) = \lambda_{kl0}(t \mid \mathcal{Z}_i(t^-); \alpha_{kl}) \exp(x_i' \beta_{kl} + w_i' \gamma_{kl}) \tag{7.30}$$

for $k \neq l$ where $\mathcal{H}_i(t) = \{Z_i(s), 0 \leq s \leq t; X, W\}$ and $\mathcal{Z}_i(t) = \{Z_i(s), 0 \leq s \leq t\}$. Letting $\theta_{kl} = (\alpha'_{kl}, \beta'_{kl}, \gamma'_{kl})'$ and assuming that no parameters are shared for different transition intensities, we have from (3.2) that the log-likelihood contributions based on complete observation of individuals $i = 1, \ldots, N$ are

$$\ell_{ikl}(\theta_{kl}) = \int_0^\infty \bar{Y}_{ik}(u) \left\{ \log \lambda_{kl}(u \mid \mathcal{H}_i(u^-); \theta_{kl}) \, dN_{ikl}(u) - \lambda_{kl}(u \mid \mathcal{H}_i(u^-); \theta_{kl}) \, du \right\}.$$

The score function contributions $U_{ikl}(\theta_{kl}) = \partial \ell_{ikl}(\theta_{kl})/\partial \theta_{kl}$ take the form

$$\int_0^\infty \bar{Y}_{ik}(u) \left\{ \frac{\partial \log \lambda_{kl}(u \mid \mathcal{H}_i(u^-); \theta_{kl})}{\partial \theta_{kl}} \, dN_{ikl}(u) - \frac{\partial \lambda_{kl}(u \mid \mathcal{H}_i(u^-); \theta_{kl})}{\partial \theta_{kl}} \, du \right\}. \quad (7.31)$$

We consider Markov models where $\lambda_{kl0}(t \mid \mathcal{Z}_i(t^-); \alpha_{kl}) = \lambda_{kl0}(t; \alpha_{kl})$ in (7.30), but other types of models can also be handled. In this case (7.31) gives score functions for α_{kl}, β_{kl} and γ_{kl}, respectively, as

$$U_{ikl1}(\theta_{kl}) = \int_0^\infty \bar{Y}_{ik}(u) \left\{ \frac{\partial \log \lambda_{kl0}(u)}{\partial \alpha_{kl}} \, dN_{ikl}(u) - \frac{\partial \lambda_{kl0}(u)}{\partial \alpha_{kl}} \exp\left(x'_i \beta_{kl} + w'_i \gamma_{kl}\right) \, du \right\}$$

$$U_{ikl2}(\theta_{kl}) = \int_0^\infty \bar{Y}_{ik}(u) \, X_i \left\{ dN_{ikl}(u) - \lambda_{kl0}(u; \alpha_{kl}) \exp\left(x'_i \beta_{kl} + w'_i \gamma_{kl}\right) \, du \right\}$$

$$U_{ikl3}(\theta_{kl}) = \int_0^\infty \bar{Y}_{ik}(u) \, W_i \left\{ dN_{ikl}(u) - \lambda_{kl0}(u; \alpha_{kl}) \exp\left(x'_i \beta_{kl} + w'_i \gamma_{kl}\right) \, du \right\}.$$

We consider the use of inverse probability weighted estimating functions for the analysis of a two-phase study where the Phase 2 sample is chosen in an outcome-dependent fashion as described in the previous section. Specifically, we let

$$U^{\mathrm{w}}_{kl1}(\theta_{kl}) = \sum_{i=1}^N \frac{R_i}{\pi_i} \cdot U_{ikl1}(\theta_{kl}) \quad (7.32)$$

$$U^{\mathrm{w}}_{kl2}(\theta_{kl}) = \sum_{i=1}^N \frac{R_i}{\pi_i} \cdot U_{ikl2}(\theta_{kl}) \quad (7.33)$$

$$U^{\mathrm{w}}_{kl3}(\theta_{kl}) = \sum_{i=1}^N \frac{R_i}{\pi_i} \cdot U_{ikl3}(\theta_{kl}). \quad (7.34)$$

For models with finite-dimensional α_{kl} with specified π_i, estimates are readily obtained by solving $U^{\mathrm{w}}_{kl1}(\theta_{kl}) = U^{\mathrm{w}}_{kl2}(\theta_{kl}) = U^{\mathrm{w}}_{kl3}(\theta_{kl}) = 0$. We note that if the models correspond to hazard functions for which failure time software exists, then provided case weights are allowed, estimates can be obtained from the software. The R function `phreg` handles many common models.

We now turn to semiparametric Cox models for which the baseline intensities $\lambda_{kl0}(u)$ are not specified parametrically. We use the approach mentioned in Section 3.3.1, where we assume increments $d\Lambda_{kl0}(u)$ in $\Lambda_{kl0}(u)$ are non-zero only for times u at which a $k \to l$ transition was observed. Equating $d\Lambda_{kl0}(u)$ with a component of α in (7.32), we obtain the estimating equation

$$\sum_{i=1}^N \frac{R_i}{\pi_i} \cdot \bar{Y}_{ik}(u) \left\{ \frac{dN_{ikl}(u)}{d\Lambda_{kl0}(u)} - \exp(x'_i \beta_{kl} + w'_i \gamma_{kl}) \right\} = 0.$$

For given β_{kl} and γ_{kl}, this gives the weighted profile likelihood estimate

$$d\tilde{\Lambda}_{kl0}(u;\beta_{kl},\gamma_{kl}) = \frac{\sum_{i=1}^{N} R_i \, \pi_i^{-1} \, \bar{Y}_{ik}(u) \, dN_{ikl}(u)}{\sum_{i=1}^{N} R_i \, \pi_i^{-1} \, \bar{Y}_{ik}(u) \, \exp(x_i'\beta_{kl} + w_i'\gamma_{kl})}. \tag{7.35}$$

Inserting the estimates (7.35) into the estimating functions (7.33) and (7.34) for β_{kl} and γ_{kl} then gives the weighted profile likelihood estimating functions

$$\tilde{U}_{kl2}^{w}(\beta_{kl},\gamma_{kl}) = \int_0^{\infty} \sum_{i=1}^{N} \frac{R_i}{\pi_i} \cdot \bar{Y}_{ik}(u) \left\{ X_i - \frac{S_{k2}^{(1,w)}(u;\beta_{kl},\gamma_{kl})}{S_k^{(0,w)}(u;\beta_{kl},\gamma_{kl})} \right\} dN_{ikl}(u) \tag{7.36}$$

$$\tilde{U}_{kl3}^{w}(\beta_{kl},\gamma_{kl}) = \int_0^{\infty} \sum_{i=1}^{N} \frac{R_i}{\pi_i} \cdot \bar{Y}_{ik}(u) \left\{ W_i - \frac{S_{k3}^{(1,w)}(u;\beta_{kl},\gamma_{kl})}{S_k^{(0,w)}(u;\beta_{kl},\gamma_{kl})} \right\} dN_{ikl}(u), \tag{7.37}$$

for β_{kl} and γ_{kl}, respectively, where

$$S_{k2}^{(1,w)}(u;\beta_{kl},\gamma_{kl}) = \sum_{i=1}^{N} \frac{R_i}{\pi_i} \cdot \bar{Y}_{ik}(u) \, x_i \, \exp(x_i'\beta_{kl} + w_i'\gamma_{kl})$$

$$S_{k3}^{(1,w)}(u;\beta_{kl},\gamma_{kl}) = \sum_{i=1}^{N} \frac{R_i}{\pi_i} \cdot \bar{Y}_{ik}(u) \, w_i \, \exp(x_i'\beta_{kl} + w_i'\gamma_{kl})$$

and

$$S_k^{(0,w)}(u;\beta_{kl},\gamma_{kl}) = \sum_{i=1}^{N} \frac{R_i}{\pi_i} \cdot \bar{Y}_{ik}(u) \, \exp(x_i'\beta_{kl} + w_i'\gamma_{kl}).$$

We obtain estimates $\hat{\beta}_{kl}$, $\hat{\gamma}_{kl}$ by setting (7.36) and (7.37) equal to zero, and then from (7.35) we obtain an estimate of $\Lambda_{kl0}(u)$ by inserting $\hat{\beta}_{kl}$, $\hat{\gamma}_{kl}$.

Estimation via (7.36) and (7.37) is equivalent to maximizing the weighted pseudo-partial likelihood function

$$L_{pp}(\beta_{kl},\gamma_{kl}) = \prod_{i=1}^{N} \prod_{u \in \mathcal{D}_{ikl}} \left\{ \frac{\exp(x_i'\beta_{kl} + w_i'\gamma_{kl})}{S_k^{(0,w)}(u;\beta_{kl},\gamma_{kl})} \right\}^{R_i/\pi_i}, \tag{7.38}$$

where $\mathcal{D}_{ikl}$ is the set of times u at which $\bar{Y}_{ik}(u)\,dN_{ikl}(u) = 1$, that is, where $k \to l$ transitions are observed. If $\mathcal{D}_{ikl}$ is empty for a specific individual, the corresponding term in (7.38) equals one. The function (7.38) is one used for case-cohort failure time studies, where $\mathcal{D}_{ikl}$ contains either one time or more. Samuelsen et al. (2007) and Borgan and Samuelsen (2014) have shown how the Cox model software **coxph** in R and S-PLUS can be used to obtain not only estimates $\hat{\beta}_{kl}$, $\hat{\gamma}_{kl}$, $\hat{\Lambda}_{kl0}(u)$, but also variance estimates for $\hat{\beta}_{kl}$ and $\hat{\gamma}_{kl}$.

7.2.4 Design Issues

Primary issues for two-phase studies are (a) the choice of strata $S_{(1)},\ldots,S_{(M)}$ used to specify differential Phase 2 sampling, and (b) the selection probabilities or sample sizes for each stratum. This has received a great deal of attention in the case of failure time studies employing case-control or case-cohort designs. Surveys of types of designs can be found in Samuelsen et al. (2007) and Borgan and Samuelsen (2014). Much of the focus is on settings where failures are rather rare in the population or phase 1 cohort, and on logistical aspects of Phase 2 selection over an extended period of time. It is found in the case of rare failures that selection of all cases (failed individuals) and of three or four non-cases (unfailed individuals) for each case provides estimates of expensive covariate effects that are close in efficiency to ones based on the full cohort. Weighted pseudo-likelihood estimation for Cox models is widely used (e.g. Samuelsen et al., 2007). and formulas for sample size determination in hypothesis testing contexts have been developed (e.g. Cai and Zeng, 2004; Cai and Zeng, 2007).

For multistate models, there has been little study of design issues, in part because covariate effects vary across different transition intensities. For intensity-based analysis of expensive covariate effects in Cox models described in Section 7.2.3, results for case-cohort failure time studies might possibly be applied if we are able to estimate $E(\bar{Y}_{ik}(u))$ in expressions (7.32)–(7.34). In general, however, simulation of data using alternative designs seems the best approach; as always we require preliminary estimates of model parameters to do this. In settings where entry to some specific state is of interest, we may prefer to base design decisions on reduced models for estimation of prevalence (occupancy) at some designated time, or on time to entry to the state.

Example 7.2.2: Diabetic Retinopathy and an Expensive Covariate

Stratified sampling of Phase 2 individuals can provide a more dispersed set of X-values and thus increase efficiency of tests or estimates for expensive covariate effects, and it can also improve estimation of the intensities for less common transitions. However, an assessment of designs requires estimates of model parameters and decisions about their relative priorities. For illustration we consider the 5-state progressive model in Figure 5.5 and suppose there are no covariates besides the expensive X. Motivated by the diabetic retinopathy process in the Diabetes Control and Complications Trial (DCCT) discussed in Sections 1.2.2 and 5.2.4, we suppose that X represents a binary genomic marker and pattern our investigation on a model M3 fitted to the conventional therapy (CT) group and reported on in Table 4 of Cook and Lawless (2014). The assumed multistate model is Markov with piecewise-constant intensities

$$\lambda_{k,k+1}(t \mid x) = \lambda_{k0}(t)\exp(\beta x), \quad k = 1,2,3,4 \tag{7.39}$$

where $\lambda_{k0}(t)$ for $0 \le t \le 4$ equals 0.251, 0.133, 0.038, 0.038 ($k = 1,2,3,4$) and for $t > 5$ equals 0.512, 0.235, 0.051, 0.051. We assume X is Bin(1,0.1) and that $\beta \ge 0$, so the rarer value $X = 1$ is associated with faster progression of retinopathy.

We assume for simplicity that a cohort of $N = 1000$ persons are all observed at times $(0,0.5,1,\ldots,8)$ years and that Phase 2 individuals are selected at the end of year 8. We consider a Phase 2 sample of size $n = 500$ and two sampling plans: (a) a simple random sample, and (b) a stratified sampling plan with strata $S_{(1)} = \{i : Z_i(8) = 4 \text{ or } 5\}$ and $S_{(2)} = \{i : Z_i(8) = 1,2 \text{ or } 3\}$. Letting N_1, N_2 be the number of cohort members in $S_{(1)}$ and $S_{(2)}$, we take all $n_1 = N_1$ persons from $S_{(1)}$ and $n_2 =$

$500 - n_1$ persons from $S_{(2)}$. Estimation is based on maximization of the weighted log-likelihood corresponding to the estimating functions (7.26). This is

$$\ell_w(\theta) = \sum_{i=1}^{N} \frac{R_i}{\widehat{\pi}_i} \cdot \ell_i(\theta), \qquad (7.40)$$

where θ denotes the full parameter set and $\ell_i(\theta) = \log L_i(\theta)$, with

$$L_i(\theta) = \prod_{j=1}^{16} P(Z_i(a_j) \mid Z_i(a_{j-1}), X_i),$$

and $(a_0, a_1, a_2, \ldots, a_{16}) = (0, 0.5, 1, \ldots, 8)$. As in Section 5.2.4, we constrain λ_{30} and λ_{40} to be equal. The sampling probabilities in (7.40) are

$$\widehat{\pi}_i = I(i \in S_{(1)}) + (n_2/N_2) I(i \in S_{(2)}),$$

and variance estimation for $\widehat{\theta}$ with observed data can be based on (7.25). Our focus here is on efficiency and so we consider only empirical standard deviations (ESDs) of estimators based on 500 simulation replicates, each consisting of $N = 1000$ processes and sub-selection of a Phase 2 sub-sample of size $n = 500$. Estimates for a given sample were obtained using the R `maxLik` package, which contains functions that also allow the calculation of variance estimates via (7.28) with terms involving α omitted.

Table 7.5: Empirical standard deviations of estimators based on (a) simple random sampling (SRS) and (b) stratified sampling (SS) in Phase 2. Results are based on 500 simulation replicates for a cohort of $N = 1000$, with Phase 2 sample size $n = 500$.

e^β	\multicolumn{4}{c}{(a) SRS}	\multicolumn{4}{c}{(b) SS}
	$\widehat{\beta}$	$\widehat{\lambda}_{10}^\dagger$	$\widehat{\lambda}_{20}$	$\widehat{\lambda}_{30}$	$\widehat{\beta}$	$\widehat{\lambda}_{10}^\dagger$	$\widehat{\lambda}_{20}$	$\widehat{\lambda}_{30}$
1.5	0.114	0.054	0.104	1.120	0.105	0.056	0.108	0.338
		0.069	0.055	0.120		0.067	0.053	0.112
2.0	0.106	0.053	0.104	0.445	0.099	0.056	0.102	0.300
		0.068	0.053	0.118		0.069	0.051	0.108
4.0	0.094	0.055	0.093	0.276	0.087	0.055	0.095	0.211
		0.071	0.054	0.114		0.070	0.057	0.104

$\dagger$ Top row for $\widehat{\lambda}_{10}, \widehat{\lambda}_{20}, \widehat{\lambda}_{30}$ – interval $0 \le t \le 4$; bottom row – $t > 4$.

Table 7.5 shows ESDs of estimators under Phase 2 sampling plans (a) and (b). We see that the stratified sampling plan (b) improves on estimation of β only slightly, even for $e^\beta = 4$. Improvements in estimation of λ_{30} over $0 \le t \le 4$ are more substantial. These results reflect the fact that plan (b) increases the number of individuals with $X = 1$ in the Phase 2 sub-sample only slightly unless β is large but that it increases more substantially the numbers of persons making transitions to states 4 and 5. In particular, with $e^\beta = 1.5$, 2.0 and 4.0, the average number of persons with $X = 1$ in the Phase 2 sample are 49.9, 50.1, 4.9 for plan (a) and 54.8, 60.2, 77.8 for plan (b).

7.2.5 Checks on Independent Follow-up Assumptions

The methods of analysis discussed in this book mostly depend on independence or coarsening at random assumptions concerning observation of the life history processes. The two main types are the independent random censoring assumption (see Section 2.2.2) and the conditionally independent visit process (CIVP) assumption for intermittent observation schemes (see Sections 5.1 and 5.4). These assumptions are often plausible when observed process histories contain rich enough data on factors that affect censoring or observation times along with the outcomes of interest. Nevertheless, there are situations where this is not the case. For example, in the cerebrospinal fluid shunt failure study in Section 4.1.5, there were long gaps between the date last seen and the administrative end-of-follow-up time for many patients, and it is a concern that those individuals may have lower than expected rates of shunt failure. A second example is provided by the study of persons with psoriatic arthritis (PsA) in Section 5.4.5, where it can be seen that the gap times between clinic visits depend on disease status and other factors. The CIVP assumption requires that the visit process depend only on previous observed disease history, but the possibility exists that patients' unobserved, more recent, history may also play a role.

Conditional independence assumptions cannot be checked solely on the basis of the observed data. This has resulted in the development of "dependent" observation models that can be used for sensitivity analyses, even though they are not estimable from the observed data. The work in this area has been conducted primarily in the setting of failure time models and to some extent for other multistate models. For example, Barrett et al. (2011) consider a model for cognitive impairment (CI) and death that is represented in Figure 7.10. The figure also includes a lost-to-follow-up (LTF) state. The CI status of individuals is ascertained at intermittent visits, but times of death are known exactly. Many individuals are lost to follow-up, but deaths for such persons are observable provided they occurred before the administrative end-of-follow-up time.

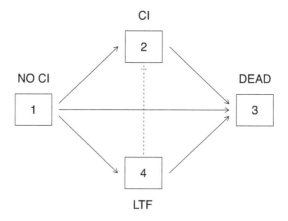

Figure 7.10: A model for cognitive impairment (CI), loss-to-follow-up (LTF) and death. The solid arrows indicate observable transitions under standard follow-up.

The solid arrows in Figure 7.10 show the observable transitions, if exact time at entry to the CI state and LTF state were observable. In fact, a person might move from the LTF state to CI, but we cannot observe this since it is not possible to obtain information on CI status at the time of death for all persons. If LTF times are independent of the process for states 1, 2 and 3, then $\lambda_{12}(t|\mathcal{H}(t^-))$ and $\lambda_{42}(t|\mathcal{H}(t^-))$ would be the same; if they are not, however, then estimation of $\lambda_{12}(t|\mathcal{H}(t^-))$ based on treating LTF times as independent may be biased. A sensitivity analysis based on parametric assumptions for $\lambda_{42}(t|\mathcal{H}(t^-))$ could be considered; this would examine the effect of $\lambda_{42}(t|\mathcal{H}(t^-))$ on the estimation of $\lambda_{12}(t|\mathcal{H}(t^-))$ based on standard methods.

A preferable alternative to sensitivity analysis is to collect auxiliary data that can be used to assess conditional independence assumptions. With intermittent visits, good practice is to ask individuals how their current visit time was set. Additional information can also be acquired by a type of two-phase study, and although this increases costs, it provides ways to assess and mitigate bias. We mention three types of such study design. The first is to randomly select a group of individuals at enrollment who will be followed closely through to administrative end-of-follow-up. This ensures their observation process is independent of the multistate process of interest, and estimates based on them alone may be compared with estimates from the remainder of the cohort. Baker et al. (1993), for example, used such an approach for a failure time study. A second related approach is to identify and measure time-dependent covariates $X(t)$ for a randomly selected group of individuals, the idea being to include factors that may be related to both the observation and multistate processes. If that is the case, we have a violation of independent LTF or CIVP assumptions for individuals for whom $X(t)$ is not measured.

A third approach that is especially useful in the case of losses to follow-up is to carry out a tracing study. This consists of identifying a group of persons LTF, and then tracing them so as to obtain auxiliary information up to a defined administrative end of follow-up time. For the process shown in Figure 7.10, for example, we could trace persons alive and in the LTF state (state 4) at that time, determine whether they had CI or not and, if possible, when they were diagnosed with it. Tracing has mainly been considered in failure time studies (e.g. Frangakis and Rubin, 2001; Farewell et al., 2003), but can be applied more generally. For example, suppose that an individual has administrative end-of-follow-up (censoring) time C_i^A, but that they may be lost to follow-up at an earlier random time $C_i^R < C_i^A$. It may be a concern that C_i^R is not independent of future process history $\{Z_i(t), t > C_i^R\}$, conditional on the observed process history $\mathcal{H}(C_i^R)$. For example, a person might stop reporting regularly to a clinic if their disease status is favourable. A tracing study in this case could consist of obtaining data on $\{Z_i(t), C_i^R \leq t \leq C_i^A\}$ for selected individuals.

It is difficult to give general methodology for sensitivity analysis or for incorporating data from tracing studies, given the various types of auxiliary data collection and loss to follow-up. We provide here two illustrations that represent common situations. An example involving intermittent visit processes is given in Problem 7.8.

7.2.5.1　Failure Time Analysis with Intermittent Data Collection

In many contexts we are interested in the time T to some event ("failure", for convenience), but failure status is ascertained at successive prespecified observation times $a_1, \ldots, a_M$. For example, in a longitudinal study, T might denote the time or age when some mental or physical condition is first detected. It is common for some individuals to be lost to follow-up before the administrative end of follow-up time a_M; this occurs when an individual is last seen at $a_m < a_M$. A concern in cases where covariate information that might be related to both dropout and T is limited, is that the independent LTF assumption may not hold. We illustrate how this might arise by considering the 3-state model in Figure 7.11, where state 1 is "unfailed", state 2 is "failed", and state 3 is premature LTF. We wish to make inferences about $\lambda_T(t)$, the hazard function for T. For simplicity, we have not included a death state; state 2 here represents an event that does not terminate follow-up.

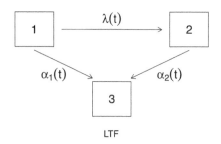

Figure 7.11: A model for failure ($1 \to 2$ transition) and LTF.

Figure 7.11 shows three possible transitions. In some contexts, discussed later, we may also want to include transitions from state 3 to state 2. Here, however, we focus on the case where $\lambda(t) = \lambda_T(t)$, which holds under independent censoring. There is, however, the possibility of bias in the estimation of $\lambda(t)$ in cases where failure times T_i that occur in time intervals $(a_{m-1}, a_m]$ can only be observed when individual i is seen at time a_m. If the LTF intensities $\alpha_1(t)$ and $\alpha_2(t)$ before and after failure are different, then the independent LTF condition is violated when we treat the time last seen (say a_{m-1}) as a censoring time. This is because the specification of a_m as the LTF time, and thus a_{m-1} as the censoring time for cases where $Z(a_{m-1}) = 1$, may depend on process history beyond time a_{m-1}. As discussed by Lawless (2013), if a transition into state 2 is found out at time a_j to have occurred at time t in $(a_{j-1}, a_j]$, then we do not estimate $\lambda(t)$ but rather the intensity function from the purged process for which $Z(a_j) = 1$ or 2. This intensity is

$$\lambda^*(t) = \lambda(t) \frac{P_{22}(t, a_j)}{P_{11}(t, a_j) + P_{12}(t, a_j)}, \quad a_{j-1} < t \le a_j. \tag{7.41}$$

When $\alpha_1(t) = \alpha_2(t)$ it can be seen (see Problem 7.7 (a)) that $\lambda^*(t) = \lambda(t)$ and so estimation based on the use of observed failures up to the time last seen would not be biased. Otherwise bias exists; for example, if $\lambda(t) = \lambda$, $\alpha_1(t) = b_1\lambda$, $\alpha_2(t) = b_2\lambda$, then (Lawless, 2013)

$$R(t) = \frac{\lambda^*(t)}{\lambda} = \frac{1 + b_1 - b_2}{1 + (b_1 - b_2) \exp\{-\lambda(1 + b_1 - b_2)(a_j - t)\}}, \quad a_{j-1} < t \le a_j.$$

The bias is positive ($\lambda^*(t) > \lambda$) if $b_1 > b_2$, negative if $b_1 < b_2$, and for fixed a_j's $R(t)$ moves further away from one as λ increases. Figure 7.12 shows the plots of $R(t)$ for $b_1 - b_2 = 0.1$ (left panels) and 0.3 (right panels). Without loss of generality we take $\lambda = 1$ and show $R(t)$ curves for three sets of equi-spaced ascertainment times a_j: $\Delta a_j = a_j - a_{j-1} = 0.25$, 0.50 and 1.00 (top, middle and bottom panels). Note that $E(T) = 1$ and so the observation gap times equal $0.25E(T)$, $0.50E(T)$ and $E(T)$, respectively.

Bias is small unless b_1 and b_2 differ substantially and visit times are fairly far apart. We remark that patterns in nonparametric estimates of $\Lambda(t)$ related to those in Figure 7.12 would suggest that LTF rates should be investigated. To assess potential bias as expressed in (7.41), we require estimates of the LTF intensities $\alpha_1(t)$ and $\alpha_2(t)$. Since we can follow individuals after failure has occurred, it is in principle possible to estimate them from the observed data; see Problem 7.7. However, if the observation times a_j are too far apart there will be rather little information about the LTF intensities, unless we make parametric assumptions that cannot be fully checked. Moreover, it is possible that the loss to follow-up intensity from state 2 is non-Markov; it may depend on the time of entry to state 2. In many cases the best we may be able to do is to estimate $\alpha_1(t)$ and $\alpha_2(t)$ using tentative models suggested by contextual information. Then we can assess bias in naively estimating $\lambda(t)$ under the conditional LTF assumption, and can also use the model in Figure 7.11 to estimate $\lambda(t)$.

We make three additional remarks. First, the exact time of entry to state 2 may be unobservable, and thus interval-censored, in some situations. In that case a formula for $P_{12}(a_{j-1}, a_j)$ analogous to that in (7.41) can be obtained under Markov assumptions (see Problem 7.7 (c)). Second, in the framework here we do not observe the exact time of LTF (i.e. entry to state 3). This is standard with intermittent observation; in fact an exact LTF time may be hard to define, and all we know for certain is that a person is LTF at a visit time a_j. The intensities $\alpha_1(t)$ and $\alpha_2(t)$ can in this case be viewed as functions that define transition probabilities $P_{13}(a_{j-1}, a_j)$ and $P_{23}(t, a_j)$. Finally, we have assumed above that $\lambda(t) = \lambda_T(t)$, which holds under independent censoring in continuous time. We note that this assumption is uncheckable from the observed data even if observation of the full process is continuous; we are unable to rule out the possibility that

$$\lim_{\Delta t \downarrow 0} \frac{P(T < t + \Delta t \mid T \ge t, Z(t^-) = 3)}{\Delta t} \ne \lambda(t).$$

It is possible in many settings to trace some proportion of individuals after they become LTF, at some additional cost. In this case, we can in particular trace persons last seen in state 1, and determine whether they have entered state 2 by an administrative end-of-follow-up time. This allows us to assess the independent censoring assumption and whether $\lambda_T(t) = \lambda(t)$. Tracing has been considered for the failure

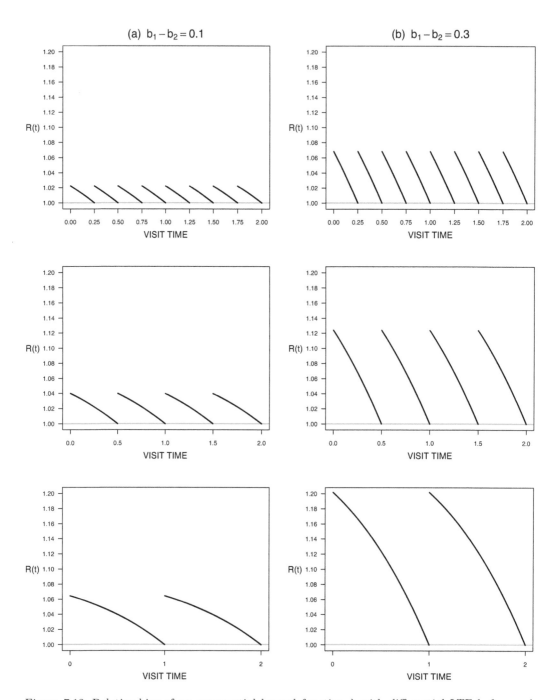

Figure 7.12: Relative bias of an exponential hazard function λ with differential LTF before and after failure.

time setting by Frangakis and Rubin (2001) and others. We discuss tracing next in the more general multistate model context.

7.2.5.2 An Illness-Death Setting with Non-Independent LTF

We consider now a situation with an illness-death process. Individuals who are alive are observed intermittently at times a_j, and it is determined whether they are in state 1 or state 2 (the "healthy" and "illness" states, respectively). Exact times of entry to state 3 are observed for all individuals, and for simplicity we assume that any delays in reporting death are small and negligible. There is an administrative end-of-follow-up time C_i^A for each individual, assumed independent of their process history, but some individuals may become LTF at a random time $C_i^R \leq C_i^A$. We consider here the concern that C_i^R and the disease process $\{Z_i(t), t > 0\}$ may not be (conditionally) independent.

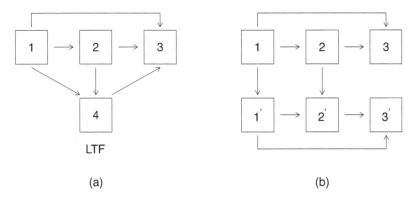

Figure 7.13: An illness-death process with LTF states.

The multistate models shown in Figure 7.13 represent this setting. The model (a) includes a single LTF state, which a person can enter only from states 1 or 2; we assume this because a person who enters state 2 before LTF will have their subsequent entry to state 3 reported. If a person becomes LTF before entry to state 2 has been observed, we may not know if they entered state 4 from state 1 or state 2. Similarly, such a person whose death is subsequently reported may or may not have entered state 2. This model allows us to assess effects associated with intermittent observation, as in the preceding section, but it does not allow us to assess the possibility or effects of non-independent LTF. Model (b) is a more detailed model that allows this; it is useful when individuals may be traced, and their histories determined, after LTF. For it we specify three states $1'$, $2'$, $3'$ for persons who are healthy, ill and dead but also LTF. Model (b) can be used to assess independence of LTF if we can obtain process histories after LTF by tracing some randomly chosen group of individuals. In particular, this would allow us to assess whether $\lambda_{kl}(t) = \lambda'_{kl}(t)$ for $(k,l) = (1,2)$, $(1,3)$ and $(2,3)$, where the $\lambda'_{kl}(t)$ are the intensities following LTF. Note that the $\lambda_{kl}(t)$ conceptually represent the transition intensities $\lambda^\circ_{kl}(t)$ that describe the basic illness-death process (in the absence of any LTF). However, if LTF is related to this process in arbitrary ways, then the $\lambda_{kl}(t)$ in model (b) do not in general equal the $\lambda_{kl}^{\circ}(t)$. If analysis suggests that $\lambda_{kl}(t) \neq \lambda_{kl}'(t)$, a question is whether the $\lambda_{kl}^{\circ}(t)$ can be estimated. If we select a random group of individuals who will be followed until their administrative end-of-follow-up time C_i^A no matter what happens, then there is in effect no LTF for such persons, and so the $\lambda_{kl}^{\circ}(t)$ can be estimated by using only data collected from them and these estimates can be compared with ones estimated from the remainder of the cohort under the independent LTF assumption. On the other hand, if LTF is allowed for all individuals, but a fraction of those LTF is traced, then we can estimate all of the intensities in Figure 7.13(b). This allows us to assess whether $\lambda_{kl}(t) = \lambda_{kl}'(t)$; if this is deemed (approximately) correct, then the "independent LTF" estimates of the $\lambda_{kl}(t)$ apply to the $\lambda_{kl}^{\circ}(t)$. However, if $\lambda_{kl}(t) \neq \lambda_{kl}'(t)$, then there is no way to estimate the $\lambda_{kl}^{\circ}(t)$. In the failure time context, marginal distributions for T have been estimated (e.g. Frangakis and Rubin, 2001) but these apply to a conceptual model where LTF occurs, and T and C^R are non-independent. This is not the same as a process where no premature LTF is possible. Finally, the discussion here has assumed the $\lambda_{kl}'(t)$ do not depend on the time of LTF; this could be untrue in some situations.

We conclude this section by outlining how estimation for the model in Figure 7.13(b) can proceed when a tracing program is used. Let C_i^A represent the administrative censoring time for individual i, and let $C_i^R < C_i^A$ represent a premature LTF time. Letting T_{i3} and T_{i3}' denote times of entry to states 3 and 3', we let $L_i = \min(T_{i3}, C_i^R, C_i^A)$ and $\Delta_i = I(L_i = C_i^R)$. We now assume that some individuals with $\Delta_i = 1$ are randomly selected for tracing; with selection indicators $R_i = I(\text{selected for tracing})$, we assume that R_i depends only on the observed process history up to C_i^R. That is, $\pi_i = P(R_i = 1 \mid \Delta_i = 1, C_i^R, \mathcal{H}_i(0, C_i^R))$, where $\mathcal{H}_i(s,t)$ stands for process history over $(s,t]$. Thus, information after time C_i^R for persons with $\Delta_i = 1$, $R_i = 0$ is missing at random, and so the following likelihood applies:

$$L = \prod P(\mathcal{H}_i(0, L_i)) \prod_{\Delta_i=1, R_i=1} P(\mathcal{H}_i(C_i^R, L_i') \mid \mathcal{H}_i(0, C_i^R), C_i^R) \qquad (7.42)$$

where $L_i' = \min(T_{i3}', C_i^A)$. This allows estimation of all intensities used to specify model (b) in Figure 7.13, with the usual caveat that parametric assumptions are needed in order to obtain estimates for all $t > 0$. Such models should naturally be checked, but this may be difficult to do comprehensively.

7.3   Bibliographic Notes

Keiding (1991) gives a detailed account of the relation between age-specific incidence and prevalence in a population based on a cross-sectional sample; see also Diamond and McDonald (1992). The need for auxilary data on mortality rates to support inferences on disease incidence, pointed out in Section 7.1, has been noted often including Keiding et al. (1989). Keiding (2006) considers the issue of modeling life history processes observed under complex sampling schemes with particular reference to processes with different time scales and types of trends. The utility of the Lexis diagram for discussing the impact of response-based sampling schemes for a population of multistate processes has been pointed out by several authors including

Lund (2000); see Keiding (2011) for a historical account. For a recent survey of length-biased observation in survival analysis see Shen et al. (2017). Boudreau and Lawless (2006) and references therein consider survival analysis when individuals are sampled within a complex survey, and Hajducek and Lawless (2012, 2013) consider duration analysis in longitudinal surveys.

Klein and Moeschberger (2003) discuss methods for dealing with truncated data in the analysis of multistate processes. Turnbull (1976) and Frydman (1994) develop maximum likelihood estimation of a survival distribution based on interval-censored and interval-truncated data. Alioum and Commenges (1996) extend these methods to deal with semiparametric regression. Pan and Chappell (1998) consider computational aspects for fitting the Cox model. Kalbfleisch and Lawless (1989) consider the implication of right-truncation schemes on the estimation of onset and disease duration distributions. Gill and Keiding (2010) and Kvist et al. (2010) discuss duration distribution estimation in a variety of settings involving history- and state-dependent sampling schemes. Aalen and Husebye (1991), Hamerle (1991), Lawless and Fong (1999) and references cited by these authors discuss the implications of selection conditions on models accommodating heterogeneity and point out how to construct likelihoods for inference.

Two-phase studies including outcome-dependent sampling were discussed by Lawless et al. (1999), who described both likelihood and weighted estimating function methodology. For more on maximum likelihood, see, for example, Zhang and Rockette (2005), Zhao et al. (2009) and Zeng and Lin (2014), and for weighted estimating functions, Whittemore (1997) and Lumley et al. (2011). There is a large literature on failure time analysis based on case-cohort and case-control designs, with much of it focused on the Cox model (e.g. Prentice, 1986; Kalbfleisch and Prentice, 2002, Section 11.4; Samuelsen et al., 2007; Breslow et al., 2009; Borgan and Samuelsen, 2014; Ding et al., 2017); Zeng and Lin (2014), Kim et al. (2016) and others consider transformation models. Saarela et al. (2008) and von Cube et al. (2017) consider case-cohort designs with multiple event times and multistate models, respectively. The collection of auxiliary data for assessing dependent loss to follow-up (LTF) through Phase 2 sampling has mainly been applied to failure time studies (e.g. Baker et al., 1993; Frangakis and Rubin, 2001; Farewell et al., 2003). Likewise, joint models for LTF and failure that allow assessment of independence and sensitivity analyses under non-independence have been proposed (e.g. Slud and Rubinstein, 1983; Scharfstein and Robins, 2002; Siannis, 2011). Lange et al. (2015) consider a model that includes non-CIVPs for intermittent observation. The discussions here of general multistate models and LTF are extensions of previous approaches (e.g. Barrett et al., 2011; Lawless, 2013) and recent work (Lawless and Cook, 2017a,b).

7.4   Problems

Problem 7.1   Suppose that the onset time distribution for U_0 in Section 7.1.3 is uniform over (B_0', B_0) for persons selected at B_0. Show in this case that

$$L_1 = \frac{\mathcal{F}(B_0 - u_0)}{\int_{B_0'}^{B_0} \mathcal{F}(B_0 - u_0)\, du_0}, \qquad L_2 = \frac{f(\tilde{t})^{\delta_2}\, \mathcal{F}(\tilde{t})^{1-\delta_2}}{\mathcal{F}(B_0 - u_0)}.$$

Show also that if $B_0' \to -\infty$ then

$$L = \frac{f(\tilde{t})^{\delta_2}\, \mathcal{F}(\tilde{t})^{1-\delta_2}}{\mu},$$

where $\mu = E(T)$. Finally, show that the marginal density for T, given that a person is sampled, is $P(T = t \mid B_0 < u_1) = t\, f(t)/\mu$. This is called a length-biased density and reflects the fact that persons with longer times between onset and failure have a greater probability of being sampled.

(Section 7.1)

Problem 7.2   Kessing et al. (2004) consider data from the Danish Psychiatric Central Registry which has information on all admissions from January 1, 1994 to December 31, 1999, restricted to individuals who had a first discharge with a primary diagnosis of affective disorder. The sample was comprised of 10,523 patients who ultimately experienced an average of 1.6 readmissions over this period. If B is the calendar time of birth, T_{kA} and T_{kD} are the ages of the kth admission and discharge from hospital, then the entry criterion is $T_{1D} \in [B_L - B, B_R - B]$, where B_L and B_R denote January 1, 1994 and December 31, 1999, respectively (see Figure 7.14). Data on subsequent hospitalizations and death are available over this study period.

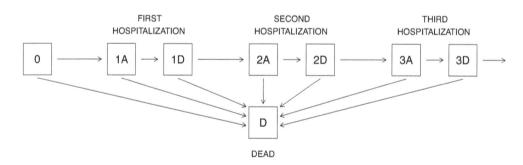

Figure 7.14: Multistate diagram for recurrent hospitalization and death for individuals with affective disorder.

(a) Write the likelihood accounting for the selection condition.

(b) Psychiatrists are interested in the possibility that the rate of hospital readmission may increase with longer duration of the disorder but as discussed in Chapter 6 it can be difficult to distinguish between heterogeneity in the course of the disorder and trends of this sort. Discuss modeling strategies to investigate this question.

(c) Comment on how an effect of the disorder on mortality over the disease course could further complicate the study of this phenomenon.

(Sections 7.1, 7.2)

Problem 7.3 Consider individuals in the University of Toronto Psoriatic Arthritis Clinic (UTPAC) discussed in Section 7.1.1. Suppose a 3-state model shown in Figure 7.15 is used to model the onset of psoriatic arthritis and death, and let $\{Z_0(s), s \geq 0\}$ denote this process for an individual labeled "0" in the registry; T_{0k} denotes the entry time to state k, $k = 1, 2$. Suppose this individual was aged A_0 at the time they were screened for recruitment; it is necessary that $Z_0(A_0) = 1$ as they would otherwise not have been selected, and we suppose T_{01} is available retrospectively for such a recruited individual. We further suppose that individuals in the registry are followed prospectively over a period (A_0, C_0) of their lifetime.

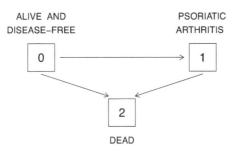

Figure 7.15: A 3-state illness-death model for modeling the onset of psoriatic arthritis and death in an individual.

Preliminary studies about the genetic basis for disease often involve random selection of a diseased individual, called a proband, from such a registry and recruitment of some (say m) of their family members to obtain information on their health and disease course. The goal is to examine the within-family association in disease status. Let T_{jk} denote the entry time to state k for the jth member in a family of $m + 1$ individuals $j = 0, 1, 2, \ldots, m$. Let C_j be the age of family member j at the time the family study was conducted. We then observe $\{Z_0(s), 0 \leq s < C_0\}$ subject to $Z_0(A_0) = 1$ for the proband sampled from the registry, and $(Z_j(C_j), C_j)$ for their non-proband family member j, subject to $Z_0(A_0) = 1$, $Z_j(C_j) \neq 2$, $j = 1, \ldots, m$ as they must be alive to be examined.

(a) Write the likelihood contribution for a family recruited to this study, conditional on $C_1, \ldots, C_m$ and the information in the proband.

(b) Consider approaches to modeling within-family association in the onset-time distribution for PsA.

(c) This sampling scheme offers no information about disease-free mortality rates (i.e. the $0 \to 2$ transition rate in Figure 7.15). What additional kinds of data would be useful to estimate parameters in the joint model you consider most suitable in (b)?

(Section 7.1)

Problem 7.4   Consider the model of Figure 7.2. Interest lies in assessing how the risk of psoriatic arthritis depends on the onset time of psoriasis through specification of the $1 \to 2$ intensity as $\lambda_{12}(t|\mathcal{H}(t^-)) = \lambda_{12}(t|t_1)$. Suppose a population is screened, and for individuals found to have psoriatic arthritis we are able to retrospectively obtain age of onset of both psoriasis and psoriatic arthritis. Consider $\lambda_{12}(t|t_1) = \lambda_{12}(t) \exp(\beta\, g(t_1))$ and assume values for $\lambda_{k3}(t)$, $k = 0,1,2$ exist from external sources. Compare the relative efficiency of estimates of β from likelihoods L_2 and $L = L_1 \cdot L_2$ of Section 7.1.3.

(Section 7.1.3)

Problem 7.5   Consider a 3-state progressive process with conditional transition intensities $\lambda_{12}(t|v) = v\,\lambda_{12}(t)$ and $\lambda_{23}(t|v,t_2) = v\,\lambda_{23}(t|t_2)$, where v is an unobserved random effect and t_2 is the time of entry to state 2.

(a) If individuals selected for a study must be in state 2, determine the form of the likelihood based on prospective follow-up for C years, or until entry to state 3. Consider the cases where

(i) t_2 is known,
(ii) t_2 is unknown.

(b) If t_2 is observable, $\lambda_{23}(t|t_2) = \lambda_{23} \exp(\beta(t - t_2))$ and $V \sim \text{Gamma}(\text{mean} = 1,\ \text{variance} = \phi)$, obtain the likelihood for λ_{23}, β and ϕ and examine the information concerning β and ϕ.

(Section 7.1.5)

Problem 7.6   Consider the setting of Example 7.2.2, involving a 5-state progressive model, and a two-phase study. Let p_1 and p_2 denote the phase 2 sampling probabilities for individuals in strata $S_{(1)}$ and $S_{(2)}$.

(a) Determine the sampling probabilities conditional only on X_i, i.e. $P(R_i = 1|X_i)$.
(b) Determine the conditional likelihood function

$$L_C = \prod_{i:R_i=1} P(\mathcal{H}_i(C_i^A) \mid X_i, R_i = 1),$$

where here, $C_i^A = 8$ years for all individuals. This conditional likelihood does not depend on a model for the covariate X_i, unlike the full likelihood.

(Section 7.2)

Problem 7.7

(a) For the LTF model described in Figure 7.11 of Section 7.2.5.1, derive (7.41) and then show that $\lambda^*(t) = \lambda(t)$ if and only if the censoring rates $\alpha_1(t)$, $\alpha_2(t)$ are equal.
(b) Generalize the discussion to the case where only the states $Z(a_j)$ are observed, and not the exact transition times.

(c) Discuss how $\alpha_1(t)$ and $\alpha_2(t)$ may be estimated in the situation here.

(Section 7.2.5.1)

Problem 7.8 A form of auxiliary data when observation is intermittent is to obtain information on the reasons why an individual had a visit at time a_j. In a very simple situation, we might be able to classify a visit at a_j as either scheduled (S) or disease driven (DD). Scheduled visits satisfy the CIVP conditions of Section 5.4, but DD visits usually would not. Suppose that DD visits can be modeled using the process in Figure 5.4; for simplicity consider the case where $K = 3$, and suppose that all transition intensities are of time-homogeneous form λ_{kl} $(k \neq l)$.

(a) Explain why this model includes CDVP cases.

(b) Give a sufficient condition for the visit process to be a CIVP.

(c) Note that visit times A_j $(j = 1, 2, ...)$ correspond to the times of downward transitions in Figure 5.4. Use this to write a likelihood function based on observed data $\mathcal{Z}(a_m) = \{(a_j, Z(a_j)), j = 1, ..., m\}$.

(Sections 5.4.1, 7.2; Lange et al., 2015)

Problem 7.9 Consider the disease process of Figure 7.2 and suppose interest lies in estimating the effect of a binary genetic marker X on the risk of psoriatic arthritis among individuals with psoriasis; we simplify the problem by setting the mortality rates to zero and ignore the complications from deaths. We consider a simple model with $\lambda_{12}(t|x) = \lambda_2$ and $\lambda_{23}(t|x) = \lambda_3 \exp(\beta x)$. Suppose a random sample from the population is screened and a sub-sample of the individuals found to have psoriasis are recruited to a registry of psoriasis patients in which they retrospectively report their age at psoriasis onset, provide a sample for genetic testing, and are examined annually for 10 years for the development of psoriatic arthritis. A sub-sample of the individuals screened and found to have psoriatic arthritis are recruited to a psoriatic arthritis registry in which they retrospectively report their ages at onset of psoriasis and psoriatic arthritis and provide a sample for genetic testing. Suppose $\lambda_2 = \log 2/40$ so the median age of onset is 40, and let the intensity for the transition from the psoriasis state to the psoriatic arthritis state be $\lambda_3 = R \times \lambda_2$.

(a) Let n_k denote the number of individuals found to be in state k at screening and recruited to the study, $k = 1, 2$. Write the likelihood under this sampling scheme making use of the data provided by the recruited individuals.

(b) Suppose the cost of recruiting an individual is D_r for both registries, the cost of the genetic test is D_g, and each annual assessment costs D_a. With a sample size of $n = n_1 + n_2 = 5000$ and a total budget D, what fraction of individuals should be recruited to the psoriasis registry to optimize the efficiency of the estimator for β?

(Section 7.1)

Chapter 8

Additional Topics

8.1 Analysis of Process-Related Costs and Benefits

Most life history processes have costs or benefits associated with them. For example, individuals with chronic disease require medicine, surgical interventions or hospitalization, each of which lead to financial costs to the health care system or health insurance agencies. Moreover, the pain or functional impairment arising from chronic disease may affect an individual's ability to carry out activities of daily living and their ability to maintain employment. Processes related to education and employment have costs for education or training and benefits from employment. In this section we describe some basic settings and analysis based on multistate models.

8.1.1 Individual-Level Models

We assume a life history process of interest is represented by a multistate model, with $\{Z(s), s > 0\}$ the history of states occupied for a given individual. We assume there is a related process $\{\mathcal{U}(s), s > 0\}$ that depends on the life history process; specifically, $\mathcal{U}(t)$ represents accumulated costs or benefits that accrue from the life history process. In many settings, cost accumulates roughly linearly according to the state occupied; in particular, there is a given cost rate u_k when an individual is in state k, and so

$$\mathcal{U}(t) = \int_0^t \sum_{k=1}^K u_k \, Y_k(s) \, ds \,, \tag{8.1}$$

where $Y_k(t) = I(Z(t) = k)$. Two features of interest for the cost process are the mean cost function and the distribution function of total cost up to a given time, denoted respectively, as

$$\mu(t) = E\{\mathcal{U}(t)\} = \int_0^t \sum_{k=1}^K u_k \, P_k(s) \, ds \tag{8.2}$$

$$F_{\mathcal{U}(t)}(w) = P(\mathcal{U}(t) \leq w) \,. \tag{8.3}$$

For simplicity, we assume that $Z(0) = 1$ so that $P_k(s) = P_{1k}(0, s)$, but this is easily modified. Fixed covariates may also be introduced in which case we write $\mu(t|x)$, $F_{\mathcal{U}(t)}(w|x)$ and so on.

Given state occupancy probability estimates $\widehat{P}_k(s)$, $0 < s \leq t$, we can obtain an estimate $\hat{\mu}(t)$. This was illustrated in Section 4.2.3 in a context that we revisit in the next section. The variance or asymptotic distribution of $\hat{\mu}(t)$ is difficult to obtain in general, so we use a nonparametric bootstrap for variance estimates and confidence intervals. For $F_{\mathcal{U}(t)}(w)$, even computing an estimate is intractable in most cases, but simulation can be employed. By generating B realizations of $\{Z(t), t > 0\}$ using the fitted multistate model we can estimate $F_{\mathcal{U}(t)}(w)$ as

$$\tilde{F}_{\mathcal{U}(t)}(w) = \frac{1}{B} \sum_{b=1}^{B} I(\mathcal{U}_b(t) \leq w), \qquad (8.4)$$

where $\mathcal{U}_b(t)$ is obtained from (8.1) using the bth simulated process history.

We remark that more complicated cost processes may be needed in some settings. For example, there may be a fixed cost u_{kl} associated with $k \to l$ transitions. In that case, we have

$$\mathcal{U}(t) = \sum_{k=1}^{K} u_k S_k(t) + \sum_{k \neq l} u_{kl} N_{kl}(t), \qquad (8.5)$$

where $S_k(t) = \int_0^t Y_k(s)\, ds$ is total sojourn time spent in state k and $N_{kl}(t)$ is the number of $k \to l$ transitions up to time t. In simple cases $E\{N_{kl}(t)\}$ and thus $E\{\mathcal{U}(t)\}$ are easily calculated; more generally, simulation can be used to approximate $E\{\mathcal{U}(t)\}$. There are also many situations where costs u_k or u_{kl} might vary across individuals. In that case, we can adopt distributions, with costs u_{ik} and u_{ikl} for individual i having distributions $G_k(u)$ and $G_{kl}(u)$, respectively. If u_{ik} and u_{ikl} are independent of the multistate process $\{Z_i(s), s > 0\}$, expected costs are still readily calculated. Still more complex situations where costs are related to prior as well as current life history are beyond our scope here, but in principle can be considered through joint processes with

$$P(\Delta N(t), \Delta \mathcal{U}(t) \mid \mathcal{H}(t^-)) = P(\Delta N(t) \mid \mathcal{H}(t^-)) \, P(\Delta \mathcal{U}(t) \mid \mathcal{H}(t^-), \Delta N(t)),$$

where $N(t)$ counts transitions in the multistate process and $\mathcal{H}(t) = \{N(s), \mathcal{U}(s), 0 \leq s \leq t\}$.

8.1.2 *Quality of Life Analysis and Breast Cancer Treatment*

We briefly revisit the therapeutic breast cancer trial described in Section 4.2.3. Patients were randomly assigned to either short- or long-duration chemotherapy, and the 4-state model in Figure 4.4 was used to conduct a quality of life (QOL) analysis. This was based on different QOL scores u_1, u_2, u_3, u_4 associated with the four states. While undergoing treatment, a person is in the "toxicity" state 1, and this and the "relapse" state 3 have lower scores than the "remission" state 2. State 4, death, has a zero score. Expression (4.27) is a version of (8.2) with treatment covariate X added, with $X = 0$ and $X = 1$ denoting short- and long-duration chemotherapy, respectively. Figure 4.6 showed estimates $\hat{\mu}(t|0)$ and $\hat{\mu}(t|1)$ that were based on methods of occupancy probability estimation introduced in Section 4.2. The data consist of process histories for 413 persons assigned short-duration and 816 persons assigned long-duration therapy, and the QOL scores used were $u_1 = 0.1$, $u_2 = 0.5$, $u_3 = 0.1$, $u_4 = 0$. Here we fit Markov models to the data. First, separate Aalen-Johansen nonparametric estimates $\widehat{P}_k(0, s)$ are obtained using the R etm package described in Section 3.4.3 for each treatment group ($X = 0, 1$). These give estimates $\widehat{\mu}(t|0)$ and $\widehat{\mu}(t|1)$ based on (8.2) that are indistinguishable from the ones in Figure 4.6.

We recall that subjects in the short-duration group have higher cumulative QOL initially but after about 48 months the long-duration group's QOL is higher. This is attributable to the fact that the short-duration group spends less time in state 1 but also that a higher proportion of that group experience relapse and death. Bootstrap standard errors given in Section 4.2.3 indicate that the difference in $\mu(t|0)$ and $\mu(t|1)$ at $t = 84$ months is statistically significant. A more detailed comparison of the treatment groups can be made by fitting multiplicative Markov models,

$$\lambda_{k,k+1}(t \mid X) = \lambda_{k0}(t) \exp(\beta_k X) \quad k = 1, 2, 3 \tag{8.6}$$

using the R/S-PLUS function coxph (see Section 3.3). Table 8.1 shows the estimates $\widehat{\beta}_k$. We see that the long-duration group has lower rates of transition into states 2 and 3 but, interestingly, a higher rate of transition from relapse to death.

Table 8.1: Treatment effects on transition intensities.

Transition	EST ($\widehat{\beta}_k$)	SE($\widehat{\beta}_k$)	RR[†]	95% CI for RR
$1 \to 2$	-2.190	0.101	0.11	(0.09, 0.14)
$2 \to 3$	-0.442	0.081	0.64	(0.55, 0.75)
$3 \to 4$	0.373	0.093	1.45	(1.21, 1.74)

[†] RR $= \exp(\widehat{\beta}_k)$; 95% CI based on $\exp(\widehat{\beta}_k \pm 1.96 \, \text{s.e.}(\widehat{\beta}_k))$.

8.1.3 Individual-Level Decision-Making

In some settings, decisions concerning interventions for individuals can be guided by estimation of life history paths expected under alternative actions (interventions). Covariates that may help to differentiate among potential outcomes are then useful. A simple but important example concerns individuals wait-listed for a kidney transplant (Wolfe et al., 2008). When an individual is on the wait list, they may at any point in time (a) receive a transplant or (b) die; this is portrayed in Figure 8.1. A third possibility is that they are taken off the wait list; we ignore this since our interest is in the effect of receiving a transplant at a given time. In particular, suppose that a donated kidney becomes available at time t. A quantity that can help inform the decision as to which person on the wait list should be first offered the transplant is the difference between ELT = expected remaining lifetime if the transplant is received at time t and ELW = expected remaining lifetime if the person remains on the wait list (does not receive the transplant). Persons for whom ELT $-$ ELW is large might be viewed more favourably, though other factors are also relevant. To calculate this quantity we need to know transition intensities, and these should incorporate covariates related to the patient and the donated organ. For the case where the patient remains on the wait list, we want to incorporate factors related to their time to death and time of being offered another donated organ. Problem 8.1 gives additional examples related to alternative forms of treatment for disease.

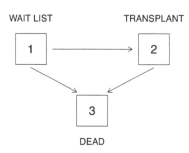

Figure 8.1: A model for persons wait-listed for a transplant.

8.1.4 Population-Level Cost Analysis

Multistate models are often used to assess aggregate public health costs associated with chronic conditions. One approach is through individual-level models combined with information about the age structure of a population. For example, the model shown in Figure 8.2 has been used to forecast the burden from dementia and Alzheimer's disease in different parts of the world (Brookmeyer et al., 2007). At the individual level, the model would be formulated in terms of age, using age-specific intensities relevant to a specific region or country. Such models could include risk factors or stratification variables, provided there were sufficiently accurate information on which to base the intensities. Given the numbers of persons of different ages in the population, one could then forecast the numbers of persons in each state at some future time. For example, let $Z(t)$ denote an individual's state at calendar time t, and let $A(t)$ denote their age. Then, a Markov model for the process in Figure 8.2 allows us to compute $P(Z(t_1) = k | Z(t_0), A(t_0))$, where $t_0 < t_1$. In a population of N individuals, we can then calculate the expected number of persons in each state k at time t_1, given conditions at time t_0. This is

$$E_k(t_1 \mid \mathcal{H}(t_0)) = \sum_{i=1}^{N} P(Z_i(t_1) = k \mid Z_i(t_0), A_i(t_0)), \qquad (8.7)$$

where $\mathcal{H}(t_0)$ refers to the information used at time t_0 and for simplicity we suppress notationally covariates or stratification factors. We would not generally have information on all persons in a population, so in practice (8.7) is normally replaced with something like

$$E_k(t_1 \mid \mathcal{H}(t_0)) = \sum_{r=1}^{R}\sum_{l=1}^{K} P(Z(t_1) = k \mid Z(t_0) = l, A(t_0) \in A_r)\, n_{rl}(t_0), \qquad (8.8)$$

where $\{A_1,\ldots,A_R\}$ is a partition of the age range for the population, and $n_{rl}(t_0)$ is the number of persons in age interval A_r who are in state l at time t_0.

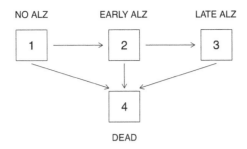

Figure 8.2: A model for Alzheimer's disease.

Given good estimates of the prevalences $n_{rl}(t_0)$ for the states 1–3 shown in Figure 8.2, (8.8) can be obtained. Such prevalence estimates are often available from health surveys or administrative data, but their accuracy should be considered carefully, and the sensitivity of projections to variation in them examined. Costs can also be incorporated; this is most simply done by assigning a fixed cost to each combination (r, k, l) for the time interval (t_0, t_1). A valuable aspect of the approach here is that we can also examine the effect of public health programs or interventions that would reduce one or more intensities, such as those in Figure 8.2.

There is a large literature on "microsimulation", which considers the generation of life history paths for individuals within populations (e.g. Lymer et al., 2016). Much of this is based on discrete-time models, but see also Zinn (2014). In the multistate context, this typically involves the consideration of individuals at discrete ages such as 5, 10, 15, ... years and then multinomial generalized linear models are used for $Z(a_r)$, $r = 1, 2, \ldots$, given $Z(a_{r-1})$ and any other factors considered. Simulation for the continuous-time case was considered in Section 8.1.1 and more generally in Section 2.5. Confidence or prediction limits for state prevalences or related costs can also be obtained via simulation, as discussed in Section 8.1.1.

8.2 Prediction

Multistate models are often used for prediction; this was discussed briefly in Section 3.5.3. As illustrated in Sections 8.1.3 and 8.1.4, a fundamental problem is as follows: given an individual's observed history $\bar{\mathcal{H}}^\circ(t_0) = \{\mathcal{Z}(t_0), \mathcal{X}(t_0)\}$ of states and covariates up to time t_0, predict their future life history $\{Z(s), s > t_0\}$. Often a specific time $t_1 > t_0$ is singled out, in which case we wish to predict $Z(t_1)$. This can be done by specifying a model $\tilde{P}(Z(t_1) = z | \bar{\mathcal{H}}^\circ(t_0))$; we use the tilde to indicate that the model is in all practical cases an estimate based on analysis of data, model fitting and contextual information. This distinguishes it from the true probability distribution $P(Z(t_1) = z | \bar{\mathcal{H}}^\circ(t_0))$, which refers to the population of processes under consideration; we note that this population is usually conceptual and hard to specify rigorously. For example, in the context of Alzheimer's disease portrayed in Figure 8.2 and discussed in Section 8.1.4, we may wish to predict an individual's state at age 80, given their life history up to age 70. In some settings we might focus on a specific absorbing state, say k, and thus, $P(Z(t_1) = k|\bar{\mathcal{H}}^\circ(t_0))$; the time T_k of entry to state k can also be considered, with $P(T_k \leq t|\bar{\mathcal{H}}^\circ(t_0)) = P(Z(t) = k|\bar{\mathcal{H}}^\circ(t_0))$.

When used for prediction, we refer to a model $\tilde{P}(Z(t) = z|\bar{\mathcal{H}}^\circ(t_0))$ as a predictive model. How to assess the predictive power or performance of such models is a topic that has received much attention. Two aspects of performance are termed *calibration* and *sharpness* (Gneiting, 2014). Calibration refers to whether the probabilities $\tilde{P}$ from a predictive model closely approximate the corresponding true probabilities P, either marginally or conditional on certain factors associated with X. Sharpness on the other hand refers to whether covariates or initial conditions $\bar{Z}(t_0)$ are highly predictive of which state a person will be in at some later time. Calibration is important so that models do not give systematically biased predictions, but sharpness is elusive in many life history processes, since it requires that the true processes have tightly linked causal mechanisms with identifiable factors. In the Alzheimer's disease example of Figure 8.2, for example, predictive probabilities $\tilde{P}(Z(80) = 4|Z(70) = k, X)$ of death by age 80 given a person's state $(k = 1, 2, 3)$ at age 70 and their sex, $X = I(\text{female})$, would be well-calibrated if they are close to the actual proportions of females and males dying by age 80 in the population. However, they would not necessarily be "sharp" in the sense of identifying which persons at age 70 were more or less likely to die by age 80. A better model for doing this should include additional covariates.

The performance of predictive models can be measured using scoring rules (Gneiting, 2014), which combine the aspects of calibration and sharpness. A number of scoring rules or functions are commonly used, depending on the type of variable or outcome being predicted. The Brier score was introduced in Section 3.5.3; it is widely used with binary response variables Y. We focus here on the widely applicable logarithmic score, defined as follows: consider a (possibly multivariate) variable Y and associated predictive probability density or mass function $\tilde{f}(y|x)$, where x denotes covariates or other factors used in the model. The logarithmic score that is assigned when the outcome y is observed is

$$\text{LS}(\tilde{f}, y) = -\log \tilde{f}(y \mid x), \qquad (8.9)$$

where for convenience we consider y discrete, so that $0 \leq \tilde{f}(y|x) \leq 1$. Note the intuitive appeal of (8.9): the LS is large (and positive) when $\tilde{f}(y|x)$ is small and LS is small (close to zero) when $\tilde{f}(y|x)$ is close to one. Thus, (8.9) can be thought of as a loss function, with larger losses corresponding to cases where an observed outcome y had a small predictive probability. We note that poorly calibrated predictive distributions tend to have larger scores or losses, but that in order to obtain very small losses, the model must be "sharp". In other words, the information in x must be highly predictive of the outcome Y. This is related to the concept of explained variation; see, for example, Lawless and Yuan (2010).

The proper use of (8.9) or the Brier score is on new cases that are separate from any data used in the development of $\tilde{f}(y|x)$. This includes situations where a model developed on some group of individuals is used on a new group. However, in life history contexts it can also be used to assess prediction within a group of individuals, based on information already observed. For a given individual i and multistate model $\tilde{P}$, we have

$$\mathrm{LS}(\tilde{P}, z_i(t_1)) = -\log \tilde{P}(Z_i(t_1) = z_i(t_1) \mid \bar{\mathcal{H}}_i^\circ(t_0)), \qquad (8.10)$$

where $z_i(t_1)$ is the observed state at time t_1. Writing (8.10) as a random function

$$\mathrm{LS}(\tilde{P}, Z_i(t_1)) = -\sum_{z=1}^{K} I(Z_i(t_1) = z) \log \tilde{P}(Z_i(t_1) = z \mid \bar{\mathcal{H}}_i^\circ(t_0)),$$

we see that models $\tilde{P}$ for which observed states tend to have higher probabilities yield smaller LS values.

The main use of scoring rules is in comparing alternative models, say $\tilde{P}_1$ and $\tilde{P}_2$. When this is done on new data we have an objective measure for which models giving smaller values are favoured. Nevertheless, LS is but one measure and if there were economic costs or benefits associated with predictions, we might prefer an alternative measure. This would be the case, for example, when a predictive model is used to make a decision at time t_0, with consequences assessed at time t_1. Problem 8.4 considers some alternative approaches when decisions are to be made. An important additional observation when a prediction addresses some specific state is that different models can be compared, as long as they each include the state in question. We provide an illustration in Section 8.2.1 below.

Logarithmic scores, also known as Kullback-Leibler (KL) scores, are often used to compare models within the same dataset used to develop the models. This is more properly viewed as estimation of the expected values of score $\mathrm{LS}(\tilde{f}, Y, X) = -\log \tilde{f}(Y|X)$, where the expectation is with respect to (Y, X) and the data $D = \{(Y_i, X_i), i = 1, \ldots, n\}$ from which $\tilde{f}(y|x)$ was developed (e.g. Lawless and Yuan, 2010). To do this "honestly" we should therefore use some form of cross-validation, in which the model $\tilde{f}(y|x)$ used to "predict" Y_i on the basis of $X_i = x_i$ is not based even partially on (y_i, x_i). This was described briefly in Section 3.5.3, and we add a few remarks here before providing an illustration. Leave-one-out cross-validation (LOOCV) uses the model $\tilde{f}_{(-i)}(y|x)$ to predict Y_i, for example, giving the total LS or KL score

$$\mathrm{LS} = -\sum_{i=1}^{n} \log \tilde{f}_{(-i)}(y_i \mid x_i). \qquad (8.11)$$

Here $\tilde{f}_{(-i)}$ refers to the model obtained from D but with (y_i, x_i) dropped. It may be noted that if we simply used the model $\tilde{f}(y_i|x_i)$ based on the full data D instead of $\tilde{f}_{(-i)}(y_i|x_i)$, then (8.11) would be the negative log-likelihood.

8.2.1 Viral Rebounds among Persons with HIV

In Section 1.2.4 we introduced a model, shown in Figure 1.4, that is useful when tracking viral load (VL) measurements in HIV-positive individuals, and in examining the occurrence of viral rebounds (VR), defined as the event when VL exceeds 999 copies of the virus per mL of blood. All individuals start their process when their initial combined anti-retroviral therapy (cART) achieves viral suppression (VL < 50 copies/mL), so $Z(0) = 1$ in each case. By fitting models $\tilde{P}$ to the process of

Figure 1.4 we can then compute predictive probabilities for the time T to viral rebound as

$$\tilde{P}_T(T \leq t \mid x) = \tilde{P}(Z(t) = 3 \mid x). \tag{8.12}$$

In (8.12) we use $\tilde{P}_T$ to refer to a model for T and $\tilde{P}$ for a multistate model; x refers to covariates known at time $t = 0$. Section 1.6.3 briefly describes the data available from a Canadian HIV cohort.

Here we consider models discussed by Nazeri Rad (2014) and Lawless and Nazeri Rad (2015) that were fitted to a cohort of 1035 men who started cART on or after January 1, 2005. The data are based on follow-up to February 28, 2010; cohort members were nominally seen every 3 or 4 months and viral load was measured, though the actual times between visits varies substantially. Analysis showed that Markov regression models with piecewise-constant intensities that changed at $t = 2$ years, as described in Section 2.3.1, fitted the data well. Covariates in the model were a person's age at baseline (time of cART initiation), whether they were injection drug users, and the type of cART received (protease inhibitor (PI) versus non-nucleoside reverse transcriptase inhibitor (NNRTI)). A time-dependent covariate that indicated whether a person had experienced an observed pattern (state 1, state 2, state 1) over three consecutive prior visits was also considered. This pattern was referred to as a "viral blip", and was found to be associated with a prolonged tendency for visits to state 2, which represents a detectable but moderate viral load. This may reflect the presence of unmeasured factors related to the variability of a person's viral load, for example, compliance with their treatment regimen. The baseline covariates also had significant effects, with younger age, injection drug use, and NNRTI treatment associated with higher state $2 \to 3$ intensity.

The estimated probabilities $\hat{P}_{13}(2|x)$ of a viral rebound within 2 years ranged from about 0.11 for non-injection drug users over 45 years of age and on PI treatment to about 0.31 for injection drug users under 45 at baseline, and on NNRTI treatment. We can compare the predictive power of alternative models, and for illustration we compare the 3-state model discussed above and a semiparametric Cox failure time model for the distribution of time T to viral rebound (VR – entry to state 3), using the same covariates as in the multistate model. We do this by predicting for each cohort member their VR status at each visit time a_{ij}, given observed covariate and process information up to time $a_{i,j-1}$ $(j = 1, \ldots, m_i)$, with $a_{i0} = 0$. We define $R_{ij} = I(T_i \leq a_{ij})$, the VR status at a_{ij}, and denote

$$\tilde{P}(R_{ij} = 1 \mid \bar{\mathcal{H}}_i^\circ(a_{ij}), x_{ij}) = \tilde{P}_{ij}, \tag{8.13}$$

where x_{ij} is the vector of baseline covariates plus the blip covariate at time $a_{i,j-1}$. We note that $\tilde{P}_{ij} = 1$ if $Z_i(a_{i,j-1}) = 3$ and so we define the logarithmic scores

$$\text{LS}(\tilde{P}_{ij}, R_{ij}) = -R_{ij} \log \tilde{P}_{ij} - (1 - R_{ij}) \log(1 - \tilde{P}_{ij}),$$

with $0 \log(0)$ defined as zero. The total logarithmic score for individual i is then

$$\text{LS}_i = \sum_{j=1}^{m_i} \text{LS}(\tilde{P}_{ij}, R_{ij}).$$

The $\tilde{P}_{ij}$ in (8.13) are based on a specific predictive model. As indicated, we consider the Markov multistate model discussed above, in which case $\tilde{P}_{ij} = \tilde{P}(Z_i(a_{ij}) = 3|Z_i(a_{i,j-1}), x_{ij})$ and the Cox model, in which case $\tilde{P}_{ij}$ takes the form

$$\tilde{P}(T_i \le a_{ij} \mid R_{i,j-1} = 0, x_{ij}) = 1 - \exp\Big((\tilde{\Lambda}_0(a_{ij}) - \tilde{\Lambda}_0(a_{i,j-1})) \exp(x'_{ij}\tilde{\beta})\Big)$$

and

$$\tilde{P}(T_i \le a_{ij} \mid R_{i,j-1} = 1, x_{ij}) = 1\,,$$

where $\tilde{\Lambda}_0(t)$ is the estimated baseline cumulative hazard function.

To obtain an honest predictive score, we use cross-validation (CV), for which the model giving the $\tilde{P}_{ij}$ $(j = 1, \ldots, m_i)$ does not use information from individual i. Nazeri Rad (2014) used 10-fold CV, where the 1035 persons were randomly split into 10 groups, $S_1, \ldots, S_{10}$, of size 103 or 104. Ten models of each type (multistate or Cox) were then fitted, using 10 datasets, with individuals in S_l dropped from dataset l $(l = 1, \ldots, 10)$. We denote the 10 models obtained as $\tilde{P}^{(-l)}$ $(l = 1, \ldots, 10)$ and then a total LS is computed for a given model as

$$\mathrm{LS_{CV}} = \frac{1}{10} \sum_{l=1}^{10} \sum_{i \in S_l} \sum_{j=1}^{m_i} \mathrm{LS}(\tilde{P}_{ij}^{(-l)}, R_{ij})\,. \tag{8.14}$$

This gave, for a single random partition of the cohort, the values $\mathrm{LS_{CV}} = 614.94$ and 617.72 for the multistate model and Cox model, respectively (Nazeri Rad, 2014). This favours the multistate model. However, it should be stressed that there is random variation in $\mathrm{LS_{CV}}$ in this situation, and that a different 10-fold partition would yield different $\mathrm{LS_{CV}}$ values. A way to avoid this is to use leave-one-out CV, but this requires fitting 1035 models of each type and so is computationally more onerous. Another procedure would be to partition time on study so as to consider real prediction. For example, we could use a model based only on observed data up to 1 year for each individual to predict outcomes at some later time. This approach requires parametric models in order to extrapolate beyond one year; a Cox model cannot do this.

8.3   Joint Modeling of Marker and Event Processes

8.3.1   Roles of Markers in Disease Modeling

An individual under study may generate several stochastic processes. In addition to a primary multistate process, an additional auxiliary process may seemingly play an important role and exploration of the relation between these processes may be of interest. Measurements on the realization of such an auxiliary process may be used, for example, to form internal time-dependent covariates. In the survival setting such measurements are only possible on the individual when they are alive and as such, the fact that measurement can be made at time t conveys the fact that the individual survived to $t > 0$; see Section 6.3.2 of Kalbfleisch and Prentice (2002). In the multistate setting, internal covariates may perhaps only be observable when the process is in a transient state and so the availability of a measurement on such a process may similarly convey information about the possible states of the response process.

Likelihood construction was discussed in Section 2.2.2 for settings with multistate processes and internal time-dependent covariates under right-censoring schemes. There we pointed out that if interest lies primarily in the effect of time-dependent covariates on the multistate process, then attention may be restricted to a partial likelihood, which is functionally dependent on the intensity of the multistate process alone. Internal covariate processes take an elevated importance, however, in situations where their values are closely tied to the underlying disease process of interest. In such settings the dynamic covariates are called *markers* and the stochastic process generating them is called a *marker process*. Joint models for markers and the multistate process may be of interest in such settings. There is a large literature on joint models, especially for failure time processes, but the complexity of marker processes and issues concerning their measurement render most approaches problematic. Space does not permit a thorough discussion of this challenging area, and we will only review some major points.

We first discuss some specific health research settings in which information from marker processes is relevant to scientific objectives.

1. In some contexts, marker values reflect the *health of an organ or organ system*. When the organ, say, is diseased then the marker value may characterize how well it is functioning. As an example, serum creatinine level is a measure of kidney function which becomes elevated once kidney damage has occurred. It can be useful for monitoring patients' health and the adequacy of their treatment once the disease has advanced to the point, for example, that they are on dialysis, but it does not play as useful a role in the early stages of nephropathy. In liver disease, alanine aminotransferase (ALT) is an enzyme that can be measured in the blood and is reflective of liver function; it can be useful in characterizing liver health over the full course of liver disease.

2. Markers often reflect *disease activity* (i.e. the extent to which affected body systems are in a period where the disease manifestations are more acute), and the relationship between markers and complications or disease outcomes is of key interest. The erythrocyte sedimentation rate (ESR) is the rate at which red blood cells sediment in one hour, and is a reflection of the extent of systemic inflammation of the body. It is used as a quantitative measure of inflammation in autoimmune disorders such as rheumatoid arthritis. High ESR values are known to be associated with more rapid joint damage, so control of inflammation through ESR-directed therapy is aimed at both reducing short-term joint swelling and pain, and prevention of joint destruction and functional impairment.

3. Markers often reflect the *burden of disease*. In patients with skeletal metastases, for example, one individual may have many small metastatic lesions while another may have a smaller number of larger lesions amounting to the same total volume of affected bone. Because the lesions can vary greatly in size, the lesion count does not adequately reflect the extent of bone disease, and it is challenging and prohibitively costly to measure total volume of each bone lesion.

N-telopeptide of type I collagen (Ntx) is a marker of bone resorption (and hence bone destruction) that is associated with the presence and volume of skeletal metastases in breast cancer patients (Lipton et al., 2000), and high values are also associated with increased risk of fractures and other skeletal complications (Coleman et al., 1997). Bone-specific alkaline phosphatase (BALP) is a marker of bone formation (growth), which can be elevated when the body aims to repair damage to the skeleton due to metastatic bone lesions; it is also reflective of the extent of bone disease in prostate cancer patients (Smith et al., 2011). Scientists are interested in understanding how the levels of bone formation and resorption markers are related to the occurrence of skeletal complications to learn how they might be used to guide therapy and to evaluate the impact of therapy; we consider an application in this field in Section 8.3.4.

4. In other settings, markers may be more directly related to immediate risk of important clinical events. Thrombocytopenia is a hematological disorder in which patients' platelet counts vary about much lower levels than are seen in healthy individuals. While there is considerable temporal variation in platelet counts in both healthy and diseased individuals, risk of potentially fatal bleeds increases when counts drop below $10 \times 10^9/L$, so platelet transfusions or other interventions are commonly administered at such times. In an analogous way, CD4+ cell counts are markers of immune health in HIV-infected individuals. Given the critical importance of the immune system in staving off potentially fatal infections, understanding the dynamic features of the marker process is important. The Centers for Disease Control and Prevention's (CDC) definition for AIDS includes the condition of a CD4+ cell count below 200 cells per cubic millimeter of blood. Joint models of CD4+ cell counts and events such as AIDS diagnosis and death are often of interest.

5. Markers can also play a useful role in understanding the causal mechanism of treatment effects through mediation analysis. This is most evident when considering scenario 2 above; treatment effects on longer-term clinically important outcomes can be explored by decomposing them into direct and (marker-mediated) indirect effects. The DCCT trial discussed in Section 5.2.4 offers an example where the effect of intensive blood glucose control on retinopathy can be decomposed into an indirect effect via control of blood glucose level or Hba1C, and a direct effect on ETDRS. Multistate models offer a useful framework for carrying out such analyses. Aalen et al. (2008) discuss this in detail.

6. Finally, as discussed in Section 8.2, markers can be useful for prediction and decisions about disease management and treatment. For example, serial measurements of prostate-specific antigen may be used in risk prediction models for prostate cancer recurrence or death (Taylor et al., 2005).

While the reasons for interest in markers may differ across settings, scientific goals are often best addressed through joint models for marker processes and multistate processes. Such models can describe the joint evolution of the two processes and help to advance understanding. We begin by considering the probability of a particular marker and event path, focussing on continuous markers and using the idea of product integration (Section 2.2), setting aside the issue of censoring or otherwise incomplete observation; this is dealt with in Section 8.3.3.

8.3.2 Models for Markers and Life History Processes

For a K-state process $\{Z(s), s \geq 0\}$, we let $\{N(s), s \geq 0\}$ denote the right-continuous multivariate counting process recording the transitions of the process. As in Section 2.2 we let $\mathcal{A}$ denote the set of absorbing states and we presume here the marker process terminates when the multistate process enters an absorbing state. The marker value at time s is denoted by $X(s)$ and $\{X(s), s \geq 0\}$ is the left-continuous stochastic marker process. The full process history at time t is then $\mathcal{H}(t) = \{Z(s), X(s), 0 \leq s \leq t\}$.

Suppose there is a time interval of interest denoted by $[0, C^A]$ and let $0 = u_0 < u_1 < \cdots < u_R = C^A$ partition this interval. Then $\Delta N(u_r) = N(u_r^-) - N(u_{r-1}^-)$ records the occurrence and nature of transitions over $[u_{r-1}, u_r)$, and $\Delta X(u_r) = X(u_r) - X(u_{r-1})$ records the change in marker value over this interval for $r = 1, \ldots, R$. As before, we will ultimately take the limit as $R \to \infty$, so we assume that at most one transition can occur in any interval $[u_{r-1}, u_r)$. If $H(u_r) = \{(N(u_s), X(u_s)), s = 1, \ldots, r; Z(0)\}$, then the probability of a particular path over $[0, C^A]$ is obtained as

$$\lim_{R \to \infty} \prod_{r=1}^{R} \Big\{ P(\Delta N(u_r) \mid H(u_{r-1}), Z(u_{r-1}^-) \notin \mathcal{A}) \tag{8.15}$$

$$\times \ P(\Delta X(u_r) \mid H(u_{r-1}), Z(u_r^-) \notin \mathcal{A})^{I(Z(u_r^-)\notin\mathcal{A})} \Big\}^{I(Z(u_{r-1}^-)\notin\mathcal{A})} .$$

If there is a contribution in interval $[u_{r-1}, u_r)$, and if a transition is made into an absorbing state, then we presume that the change in the marker value is not measurable. This would be the case in a competing risks setting where a change in a marker value is not defined over intervals in which a death occurs. In this case we note that $P(Z(t) \notin \mathcal{A} \mid X(s), 0 \leq s \leq t) = 1$ since $X(t)$ is only observed if $Z(t) \notin \mathcal{A}$. In the survival setting, this corresponds to the case in which marker measurements (e.g. systolic blood pressure) cannot be made for an individual who has died.

As an illustration, consider a joint model for a continuous scalar marker process and an illness-death process. Let $\{X_1(s), s \geq 0\}$ denote the marker process and assume there is a fixed Bernoulli covariate X_2 with $P(X_2 = 1) = 0.5$; we let $X(s) = (X_1(s), X_2)'$. We take $Z(0) = 1$ with probability 1 and consider a modulated Markov illness-death model with intensities

$$\lambda_{12}(t \mid \mathcal{H}(t^-), Z(t^-) = 1) = \lambda_{12} \exp(X'(t) \alpha_{12}), \tag{8.16}$$

where $\alpha_{12} = (\alpha_{121}, \alpha_{122})'$ and

$$\lambda_{k3}(t \mid \mathcal{H}(t^-), Z(t^-) = k) = \lambda_{k3} \exp(X'(t) \alpha_{k3}), \tag{8.17}$$

where $\alpha_{k3} = (\alpha_{k31}, \alpha_{k32})'$, $k = 1, 2$.

Biological markers are typically continuous or, in the case of counts with very wide ranges, may be effectively treated as such. Models used for continuous markers are mainly of two types. The first are diffusion processes in which increments $\Delta X(t)$ are modeled in terms of features of $\mathcal{H}(t^-)$ and Gaussian process increments. The second type are random coefficient models for mean individual-specific trajectories $\mu_i(t)$ with Gaussian errors or perturbations; we discuss these in Section 8.3.3. The fact that markers are only measured at a set of discrete times poses challenges for general model fitting in the former case, while the latter are more readily handled. Models in which $X_i(t)$ is categorical can also be considered, including cases where the categorical values are based on intervals of a continuous marker; we discuss this in Section 8.3.4 and show how joint models for markers and life history models can deal with intermittent observation of both markers and the response process. Finally, we reiterate that a primary goal is to discover what features of a marker process affect transition intensities. The nature of the relationship can also play a role in the choice of the marker model. We discuss this in subsequent sections.

8.3.3 Intermittent Measurement of Markers and Censoring

Joint model fitting can be challenging when the marker values are only observed intermittently, which is routinely the case. We consider the case where transitions of a multistate process are observed subject to right censoring. Let $C^R < C^A$ denote a random drop-out time so that transitions are recorded over $[0, C]$ where $C = \min(C^R, C^A)$; as before we let $Y(s) = I(s \leq C)$, $d\bar{N}(s) = Y(s)dN(s)$, and $\bar{N}(t) = \int_0^t d\bar{N}(s)$. Let a_j, $j = 0, 1, \ldots, A(C)$ denote the assessment times at which the marker values are recorded over $[0, C]$ for an individual of interest, where $\{A(s), s > 0\}$ is the counting process for assessments. The observed history at t is then

$$\bar{\mathcal{H}}^\circ(t) = \{Y(u), A(u), dN(u), 0 \leq u \leq t; \; (a_j, X(a_j)), j = 0, \ldots, A(\min(t, C)), Z(0)\}. \quad (8.18)$$

Gaussian processes for continuous $X(t)$ are convenient to fit to observations $\{(a_j, X(a_j)), j = 0, 1, \ldots, A(C)\}$, provided the assessment times a_j arise from a conditionally independent visit process. In very special cases, it may be possible to fit transition intensity models like (8.16) and (8.17). Generally, however, it is difficult to marginalize over unobserved marker values for likelihood construction, and so intermittent observation of markers usually leads to considerable model simplification. The transition intensities are often specified to depend only on the observed marker values, which can be viewed as an approximation to a potentially more realistic model that may involve the unobserved marker values. For example, analogous to (8.16) and (8.17), we could consider an illness-death process with

$$\bar{\lambda}_{12}(t \mid \bar{\mathcal{H}}^\circ(t^-)) = \lim_{\Delta t \downarrow 0} \frac{P(\Delta \bar{N}_{12}(t) = 1 \mid \bar{\mathcal{H}}^\circ(t^-), Z(t^-) = 1, Y(t) = 1)}{\Delta t}$$

and

$$\bar{\lambda}_{k3}(t \mid \bar{\mathcal{H}}^\circ(t^-)) = \lim_{\Delta t \downarrow 0} \frac{P(\Delta \bar{N}_{k3}(t) = 1 \mid \bar{\mathcal{H}}^\circ(t^-), Z(t^-) = k, Y(t) = 1)}{\Delta t}, \quad k = 1, 2.$$

When considered in the context of the examples discussed in Section 8.3.1 these models may not be particularly appealing because as the time since the last assessment $(t - a_{A(t)})$ increases, the marker value $X(a_{A(t)})$ becomes a poorer approximation to $X(t)$. Motivated by problems in a liver cirrhosis study, Andersen and Listøl (2003) explore the extent of attenuation in regression coefficient estimates arising from infrequent measurement of time-dependent covariates and propose simple methods for dealing with this; see also de Bruijne et al. (2001).

Much of the recent work on joint modeling of longitudinal and survival data is based upon linear mixed effect models. This approach, which can be extended to general multistate models, is computationally tractable in some cases but has some features that affect applicability and interpretability of analyses. We discuss this here, focussing for convenience on survival processes represented by a 2-state irreversible model. Joint models are formulated through shared dependence on random effects. The basic model is described in Chapter 4 of Rizopoulos (2012); we present a version here in our notation.

Let $V(t)$ denote a $p \times 1$ vector, whose elements may be defined functions of time or fixed or time-dependent covariates, and let B denote a latent $p \times 1$ random vector with a distribution $G(b)$, where b denotes a realized value. Given $B = b$, suppose the marker distribution at time t for a particular individual is of the form

$$X(t) = V'(t)\beta + W'(t)b + \epsilon(t) \tag{8.19}$$

where β is a $p \times 1$ vector of fixed parameters, $W(t)$ is a $p \times 1$ vector, and $\epsilon(t) \sim N(0, \sigma^2)$ is an error term where $\epsilon(s) \perp \epsilon(t)$ for $s < t$. The vector $W(t)$ is introduced to allow individual-specific levels and trends in mean marker values and effects of covariates; each element of $W(t)$ is typically either zero or the corresponding element of $V(t)$. Suppose, for example, that $B \sim N_p(0, \Sigma)$, where Σ is a unstructured $p \times p$ covariance matrix. If $V(t) = W(t)$ then (8.19) can be written as

$$X(t) = V'(t)B^* + \epsilon(t)$$

where $B^* \sim N_p(\beta, \Sigma)$ is a vector of random coefficients with $E(B^*) = B$. If on the other hand, $W(t)$ is a $p \times 1$ vector of zeros, then all individuals have a common mean $E(X(t)) = V'(t)\beta$ and values of $X(t)$ deviate from this solely due to the random error $\epsilon(t)$. More generally, some elements of $W(t)$ may be non-zero to accommodate serial dependence in marker values and heterogeneity in trends and/or covariate effects.

Consider a simple example with a single defined covariate of interest with $V_i(t) = (1, V_{i1}(t))'$ for individual i, where $V_{i1}(t) = t$. We let $\beta = (\beta_0, \beta_1)'$, $W_i(t) = (1, t)'$ and assume $B_i = (B_{i0}, B_{i1})' \sim BVN(0, \Sigma)$ where Σ has σ_0^2 and σ_1^2 as the diagonal and σ_{01} in the off-diagonal; $\epsilon_i(t) \sim N(0, \sigma^2)$. This is a linear random effects model with an intercept and slope unique to each individual. If $B_i^* = \beta + B_i$, given $B_i^* = b_i^*$, this can be rewritten as

$$X_i(t) = \mu_i(t) + \epsilon_i(t) \tag{8.20}$$

with $\mu_i(t) = b_{i0}^* + b_{i1}^* t$.

In the general formulation the function $\mu_i(t) = V_i'(t)\beta + W_i'(t)b_i$ is often viewed as representing the "true" but unobserved marker value at time t for individual i, and hence interest lies in modeling its effect on the hazard for death or on other transition intensities. This is naturally an idealization, and the implied assumption that the hazard say, is independent of $\bar{X}_i(t)$, given $B_i = b_i$, is a strong one. The model, however, has utility in certain situations where the idea of a smoothly varying underlying process $\mu_i(t)$ is plausible. The hazard or intensity for death might then be taken to have the form

$$\lambda_{12}(t \mid \mathcal{H}_i(t^-), \bar{\mu}_i(t)) = \lambda_{120}(t) \exp(\alpha \mu_i(t))$$

where $\lambda_{120}(t)$ is a baseline hazard and $\bar{\mu}_i(t) = \{\mu_i(s), 0 \leq s \leq t\}$ is the history of the latent marker values over $[0, t]$. Here then, the effect of a one-unit increase in $\mu_i(t)$ is to multiply the baseline hazard for death by a factor $\exp(\alpha)$. More generally, $\mu_i(t)$ in the hazard function could be replaced with a vector of functions based on $\bar{\mu}_i(t)$. The probability of being alive given the latent marker path is computed as

$$P(Z(t) = 1 \mid Z(0) = 1, \bar{\mu}_i(t), B_i = b_i) = \exp\left(-\int_0^t \lambda_{120}(s) \exp(\alpha \mu_i(s)) \, ds\right),$$

since $\mu_i(t)$ is a fixed function given $B_i = b_i$. Of course, if $V_i(t)$ includes time-dependent covariates that are observed only intermittently, then once again we are forced to make simplifying assumptions or to impute missing values.

The joint model for the observable marker and survival process is obtained by marginalizing over the random effects. For discussion, we continue with the model above. If $a_{i0} < a_{i1} < \cdots < a_{im_i}$ are the times the marker $X_i(t)$ was observed for individual i, joint outcome probabilities are given by

$$\int \Big\{ \lambda_{12}(s_i \mid T_i \geq s_i, \mu_i(s_i))^{\delta_i} \exp\left(-\int_0^\infty \bar{Y}_i(u) \lambda_{12}(u \mid T_i \geq u, \mu_i(u)) \, du\right)$$
$$\times \prod_{j=1}^{m_i} P(X_i(a_{ij}) \mid V_i(a_{ij}), B_i = b_i) \Big\} dG(b_i)$$

where $S_i = \min(T_i, C_i)$, $\delta_i = I(S_i = T_i)$ and s_i is the realized value of S_i. The first term in the integrand comes from the distribution of survival time given the mean marker function $\mu_i(t)$, and the second from the measurements on $X_i(t)$. Here,

$$\prod_{j=1}^{m_i} P(X_i(a_{ij}) = x_{ij} \mid V_i(a_{ij}), B_i = b_i) = \prod_{j=1}^{m_i} f_i(x_{ij})$$

where $f_i(x_{ij})$ is a normal density with mean $\mu_i(a_{ij}; b_i)$ and variance σ^2. In a more general multistate process, we may compute the joint probability of transition times and observed marker values as

$$\int \Big\{ \prod_k \prod_l \Big[\prod_{t_{ir} \in \mathcal{D}_{ikl}} \lambda_{kl}(t_{ir} \mid \mathcal{Z}_i(t), \bar{\mu}_i(t)) \exp\left(-\int_0^\infty \bar{Y}_{ik}(u) \lambda_{kl}(u \mid \mathcal{Z}_i(u), \mu_i(u)) \, du\right) \Big]$$
$$\times \prod_{j=1}^{m_i} P(X_i(a_{ij}) \mid V_i(a_{ij}), B_i = b_i) \Big\} dG(b_i)$$

where $\mathcal{D}_{ikl}$ is the set of $k \to l$ transition times for individual i, and $\mathcal{Z}_i(t)$ is the history of the multistate process.

The desire to obtain graphical or numerical summaries of marker trajectories presumably motivates, at least in part, this formulation; however, remarks are warranted. First, the joint models accommodate dependence between the marker process and the life history process by modeling the dependence of the life history intensities on the latent marker process $\mu_i(t;b_i)$. As noted, this is a strong constraint. Second, this joint model does not imply that the marker process terminates upon failure. This may be reasonable in settings where mortality is negligible and failure represents a non-fatal event; in other settings this seems problematic. Third, while the models for the marker paths may appeal in their apparent simplicity, the parameters and associated estimates do not have a clear interpretation when marker processes terminate at failure. The expected marker profiles over $[0,t]$ given $T > t$ do not, for example, have the simple form in (8.19), and the expected increment over $[u_{r-1}, u_r)$ given $u_r < T$ is complex. That is, while the conditional independence assumption simplifies the construction of the joint model and

$$E\{X_i(u_r) - X_i(u_{r-1}) \mid B_i = b_i, T > u_r\} = E\{X_i(u_r) - X_i(u_{r-1}) \mid B_i = b_i\},$$

the distribution of $\Delta X_i(u_r) \mid T > u_r$ does not retain this simple form. The increment in the marker value over $[t, t+\Delta t)$ given survival to $t+\Delta t$ has probabilities

$$P(\Delta X_i(t) \mid T_i \geq t+\Delta t, \bar{X}_i(t)) = \frac{P(T_i \geq t+\Delta t, \Delta X_i(t) \mid T_i \geq t, \bar{X}_i(t))}{P(T_i \geq t+\Delta t \mid T_i \geq t, \bar{X}_i(t))}.$$

Thus, $\Delta X_i(t)$ or $V_i(t)$ do not refer to the population of individuals alive at time t, or indeed to any real population. Finally, model checking has received little attention in this framework and methods that utilize estimated random effects $\hat{b}_i$ are not well understood.

8.3.4   A Joint Multistate and Discrete Marker Process Model

In some contexts the marker values are discrete by nature, and in other cases continuous markers may be categorized according to established thresholds defining, for example, normal ranges, or more generally intervals representing meaningfully different levels; this was done in Section 8.2.1. A joint model for a categorized marker process $\{X(s), s \geq 0\}$ and the multistate process $\{Z(s), s \geq 0\}$ can be formed by considering a composite process $\{Z^*(s), s \geq 0\}$, where $Z^*(s) = (Z(s), X(s))$; see Section 6.2.2. We suppose for discussion that $X(s)$ has G ordered categories or states $g = 1, \ldots, G$. Then, for example, we may consider a joint model for the marker process and an illness-death process, with a state space diagram given in Figure 8.3. The states in the middle two columns are identified by two numbers with the first reflecting event (illness) status in the original illness-death model (1 = alive and event-free, 2 = alive post-event). The second number reflects the value of $X(s)$. States 3 and 4 are death states that allow one to distinguish between deaths among individuals who were event free and individuals who had experienced the event prior to death. The vertical transitions are made when the marker states change and transitions from state $(1,g)$ to $(2,g)$ occur for some g, when the event occurs. Death states 3 and 4 are absorbing.

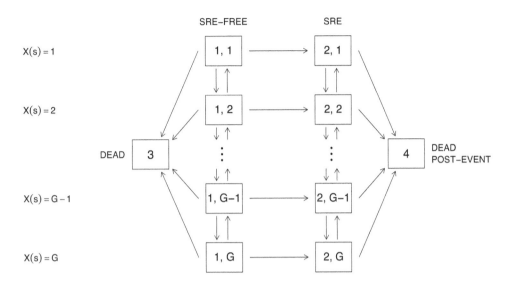

Figure 8.3: A joint model for a discrete marker with G categories, a non-fatal event (SRE - skeletal-related event) and death.

If $\mathcal{H}(t) = \{Z^*(s), 0 \le s \le t\}$ then following the approach of Section 6.2.2 we may define a Markov intensity for the non-fatal event as

$$\lambda_{1l;2l}(t \mid \mathcal{H}(t^-)) = \lim_{\Delta t \downarrow 0} \frac{P(Z^*(t+\Delta t^-) = (2,l) \mid Z^*(t^-) = (1,l))}{\Delta t} \tag{8.21}$$

for $l = 1, \ldots, G$ and Markov death intensities as

$$\lambda_{kl;j}(t \mid \mathcal{H}(t^-)) = \lim_{\Delta t \downarrow 0} \frac{P(Z^*(t+\Delta t^-) = j \mid Z^*(t^-) = (k,l))}{\Delta t}, \tag{8.22}$$

with $j = 3$ for event-free death and $j = 4$ for death following event occurrence. The dynamics of the discrete marker process are specified by

$$\lambda_{kl;kl'}(t \mid \mathcal{H}(t^-)) = \lim_{\Delta t \downarrow 0} \frac{P(Z^*(t+\Delta t^-) = (k,l') \mid Z^*(t^-) = (k,l))}{\Delta t}, \tag{8.23}$$

where $l' \in \{l-1, l+1\}$ for $l = 2, \ldots, G-1$, $l' = 2$ if $l = 1$, and $l' = G-1$ if $l = G$. While we have omitted consideration of fixed or time-dependent covariates here for exposition, they can of course be included. Comparison of the intensities $\lambda_{1l;2l}(t \mid \mathcal{H}(t^-))$ for different values of l, or the corresponding cumulative intensities conveys the impact of higher marker values on event risk; this is often done parsimoniously by specifying multiplicative models for $\lambda_{1l;2l}(t \mid \mathcal{H}(t^-))$ treating $X(t)$ as a time-dependent covariate. The same can be done for $\lambda_{kl;j}(t \mid \mathcal{H}(t^-))$.

The joint approach described here, which involves modeling assumptions for the marker process (see (8.23)) enables one to make inferences about the dynamics of the marker process. Importantly, the multistate formulation also provides useful guidance on what estimands are reasonable. For example, because the death states are absorbing and the marker process ceases to be defined upon death, probability statements regarding the marker values at a given time incorporate survival status. If $P(X(0) = g) = \pi_g$, $g = 1, \ldots, G$ is the initial distribution for the categorical marker, for example, the probability that an individual is alive at time s and $X(s) = g$ is

$$P(X(s) = g \mid Z(0) = 1) = \sum_{g_0=1}^{G} \sum_{k=1}^{2} P(Z^*(s) = (k,g) \mid Z^*(0) = (1,g_0)) \pi_{g_0},$$

where $P(Z(0) = 1 \mid X(0) = g) = 1$. The conditional probability of $X(s) = g$ given survival to s is then

$$P(X(s) = g \mid Z(0) = 1) / P(Z^*(s) \notin \{3,4\} \mid Z(0) = 1)$$

where

$$P(Z^*(s) \notin \{3,4\} \mid Z(0) = 1) = \sum_{g_0=1}^{G} \sum_{k=1}^{2} \sum_{l=1}^{G} P(Z^*(s) = (k,l) \mid Z^*(0) = (1,g_0)) \pi_{g_0}.$$

Intensity-based models can provide a detailed description of the marker dynamics and the marker effect on event risk. However, when Markov assumptions are adopted for the transition rates in Figure 8.3 and data are subject to independent right censoring, the robustness of the Aalen-Johansen estimates (Section 3.4.1) mean that the state occupancy probabilities can be consistently estimated for more general processes. Moreover, Markov models allow convenient model fitting based on data from intermittent observation, as discussed in Chapter 5. We illustrate this further in the application that follows.

Example 8.3.1: Bone Alkaline Phosphatase in Metastatic Prostate Cancer
Here we consider data from a trial of prostate cancer patients with skeletal metastases, where the goal is to evaluate the effect of a bisphosphonate on the prevention of skeletal complications (Saad et al., 2004). This trial is broadly similar in design and objectives to the trial of breast cancer patients discussed in Section 1.6.1 and in Chapter 3. We restrict attention to individuals who had at least one bone marker assessment giving a total of 421 individuals randomized to receive monthly infusions of zoledronic acid and 201 randomized to receive a placebo infusion. We use this data to illustrate a joint model for the level of a bone formation marker, bone-specific alkaline phosphatase (BALP), and the occurrence of a skeletal complication and death. The skeletal complications that may arise include pathological and vertebral fractures, acute bone pain requiring radiotherapy, and need for surgery to treat or prevent fractures. We use the term *skeletal-related event* (SRE) when referring to this composite event. Figure 8.4 shows the values of BALP for two prostate cancer patients in the study; also indicated on the horizontal axes are the times of the first skeletal event (T_2), death (T_3 or T_4) and end of follow-up (C). It is apparent that there can be considerable variation in the bone marker values within patients over

Table 8.2: Frequencies of consecutive pairs of known states (or censoring) among 622 prostate cancer patients in Saad et al. (2004).

											DEAD		
		SRE-FREE				SRE[†]					SRE-FREE	SRE	
		1, 1	1, 2	1, 3	1, 4	2, 1	2, 2	2, 3	2, 4	$\mathcal{S}_2$	3	4	CENS
FROM													
SRE-FREE	1, 1	648	99	17	4	0	0	0	0	52	15	0	83
	1, 2	105	235	84	12	0	0	0	0	54	13	0	54
	1, 3	6	63	183	103	0	0	0	0	66	12	0	44
	1, 4	4	4	38	293	0	0	0	0	94	35	0	100
SRE	2, 1	0	0	0	0	87	22	5	1	0	0	8	28
	2, 2	0	0	0	0	12	44	27	7	0	0	4	38
	2, 3	0	0	0	0	1	16	36	23	0	0	15	32
	2, 4	0	0	0	0	0	4	15	104	0	0	28	68
	$\mathcal{S}_2$	0	0	0	0	51	46	40	84	0	0	12	33

[†] SRE is an acronymn for skeletal-related event.

time. Also apparent is the variation in the period over which markers are measured due to incomplete assessments, study withdrawal or death.

A discrete marker is constructed by discretizing the BALP value according to the cut-points 150.25, 267.50 and 529.75 IU/L giving $G = 4$ categories. The resulting process can be represented by the multistate diagram in Figure 8.3. A challenge in this dataset is that the skeletal event and death times are subject to right censoring but the bone marker is only measured periodically. When an SRE is observed a transition is made from the SRE-FREE to SRE column of states, but it is not known from which state the individual made the transition nor what precise state was entered. As a result it is only known that they moved from one of the states in $\mathcal{S}_1 = \{(1, g), g = 1, \ldots, G\}$ to one of the states in $\mathcal{S}_2 = \{(2, g), g = 1, \ldots, G\}$. Table 8.2 gives the raw counts of what was known at consecutive times. The marker value at the time of an SRE is unknown, which is reflected by the counts in the column headed $\mathcal{S}_2$; the following marker assessments determine the particular state occupied in $\mathcal{S}_2$ which is reflected in the last row of the table.

Let $a_0 < a_1 < \cdots < a_m$ denote the marker assessment times over $(0, C]$ for an individual. If they are event-free over $(0, C]$, then given $Z^*(a_0)$ the likelihood contribution is

$$\prod_{j=1}^{m} P(Z^*(a_j) \mid Z^*(a_{j-1})) \, P(Z(C) \in \mathcal{S}_1 \mid Z^*(a_m)),$$

where $Z^*(a_j) \in \mathcal{S}_1 = \{(1, g), g = 1, \ldots, G\}$ and the last term simply reflects the fact that they remained event-free over $(a_m, C]$. If a skeletal event was observed, but

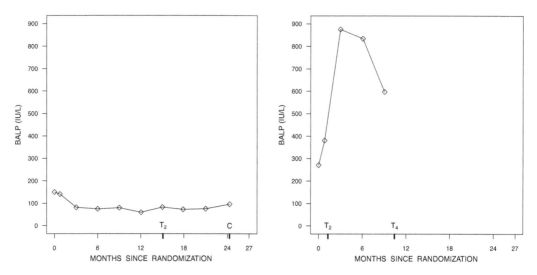

Figure 8.4: Two profiles of bone alkaline phosphatase (BALP) along with indications of the time of first skeletal events (T_2), death (T_4) and end-of-follow-up (C) for two prostate cancer patients from Saad et al. (2004).

they survived over $(0, C]$, then if $a_{r-1} < t_2 < a_r$ for some $r = 1, \ldots, m$, we write

$$\prod_{j=1}^{r-1} P(Z^*(a_j) \mid Z^*(a_{j-1})) \left\{ \sum_{l=1}^{G} P(Z^*(t_2^-) = (1, l) \mid Z^*(a_{r-1})) \lambda_{1l;2l}(t_2) \right.$$

$$\times \left. P(Z^*(a_r) \mid Z^*(t_2) = (2, l)) \right\} \prod_{s=r+1}^{m} P(Z^*(a_s) \mid Z^*(a_{s-1})) P(Z^*(C) \in \mathcal{S}_2 \mid Z^*(a_m)).$$

If the non-fatal event occurred over $(a_{r-1}, a_r]$ and they died at $t_4 \in (a_s, a_{s+1}]$ for $s \geq r$, then the contribution is

$$\prod_{j=1}^{r-1} P(Z^*(a_j) \mid Z^*(a_{j-1})) \sum_{l=1}^{G} \left\{ P(Z^*(t_2^-) = (1, l) \mid Z^*(a_{r-1})) \lambda_{1l;2l}(t_2) \right.$$

$$\times \left. P(Z^*(a_r) \mid Z^*(t_2) = (2, l)) \right\} \prod_{u=r+1}^{s+1} P(Z^*(a_u) \mid Z^*(a_{u-1}))$$

$$\times \sum_{l'=1}^{G} \left\{ P(Z^*(t_4^-) = (2, l') \mid Z^*(a_s)) \lambda_{2l',4}(t_4) \right\}$$

The likelihood contributions for other observations can similarly be derived.

The estimates of the baseline intensities for the clinical events are displayed in the bottom of Table 8.3. The increasing risk of an SRE with the marker states defined in terms of higher BALP values is apparent. The same is broadly true for transitions into state 3 (death without prior SRE). The treatment effects reported in the last columns suggest a higher rate of transitions into the lower marker states for individuals receiving zoledronic acid among individuals who are SRE-free; there is no evidence of this from individuals post-SRE. If individuals are put into strata

Table 8.3: Parameter estimates obtained from fitting a multistate model with $G = 4$ shown in Figure 8.3.

SRE STATUS	FROM (l)	TO (l')	BASELINE INTENSITY ($\times 10^{-3}$) $\widehat{\lambda}_{kl;kl'}$	95% CI	TREATMENT EFFECT RR	95% CI	p
SRE-FREE ($k=1$)	[0, 150.25)	[150.25, 267.50)	2.41	(1.57, 3.70)	1.21	(0.74, 1.99)	0.454
	[150.25, 267.60)	[267.60, 529.75)	4.62	(3.10, 6.90)	1.61	(0.98, 2.64)	0.059
	[267.60, 529.75)	[529.75, ∞)	7.53	(5.33, 10.64)	1.31	(0.84, 2.05)	0.230
SRE-FREE ($k=1$)	[150.25, 267.50)	[0, 150.25)	2.83	(1.68, 4.76)	3.03	(1.69, 5.44)	< 0.001
	[267.60, 529.75)	[150.25, 267.50)	2.75	(1.49, 5.08)	2.67	(1.32, 5.39)	0.006
	[529.75, ∞)	[267.60, 529.75)	2.17	(1.13, 4.16)	1.83	(0.85, 3.94)	0.124
SRE ($k=2$)	[0, 150.25)	[150.25, 267.50)	5.41	(2.91, 10.05)	0.60	(0.28, 1.30)	0.197
	[150.25, 267.60)	[267.60, 529.75)	5.81	(3.25, 10.39)	1.41	(0.69, 2.86)	0.345
	[267.60, 529.75)	[529.75, ∞)	4.51	(2.25, 9.05)	1.64	(0.75, 3.59)	0.213
SRE ($k=2$)	[150.25, 267.50)	[0, 150.25)	2.47	(1.10, 5.55)	1.45	(0.54, 3.91)	0.461
	[267.60, 529.75)	[150.25, 267.50)	6.12	(3.15, 11.89)	0.79	(0.32, 1.94)	0.606
	[529.75, ∞)	[267.60, 529.75)	1.79	(0.87, 3.67)	1.32	(0.52, 3.34)	0.552

BALP LEVEL	STATE FROM (l)	TO (l')	BASELINE INTENSITY ($\times 10^{-3}$) $\widehat{\lambda}_{kl;kl'}$ or $\widehat{\lambda}_{kl;j}$	95% CI	TREATMENT EFFECT RR	95% CI	p
[0, 150.25)	SRE-FREE ($k=1$)	SRE ($k=2$)	0.98	(0.69, 1.39)	0.91	(0.71, 1.18)	0.488
	SRE-FREE ($k=1$)	DEAD ($j=3$)	0.19	(0.08, 0.42)	1.34	(0.81, 2.23)	0.257
	SRE ($k=2$)	DEAD ($j=4$)	0.54	(0.20, 1.47)	0.90	(0.54, 1.48)	0.675
[150.25, 267.50)	SRE-FREE ($k=1$)	SRE ($k=2$)	2.45	(1.73, 3.46)			
	SRE-FREE ($k=1$)	DEAD ($j=3$)	0.48	(0.20, 1.14)			
	SRE ($k=2$)	DEAD ($j=4$)	0.47	(0.12, 1.83)			
[267.60, 529.75)	SRE-FREE ($k=1$)	SRE ($k=2$)	3.09	(2.18, 4.38)			
	SRE-FREE ($k=1$)	DEAD ($j=3$)	0.29	(0.06, 1.47)			
	SRE ($k=2$)	DEAD ($j=4$)	0.84	(0.23, 3.09)			
[529.75, ∞)	SRE-FREE ($k=1$)	SRE ($k=2$)	3.88	(3.01, 5.01)			
	SRE-FREE ($k=1$)	DEAD ($j=3$)	1.19	(0.72, 1.96)			
	SRE ($k=2$)	DEAD ($j=4$)	2.13	(1.33, 3.39)			

based on the BALP value at the time of randomization, then one can estimate the cumulative risk of death based on Kaplan-Meier estimates for each stratum. Standard unweighted Kaplan-Meier estimates are sensitive to a dependent censoring scheme as discussed in Section 3.4.3, but we plot these in Figure 8.5 for the zoledronate treated patients, along with estimates based on the fitted multistate model. Weighted estimates which accommodate SRE-dependent censoring are also shown but are in good agreement with the unweighted estimates. We stress that these are descriptive analyses that can be useful for summarizing the dynamics of the joint processes, but this type of analysis does not admit estimates of treatment effect with a causal interpretation. We discuss this more in the next section.

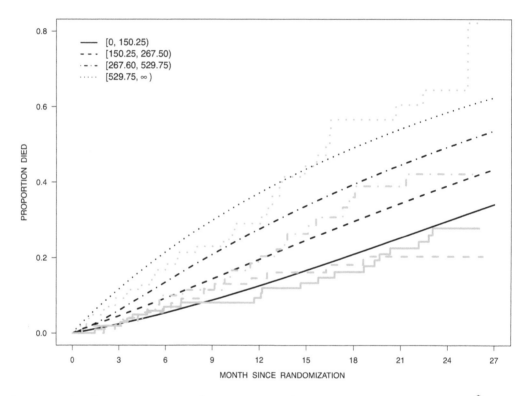

Figure 8.5: Kaplan-Meier estimates (grey stepped line) for the probability of death and $\widehat{P}(Z(t) \in \{3,4\}|Z(0) = 1l)$ for $l = 1,2,3,4$ based on a multistate model (black line) by quartiles of serum BALP (IU/L) for zoledronate arm.

8.4   Remarks on Causal Inference with Life History Processes

Causal inference is an implicit goal in most epidemiological studies about the effects of risk factors and interventions on chronic disease processes. The Canadian Longitudinal Study on Aging (Raina et al., 2009), for example, aims to learn about behavioural and environmental risk factors for disease onset with a long-term view that interventions changing these will reduce disease occurrence. Much of the early work in formalizing the probabilistic framework for causal reasoning was developed in the context of fixed exposure variables, confounders, and a response implicitly measured subsequent to these. With life history processes involving the complications of truncation, selection effects, and censoring, challenges are more substantial. A rigorous framework for causal inference developed over the past few decades has been authoritatively reported on in Hernán and Robins (2015). Chapter 9 of Aalen et al. (2008) discusses causal inference for life history processes while emphasizing the importance of appropriately dealing with time through the use of dynamic path analysis. We refer readers to these books and references therein for further reading, but make brief informal remarks here beginning with the discussion of a simple setting.

Suppose the goal is to estimate the effect of an intervention on a univariate response. Let X denote a binary variable that indicates whether an individual receives the intervention of interest ($X = 1$) or not ($X = 0$), and let Y denote the univariate response. Consider the mean model

$$E(Y \mid X, V) = \alpha_0 + \alpha_1 X + \alpha_2 V, \tag{8.24}$$

where V is an auxiliary variable. In general,

$$E(Y \mid X) = \alpha_0 + \alpha_1 X + \alpha_2 E(V \mid X) = \alpha_0^{\dagger} + \alpha_1^{\dagger} X$$

with $\alpha_0^{\dagger} = \alpha_0 + E(V \mid X = 0)$ and

$$\alpha_1^{\dagger} = \alpha_1 + \alpha_2 [E(V \mid X = 1) - E(V \mid X = 0)],$$

where $\alpha_1^{\dagger} - \alpha_1 = \alpha_2 [E(V \mid X = 1) - E(V \mid X = 0)]$ represents the confounding effect of omitting V on inferences regarding the $X - Y$ association. If $X \not\perp V$ then a naive analysis of $Y \mid X$ will yield estimators consistent for $\alpha_0^{\dagger}$ and $\alpha_1^{\dagger}$.

To discuss causal estimands more formally, Rubin (1974) used the framework of potential outcomes. In this framework with a binary treatment variable X, the response of each individual in a population is considered under two conditions: one in which they receive the treatment of interest ($X = 1$), and one in which they do not receive the treatment ($X = 0$). Custom notation is used to denote the outcomes under these two scenarios as $Y^{[1]}$ and $Y^{[0]}$, respectively, and we write $Y^{[x]}$ generically. The causal effect of the treatment for an individual is defined as $Y^{[1]} - Y^{[0]}$. A marginal structural model (Robins et al., 2000) for the treatment effect on the mean response can be defined as

$$E(Y^{[x]}) = \mu^{[x]} = \gamma_0 + \gamma_1 x \tag{8.25}$$

where $\mu^{[x]}$ is the average response in the population if all individuals had $X = x$, $x = 0, 1$. In this case γ_0 is the average response if no individuals received the intervention and $\gamma_0 + \gamma_1$ is the average response if all individuals received the treatment. Note that $\mu^{[x]}$ in (8.25) may be equivalently viewed in terms of

$$\mu^{[x_o]} = E_V \{ E(Y \mid X = x_o, V) \}$$

where the outside expectation is taken simply with respect to the marginal distribution of V. The average causal treatment effect is then defined as $\mu^{[1]} - \mu^{[0]} = \gamma_1$. In most settings individuals are observed under either the treatment or standard condition, so a response $Y^{[1]}$ or $Y^{[0]}$ is available. Well-conducted randomized experiments, however, facilitate estimation of average causal treatment effects.

Having identified the causal estimand in this framework, the challenge is to analyse observational data in which $X \not\perp V$ to estimate γ. This can be achieved by weighting standard estimating functions to create a pseudo-sample representative of a population in which $X \perp V$. Weights that are proportional to the inverse of the probability of treatment given the confounders V will achieve this. Let x_o denote the value for the treatment variable for an individual and consider their contribution to the inverse probability of treatment weighted estimating function:

$$U(Y, X, V; \mu^{[x_o]}) = \frac{I(X = x_o)}{P(X = x_o \mid V)} \left(Y - \mu^{[x_o]} \right). \tag{8.26}$$

To see that (8.26) is unbiased for $\gamma = (\gamma_0, \gamma_1)$, note that

$$E_V \left(E_X \left\{ E_Y \left[U(Y, X, V; \mu^{[x_o]}) \mid X, V \right] \mid V \right\} \right) \tag{8.27}$$

$$= E_V \left(E_X \left\{ \frac{I(X = x_o)}{P(X = x_o \mid V)} \left(E(Y \mid X, V) - \mu^{[x_o]} \right) \mid V \right\} \right)$$

$$= E_V \left(\sum_{x=0}^{1} \frac{P(X = x \mid V) I(x = x_o)}{P(X = x_o \mid V)} \left(E(Y \mid x, V) - \mu^{[x_o]} \right) \right)$$

$$= E_V \left(E(Y \mid X = x_o, V) - \mu^{[x_o]} \right) = 0.$$

A model for the propensity score $P(X|V)$ must be fitted in practice, but provided a consistent estimate of $P(X|V)$ is used in (8.26), consistent estimates of γ will still be obtained.

Causal inference is considerably more challenging for life history processes where intensity-based models tend to play a key role in modeling and inference. Consider first a simple 2-state failure time process in which state 1 represents the condition of being event-free and state 2 is entered at $T = T_2$ upon occurence of the event. Rigorously defining the causal effect of an intervention in the common framework of Cox regression is challenging, since known and unknown variables prognostic for the failure time are usually modulating the underlying event intensity. Consider a simple setting analogous to (8.24) involving a fixed categorical vector V and suppose

$$\lambda_{12}(t \mid X = x, V = v) = \lambda_{120}(t) \exp(\beta_1 x + \beta_2 v) \tag{8.28}$$

represents the true event intensity. In this framework, a parsimonious definition of a causal effect of X on T is elusive for two reasons. First, since

$$P(X, V \mid T \geq t) = \exp(-\Lambda_{12}(t \mid X, V)) P(X, V) / P(T \geq t) \tag{8.29}$$

where $\Lambda_{12}(t|X, V)$ is the cumulative intensity for $1 \to 2$ transitions, $X \not\perp V|T \geq t$. So even if $X \perp V$ at $t = 0$ as it would be in a randomized trial, omission of V has a confounding effect on inferences about the association between X and T beyond the first failure time in the sample. While the partial likelihood estimating functions given in Chapter 3 can in principle be re-weighted with time-dependent propensity scores $P(X = 1|T \geq t, V)$ to create pseudo-samples for which $X \perp V \mid T \geq t$, this pseudo-sample would not correspond to any real-world situation, making it difficult to interpret the results. A second challenge is that under (8.28) the distribution of $T|X$ has marginal intensity

$$\lambda_{12}^m(t \mid X = x) = E_V\{\lambda_{12}(t \mid X = x, V) \mid T \geq t, X = x\},$$

which does not satsify the proportional hazards assumption if $\beta_2 \neq 0$. Taken together these points imply that if there are any prognostic variables V which act as in (8.28), marginal hazard-based causal inferences about X are not possible (Hernán, 2010; Aalen et al., 2015a).

Such issues are at play when $\{Z(s), s \geq 0\}$ is a general K-state process with $\mathcal{H}(t) = \{Z(s), 0 \leq s \leq t; X, V\}$ and $\bar{Z}(t) = \{Z(s), 0 \leq s \leq t\}$. In this case, if

$$\lambda_{kl}(t \mid \mathcal{H}(t^-)) = \lambda_{kl0}(t) \exp(\beta_{kl1}x + \beta_{kl2}v),$$

then the "marginal" intensity $\lambda_{kl}^m(t|X = x, \bar{Z}(t^-))$ has the form

$$Y_k(t^-) \lambda_{kl0}(t) \exp(\beta_{kl1}x) E_V\{\exp(\beta_{kl2}V) \mid Y_k(t^-) = 1, X)\},$$

where $P(V|X, Y_k(t^-) = 1) = P(X, V|Y_k(t^-) = 1)/P(X|Y_k(t^-) = 1)$ and

$$P(X, V \mid Y_k(t^-) = 1) = \frac{P(Y_k(t^-) = 1 \mid X, V) P(X, V)}{P(Y_k(t^-) = 1)}.$$

Conditioning on the event $Y_k(t^-) = 1$ likewise induces a confounding effect regarding the effect of X on the $k \to l$ transition time.

Despite issues with multiplicative intensity-based models, they can play a role in thinking about causal issues for complex processes. First, they are crucial in the "dynamic" view of causality (Aalen et al., 2008) as we discuss below. In simple settings where exposure and auxiliary covariates do not change over time, marginal features of multistate processes are more amenable to causal inference. The state k occupancy probability at a time t_o, for example, can be viewed as the expectation of $Y_k(t_o)$, enabling one to consider a model analogous to (8.24) or (8.25). In Section 4.1.2 we considered the competing risks problem with K possible causes of failure (see Figure 4.1). We may choose to examine the effect of a treatment on failure due to cause 1 via a model

$$g(P_1(t \mid x)) = \alpha_1(t) + \beta_1 x,$$

where $P_1(t|x) = P(Z(t) = 1|Z(0) = 0, x)$ and $g(\cdot)$ is a strictly monotonic differentiable function. Note, however, that while $P_1(t_o|x)$ is a marginal feature that can be robustly estimated under independent censoring, it is not a sufficient basis for the formal assessment of treatment effects in randomized trials. The primary reason is that if $Y_1(t_o) \neq 1$, then the process could be in state 0 (event-free) or it could have failed for any of the other causes $2, \ldots, K$. These states $(0, 2, \ldots, K)$ would be treated as equivalent in the sense that they are not state 1, but they have very different meanings and values. The parsimonious summaries of treatment effects are very challenging in such settings, and it can be misguided to summarize causal treatment effects in terms of a scalar parameter when the problem is inherently multidimensional.

A more appealing use of the Aalen-Johansen estimator is to estimate other marginal features such as the mean total utility as illustrated in Sections 4.2.3 and 8.1. There utilities of u_1, u_2 and u_3 were specified for each unit of time in the states 1−3 of Figure 4.4 and interest was in the mean cumulative utility over a period of time. The marginal expected cumulative utilities at C^A given by

$$\mu(C^A \mid x) = \sum_{j=1}^{3} \int_0^{C^A} u_j \, P(Z(s) = j \mid Z(0) = 1, x) \, ds \qquad (8.30)$$

were then estimated and tests of H_0: $\mu(C^A|1) - \mu(C^A|0)$ carried out; any differences could be given a causal interpretation in this setting if X was randomized. Of course the challenge here is to ensure the states chosen adequately distinguish different health conditions and that the adopted utilities will be widely accepted. This general utility-based approach has received increased attention in recent years in studies of quality of life as well as for the simple comparisons of survival distrbutions based on restricted mean lifetime (Chen and Tsiatis, 2001; Andersen et al., 2004; Zhao et al., 2016); the latter may be viewed as a special case in which $K = 2$, $u_1 = 1$ and $u_2 = 0$. Utility-based analyses also have appeal when the goal is to estimate a causal effect of treatment in settings with recurrent events and death (Section 3.4.3), or when there are several marker-based states of interest but mortality rates are non-negligible (Section 8.3.4).

In summary, causal effects are typically expressed in terms of simple marginal features that can be estimated robustly. With complex processes, much information is lost when only reporting intervention effects on marginal features such as the expected utility, state occupancy probability at a particular time, the mean sojourn time in a particular state, the probability of experiencing a particular transition, and so on. As we have stressed throughout this book, scientific understanding about the dynamics of a disease process and the effect of interventions are often best obtained via intensity-based modeling. This is the approach to dynamic path analysis put forward by Fosen et al. (2006); see also Section 9.3 of Aalen et al. (2008). The challenge with causal inferences in this framework is the adequate specification of models. Models based on misspecified intensity functions will yield estimators that are uninterpretable (see Section 3.5.4) and therefore not suitable as a basis for causal inference, and with complex observation schemes featuring delayed entry, random censoring and intermittent observation, our ability to check models adequately can be limited. There is therefore a tension between the scientific goal of formulating models that adequately describe multistate processes to enhance understanding and the desire to make robust causal statements about intervention effects. Intensity-based models also play a crucial role in mediation analysis where the goal is to decompose overall treatment effects on a given outcome into direct and indirect effects; here multistate models should have a central role when aiming to understand process dynamics.

8.5 Bibliographic Notes

The discussion of process-related costs and benefits here is based on the framework of Cook et al. (2003). Cook and Lawless (2007, Sections 8.2, 8.9) consider the special case of recurrent event processes, and note some alternative approaches. The discussion of probabilistic prediction in Section 8.2 is closely related to those of Gneiting and Raftery (2007), Gneiting et al. (2007), and Gneiting and Katzfuss (2014). Lawless and Yuan (2010) discuss loss functions and prediction error for point predictions, stressing that these are typically estimated with some degree of imprecision. References to the extensive literature on prediction can be found in these articles. For ROC curves and related ideas, see Pepe (2003), Krzanowski and Hand (2009) and van Houwelingen and Putter (2012, Chapter 3). Prediction in the context of disease processes is discussed at length by van Houwelingen and Putter (2012), who emphasize "landmark" models that focus on information available at the time predictions are to be made. Their Chapter 9 has an illustration based on multistate models. Steyerberg (2009) has rather few technical details but good discussions of a wide range of issues concerning prediction in clinical settings.

The literature on joint modeling of longitudinal and survival data has grown rapidly since Wulfsohn and Tsiatis (1997), and there are several review papers (e.g. Tsiatis and Davidian, 2004; Ibrahim et al., 2010). Jewell and Kalbfleisch (1996) discuss models for markers that lead to fairly tractable determination of marginal survival probabilities. Tsiatis and Davidian (2004) critically discuss the approaches based on the mixed-effects models for longitudinal profiles and the increments-based models for stochastic processes (see their Section 2.2). Rizopoulos and Ghosh (2011) consider multivariate longitudinal data, and Chi and Ibrahim (2006) extend the standard mixed model formulation to accommodate multivariate longitudinal and multivariate time to event data. Rizopoulos (2012) gives a thorough account of the mixed models with failure time hazard based on mean profiles along with illustrations on computing in R. Król et al. (2017) review many types of joint random effects models for failure times and for recurrent events, and illustrate analysis based on the R `frailtypack` package. Proust-Lima et al. (2014) review joint models based on a latent class formulation, and Proust-Lima and Taylor (2009) extend the approach to deal with multivariate longitudinal data. Joint models for multistate processes are discussed in Dantan et al. (2001); failure times in the competing risks framework are considered by Williamson et al. (2008). Tom and Farewell (2011) consider the use of joint multistate models for discrete time-dependent covariates under intermittent observation. Maziarz et al. (2017) compare methods of prediction for event times based on joint models with one based on conditioning on observed marker values as in the "landmark" approach (van Houwelingen and Putter, 2012).

Aalen et al. (2008) give a thoughtful account of the various frameworks for causal reasoning with life history processes. They highlight the different targets of estimation in the contexts of marginal structural modeling (e.g. Robins et al., 2000) and dynamic path analysis; the latter is a more natural framework for life history processes (Fosen et al., 2006). Martinussen and Vansteelandt (2013) and Aalen et al. (2015a) discuss collapsibility and marginal effects for failure times. Propensity score matching offers an alternative approach considered by Austin (2014).

Aalen (2012) discusses the use of dynamic path analysis as a basis for studying indirect and direct effects of exposure variables in mediation analysis; see also Aalen et al. (2016). Hvidtfeldt et al. (2012) report on the results of a mediation analysis studying the role of estrogen and/or insulin in the relationship between obesity, physical activity, alcohol consumption and risk of post-menopausal breast cancer. The analysis used additive hazards models to examine indirect and direct effects using data from two case-cohort studies involving data from the Women's Health Initiative Observational Study. See also Martinussen et al. (2011) and Aalen et al. (2018) concerning additive hazards models.

Marginal features of multistate processes such as state occupancy probabilities or expected utilities may be targets of causal inference and can often be estimated robustly using Aalen-Johansen estimates (Aalen et al., 2001) or more directly using the methods of Sections 4.1 and 4.2. Jensen et al. (2016) discuss the use of inverse probability weights for marginal structural models for recurrent event data. Keiding et al. (2001) used a multistate model for a mediation analysis and used the model to assess the consequences of intervening on a multistate process. A similar approach was considered by Gran et al. (2015) who also discussed the use of inverse probability of exposure weights for estimation.

8.6   Problems

Problem 8.1   Consider the organ transplantation scenario portrayed in Figure 8.1 and suppose that at age a an individual is a potential transplant recipient. Let x denote a vector of covariates related to the individual and to the donated organ. Suppose that transition intensities in Figure 8.1 may be modeled as $\lambda_{12}(t|x)$, $\lambda_{23}(t|x)$ and $\lambda_{13}(t|x_1)$, where t denotes age and x_1 is the subset of x when donated organ characteristics are dropped. This model does not explicity consider rejection of the donated organ; for simplicity we ignore this but we can think of it as an internal factor affecting the $2 \to 3$ transition intensity.

(a) Show that the expected residual (remaining) lifetimes EL_2 if the person receives the transplant and EL_1 if they remain on the wait list are

$$EL_2 = \int_a^\infty P_{22}(t \mid x)\, dt$$

$$EL_1 = \int_a^\infty [P_{11}(t \mid x) + P_{12}(t \mid x)]\, dt\,.$$

(b) For a given individual we can suppress the dependence on x and just write $\lambda_{12}(t)$, $\lambda_{23}(t)$ and $\lambda_{13}(t)$. If $\lambda_{12}(t) = 2$ with age t in years, give an expression for $EL_2 - EL_1$.

(c) The calculation in (b) ignores possible variation in the quality of a future organ donation that "matches" the individual on the wait list. How could this be incorporated?

(Section 8.1; Wolfe et al., 2008)

Problem 8.2   For the scenario portrayed in Figure 8.2, suppose that average excess (due to ALZ) public health costs of D_2 and D_3 dollars per year could be ascribed to persons in states 2 and 3, respectively. If a program costing D dollars per year could reduce the $2 \to 3$ transition intensity by 5%, outline how you could assess the net effect of the program.

(Section 7.1)

Problem 8.3   The assessment of prediction rules for binary outcomes has received much study. This can be applied to an arbitrary life history process by defining binary outcomes related to the process, for example, $Y_k(t) = I(Z(t) = k)$ or $Y_k^e(t) = I(T_k^{(1)} \le t)$. Let us consider some aspects of binary outcomes by letting Y denote a binary outcome associated with a specified time t, X a vector of related covariates, and $\mu(X) = E(Y|X) = P(Y = 1|X)$.

(a) For a specified model $\tilde{\mu}(x)$, the squared error of prediction $(Y - \tilde{\mu}(x))^2$ is often called the Brier score. Suppose $Y \sim \text{Bin}(1, \mu(x))$, where $\mu(x)$ is the true value of $P(Y = 1|x)$. Show that the average Brier score is

$$\text{EBS} = E_{YX}\{(Y - \tilde{\mu}(X))^2\}$$
$$= E_X\{\mu(X)(1 - \mu(X))\} + E_X\{(\tilde{\mu}(X) - \mu(X))^2\}\,.$$

The first term represents sharpness and the second calibration. In particular, if $\tilde{\mu}(x) = \mu(x)$ then only the first term remains, and its magnitude depends on the form of $\mu(x)$, and on the distribution of X in the population of interest.

(b) Compare EBS and the expected logarithmic score in (8.9):

$$\text{ELS} = E_{YX}\{-\log \tilde{f}(Y \mid X)\},$$

noting that $\tilde{f}(Y|X)$ is assumed to be binomial. Consider a very simple case where $X = 0$, 1 only, with probabilities p_0, p_1 in the population, and where $\tilde{\mu}(x) = \mu(x)$ for $x = 0$, 1.

(Sections 3.5.3, 8.2)

Problem 8.4 Consider a binary outcome Y as in the preceding problem, and suppose X is a continuous scalar covariate; it could also be a "risk score" such as $\tilde{\beta}'X^*$ from some regression model with covariate vector X^*. For convenience we assume X takes on values in the interval $(0,1)$. Suppose that our objective is to make a prediction or to classify individuals with respect to their as yet unobserved Y, based on knowledge of their X-value. Consider a rule of the following form: for some specified value x_0, predict $Y = 1$ if $X \geq x_0$ and predict $Y = 0$ if $X < x_0$. This assumes that larger values of X are related to the outcome $Y = 1$. Error rates can then be defined:

$$\text{FPR}(x_0) = P(X \geq x_0 \mid Y = 0)$$
$$\text{FNR}(x_0) = P(X < x_0 \mid Y = 1)$$

are called the false-positive rate and false-negative rate, respectively. The receiver operating characteristic (ROC) curve is a plot of points $(\text{FPR}(x_0), \text{TPR}(x_0))$ as x_0 varies, where $\text{TPR}(x_0) = P(X \geq x_0|Y = 1)$ is called the true-positive rate (Pepe, 2003).

(a) If X and Y are independent, show that the ROC curve is the straight line from $(0,0)$ to $(1,1)$.

(b) A "good" predictor X is one for which there exist values x_0 such that $(\text{FPR}(x_0), \text{TPR}(x_0))$ is close to $(0,1)$. Why is this the case?

(c) Performance of the classification rule can also be considered through the distribution of Y given X. The positive predictive value (PPV) is defined as $\text{PPV} = P(Y = 1|X \geq x_0)$ for a specified x_0 and, likewise, the negative predictive value (NPV) is defined as $\text{NPV} = P(Y = 0|X < x_0)$. For a given x_0 define $V = I(X \geq x_0)$ and let

$$\pi(y,v) = P(Y = y, V = v) \qquad y = 0,1; \quad v = 0,1,$$

be joint probabilities for Y, V. Write FPR, FNR, PPV and NPV in terms of the $\pi(y,v)$.

(d) The predictive performance measures in Problem 8.3 are based on the joint distribution of Y and X. For prediction of Y based on $V = I(X \geq x_0)$, obtain the expected Brier score and expected logarithmic score in terms of $\pi(y, v)$. How do conditional probabilities such as TPR and PPV relate to these performance measures?

(e) Suppose costs $C(1,0)$ and $C(0,1)$ are associated with the outcomes $Y = 1$, $V = 0$ and $Y = 0$, $V = 1$. How might they be incorporated into the choice of x_0?

(f) The area under the ROC curve (AUC) is often used as a measure of predictive power for a continuous scalar covariate or "risk score" X. Show that if independent random variables X_1 and X_0 have distribution $P(X|Y = 1)$ and $P(X|Y = 0)$, respectively, then AUC $= P(X_1 > X_0)$. If $P(Y = 1) = \pi_1$ and $P(Y = 0) = \pi_0$, can you relate AUC to the performance measures EBS and LS?

(Sections 3.5.3, 8.2; Pepe, 2003)

Problem 8.5   Collapsibility and Marginal Effects

Suppose we have a model for an outcome Y, conditional on covariates X and Z, where X and Z are independent. If Z has distribution function $G(z)$, the model for $Y|(X, Z)$ is called collapsible with respect to Z if the model for $Y|X$ has the same form.

(a) Suppose Y is a failure time with conditional survivor function $\mathcal{F}(Y|X, Z)$ of additive hazards form:

$$\mathcal{F}(y \mid x, z) = \exp\{-\Lambda_0(y) - B_x(y)\,x - B_z(y)\,z\}, \quad y > 0.$$

Show that the survivor function for $Y|X$ has the form

$$\mathcal{F}_1(y \mid x) = \exp\{-\Lambda_0^*(y) - B_x(y)\,x\},$$

where $\Lambda_0^*(y) = \Lambda_0(y) + \log\{\int \exp(-B_z(y)\,z)\,dG(z)\}$. Thus, $B_x(y)$ represents the effect of x in both the model for $Y|X, Z$ and the model for $Y|X$, which are both of additive hazards form.

(b) Show that this does not happen in general if $Y|X, Z$ has a proportional hazards form with

$$\mathcal{F}(y \mid x, z) = \exp\{-\Lambda_0(y)\,e^{\beta_x x + \beta_z z}\}.$$

That is, the distribution of $Y|X$ is not in general of proportional hazards form, nor does β_x represent the effect of X in this distribution.

(Section 8.4; Aalen et al., 2015a; Martinussen and Vansteelandt, 2013)

Problem 8.6   Collider Effect

Consider a progressive 3-state Markov model for a chronic disease process in which $1 \to 2$ and $2 \to 3$ transitions are possible. Suppose that given a binary genetic marker

X the process is Markov and the marker is prognostic for entry to state 3 through the model

$$g(P_{13}(t) \mid x) = \alpha_3(t) + \beta x \qquad (8.31)$$

as in equation (4.8) where $g(u) = \log(-\log(1-u))$.

(a) Show that (8.31) is a proportional hazards model for T_3 given $X = x$.

(b) Let $\mathcal{H}_i(t) = \{Z_i(s), 0 \le s \le t, X_i\}$ denote the process history for individual i at t in a sample of size n subject only to administrative censoring at C^A. Suppose interest lies in assessing the effect of X on the individual transition intensities by fitting models of the form

$$\lambda_{12}(t \mid \mathcal{H}(t^-)) = \lambda_1(t) \exp(\alpha_1 x) \quad \text{and} \quad \lambda_{23}(t \mid \mathcal{H}(t^-)) = \lambda_2(t) \exp(\alpha_2 x)$$

Using the results in Section 3.3 describe how you would evaluate the limiting value of $\hat{\alpha}_2$ denoted $\alpha_2^\dagger$.

(c) Let $C^A = 1$, $\lambda_1(t) = \lambda_1$ and $\lambda_2(t) = r \times \lambda_1$ and determine λ_1 such that $P(T_3 < 1 | X = 0) = \pi$, $\pi = 0.40, 0.80$ for $r = 1, 2, 5$ and 10. Plot $\alpha_2^\dagger$ versus β for $\beta > 1$. What do you conclude about the sensitivity of inferences about covariate effects to the omission of prognostic covariates?

(Section 8.4)

Problem 8.7   Consider a progressive 3-state Markov model with transition intensities

$$\lambda_{12}(t \mid x_1) = \lambda_1 \exp(\beta_1 x_1) \quad \text{and} \quad \lambda_{23}(t \mid x_1) = \lambda_2 \exp(\beta_2 x_1),$$

where X_1 is a binary covariate with $P(X_1 = 1) = 1 - P(X_1 = 0) = p$. Suppose the process is observed over $(0, C^A]$ where $C^A = 1$, and that $\beta_1 = \log 1.1$, $\beta_2 = \log 1.25$, $\lambda_2/\lambda_1 = 2$ and λ_1 is selected so that $P(T_3 < 1|X_1 = 0) = 0.8$.

(a) Write a function to derive the value of λ_{10} satisfying the constraint above.

(b) The Nelson-Aalen estimate in (3.13) can be obtained as the solution to the "score" equation by treating $d\Lambda_{kl}(t)$ as a parameter and differentiating the log of the partial likelihood

$$L_{kl} \propto \prod_{i=1}^{n} \left\{ \left[\prod_{t_r \in \mathcal{D}_{ikl}} d\Lambda_{kl}(t_{ir}) \right] \exp\left(- \int_0^\infty \bar{Y}_{ik}(u) \, d\Lambda_{kl}(u) \right) \right\}$$

where $\mathcal{D}_{ikl}$ contains the set of all $k \to l$ transition times experienced by individual i over $(0, C^A]$, $i = 1, \ldots, n$. This gives

$$U_{kl}(d\Lambda_{kl}(u)) = \sum_{i=1}^{n} \bar{Y}_{ik}(u) \left[dN_{ikl}(u) - d\Lambda_{kl}(u) \right] = 0$$

with solution $d\hat{\Lambda}_{kl}(u) = d\bar{N}_{.kl}(u)/\bar{Y}_{.k}(u)$ where $d\bar{N}_{.kl}(u) = \sum_{i=1}^{n} \bar{Y}_{ik}(u) \, dN_{ikl}(u)$ and $\bar{Y}_{.k}(u) = \sum_{i=1}^{n} \bar{Y}_{ik}(u)$. As discussed in Section 3.3, the Nelson-Aalen estimate $\widehat{\Lambda}_{kl}(t)$ is consistent for $\Lambda_{kl}^\dagger(t) = \int_0^t d\Lambda_{kl}^\dagger(u)$ where $d\Lambda_{kl}^\dagger(u)$ solves $E\{U_{kl}(d\Lambda_{kl}(u))\} = 0$. Derive the limiting value of the estimators

$$\widehat{\Lambda}_{12}(t) = \int_0^t d\widehat{\Lambda}_{12}(u) \quad \text{and} \quad \widehat{\Lambda}_{23}(t) = \int_0^t d\widehat{\Lambda}_{23}(u)$$

where the covariate X_1 is not accounted for.

(c) Suppose a sample of individuals is recruited at $t = 0$ where individuals are randomized to an experimental or control condition. Let $X_{i2} = 1$ if individual i is assigned to the experimental treatment and $X_{i2} = 0$ otherwise, where $P(X_{i2} = 1) = 0.5$. Suppose then that the intensities become $\lambda_{12}(t \mid x_1, x_2) = \lambda_1(t) \exp(\beta_1 x_1 + \gamma_1 x_2)$ and $\lambda_{23}(t \mid x_1, x_2) = \lambda_2 \exp(\beta_2 x_1 + \gamma_2 x_2)$. Using results from Section 3.3 derive the limiting value $\widehat{\gamma}_2^\dagger$ if X_1 is omitted and a model is fitted of the form

$$\lambda_{23}(t \mid x_2) = \lambda_{20}(t) \exp(\gamma_2 x_{i2}).$$

(Section 8.4)

Appendix A

Selected Software Packages

A.1 Software for Time to Event Data

A.1.1 Parametric Analyses

In Section 2.2.2 we pointed out that the likelihood for a multistate analysis can be written as a product of components with each having the form of a likelihood for the analysis of left-truncated survival data. The ability to accommodate left truncation is needed to deal with the fact that individuals cannot be at risk of transitions out of a particular state until they have entered it; the term *delayed entry* is used when considering this phenomenon in the multistate setting. This close connection between likelihoods for multistate and survival analysis means that routines for survival analysis can be exploited for multistate analysis if dataframes are suitably constructed. In S-PLUS the key function for parametric analysis is censorReg, which uses the location-scale parameterization; a location-scale model with an extreme value error distribution is also a proportional hazards model with a Weibull distribution (Cox and Oakes, 1984). A broader class of parametric proportional hazards models can be fitted using the phreg function in the eha library (Broström, 2012); these include models with Weibull, log-normal, log-logistic and Gompertz forms as well as more flexible models with piecewise-constant baseline hazards. This has been illustrated in Section 3.3.2 in a parametric analysis based on proportional cause-specific hazards.

Functions that have the facility to handle left truncation may also accommodate parametric regression modeling with time-dependent covariates provided the covariates change value at a finite number of timepoints. In this case a separate contribution to the dataframe is required every time either a transition is observed or a covariate changes value.

A.1.2 Semiparametric Analyses

The *coxph* function is named for its ability to fit Cox regression models for censored failure time data, but the code is written in such a general way that it can be used to fit models for a remarkably broad range of processes; see Therneau and Grambsch (2000). Dataframes using the counting process specification of the periods of risk if failure are most useful for the analysis of multistate processes, since they naturally deal with delayed entry as well as the possibility of repeat visits to a state. Robust standard errors of regression coefficients can also be obtained. Models with univariate multiplicative random effects can be specified with fitting carried out using penalized likelihood. Models with higher-dimensional Gaussian random effects can be fitted with the package `coxme`, which uses a Laplace approximation. The `frailtypack` package in R (Król et al., 2017) fits a variety of shared random effects models for failure times, recurrent events, and time-dependent covariates, with splines used to model baseline hazard or intensity functions.

The package `timereg` enables one to fit semiparametric additive regression models; see Martinussen and Scheike (2006) for a comprehensive account of the associated models, theory and illustrative applications. Aalen et al. (2001) point out that additive intensities can be adopted in multistate analyses and illustrate how transition probability matrices can be computed from separate fits.

A.2   Selected Software for Multistate Analyses

A.2.1   Multistate Software

Several packages have been written expressly for the analysis of multistate data. The package `msSurv` (Ferguson et al., 2012) provides nonparametric estimates of state occupancy probabilities, and of state entry and exit time distributions, for general multistate processes. The `cmprsk` library contains a suite of functions for competing risks analysis. Nonparametric estimates of the cumulative incidence functions (see (4.3) of Section 4.1.1) can be obtained along with confidence intervals. We illustrated its use first in Section 4.1.5. The function `crr` within the `cmprsk` library can be used to fit regression models based on a transformation of the cumulative incidence function. In Section 4.1.5.3, we illustrate its use with the default complementary log-log transformation. Such analyses use inverse probability of censoring weighted estimating equations as developed by Fine and Gray (1999). Intensity or cause-specific hazards-based analysis can be based on survival analysis software.

The package `mvna` can be used to obtain the nonparametric Nelson-Aalen estimates of the cumulative transition rates for general multistate models. This can be useful for graphical assessment of the plausibility of a proportional intensity assumption; two variance estimates are available, including one based on the optional variation process and one based on the predictable variation process given in Andersen et al. (1993) in their equations (4.1.6) and (4.1.7), respectively. The Aalen-Johansen estimate of the transition probability matrix can be computed from the Nelson-Aalen estimates of the transition intensity functions as discussed in Section 3.4.1; such estimates are conveniently obtained from the `etm` package (Beyersmann et al., 2012), which provides a Greenwood-type estimator of the covariance matrix (see Andersen et al., 1993). While Glidden (2002) has derived a robust covariance matrix, this has yet to be implemented to our knowledge.

The R `mstate` library has many useful functions for data manipulation and analysis of right-censored multistate data. It permits nonparametric estimation as well as fitting semiparametric models to transition intensities. We comment more on this function in Section A.2.3, where we show how it can be used to create dataframes for analysis using several different functions.

A.2.2 Methods Based on Marginal Features

The Fine and Gray approach to modeling covariate effects on transforms of the cumulative incidence function is directed at marginal inferences regarding a particular cause of failure. More generally, marginal semiparametric models can be specified through transformations of the state occupancy probabilities, which can be fitted by the use of pseudo-values; Andersen, Klein and colleagues have developed a library of R functions **pseudo** to support this. The function **pseudoci**, for example, is useful for formulating regression models based on the cumulative incidence function; this was illustrated in Section 4.1.5.3. When pseudo-values from multiple time-points are used, robust variance estimation must be carried out to address the dependence between the pseudo-values at the distinct time-points. This is facilitated by use of the **geese** function in the **geepack** library; see Section 4.1.5.3. Pseudo-values have much broader utility than just for competing risks analysis; we use this approach in Section 4.3 for the evaluation of a treatment effect on the expected total sojourn time in a symptomatic state of a 2-state process. Direct binomial regression offers another approach to modeling covariate effects on transformations of marginal state occupancy probabilities. The **comp.risk** function in the **timereg** library is used in Section 4.1.5.2 for illustration.

A.2.3 Dataframe Construction with the mstate Package

The **mstate** package has functions for converting life history data into the standard counting process form used in databases in this book. Here we illustrate the use of the function **msprep** in the **mstate** library. This function is used when a rectangular wide-format dataframe is available, containing one line per individual with the times (and censoring or event-type indicators) for various events. For illustration we consider a dataset reported in Putter et al. (2007), comprised of 2204 individuals from the European Blood and Marrow Transplant (EBMT) registry who received a bone marrow transplant between 1995 and 1998. Platelet counts are greatly depressed due to non-functioning bone marrow in such patients and take time to recover following transplant. Patients are also at risk of relapse and death, and like Putter et al. (2007) we define a composite event of relapse/death.

The rectangular version of the dataframe **ebmt3** comprised of the following variables is provided in the **mstate** package. The day of the bone marrow transplant is the time origin. The variable **prtime** records the minimum of the day of platelet recovery or censoring, with **prstat** = 1 if platelet recovery was observed and **prstat** = 0 otherwise. The variable **rfstime** records the time of relapse/death or censoring with **rfsstat** = 1 if relapse/death is observed and **rfsstat** = 0 otherwise. Covariates include the disease subclassification (**dissub**), patient age at transplant (**age**), donor-recipient gender match status (**drmatch**), and whether or not there is a T-cell depletion (**tcd**).

```
> library(mstate)
> data(ebmt3)
> head(ebmt3)
  id prtime prstat rfstime rfsstat dissub  age            drmatch    tcd
   1     23      1     744       0    CML  >40  Gender mismatch No TCD
```

```
2     35      1      360      1      CML   >40 No gender mismatch No TCD
3     26      1      135      1      CML   >40 No gender mismatch No TCD
4     22      1      995      0      AML 20-40 No gender mismatch No TCD
5     29      1      422      1      AML 20-40 No gender mismatch No TCD
6     38      1      119      1      ALL   >40 No gender mismatch No TCD
```

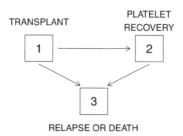

Figure A.1: A 3-state model for platelet recovery and the composite event of relapse or death.

The 3-state model is illustrated in Figure A.1 where state 1 is occupied at the time of transplant, state 2 represents platelet recovery and state 3 is entered upon relapse or death. To construct a counting process dataframe for a multistate analysis, we first define the structure of the transition intensity matrix, **tmat**; the numbered entries correspond to admissible transitions, while NA appears in entries that do not correspond to transitions.

```
> tmat <- transMat(x=list(c(2,3),c(3),c()),
                names=c("TX","PLT Recovery","Relapse/Death"))
> tmat
                  to
 from             TX PLT Recovery Relapse/Death
    TX            NA           1             2
    PLT Recovery  NA          NA             3
    Relapse/Death NA          NA            NA
```

Note that since the process begins at the time of the transplant (in state 1) and no transitions are possible into this state, there is no column in the **ebmt3** dataframe for state 1. The **time** and **status** specifications in the **msprep** function therefore have an "NA" entry.

```
> ebmt <- msprep(data = ebmt3, trans = tmat,
          time = c(NA, "prtime","rfstime"), status = c(NA, "prstat","rfsstat"),
          id=ebmt3$id, keep = c("dissub","age","drmatch","tcd"))
> ebmt[ebmt$id %in% c(1,2,3),]
 id from to trans Tstart Tstop time status dissub age        drmatch    tcd
  1    1  2     1      0    23   23      1    CML >40    Gender mismatch No TCD
  1    1  3     2      0    23   23      0    CML >40    Gender mismatch No TCD
  1    2  3     3     23   744  721      0    CML >40    Gender mismatch No TCD
  2    1  2     1      0    35   35      1    CML >40 No gender mismatch No TCD
  2    1  3     2      0    35   35      0    CML >40 No gender mismatch No TCD
  2    2  3     3     35   360  325      1    CML >40 No gender mismatch No TCD
  3    1  2     1      0    26   26      1    CML >40 No gender mismatch No TCD
  3    1  3     2      0    26   26      0    CML >40 No gender mismatch No TCD
  3    2  3     3     26   135  109      1    CML >40 No gender mismatch No TCD
```

A.3 Software for Intermittently Observed Multistate Processes

When multistate processes are subject to intermittent observation a conditionally independent visit process (CIVP) is typically assumed; see Section 5.4.1. Parametric Markov models with time-homogeneous or piecewise-constant intensities are usually adopted; code is provided in Section C.1 for fitting models M1B and M2B in the diabetic retinopathy example of Section 5.2.4. The suite of functions in the `msm` library can collectively accommodate a range of models and data types (Jackson, 2011). The algorithms for estimation are primarily based on direct maximization of the observed data likelihood and allow for the possibility that transitions into some states (typically absorbing states) are observed exactly subject to right censoring, and also that at some assessment times it is only known that the process is in one of a set of possible states. The study of bone progression and death in Section 5.6.3.2 involves intermittent observation of bone progression status along with exact observation of death times. The `msm` package can also be used to fit hidden Markov models. In Section 6.4.2 we model diabetic retinopathy with a progressive underlying process and the possibility of misclassification of states; this accommodates data in which transitions to lower states are observed. Section C.1.3 gives the associated code.

Joly et al. (2002) describe a package `SmoothHazard` that enables one to fit parametric or "semiparametric" intensity models for survival data or data from the 3-state illness-death model in which the transition time to the intermediate state may be interval censored. The semiparametric approach is based on M-splines, and estimation is based on penalized likelihood.

A.4 Miscellaneous Functions Useful for Multistate Analysis

In the following sections, we revisit some examples in Section 1.2 and introduce new ones to provide more information on how dataframes can be constructed in preparation for analyses, and how models and data can be graphically displayed.

A.4.1 Timeline Plots

Inspection of raw data is often an informative first step in analysis, and graphical methods can be appealing. The `event.chart` function in the `Hmisc` library of R functions is useful for displaying individual life histories in a timeline chart (see Figure A.2), or "event chart" (Lee et al., 2000; Dubin et al., 2001).

We illustrate this here using data from 8 individuals selected from the `rhDNase` dataframe describing the recurrent exacerbations in patients with cystic fibrosis. These exacerbations arise from a simple alternating 2-state process and vary in frequency and duration. In the following dataframe, event types `etype` = 1 and 2 denote transitions into the exacerbation and non-exacerbation states, respectively.

```
> rhDNase[rhDNase$id %in% c(493303,493313,589303,589310,601320,765303,766303,768309),
          c("id","time1","time2","status","etype","enum")]
     id time1 time2 status etype enum
 493303     0   169      0     1    1
 493313     0    90      1     1    1
```

493313	90	104	1	2	2
493313	104	166	0	1	3
589303	0	60	1	1	1
589303	60	74	1	2	2
589303	74	83	1	1	3
589303	83	124	1	2	4
589303	124	169	0	1	5
:	:	:	:	:	:

To make use of the `event.chart` function, we first construct a dataframe with one line per individual, shown below. The variables `dstart` and `dstop` contain the times of the start and end of the observation period. The variables `e1` to `e3` record the times the first three exacerbations started, and variables `ef1` to `ef3` record times when the respective exacerbations were resolved. The variables `etime1` to `etime12` specify different line types to distinguish exacerbation and exacerbation-free periods.

```
> copdEC
  old.id id  e1 e2  e3 ef1 ef2 ef3 dstart dstop etime1 etime2 etime3
1 493303  1  NA NA  NA  NA  NA  NA      0   169      0    169     NA
2 493313  2  90 NA  NA 104  NA  NA      0   166      0     90     90
3 589303  3  60 83  NA  74 124  NA      0   169      0     60     60
4 589310  4  35 71  NA  64 108  NA      0   169      0     35     35
5 601320  5  13 51 166  35  77 169      0   169      0     13     13
6 765303  6  69 NA  NA  75  NA  NA      0   171      0     69     69
7 766303  7  59 NA  NA  63  NA  NA      0   173      0     59     59
8 768309  8 122 NA  NA 141  NA  NA      0   170      0    122    122

  etime4 etime5 etime6 etime7 etime8 etime9 etime10 etime11 etime12
1     NA     NA     NA     NA     NA     NA      NA      NA      NA
2    104    104    166     NA     NA     NA      NA      NA      NA
3     74     74     83     83    124    124     169      NA      NA
4     64     64     71     71    108    108     169      NA      NA
5     35     35     51     51     77     77     166     166     169
6     75     75    171     NA     NA     NA      NA      NA      NA
7     63     63    173     NA     NA     NA      NA      NA      NA
8    141    141    170     NA     NA     NA      NA      NA      NA
```

In the `event.chart` function, we specify the `line.add` option to read in individual data from these 12 columns as a 2×6 matrix so that up to 6 line segments can be represented in the event chart. The `line.add.lty`, `line.add.col` and `line.add.lwd` option controls the line type, color and thickness, respectively. The `subset.c` option indicates the data we want to include in the event chart. We specify `y.idlabels = "id"` to display patient ID in the first column of the dataframe on the y-axis, and `sort.by = "id"` will sort the data by patient ID. The `point.pch` option allows us to specify different symbols for each type of events, and `point.cex` option controls the size of the symbols. Figure A.2 shows the event chart of these 8 individuals.

```
> library(Hmisc)
> event.chart(copdEC, subset.c=c(paste("e",1:3,sep=""), paste("ef",1:3,sep=""),
                    "dstart", "dstop", paste("etime",1:12,sep="")),
  sort.by="id", y.idlabels="id", x.julian=TRUE, y.lim.extend=c(0,0.5),
  point.cex=c(rep(1.2,6), 1.5, 1,5, rep(0,12)),
```

```
point.pch=c(rep(19,3), rep(1,3), 3, 3, rep(3,12)),
x.lab="DAYS SINCE RANDOMIZATION", y.lab="PATIENT ID",
titl ="EVENT CHART FOR 8 SELECTED INDIVIDUALS",
line.lty=2, line.lwd=1, cex=1, legend.plot=TRUE, legend.location="i",
line.add = matrix(paste("etime",1:12,sep=""),2,6),
line.add.lty=rep(c(1,2),3), line.add.col=rep("black",6), line.add.lwd=rep(c(3,1),3),
legend.point.at=list(x=c(0,50), y=c(9,9.5)), legend.point.pch=c(19,1),
legend.point.text=c("EXACERBATION","EXACERBATION FREE"),
legend.cex=1, legend.bty="n")
```

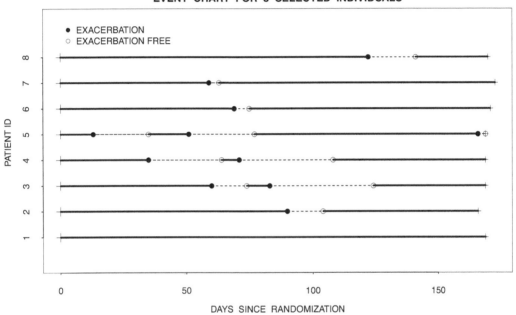

Figure A.2: An event chart for the onset and resolution of exacerbations in a sample of 8 individuals in a cystic fibrosis trial (Therneau and Hamilton, 1997).

A.4.2 Lexis Diagrams

Lexis diagrams are useful for both displaying data and considering the implications of sampling schemes on life history analysis. The **Lexis** function in the R package **Epi** can create such plots, and we illustrate it briefly here. For a more detailed description and illustration see Plummer and Carstensen (2011) and Carstensen and Plummer (2011).

Here we consider the data from the University of Toronto Psoriasis Registry and the University of Toronto Psoriatic Arthritis Registry discussed in Example 7.1.3 of Section 7.1.3. We aim to display data related to the 4-state process of Figure 7.2 and the recruitment time. We use the first 10 lines of the **incPsA** dataframe of Section D.9 for illustration, where all 10 individuals were enrolled in the PsA registry. In the dataframe below the calendar year of birth is recorded along with the ages of the individuals at key points in the disease process and at entry to the registry; the **death.status** variable indicates whether follow-up was terminated by death.

```
> events
   id bday age.entry age.ps age.psa age.last.contact death.status
    1 1942    39.461     18      33            74.853            0
    2 1904    78.689     22      31            86.746            1
    3 1965    15.540     15      15            52.107            0
    4 1910    70.286     50      50            80.821            1
    5 1923    60.011     52      52            74.229            1
    6 1952    28.753     12      19            65.227            0
    7 1938    42.705     37      42            63.184            1
    8 1921    59.398     42      42            83.077            1
    9 1925    55.680     48      48            59.469            1
   10 1949    33.667     15      28            45.990            1
```

The following commands produce a Lexis diagram for the 10 individuals.

```
> library(Epi)
> followup <- Lexis(entry=list(cal = age.entry + bday, age = age.entry),
     exit=list(cal = age.last.contact + bday), exit.status = death.status,
     id=id, data=events)

> followup[,c("cal","age","lex.dur","lex.Cst","lex.Xst","lex.id")]
       cal      age lex.dur lex.Cst lex.Xst lex.id
  1981.461 39.461  35.392       0       0      1
  1982.689 78.689   8.057       0       1      2
  1980.540 15.540  36.567       0       0      3
  1980.286 70.286  10.535       0       1      4
  1983.011 60.011  14.218       0       1      5
  1980.753 28.753  36.474       0       0      6
  1980.705 42.705  20.479       0       1      7
  1980.398 59.398  23.679       0       1      8
  1980.680 55.680   3.789       0       1      9
  1982.667 33.667  12.323       0       1     10

> events$age.exit <- ifelse(events$age.entry <= events$age.ps,
                            events$age.ps + 0.5, events$age.entry)
> psc.to.entry <- Lexis(entry=list(cal = age.ps + bday, age = age.ps),
     exit=list(cal = age.exit + bday), exit.status = 1, id=id, data=events)

> events$age.exit <- ifelse(events$age.ps == events$age.psa,
                            events$age.psa + 0.5, events$age.psa)
> psc.to.psa <- Lexis(entry=list(cal = age.ps + bday, age = age.ps),
     exit=list(cal = age.exit + bday), exit.status = 1, id=id, data=events)

> plot(0,0,type="n",axes=F,xlim=c(1920,2020),ylim=c(0,100),xlab="",ylab="")
> axis(side=1, at=seq(1920,2020,by=10), label=T)
> axis(side=2, at=seq(0,100,by=20), label=T, las=2, adj=1)
> mtext("CALENDAR YEAR", side=1, line=2.6)
> mtext("AGE OF INDIVIDUAL", side=2, line=2.8)
> lines(psc.to.psa, lty=1, lwd=2)
> lines(psc.to.entry, lty=2, lwd=1); points(psc.to.entry, pch=1, cex=1)
> lines(followup, lty=2, lwd=1)
> points(subset(followup, status(followup) %in% 1), pch=19, cex=1)
> box()
```

Figure A.3 contains the resulting plot.

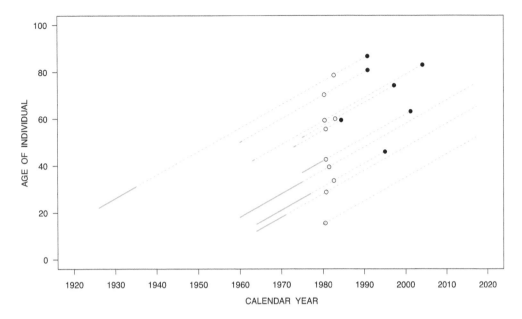

Figure A.3: A Lexis diagram for a sample of 10 individuals from the PsA registry with lines starting at the onset of psoriasis and changing to a dashed line upon the onset of PsA and terminating upon death (circle) or last contact; the open circle corresponds to the recruitment time to the PsA registry.

A.5   Drawing Multistate Diagrams with the Epi R Package

The Epi package also has a function boxes.Lexis for creating multistate diagrams, and we illustrate this here for a simple illness-death model; see Carstensen and Plummer (2011) for more details. This first involves specifying the structure of the transition intensity matrix qmat. We then call the boxes.Lexis function to create a diagram with states labeled as State 0 to 2 and arrows labeled as λ_{01}, λ_{02} and λ_{12}. Additional functionality can be exploited to refine and customize multistate diagrams; see the Epi help manual (Carstensen et al., 2016). The resulting figure is given in Figure A.4(a).

```
> qmat <- rbind(c(NA,1,1),c(NA,NA,1),c(NA,NA,NA))
> rownames(qmat) <- colnames(qmat) <- c("State 0","State 1","State 2")
> boxes.Lexis(qmat, boxpos=TRUE, wmult=3, hmult=3, cex=1.2,
             txt.arr=c(expression(lambda['01']), expression(lambda['02']),
                       expression(lambda['12'])) )
```

For LaTeX users, it is possible to use the package tikz to create multistate diagrams directly. The following is sample LaTeX code for the creation of an illness-death diagram using the tikZ package; the resulting figure is in Figure A.4(b).

```
\usepackage{tikz}
\usetikzlibrary{automata, positioning, arrows, shapes.geometric}

\begin{document}
\tikzset{square/.style={draw, regular polygon, regular polygon sides=4,
                minimum size=2cm}}
```

```
\begin{figure}
\centering
\begin{tikzpicture}
\node[square] at (0,0)  (0) {$0$};
\node[square] at (6,0)  (1) {$1$};
\node[square] at (3,-4) (2) {$2$};
\draw[->, line width=2pt, >= triangle 45, auto]
(0) edge node {$\lambda_{01}$} (1)
(0) edge node {$\lambda_{02}$} (2)
(1) edge node {$\lambda_{12}$} (2);
\end{tikzpicture}
\end{figure}
\end{document}
```

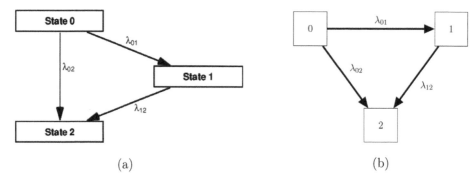

(a)   (b)

Figure A.4: A multistate diagram for an illness-death process created using the **boxes.Lexis** function in the **Epi** library (panel a) and the **tikZ** package in LaTeX (panel b).

Appendix B

Simulation of Multistate Processes

B.1 Generating a 3-State Time-Nonhomogeneous Markov Process

B.1.1 Intensities Featuring Smooth Time Trends

Here we describe the simulation of data from a time-nonhomogeneous Markov illness-death process. The incorporation of a time trend is important to realistic representations of age-dependent risk of disease onset and mortality. The illness-death process can be represented in Figure B.1.

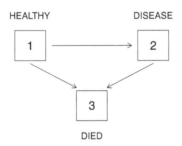

HEALTHY   DISEASE

1 ⟶ 2

3

DIED

Figure B.1: An illness-death model for disease onset and death.

We let $\lambda_{12}(t|\mathcal{H}(t^-)) = \lambda_{12}(t; \theta_{12})$ denote the intensity for disease onset at age t, $\lambda_{13}(t|\mathcal{H}(t^-)) = \lambda_{13}(t; \theta_{13})$ represent the disease-free mortality rate at age t, and $\lambda_{23}(t|\mathcal{H}(t^-)) = \lambda_{23}(t; \theta_{23})$ represent the mortality rate at age t for a diseased individual. We take the intensities to have the Weibull form for illustration, with

$$\lambda_{kl}(t; \theta_{kl}) = \alpha_{kl}\,\lambda_{kl}\,(\lambda_{kl}t)^{\alpha_{kl}-1},$$

and $\theta_{kl} = (\lambda_{kl}, \alpha_{kl})'$, for $(k,l) = (1,2), (1,3)$ and $(2,3)$. We set $\lambda_{12} = 0.0163$, $\lambda_{13} = 0.0200$ and $\lambda_{23} = 0.0224$, and let $\alpha_{12} = \alpha_{13} = \alpha_{23} = \alpha = 2$ in an illustration below. States are defined as 1, 2 or 3 with 999 used to indicate a censoring state. The following code can be used to generate data based on this model. We adopt a simple approach of generating $T_{1l} \sim \text{Weibull}(\alpha_{1l}, \lambda_{1l})$ for $l = 2, 3$, but note that $t^{(1)} = \min(t_{12}, t_{13})$ is the exit time from state 1; if $t_{13} < t_{12}$ then the process terminates in state 3 at $t^{(1)}$, but otherwise the process enters state 2 and the risk of death is characterized by the new mortality rate $\lambda_{23}(t)$. In this case, the entry time to state 3, T_{23}, is simulated according to a left-truncated Weibull distribution with conditional survivor function

$$P(T_{23} \geq t \mid T_{23} > t_{12}) = \frac{\exp(-(\lambda_{23}\, t)^{\alpha_{23}})}{\exp(-(\lambda_{23}\, t_{12})^{\alpha_{23}})}.$$

A library of functions `truncdist` is loaded for this step.

The following function implements this procedure for N individuals, with a common administrative censoring time CC.

```
> library(truncdist)
> generatedata.f <- function(N, CC, lam12, lam13, lam23, alp12, alp13, alp23) {
    outdata <- NULL
    for (i in 1:N) {
      t12 <- rweibull(1, shape=alp12, scale=(1/lam12))
      t13 <- rweibull(1, shape=alp13, scale=(1/lam13))
      t1  <- min(c(t12, t13))
      s1  <- ifelse(t12 < t13, 2, 3)
      if ((t12 < t13) && (t12 < CC)) {
        t23 <- rtrunc(1, spec="weibull", a=t1, b=Inf, shape=alp23, scale=(1/lam23))
        s23 <- 3

        estart  <- c(0, 0, t1)
        estop   <- c(t1, t1, min(t23,CC))
        estatus <- c(1, 0, ifelse(t23 < CC, 1, 0))
        from    <- c(1, 1, s1)
        to      <- c(s1, s23, s23)
        for.etm <- c(1,0,1)
      }
      else {
        estart  <- rep(0,2)
        estop   <- rep(min(t1,CC),2)
        estatus <- c(ifelse(t1 < CC, 1, 0), 0)
        from    <- rep(1,2)
        to      <- c(s1,2)
        for.etm <- c(1,0)
      }
      id <- rep(i, length(estart))
      outdata <- rbind(outdata, data.frame(id, estart, estop, estatus, from, to,
        for.etm))
    }
    outdata <- outdata[order(outdata$id, outdata$from, outdata$to),]
    return(outdata)
  }
```

A call to the above function creates a dataframe in the counting process style as displayed below. We specify for illustration a dataset of $n = 5000$ individuals with a single administrative censoring time at $t = 100$. Note that there are at least two lines contributed to the dataframe for any given individual, to ensure the times at risk in each state are represented. The `etm` package illustrated below requires less information, so the `for.etm` variable indicates the lines to be extracted for an analysis based on that function.

```
> set.seed(1000)
> simdata <- generatedata.f(N=5000, CC=100, lam12=0.0163, lam13=0.0200, lam23=0.0224,
                            alp12=2, alp13=2, alp23=2)
> simdata[simdata$id %in% c(1,3,16,181),]
```

```
 id    estart       estop estatus from to for.etm
  1   0.00000   26.26575       0    1  2        0
  1   0.00000   26.26575       1    1  3        1
  3   0.00000   49.87353       1    1  2        1
  3   0.00000   49.87353       0    1  3        0
  3  49.87353   71.84872       1    2  3        1
 16   0.00000   44.96457       1    1  2        1
 16   0.00000   44.96457       0    1  3        0
 16  44.96457  100.00000       0    2  3        1
181   0.00000  100.00000       0    1  2        0
181   0.00000  100.00000       0    1  3        1
```

We now illustrate the use of the R function `coxph` to obtain the Nelson-Aalen estimates of the cumulative intensities $\Lambda_{kl}(t)$. The Nelson-Aalen estimates are obtained with a call to the `coxph` function with the `Surv(estart, estop, estatus)` response, no covariates indicated by the "~ 1" specification and the selection of the appropriate lines of the dataframe. The resulting estimates are plotted in Figure B.2(a) along with the true cumulative intensity functions.

```
> library(survival)
> fit12 <- coxph(Surv(estart, estop, estatus) ~ 1,
    data=simdata[(simdata$from == 1) & (simdata$to == 2),], method="breslow")
> na12 <- survfit(fit12, type="aalen")

> fit13 <- coxph(Surv(estart, estop, estatus) ~ 1,
    data=simdata[(simdata$from == 1) & (simdata$to == 3),], method="breslow")
> na13 <- survfit(fit13, type="aalen")

> fit23 <- coxph(Surv(estart, estop, estatus) ~ 1,
    data=simdata[(simdata$from == 2) & (simdata$to == 3),], method="breslow")
> na23 <- survfit(fit23, type="aalen")
```

The `etm` function is then used to obtain the Aalen-Johansen estimates of the 3×3 transition probability matrix $P(0,t)$. Figure B.2(b) contains plots of the probability of being alive with disease $(P_{12}(0,t))$ and the cumulative probability of death $(P_{13}(0,t))$. These exhibit good agreement with the true values depicted by the lighter lines as would be expected with a large dataset.

```
> library(etm)
> trdata <- simdata[simdata$for.etm == 1,]
> trdata$from <- as.character(trdata$from)
> trdata$to <- ifelse((trdata$to == 3) & (trdata$estatus == 0),
                  "cens", as.character(trdata$to))
> trdata <- trdata[,c("id","from","to","estart","estop")]
> dimnames(trdata)[[2]] <- c("id","from","to","entry","exit")
> trdata[trdata$id %in% c(1,3,16,181),]
   id from   to   entry      exit
    1    1    3 0.00000  26.26575
    3    1    2 0.00000  49.87353
    3    2    3 49.87353  71.84872
   16    1    2 0.00000  44.96457
   16    2 cens 44.96457 100.00000
  181    1 cens  0.00000 100.00000
```

```
> tra <- matrix(ncol=3, nrow=3, FALSE)
> tra[1,c(2,3)] <- TRUE
> tra[2,3] <- TRUE

> tr <- etm(trdata, c("1","2","3), tra, "cens", 0)

> aj <- data.frame( cbind( tr$time, t(tr$est[1,,])) )
> dimnames(aj)[[2]] <- c("tt","P11","P12","P13")
```

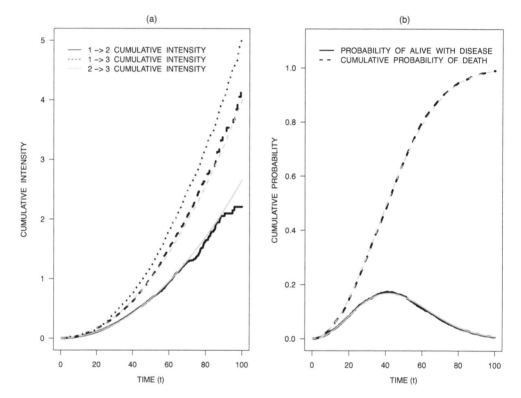

Figure B.2: Nelson-Aalen estimates (dark lines) and true cumulative transition intensities (light lines) for the illness-death model (panel a) along with the estimated cumulative incidence function for disease and cumulative probability of death (panel b), based on a sample of size $n = 5000$.

B.1.2 Processes with Piecewise-Constant Intensities

Multistate processes with piecewise-constant intensities can be generated using the sim.msm function in the msm library. Suppose that intensities are specified with common cut-points at $0 = b_0 < b_1 < \cdots < b_{R-1} < b_R = \infty$ yielding R pieces. If we define $X_j(t) = I(b_{j-1} \leq t < b_j)$, $j = 2, \ldots, R$, then the $k \to l$ intensity function at t can be written as

$$\lambda_{kl}(t) = \lambda_{kl} \exp(X'(t) \delta_{kl}) \tag{B.1}$$

where $X(t) = (X_2(t), \ldots, X_R(t))'$ and $\delta_{kl} = (\delta_{kl2}, \ldots, \delta_{klR})'$. A transition intensity matrix is specified by entering the values for the baseline transition intensities. Suppose that we consider the 3-state illness-death model again; the matrix is then given by `rbind(c(0,a,b),c(0,0,c),c(0,0,0))` where the non-zero entries are intensities that apply over $(0, b_1)$.

We need to construct a covariate matrix with one row for each cut-point b_j and one column for each covariate, having the form

$$
\begin{array}{ccccc}
 & X_2(t) & X_3(t) & \ldots & X_R(t) \\
b_0 & 0 & 0 & \ldots & 0 \\
b_1 & 1 & 0 & \ldots & 0 \\
b_2 & 0 & 1 & \ldots & 0 \\
\vdots & \vdots & \vdots & \vdots & \\
b_{R-1} & 0 & 0 & \ldots & 1
\end{array}
$$

For illustration, we take $R = 3$ and set $\delta_{12r} = (r-1)\log 4$, $r = 2, 3$ so the $1 \to 2$ intensity increases by a factor of 4 when moving from the rth to the $(r+1)$st interval. Likewise, we set $\delta_{13r} = (r-1)\log 2$ so that the disease-free mortality rate doubles moving from one interval to the next. We let $\lambda_{23}(t) = \lambda_{13}(t)$ for $r = 1$ and $\lambda_{23}(t) = \lambda_{13}(t)e^{\beta}$ for $r = 2, 3$ with $\beta = \log 1.2$.

We present the code for simulating a process with intensities having cut-points at $t = 30$ and $t = 60$. First, we initialize the baseline transition intensity matrix corresponding to the first interval, we define the cut-points at $t = 30$ and 60 and construct the matrix `bmat` containing the time-dependent covariates according to (B.1) with the form given above, and specify the matrix of regression coefficients in `beta`.

```
> library(msm)
> qmat <- rbind(c(-0.03, 0.01, 0.02), c(0, -0.02, 0.02), c(0, 0, 0))

> btime <- c(0,30,60)
> bmat <- matrix(0, nrow=3, ncol=2)
> bmat[2,1] <- bmat[3,2] <- 1

> beta <- matrix(0, nrow=2, ncol=3)
> beta[,1] <- (c(2:3) - 1)*log(4)
> beta[,2] <- (c(2:3) - 1)*log(2)
> beta[,3] <- rep(log(1.2), 2)
```

Next the `sim.msm` function is called to create sample paths for $n = 1000$ individuals. The data for the first three individuals are then displayed.

```
> set.seed(1000)
> CC <- 1000
> simdata <- NULL
> for (i in 1:1000) {
    sim <- sim.msm(qmatrix=qmat, maxtime=CC, covs=bmat, beta=beta,
               obstimes=btime, start=1, mintime=0)
    times  <- sim$times
    states <- sim$states
    if (length(times) == 3) {
      estart  <- c(times[1], times[1], times[2])
      estop   <- c(times[2], times[2], times[3])
```

```
    estatus <- c(1, 0, ifelse(times[3] == CC, 0, 1))
    from    <- c(states[1], states[1], states[2])
    to      <- c(states[2], 3, 3)
    for.etm <- c(1, 0, 1)
  }
  else {
    estart  <- rep(times[1],2)
    estop   <- rep(times[2],2)
    estatus <- c(ifelse(times[2] == CC, 0, 1), 0)
    from    <- rep(states[1],2)
    to      <- c(states[2],2)
    for.etm <- c(1, 0)
  }
  id <- rep(i, length(estart))
  simdata <- rbind(simdata, data.frame(id, estart, estop, estatus, from, to,
    for.etm))
}

> simdata[simdata$id %in% c(1,3,8),]

id    estart       estop estatus from to for.etm
 1  0.00000   36.47116       1    1  3       1
 1  0.00000   36.47116       0    1  2       0
 3  0.00000   35.96787       1    1  2       1
 3  0.00000   35.96787       0    1  3       0
 3 35.96787   81.28846       1    2  3       1
 8  0.00000   31.20764       1    1  2       1
 8  0.00000   31.20764       0    1  3       0
 8 31.20764  100.00000       0    2  3       1
```

The `sim.msm` function returns data for an individual in the form required for analysis with the `msm` function. To obtain the Nelson-Aalen estimates of the cumulative transition intensities using the `coxph` function, the data must be modified to contain the `estart`, `estop`, `status`, `from` and `to` variables.

```
> library(survival)
> fit12 <- coxph(Surv(estart, estop, estatus) ~ 1,
    data=simdata[(simdata$from == 1) & (simdata$to == 2),], method="breslow")
> na12 <- survfit(fit12, type="aalen")

> fit13 <- coxph(Surv(estart, estop, estatus) ~ 1,
    data=simdata[(simdata$from == 1) & (simdata$to == 3),], method="breslow")
> na13 <- survfit(fit13, type="aalen")

> fit23 <- coxph(Surv(estart, estop, estatus) ~ 1,
    data=simdata[(simdata$from == 2) & (simdata$to == 3),], method="breslow")
> na23 <- survfit(fit23, type="aalen")
> na23$time[1:6]
 [1] 0.399782 4.298349 4.854390 5.692744 7.184223 8.079537

> na23$cumhaz[1:6]
 [1] 0.200000 0.232258 0.261670 0.283409 0.303017 0.319966
```

The object `na23$cumhaz` contains the Nelson-Aalen estimate of the cumulative $1 \rightarrow 2$ transition intensity with the corresponding times of jumps recorded in `na23$time`. The true values of $\Lambda_{kl}(t)$ based on the piecewise-constant specifications are plotted in the left panel of Figure B.3.

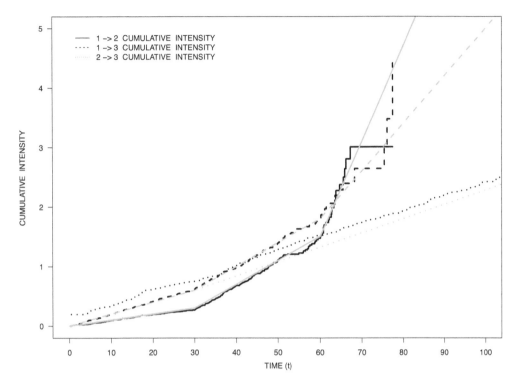

Figure B.3: Plots of the cumulative transition intensities under the piecewise-constant models along with the Nelson-Aalen estimates.

B.2   Simulating Multistate Processes under Intermittent Inspection

If interest lies in simulating data under a visit process that is completely independent of the multistate process, the visit times can be generated first and the states occupied at those times can be generated subsequently provided transition probabilities are obtainable. Consider the setting of an initial assessment at $a_{i0} = 0$, an interval of interest $[0, C^A]$ and no random censoring. Assessment times $a_{i1} < \cdots < a_{im_i} < C^A$ can be generated according to a point process with intensity $\lambda^a(t|\mathcal{H}(t^-)) = \lambda^a(t|A(u), 0 \leq u < t)$ for a completely independent assessment process; see, for example, Cook and Lawless (2007, pp. 44–45).

Let $\pi_k = P(Z_i(0) = k)$, $k = 1, \ldots, K$ so that $\pi = (\pi_1, \ldots, \pi_K)$ define the initial (multinomial) distribution for the multistate process. Then under a Markov model with transition probability matrix $P(s, t)$ having elements $p_{kl}(s, t)$, the states occupied at the assessment times may be generated as conditionally independent multinomial random variables. Specifically at a_{ij} given $Z_i(a_{i,j-1}) = k$, we simulate

$$(Y_{i1}(a_{ij}), \ldots, Y_{iK}(a_{ij})) \mid Z_i(a_{i,j-1}) = k \sim \text{multinomial}(1, p_k(a_{i,j-1}, a_{ij}))$$

where $p_k(a_{i,j-1}, a_{ij})$ is the kth row of $P(a_{i,j-1}, a_{ij})$, for $j = 1, \ldots, m_i$, $i = 1, \ldots, n$. The `rmultinom` function in the `stats` package can be used for this purpose.

To accommodate a more general conditionally independent visit process, it is sometimes preferable to first simulate the entire multistate process as described in Section B.1. The assessment time may then be generated according to any point process with the history-dependence incorporated into the intensity. Recall from Section 5.4 that we define the history of the observable process as

$$\bar{\mathcal{H}}^\circ(t) = \{Y(u), A(u), 0 \leq u \leq t; (a_j, Z(a_j)), j = 0, 1, \ldots, A(t)\}$$

and note that we may define the visit intensity at time t as dependent on $\mathcal{Z}^\circ(t^-)$ as well as $\{A(u), 0 \leq u \leq t\}$ and any fixed covariates being controlled for in the multistate model. For example, given data $\bar{\mathcal{H}}^\circ(t) = \{Y(u), A(u), 0 \leq u \leq t; (a_j, Z(a_j)), j = 0, 1, \ldots, A(t)\}$ at t, if it was thought that membership in a class of states $\mathcal{C}$ at the last assessment representing more severe disease might reduce the time to the next visit, we may set $\lambda^a(t|\bar{\mathcal{H}}^\circ(t^-))$ to

$$\lambda^a(t \mid \bar{\mathcal{H}}^\circ(t^-)) = \lambda_0^a(t) \exp(\gamma \cdot I(Z(t^-) \in \mathcal{C}))$$

for a modulated Markov visit process or

$$\lambda^a(t \mid \bar{\mathcal{H}}^\circ(t^-)) = h_0^a(B^a(t)) \exp(\theta \cdot I(Z(t^-) \in \mathcal{C}))$$

for a modulated semi-Markov visit process where $B^a(t) = t - t_{A(t^-)}$ is the time since the last assessment and $h_0^a(\cdot)$ is the baseline hazard function for the time between visits. An advantage of this approach is that, if we wished to explore the effects of a CDVP, we could let the visit intensity depend on features in the complete history $\bar{\mathcal{H}}(t^-)$ that are not in the observed history $\bar{\mathcal{H}}^\circ(t^-)$. Such a model is portrayed in Figure 5.4.

Appendix C

Code and Output for Illustrative Analyses

C.1 Illustrative Analysis of Diabetic Retinopathy

Figure C.1 gives the multistate state diagrams for the models obtained by combining states 3 to 5 in Figure 1.2. In Figure C.1 states 1, 2 and 3 represent ETDRS scores of 1, 2 or 3, and ≥ 4, respectively.

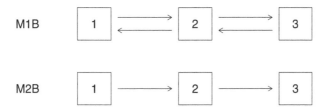

Figure C.1: Multistate diagrams for analyses of the reversible (M1B) and progressive (M2B) models for diabetic retinopathy.

What follows are a few lines of the dataframe used for the modeling of diabetic retinopathy that was reported in Sections 5.2.4 and 6.4.2. We display the lines for a single individual, with the actual data changed slightly for confidentiality reasons. The variable **enum** simply records the line in the dataframe by individual; **visit** records the nominal quarterly assessment number; **etime** is the time from randomization (years); **state** is the state occupied in the 3-state reversible model M1B at **etime**; **statep** is the state occupied under the progressive model M2B which is the highest state recorded at or before **etime**; and **trx** is the treatment indicator (0 = convential therapy, 1 = intensive therapy). So the person analyzed provided assessments for just over 9 years and was ultimately observed to enter state 3 under both models M1B and M2B. The **state** and **statep** variables are 3 at **etime** = 4.227, and **statep** remains equal to 3 thereafter.

```
> dcct
  id enum visit etime state statep trx
   1    1     0 0.000     1      1   0
   1    2     2 0.671     1      1   0
   1    3     4 1.180     1      1   0
   1    4     6 1.665     1      1   0
   1    5     8 2.166     1      1   0
   1    6    10 2.680     1      1   0
   1    7    12 3.181     2      2   0
```

```
1    8    14 3.680      2        2   0
1    9    16 4.227      3        3   0
1   10    18 4.646      2        3   0
1   11    20 5.095      3        3   0
1   12    22 5.626      2        3   0
1   13    24 6.182      2        3   0
1   14    26 6.642      2        3   0
1   15    28 7.121      3        3   0
1   16    30 7.619      3        3   0
1   17    32 8.159      3        3   0
1   18    34 8.676      2        3   0
1   19    41 9.161      3        3   0
```

Jackson (2011) gives full details on the use and functionality of the msm package, which also has an excellent user's guide on the R CRAN website (Jackson, 2016). Here we give a brief explanation of the code to produce the results of these analyses.

C.1.1   Fitting the Reversible Markov Model M1B with msm

We first load the library of msm functions and initialize the transition intensity matrix

$$
\begin{pmatrix}
-\lambda_{12} & \lambda_{12} & 0 \\
\lambda_{21} & -(\lambda_{21} + \lambda_{23}) & \lambda_{23} \\
0 & \lambda_{32} & -\lambda_{32}
\end{pmatrix} ;
$$

entries with a 0 mean that no direct transitions between the corresponding states are possible.

```
> library(msm)
> mat.q <- rbind(c(-0.1, 0.1, 0), c(0.1, -0.2, 0.1), c(0, 0.1, -0.1))
```

The fit for the conventional arm is next carried out for a Markov model with piecewise-constant intensities with cut-points at 3 and 6 years. The opt.method="optim" specifies that the general purpose optimization optim function in R be used for maximum likelihood estimation; the default method for the computation of standard errors with this specification is through numerical differentiation of the observed data log-likelihood.

```
> m1.conv <- msm(state ~ etime, subject=id, data=dcct[dcct$trx == 0,],
            qmatrix=mat.q, pci=c(3,6), center=FALSE, opt.method="optim")
> m1.conv
  Maximum likelihood estimates
  Baselines are with covariates set to 0

  Transition intensities with hazard ratios for each covariate
                    Baseline                  timeperiod[3,6)
  State 1 - State 1 -0.4425 (-0.5101,-0.3839)
  State 1 - State 2  0.4425 ( 0.3839, 0.5101) 2.1492 (1.7337,2.6642)
  State 2 - State 1  0.9716 ( 0.7962, 1.1857) 0.6617 (0.5016,0.8730)
  State 2 - State 2 -1.2965 (-1.5688,-1.0715)
  State 2 - State 3  0.3249 ( 0.2060, 0.5125) 1.6155 (0.9693,2.6923)
  State 3 - State 2  2.7160 ( 1.6037, 4.5998) 0.4288 (0.2350,0.7823)
  State 3 - State 3 -2.7160 (-4.5998,-1.6037)
```

```
                timeperiod[6,Inf)
State 1 - State 1
State 1 - State 2 2.5288 (1.6951,3.7725)
State 2 - State 1 0.3292 (0.1938,0.5591)
State 2 - State 2
State 2 - State 3 2.8182 (1.6340,4.8606)
State 3 - State 2 0.2597 (0.1369,0.4927)
State 3 - State 3

-2 * log-likelihood:   5279.873
```

The output above has one row for each non-zero entry of the transition intensity matrix. The parametrization for the piecewise-constant intensity $\lambda_{12}(t)$ is

$$\lambda_{12}(t) = \lambda_{120} \exp(\delta_{122} x_2(t) + \delta_{123} x_3(t)) \tag{C.1}$$

where $x_2(t) = I(3 \leq t < 6)$ and $x_3(t) = I(6 \leq t)$ are defined time-dependent covariates that indicate t is in $[3,6)$ and $[6,\infty)$, respectively. The first two subscripts on δ_{12r} indicates the coefficients correspond to a $1 \to 2$ transition intensity and the third represents the interval this parameter contributes in this regression model. The display of the estimates reflects this parameterization, so the maximum likelihood estimate in the row labeled State 1 - State 2 under the heading Baseline is of the $1 \to 2$ intensity for the first 3 years following randomization, and the estimates under timeperiod[3, 6) and timeperiod[6, Inf) are for δ_{122} and δ_{123}, respectively.

The qmatrix.msm function gives the estimates for the intensities themselves over the periods $[0,3)$, $[3,6)$ and $[6,\infty)$, respectively, where the values of the defined "time-dependent" covariates are specified explicitly.

```
> qmatrix.msm(m1.conv, covariates=list('timeperiod[3,6)'=0, 'timeperiod[6,Inf)'=0),
              ci="delta")
         State 1                  State 2                  State 3
State 1 -0.4425 (-0.5101,-0.3839)  0.4425 ( 0.3839, 0.5101) 0
State 2  0.9716 ( 0.7962, 1.1857) -1.2965 (-1.5688,-1.0715)  0.3249 ( 0.2060, 0.5125)
State 3 0                          2.7160 ( 1.6037, 4.5998) -2.7160 (-4.5998,-1.6037)

> qmatrix.msm(m1.conv, covariates=list('timeperiod[3,6)'=1, 'timeperiod[6,Inf)'=0),
              ci="delta")
         State 1                  State 2                  State 3
State 1 -0.9511 (-1.1067,-0.8174)  0.9511 ( 0.8174, 1.1067) 0
State 2  0.6429 ( 0.5341, 0.7738) -1.1678 (-1.3437,-1.0150)  0.5249 ( 0.4269, 0.6455)
State 3 0                          1.1645 ( 0.8943, 1.5163) -1.1645 (-1.5163,-0.8943)

> qmatrix.msm(m1.conv, covariates=list('timeperiod[3,6)'=0, 'timeperiod[6,Inf)'=1),
              ci="delta")
         State 1                  State 2                  State 3
State 1 -1.1191 (-1.6272,-0.7696)  1.1191 ( 0.7696, 1.6272) 0
State 2  0.3198 ( 0.1957, 0.5226) -1.2356 (-1.6020,-0.9530)  0.9158 ( 0.6781, 1.2367)
State 3 0                          0.7054 ( 0.4899, 1.0158) -0.7054 (-1.0158,-0.4899)
```

The confidence intervals are based on the "delta method", which first involves computation of a 95% confidence interval on the log scale and then exponentiation of the limits. So for example, the 95% CI for $\lambda_{12}(t)$ for $t \in [3,6)$ is obtained as

$$\exp((\log \widehat{\lambda}_{120} + \widehat{\delta}_{122}) \pm 1.96 \text{ s.e.}(\log \widehat{\lambda}_{120} + \widehat{\delta}_{122})).$$

The code for fitting the corresponding model to the intensive therapy arm simply involves using the corresponding subset through the specification `data=dcct[dcct$trx == 1,]`.

C.1.2   Fitting the Progressive Markov Model M2B with msm

Fitting model M2B of Figure C.1 requires initialization of a new transition intensity matrix

$$\begin{pmatrix} -\lambda_{12} & \lambda_{12} & 0 \\ 0 & -\lambda_{23} & \lambda_{23} \\ 0 & 0 & 0 \end{pmatrix}$$

since transitions to lower states are not possible.

```
> mat.q <- rbind(c(-0.1, 0.1, 0), c(0, -0.1, 0.1), c(0, 0, 0))
```

The call to the `msm` function here uses the `statep` variable which is non-decreasing over time; the command is otherwise the same as it was for the reversible model. The code below is for the analysis of the data in the conventional therapy arm.

```
> m2.conv <- msm(statep ~ etime, subject=id, data=dcct[dcct$trx == 0,],
             qmatrix=mat.q, pci=c(3,6), center=FALSE, opt.method="optim")
> m2.conv
 Maximum likelihood estimates
 Baselines are with covariates set to 0

 Transition intensities with hazard ratios for each covariate
                 Baseline              timeperiod[3,6)
 State 1 - State 1 -0.2964 (-0.34067,-0.25795)
 State 1 - State 2  0.2964 ( 0.25795, 0.34067) 1.671 (1.270,2.198)
 State 2 - State 2 -0.1211 (-0.17114,-0.08565)
 State 2 - State 3  0.1211 ( 0.08565, 0.17114) 1.765 (1.161,2.683)

                    timeperiod[6,Inf)
 State 1 - State 1
 State 1 - State 2 2.594 (1.108,6.077)
 State 2 - State 2
 State 2 - State 3 2.620 (1.491,4.603)

 -2 * log-likelihood:  2595.012

> qmatrix.msm(m2.conv, covariates=list('timeperiod[3,6)'=0, 'timeperiod[6,Inf)'=0),
                  ci="delta")
         State 1                  State 2                  State 3
 State 1 -0.2964 (-0.34067,-0.25795)  0.2964 ( 0.25795, 0.34067) 0
 State 2 0                            -0.1211 (-0.17114,-0.08565)  0.1211 ( 0.08565, 0.17114)
 State 3 0                            0                            0

> qmatrix.msm(m2.conv, covariates=list('timeperiod[3,6)'=1, 'timeperiod[6,Inf)'=0),
                  ci="delta")
         State 1                  State 2                  State 3
 State 1 -0.4953 (-0.6214,-0.3947)  0.4953 ( 0.3947, 0.6214) 0
 State 2 0                          -0.2136 (-0.2654,-0.1720)  0.2136 ( 0.1720, 0.2654)
 State 3 0                          0                          0
```

```
> qmatrix.msm(m2.conv, covariates=list('timeperiod[3,6)'=0, 'timeperiod[6,Inf)'=1),
                ci="delta")
         State 1                  State 2                  State 3
State 1 -0.7691 (-1.7814,-0.3320)  0.7691 ( 0.3320, 1.7814) 0
State 2 0                         -0.3172 (-0.4955,-0.2031)  0.3172 ( 0.2031, 0.4955)
State 3 0                          0                         0
```

C.1.3 Fitting the Hidden Markov Model with msm

For the hidden Markov model discussed in Section 6.4.2, the underlying process is taken to be a progressive 3-state Markov model, and so the initialization of the transition intensity matrix is as in Section C.1.2. In addition here we must initialize the matrix of misclassification (error) probabilities, which we denote by **emat**. Using the notation of Section 6.4.1, we write this as

$$\begin{pmatrix} 1 - \nu_{12} - \nu_{13} & \nu_{12} & \nu_{13} \\ \nu_{21} & 1 - \nu_{21} - \nu_{23} & \nu_{23} \\ \nu_{31} & \nu_{32} & 1 - \nu_{31} - \nu_{32} \end{pmatrix},$$

but note that in the initialization step the diagonal entries, which correspond to no misclassification error, are set to zero; this is in contrast to the way the transition intensity matrix is initialized where the diagonal entries are functions of the off-diagonal entries so a non-zero initial value can be specified. The following assignment to **emat** means we will accommodate all possible state misclassifications in this model.

```
> mat.q <- rbind(c(-0.1, 0.1, 0), c(0, -0.1, 0.1), c(0, 0, 0))
> emat  <- rbind(c(0, 0.1, 0.1), c(0.1, 0, 0.1), c(0.1, 0.1, 0))
```

The call to the **msm** function uses the variable **state**, which admits transitions to lower states, but with the underlying process being progressive such data are modeled jointly through the progressive Markov model and the misclassification process. The **ematrix = emat** specification is what ensures that a hidden Markov model is fitted. The R **optim** function is again exploited for maximization, which is feasible with a progressive model and a relatively small number of states. The call to the msm function and the results follow.

```
> hmm.conv <- msm(state ~ etime, subject=id, data=dcct[dcct$trx == 0,],
                qmatrix=mat.q, ematrix=emat,
                pci=c(3,6), center=FALSE, opt.method="optim")
> hmm.conv
  Maximum likelihood estimates
  Baselines are with covariates set to 0

  Transition intensities with hazard ratios for each covariate
                    Baseline                     timeperiod[3,6)
  State 1 - State 1 -0.19119 (-0.23088,-0.15832)
  State 1 - State 2  0.19119 ( 0.15832, 0.23088) 1.751 (1.259,2.435)
  State 2 - State 2 -0.06532 (-0.12907,-0.03306)
  State 2 - State 3  0.06532 ( 0.03306, 0.12907) 3.244 (1.530,6.879)
```

```
                        timeperiod[6,Inf)
  State 1 - State 1
  State 1 - State 2 1.765 (0.7627, 4.082)
  State 2 - State 2
  State 2 - State 3 5.429 (2.3936,12.314)

  Misclassification probabilities
                          Baseline
  Obs State 1 | State 1 0.9325346 (8.599e-01,0.968870)
  Obs State 2 | State 1 0.0632122 (4.913e-02,0.080983)
  Obs State 3 | State 1 0.0042532 (1.995e-03,0.009046)
  Obs State 1 | State 2 0.1846336 (1.435e-01,0.234332)
  Obs State 2 | State 2 0.7644179 (6.825e-01,0.830445)
  Obs State 3 | State 2 0.0509485 (3.470e-02,0.074227)
  Obs State 1 | State 3 0.0009228 (3.859e-07,0.688535)
  Obs State 2 | State 3 0.2783080 (1.831e-01,0.398900)
  Obs State 3 | State 3 0.7207692 (1.014e-03,0.999848)


  -2 * log-likelihood:  5068.764
```

The output above can again be reformulated in terms of the estimated transition intensities for each time period, as follows.

```
> qmatrix.msm(hmm.conv, covariates=list('timeperiod[3,6)'=0, 'timeperiod[6,Inf)'=0), ci="delta")
> qmatrix.msm(hmm.conv, covariates=list('timeperiod[3,6)'=1, 'timeperiod[6,Inf)'=0), ci="delta")
> qmatrix.msm(hmm.conv, covariates=list('timeperiod[3,6)'=0, 'timeperiod[6,Inf)'=1), ci="delta")
```

Estimates of the misclassification probabilities ν_{hk} can be expressed in matrix form, as follows. Misclassification rates can be modeled in terms of covariates and estimated rates given according to specified covariate values, but in the present example no covariates are involved.

```
> ematrix.msm(hmm.conv, covariates=0, ci="delta")
          State 1                          State 2
  State 1 0.9325346 (8.599e-01,0.968870) 0.0632122 (4.913e-02,0.080983)
  State 2 0.1846336 (1.435e-01,0.234332) 0.7644179 (6.825e-01,0.830445)
  State 3 0.0009228 (3.859e-07,0.688535) 0.2783080 (1.831e-01,0.398900)


          State 3
  State 1 0.0042532 (1.995e-03,0.009046)
  State 2 0.0509485 (3.470e-02,0.074227)
  State 3 0.7207692 (1.014e-03,0.999848)
```

C.2   Code for the Onset of Arthritis Mutilans in PsA

C.2.1   Dataframe and Fit of Intensity-Based Model

The dataframe for modeling the onset of arthritis mutilans in psoriatic arthritis is given in Section D.7. The analyses here will make use of the variable state, which contains the state occupied based on Figure 5.5 of Section 5.4.5 with state k in Figure 5.5 labeled $k+1$ here, $k = 0, 1, \ldots, 5$. The matrix of transition intensities is given as

$$\begin{pmatrix} -\lambda_{12} & \lambda_{12} & 0 & 0 & 0 & 0 \\ 0 & -\lambda_{23} & \lambda_{23} & 0 & 0 & 0 \\ 0 & 0 & -\lambda_{34} & \lambda_{34} & 0 & 0 \\ 0 & 0 & 0 & -\lambda_{45} & \lambda_{45} & 0 \\ 0 & 0 & 0 & 0 & -\lambda_{56} & \lambda_{56} \end{pmatrix}.$$

The initalization of the transition intensity matrix is provided below, followed by the call to msm listing the covariates to be included in the intensity models. Data on medication are not provided in Section D.7 and we simply regress here on the fixed covariates sex, age, and the two HLA markers B27 and C3. The covariate effects could differ for each transition intensity, but they are constrained to be the same here for all intensities through the **constraint** option. The cut-points for the piecewise-constant baseline intensities are at 6, 12 and 18 years.

```
> mutilans[mutilans$id == 9,
          c("id","sex.female","age.psa","b27","c3","times","state","status")]
  id sex.female age.psa b27 c3   times state status
   9          1      26   0  0   0.000     1      1
   9          1      26   0  0   9.489     2      1
   9          1      26   0  0  11.488     2      1
   9          1      26   0  0  13.528     2      1
   9          1      26   0  0  20.627     4      1
   9          1      26   0  0  23.072     6      1
   9          1      26   0  0  45.024   999      0

> mat.q <- rbind(c(-0.1, 0.1, 0, 0, 0, 0), c(0, -0.1, 0.1, 0, 0, 0),
                 c(0, 0, -0.1, 0.1, 0, 0), c(0, 0, 0, -0.1, 0.1, 0),
                 c(0, 0, 0, 0, -0.1, 0.1), c(0, 0, 0, 0, 0, 0))

> fitM <- msm(state ~ times, subject=id, data=mutilans[mutilans$status == 1,],
  covariates = ~ sex.female + age.psa + b27 + c3,
  constraint=list(sex.female=rep(1,5), age.psa=rep(1,5), b27=rep(1,5), c3=rep(1,5)),
  qmatrix=mat.q, gen.inits=TRUE, pci=c(6,12,18), opt.method="optim", center=FALSE,
  control=list(trace=2, fnscale=100000, reltol=1e-10, maxit=10000))
```

C.2.2 Marginal Model for Time to Entry to the Absorbing State

Here we fit a failure time model without the age, sex and HLA covariates to the time to arthritis mutilans by redefining state 1 as having ≤ 4 severely damaged joints (equivalent to states $1-5$ of the previous 6-state model) and state 2 as ≥ 5 severely damaged joints (state 6 of the 6-state model). We then use the msm function to fit this 2-state model with a piecewise-constant hazard with cut-points at 6, 12 and 18 years.

```
> mutilans$state6 <- ifelse(mutilans$state == 6, 2, 1)
> mutilans[mutilans$id == 9, c("id","sex.female","age.psa","b27","c3",
                               "times","state","state6","status")]
  id sex.female age.psa b27 c3   times state state6 status
   9          1      26   0  0   0.000     1      1      1
```

```
9              1    26  0  0  9.489    2      1    1
9              1    26  0  0 11.488    2      1    1
9              1    26  0  0 13.528    2      1    1
9              1    26  0  0 20.627    4      1    1
9              1    26  0  0 23.072    6      2    1
9              1    26  0  0 45.024  999      1    0
```

```
> mat.q <- rbind(c(-0.1, 0.1), c(0, 0))
```

```
> fitM12 <- msm(state6 ~ times, subject=id, data=mutilans[mutilans$status == 1,],
    qmatrix=mat.q, gen.inits=TRUE, pci=c(6,12,18), opt.method="optim", center=FALSE,
    control=list(trace=2, fnscale=100000, reltol=1e-10, maxit=10000))
> fitM12
Maximum likelihood estimates
Baselines are with covariates set to 0

Transition intensities with hazard ratios for each covariate
                       Baseline                      timeperiod[6,12]
State 1 - State 1 -0.008176 (-0.013408,-0.004986)
State 1 - State 2  0.008176 ( 0.004986, 0.013408) 1.917 (0.9037,4.067)

                      timeperiod[12,18]    timeperiod[18,Inf]
State 1 - State 1
State 1 - State 2 1.237 (0.5364,2.855) 1.453 (0.7402,2.853)

-2 * log-likelihood:  788.5847
```

To analyze this using standard failure time software accommodating interval censoring, we need to create a new dataframe with interval-censored and right-censored data as follows.

```
> mutilansTTE <- lapply(sort(unique(mutilans$id)), function(pid, indata) {
    datai <- indata[indata$id == pid,]
    datai <- datai[order(datai$times),]

    if ( sum(datai$state6 == 2) > 0 ) {
      estop   <- min( datai$times[datai$state6 == 2] )
      estart  <- max( datai$times[datai$times < estop] )
      estatus <- 3
    }
    else {
      estop   <- NA
      estart  <- max(datai$times)
      estatus <- 0
    }

    return(data.frame(id=pid, estart, estop, estatus))
  }, indata=mutilans[mutilans$estatus == 1,])
> mutilansTTE <- do.call("rbind", mutilansTTE)
> mutilansTTE[mutilansTTE$id %in% c(1,5,11,21),]
  id estart estop estatus
   1 31.403    NA       0
   5  0.000 26.73       3
  11 21.205    NA       0
  21 18.768    NA       0
```

The variable `estatus` = 3 if the failure time is interval censored and 0 if right censored. If `estatus` = 3, `estart` and `estop` record the start and stop time of the interval, respectively. If `estatus` = 0, `estop` records the right censoring time. We can use the `mutilansTTE` dataframe with the `kaplanMeier` function in S-PLUS (TIBCO Spotfire S+® 8.2, 2010) to obtain Turnbull estimates (Turnbull, 1976) as follows. The Turnbull estimate for time to mutilans is shown in Figure 5.6.

```
> fitNP <- kaplanMeier(censor(estart, estop, estatus) ~ 1, data=mutilansTTE, se.fit=F)
> fitNP$fits
  [[1]]:
        time1   time2      survival
   1    2.954   3.028     0.9819282
   2    3.061   3.206     0.9819282
   3    3.217   3.392     0.9819282
   4    3.417   4.041     0.9819282
   5    4.049   5.413     0.9735562
   6    5.415   5.509     0.9593133
   :    :       :         :
  82   37.585   4.9708e+01  0.6482456
  83   51.258   6.3075e+01  0.3866404
  84   69.361   1.0000e+29  0.0000000
```

C.2.3 Inverse Intensity Weighted Nonparametric Estimation

To obtain an inverse intensity weighted nonparametric estimate of the occupancy probability $P_K(t;\theta)$ given by (5.24), we first need to fit a model for the visit time process to compute weights $w_{ij} = \widehat{\lambda}_i^a(a_{ij}|v_i(a_{ij}))^{-1}$. Separate models were fitted using `coxph` for the first visit times a_{i0}, and the subsequent visit times a_{ij}, $j = 1,\ldots,m_i$. The time from disease diagnosis to first clinic visit is uncensored for all individuals and is modeled using Cox regression. The process for the second and subsequent visits can be viewed as a recurrent event process; Cook and Lawless (2007) describe how to fit such models with survival analysis software for general processes. With the modulated semi-Markov model in the arthritis mutilans example of Section 5.4.5, the intensity has the multiplicative form

$$\lambda_i^a(a_{ij} \mid v_i(a_{ij})) = \lambda_0(a_{ij} - a_{i,j-1}) \exp(x_{ij}'\beta) \tag{C.2}$$

where x_{ij} includes covariates for age at PsA onset, sex, HLA variables B27 and C03, and the number of damaged joints (0, 1, 2, 3, 4, ≥ 5) at the previous visit time $a_{i,j-1}$ for $j = 1,\ldots$. Times between successive visits may therefore be treated as a series of failure time outcomes with the time from the last assessment to the administrative censoring time (July 15, 2013) treated as a censored failure time; the `coxph` function in R can be used for model fitting. The model for the time to the first visit is of a similar form to (C.2) but without the covariates reflecting the number of damaged joints at the previous assessment. Table 5.5 displays the estimates for the resulting models.

The function below takes as input a dataframe and fits the two separate models with one for the time from disease onset to the first clinic visit, and the second a semi-Markov semiparametric Cox model using the (possibly censored) `egap` variable as the response in the `coxph` function.

```
> library(parallel)
> library(survival)

> iivw.nonpara.f <- function(intt, indata, r, cores) {
    indata$egap <- round(indata$egap, 10)

    fitWGT.GAP1 <- coxph(Surv(egap, status) ~ sex.female + age.psa + b27 + c3,
                     data=indata[indata$enum == 1,], method="breslow", x=TRUE)

    fitWGT.GAP <- coxph(Surv(egap, status) ~ sex.female + age.psa + b27 + c3 +
      factor(from),
                     data=indata[indata$enum > 1,], method="breslow", x=TRUE)

    pid <- sort(unique(indata$id))

    wFt <- Ft <- NULL
    for (tt in intt) {
      pp <- mclapply(pid, function(id, inr, intt, indata, infit1, infit) {
        datai <- indata[indata$id == id,]
        datai <- datai[order(datai$estart, datai$estop),]
        nlen <- length(datai$id)

        Iij <- ifelse((datai$estop > (intt - 0.5)) &
                    (datai$estop <= (intt + 0.5)), 1, 0)
        Yir <- ifelse(datai$to == inr, 1, 0)

        if ( sum(Iij) == 0 ) {
          top <- bot <- wtop <- wbot <- 0
        }
        else {
          top <- bot <- wtop <- wbot <- 0

          first.Yir <- 1
          for (j in 1:nlen) {
           if (Iij[j] != 0) {
             if (datai$enum[j] == 1) {
               km <- survfit(infit1, newdata=data.frame(sex.female=datai$sex.female[j],
                     age.psa=datai$age.psa[j], b27=datai$b27[j], c3=datai$c3[j]),
                     type="aalen")
             }
             else {
               km <- survfit(infit, newdata=data.frame(sex.female=datai$sex.female[j],
                     age.psa=datai$age.psa[j], b27=datai$b27[j], c3=datai$c3[j],
                     from=datai$from[j]), type="aalen")
             }
             km <- data.frame(tt=c(0, km$time), St=c(1, km$surv))

             lo.tt <- max(c(0, intt - 0.5 - datai$estart[j]))
             up.tt <- intt + 0.5 - datai$estart[j]

             lo <- km$St[sum(km$tt <= lo.tt)]
             up <- km$St[sum(km$tt <= up.tt)]

             Wij <- Iij[j]/(lo - up)
```

```
            Wij <- ifelse((lo - up) == 0, 0, Wij)

            top <- top + (Iij[j]*Yir[j])
            bot <- bot + Iij[j]

            wbot <- wbot + Wij
            wtop <- wtop + (Wij*Yir[j])
          }
        }
      }
      return( data.frame(top, bot, wtop, wbot) )
    }, inr=r, intt=tt, indata=indata[indata$status == 1,],
       infit1=fitWGT.GAP1, infit=fitWGT.GAP, mc.cores=cores)
    pp <- do.call("rbind", pp)

    Ft  <- c(Ft, sum(pp$top)/sum(pp$bot))
    wFt <- c(wFt, sum(pp$wtop)/sum(pp$wbot))
  }
  outdata <- data.frame(tt=intt, Ft, wFt)
  return(outdata)
}
```

The `mutilans` dataframe in Section D.7 is used here to construct a dataframe with this added variable as follows. A few lines of data manipulation are needed to create the `estart`, `estop`, `from` and `to` variables used by the function `iivw.nonpara.f`.

```
> mutilansMS <- mutilans
> mutilansMS$estop  <- mutilansMS$times
> mutilansMS$estart <- c(NA, mutilansMS$times[-nrow(mutilansMS)])
> mutilansMS$to      <- mutilansMS$state
> mutilansMS$from    <- c(NA, mutilansMS$state[-nrow(mutilansMS)])
> mutilansMS <- mutilansMS[mutilansMS$enum > 1,]
> mutilansMS$enum <- mutilansMS$enum - 1
> mutilansMS$egap <- mutilansMS$estop - mutilansMS$estart
> mutilansMS$from <- mutilansMS$from - 1
> mutilansMS$to   <- mutilansMS$to - 1
> mutilansMS <- mutilansMS[,c("id","sex.female","age.psa","b27","c3",
                        "enum","from","to","estart","estop","egap","status")]
> mutilansMS[mutilansMS$id == 9,]
  id sex.female age.psa b27 c3 enum from  to estart  estop   egap status
   9          1      26   0  0    1    0   1  0.000  9.489  9.489      1
   9          1      26   0  0    2    1   1  9.489 11.488  1.999      1
   9          1      26   0  0    3    1   1 11.488 13.528  2.040      1
   9          1      26   0  0    4    1   3 13.528 20.627  7.099      1
   9          1      26   0  0    5    3   5 20.627 23.072  2.445      1
   9          1      26   0  0    6    5 999 23.072 45.024 21.952      0
```

```
> nonpara <- iivw.nonpara.f(vectt=seq(1,30,by=1), indata=mutilansMS, r=5, cores=1)
> nonpara[1:10,]
  tt       Ft       wFt
   1 0.000000 0.000000
   2 0.000000 0.000000
   3 0.028571 0.012414
   4 0.015385 0.013606
   5 0.033784 0.019283
   6 0.052174 0.162046
```

```
 7 0.063492 0.070175
 8 0.069231 0.079855
 9 0.131783 0.118570
10 0.126984 0.097614
```

A call to the function `iivw.nonpara.f` will return unweighted (`Ft`) and weighted (`wFt`) nonparametric estimates. We will need to apply the `isoreg` function to make the estimates monotonically increase as below. We then plot the fitted values of y, `yf` against the ordered x values as a step function.

```
> nonpara.isoreg <- isoreg(x=c(0,nonpara$tt), y=c(0,nonpara$wFt))
> attributes(nonpara.isoreg)
  $names
  [1] "x"      "y"      "yf"     "yc"     "iKnots" "isOrd"  "ord"    "call"

> plot(nonpara.isoreg$x, nonpara.isoreg$yf, type="s")
```

Appendix D

Datasets

We describe here a few datasets that are publicly available, including some that were used for illustration in this book.

D.1 Mechanical Ventilation in an Intensive Care Unit

Grundmann et al. (2005) report on a prospective cohort study of individuals in an intensive care unit (ICU) where information on the occurrence of infections and the need for mechanical ventilation are collected along with discharge times. Here we consider data from a sample of 747 patients, with the goal of examining need for mechanical ventilation over time and the relation between mechanical ventilation status and risk of death or discharge. We consider a simple 4-state model depicted below with state 1 representing being in the ICU and off mechanical ventilation, state 2 representing being in the ICU and on mechanical ventilation, state 3 representing discharge and state 4 death. Covariates include patient age (years) and sex.

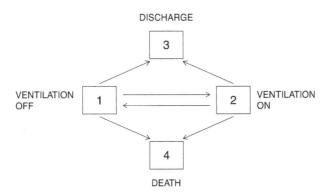

The dataframe **ventICU** described below was constructed from the **sir.adm** and **sir.cont** dataframes available in the **mvna** package; the format provided is suitable for analysis using the **etm** and **mvna** packages.

Variable	Description
id	patient ID
age	patient's age (years)
sex.female	sex: 0 if male; 1 if female

enum	cumulative number of lines by patient
from	the state from which transition may occur
to	the state to which transition may occur
	1 if ventilation OFF; 2 if ventilation ON;
	3 if discharge from ICU; 4 if death; 999 if censored
estart	the beginning of the at-risk period
estop	the end of the at-risk period

A few lines of dataframe follow.

```
> ventICU[1:9,]
```

```
      id      age sex.female enum from to estart estop
    41 75.34153          1    1    1  3      0     4
   395 19.17380          0    1    1  3      0    24
   710 61.56568          0    1    2  1      0    33
   710 61.56568          0    2    1  3     33    37
  3138 57.88038          1    1    1  3      0     8
  3154 39.00639          0    1    1  3      0     3
  3178 70.27762          0    1    1  2      0     1
  3178 70.27762          0    2    2  1      1     7
  3178 70.27762          0    3    1  3      7    24
```

The cross-tabulation of the `from` and `to` variables gives a summary of the number of observed transitions among the different states; recall 999 indicates censoring.

```
> table(ventICU$from, ventICU$to)
```

```
       1   2   3   4 999
   1   0  75 585  21   5
   2 319   0  72  55   9
```

D.2   Outcomes in Blood and Marrow Transplantation (EBMT)

We consider data on 2,279 individuals with acute lymphoid leukemia from the European Group for Blood and Marrow Transplantation (EBMT) Registry who received an allogeneic bone marrow transplant from an HLA-identical sibling donor between 1985 and 1998. These data were used in Fiocco et al. (2008), de Wreede et al. (2011), van Houwelingen and Putter (2012) and are available in the R **mstate** package as the dataframe **ebmt4**. The states are defined as state 1: alive, no recovery or adverse event (transplanted); state 2: alive in remission, recovered from treatment; state 3: alive in remission, adverse event occurred; state 4: alive, both recovered from treatment and adverse event occurred; state 5: alive in relapse (due to treatment failure); and state 6: dead (due to treatment failure). The possible transitions are shown in the figure below.

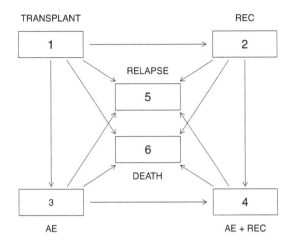

Covariates include an indicator of whether there was a match in the donor and recipient's sex (**sex.match**); whether there were any prophylactic measures for the prevention of graft versus host disease (**proph**); the calendar time of transplantation with periods 1985−1989, 1990−1994, and 1995−1998 (**year**); and a categorical variable for patient age in years with categories <=20, 20−40 and >40 years of age (**agecl**).

Variable	Description
id	patient ID
from	state from which a transition occurs
to	state to which a transition occurs
trans	numbered all possible transitions
Tstart	start time of a transition (days)
Tstop	stop time of a transition (days)
time	gap time between **Tstart** and **Tstop** (days)
status	1 if a transition is made; 0 otherwise
sex.match	donor-recipient gender match:
	no gender mismatch, gender mismatch
proph	prophylaxis: no, yes
year	year of transplantation: 1985-1989, 1990-1994, 1995-1998
agecl	patient's age (years) at transplant: <=20, 20-40, >40

This long-format dataframe is created using the function **msprep** in the **mstate** package with **ebmt4** as the input data. A few lines of the dataframe follow.

```
> ebmt[ebmt$id %in% c(1,2),]
```

id	from	to	trans	Tstart	Tstop	time	status		match	proph	year	agecl
1	1	2	1	0	22	22	1	no gender mismatch		no	1995-1998	20-40
1	1	3	2	0	22	22	0	no gender mismatch		no	1995-1998	20-40
1	1	5	3	0	22	22	0	no gender mismatch		no	1995-1998	20-40
1	1	6	4	0	22	22	0	no gender mismatch		no	1995-1998	20-40
1	2	4	5	22	995	973	0	no gender mismatch		no	1995-1998	20-40
1	2	5	6	22	995	973	0	no gender mismatch		no	1995-1998	20-40

```
1   2 6    7    22   995  973      0 no gender mismatch    no 1995-1998 20-40
2   1 2    1     0    12   12      0 no gender mismatch    no 1995-1998 20-40
2   1 3    2     0    12   12      1 no gender mismatch    no 1995-1998 20-40
2   1 5    3     0    12   12      0 no gender mismatch    no 1995-1998 20-40
2   1 6    4     0    12   12      0 no gender mismatch    no 1995-1998 20-40
2   3 4    8    12    29   17      1 no gender mismatch    no 1995-1998 20-40
2   3 5    9    12    29   17      0 no gender mismatch    no 1995-1998 20-40
2   3 6   10    12    29   17      0 no gender mismatch    no 1995-1998 20-40
2   4 5   11    29   422  393      1 no gender mismatch    no 1995-1998 20-40
2   4 6   12    29   422  393      0 no gender mismatch    no 1995-1998 20-40
```

The cross-tabulation of the **from** and **to** variables gives the aggregate counts for pairs of consecutive states.

```
> table(ebmt$from[ebmt$status == 1], ebmt$to[ebmt$status == 1])

    2   3   4   5   6
1 785 907   0  95 160
2   0   0 227 112  39
3   0   0 433  56 197
4   0   0   0 107 137
```

D.3   A Trial of Platelet Dose and Bleeding Outcomes

Heddle et al. (2009) report on an international randomized clinical trial designed to assess the impact on bleeding of routine use of a low dose of platelets ($150-300 \times 10^9/L$) versus a standard dose ($300-600 \times 10^9/L$) for prophylactic platelet transfusion in patients with thrombocytopenia. We acknowledge the BEST Collaborative for permission to release this data. The dataframe is comprised of 119 patients having received at least one platelet transfusion, with 58 assigned to the low-dose arm and 61 to the standard-dose arm. Bleeding was assessed from the day of randomization until the end of follow-up; follow-up was planned for one month but could be terminated early due to platelet function recovery (defined by a spontaneous recovery of the patient's platelet count to $50 \times 10^9/L$ or higher), patient withdrawal, physician withdrawal, a serious adverse event, death, and other or unknown reasons. The table below shows variables in the dataset, and their definitions. Several fixed covariates measured at randomization are available including the treatment assignment (**trt**), **sex**, patient blood group (**blood.grp**), Rh type (**rh.type**), height (cm) (**height**), weight (kg) (**weight**), an indicator of whether they have previously received a platelet transfusion (**history.plt.tx**), an indicator of whether they have previously received a red blood cell transfusion (**history.rbc.tx**), a baseline platelet count (**plt**), and a baseline hemoglobin value (**hb**). In addition the date of the first platelet transfusion is recorded in **ptime1** as is the last day of follow-up (**eof**) along with the status indicator at the end of follow-up (**eof.status**). Data on the days of transitions between states are recorded in **estart** and **estop** with the states involved in **from** and **to** variables.

Figure 4.9 contains a 4-state model for bleeding status (≥ 2 WHO grade), recovery of platelet function and death. To fit this model, follow-up terminated by patient or physician withdrawal, adverse event or other specified or unknown reasons can be treated as right censoring. Several lines of the dataframe for a particular individual are given below.

Variable	Description
id	patient ID
trt	treatment group: 0 = standard dose; 1 = low dose
sex	sex: 1 = male; 2 = female
blood.grp	blood group: 1 = Type O; 2 = Type A; 3 = Type B; 4 = Type AB
rh.type	Rh type: 1 = positive; 2 = negative
hgt	height (cm)
wgt	weight (kg)
history.plt.tx	history of platelet transfusion: 0 = No; 1 = Yes
history.rbc.tx	history of RBC transfusion: 0 = No; 1 = Yes
plt	baseline platelet count $(10^9/L)$
hb	baseline hemoglobin (g/L)
ptime1	first platelet transfusion day
eof	end of follow-up day
eof.status	the status at the end of follow-up day:
	1 = recovery of platelet function; 2 = death; 0 = censored
enum	cumulative number of lines by patient
from	the state from which transition may occur
to	the state to which transition may occur
	1 if no bleeding
	2 if it is a $\geq$ WHO Grade 2 bleeding
	3 if it is a platelet recovery
	4 if died
	999 indicates a censoring state
estart	the beginning of the at-risk period from randomization
estop	the end of the at-risk period from randomization

```
> SToP[SToP$id %in% c(1,2,3),]

  id trt sex blood.grp rh.type    hgt   wgt history.plt.tx history.rbc.tx
1  1   0   1         1       2 174.00 76.0              1              1
2  2   1   1         3       2 182.88' 99.5             0              1
3  2   1   1         3       2 182.88 99.5              0              1
4  2   1   1         3       2 182.88 99.5              0              1
5  3   1   1         1       1 167.64 83.0              1              1
6  3   1   1         1       1 167.64 83.0              1              1
7  3   1   1         1       1 167.64 83.0              1              1

  plt hb ptime1 eof eof.status enum from to estart estop
1  16 86      5  13          1    1    1  3      5    13
2  74 77      2  27          1    1    1  2      2    10
3  74 77      2  27          1    2    2  1     10    11
4  74 77      2  27          1    3    1  3     11    27
5  34 79      0  13          1    1    2  1      1     5
6  34 79      0  13          1    2    1  2      5     7
7  34 79      0  13          1    3    2  3      7    13
```

```
> table(SToP$from, SToP$to)

      1  2  3  4 999
  1   0 68 49  0  56
  2  69  0  5  2   3
```

D.4   Shedding of Cytomegalovirus in HIV-Infected Individuals

Betensky and Finkelstein (1999) and Goggins and Finkelstein (2000) discuss the
ACTG 181 study involving HIV-infected individuals with an opportunistic cytomegalovirus infection; we refer readers to these articles for further details on this
study. It is of interest to characterize the natural history of a CMV infection and in
particular the distributions of the times of viral shedding in the body fluids (urine
and blood) of an infected person, a precursor for the development of active CMV
disease. A more complete version of the data is used in Finkelstein et al. (2002),
but here we consider 232 individuals who had one or more follow-up assessments at
which urine or blood samples were taken. Because the samples are taken periodically
the times of viral shedding in the urine and blood are interval censored. The data in
the available file are described in the table below, where the time of viral shedding
in the urine is left censored at `urineR` if `urine.cens` = 2, right censored at `urineL`
if `urine.cens` = 0 and interval censored if `urine.cens` = 3; the censoring status
code is similar for the time to viral shedding in the blood. Note that only the dates
of the urine and blood samples necessary to convey the data on the marginal times
of shedding are provided in the data that follow; the times of all urine and blood
samples are not provided here. Cook et al. (2008) consider the joint analysis of the
time to shedding in the urine and blood based on the 4-state model in Figure 1.8.

Variable	Description
id	patient ID
urineL	days of last negative shedding in urine since earliest urine shedding date
urineR	days of first positive shedding in urine
bloodL	days of last negative shedding in blood since earliest blood shedding date
bloodR	days of first positive shedding in blood
urine.cens	shedding in urine status:
	0 if right censored; 2 if left censored; 3 if interval censored
blood.cens	shedding in blood status:
	0 if right censored; 2 if left censored; 3 if interval censored

A few lines of the dataframe follow.

```
> cmvdata[cmvdata$id %in% c(20116,20286,20309,70143,70897,140268,210066,210461),]

       id urineL urineR bloodL bloodR urine.cens blood.cens
  20116    346     NA    346     NA          0          0
  20286    264    300    350     NA          3          0
```

20309	252	280	252	336	3	3
70143	NA	1	166	NA	2	0
70897	28	56	NA	1	3	2
140268	NA	NA	NA	NA	NA	NA
210066	NA	1	172	258	2	3
210461	NA	1	NA	1	2	2

D.5 Micronutrient Powder and Infection in Malnourished Children

Lemaire et al. (2011) report on a randomized non-inferiority trial involving 263 Bangladeshi children aged 12–24 months with moderate to severe malnutrition. The children were randomized to receive either an iron-containing micronutrient powder or a placebo powder each day for a 2-month treatment period. They were assessed every 2 days for the presence of diarrhea, dysentry, lower respiratory tract infection (LRTI) and other infections over the 2-month treatment period and then weekly thereafter for 4 months; the processes are as follows:

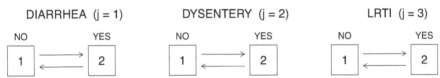

There were two phases of recruitment. In addition to the child's hemoglobin level at study entry, some additional fixed covariates were recorded. Variables in the dataset are listed in the table below.

Variable	Description
id	child ID
family.income	family monthly income (US$)
tube.well	tube well near dwelling: 0 if no; 1 if yes
size.main.room	size of dwelling's main room (m^2)
mother.literacy	literacy of mother:
	0 if cannot sign her name; 1 if only can sign;
	2 if only can read; 3 if can both read and write
hb	hemoglobin (g/L)
iron	treatment: 0 if placebo; 1 if iron
phase	1 if Phase 1; 2 if Phase 2
enum	cumulative number of lines by child
estart	start time of an interval
estop	stop time of an interval
dia.Y	1 if at risk for diarrhea; 0 otherwise
dia.status	1 if child had diarrhea; 0 if none
dys.Y	1 if at risk for dysentery; 0 otherwise
dys.status	1 if child had dysentery; 0 if none
LRTI.Y	1 if at risk for LRTI; 0 otherwise
LRTI.status	1 if child had LRTI; 0 if none

A few lines of data:

```
> sprinkles[1:10,]
```

	id	family.income	tube.well	size.main.room	mother.literacy	hb	iron	phase
1	1	3000	1	144	3	9.7	0	1
2	1	3000	1	144	3	9.7	0	1
3	1	3000	1	144	3	9.7	0	1
4	1	3000	1	144	3	9.7	0	1
5	1	3000	1	144	3	9.7	0	1
6	1	3000	1	144	3	9.7	0	1
7	1	3000	1	144	3	9.7	0	1
8	1	3000	1	144	3	9.7	0	1
9	1	3000	1	144	3	9.7	0	1
10	1	3000	1	144	3	9.7	0	1

	enum	estart	estop	dia.Y	dia.status	dys.Y	dys.status	LRTI.Y	LRTI.status
1	1	0	2	1	0	1	0	1	0
2	2	2	4	1	0	1	0	1	0
3	3	4	6	1	0	1	0	1	0
4	4	6	8	1	0	1	0	1	0
5	5	8	10	1	0	1	0	1	0
6	6	10	12	1	0	1	0	1	1
7	7	12	14	1	0	1	0	1	1
8	8	14	16	1	0	1	0	1	0
9	9	16	18	1	0	1	0	1	0
10	10	18	20	1	0	1	0	1	0

D.6   The Dynamics of *Giardia lamblia* Infection in Children

Nagelkerke et al. (1990) report on an observational field study regarding infection with the *Giardia lamblia* parasite in the Kiambu district of Kenya. This parasite, when present in the small intestine, can cause serious gastrointestinal health problems in infants. The study involved 84 children aged 11 to 18 months who were to provide weekly stool samples to be tested for the presence of *Giardia lamblia*. Follow-up was to take place for 44 weeks but early study withdrawal was common yielding highly variable follow-up. The infection status for the selected 58 children is given in Table D.6, where 1 denotes a negative stool sample and 2 indicates a positive test result for the *Giardia lamblia* parasite.

An alternating 2-state model with states representing a negative (state 1) and positive (state 2) stool sample test is of interest.

NEGATIVE TEST   POSITIVE TEST

Table D.6: Weekly infection with *Giardia lamblia* among 58 children in Kenya (Nagelkerke et al., 1990).

Child	Weekly Infection (w1 w2 w3 ... w44)
1	2 1 2 2 1 1 2 2 2 2 2 1
2	1 1 1 1 2 1 1 1 1 1 1 1 2 1 1 2 2 2 1 1 1 2 1 2
3	2 2 1 1 1 1 1 1 1 1
4	1 1 1 2 2 2 2 2 2 2 2
5	2 2 2 2 2 2 1 1 2 2 2 2 2 2 2 2 1 2 2 2 2 2 2 2
6	1 1 2 1 2 1 1 1 2 1 2 2 1 1 1 1 2 1 1 1 1 1 1 1
7	2 2 2 2 2 1 1 2 2 2 2 2
8	2 1 2 1 2 2 2 1 1 2 1 1
9	1 1 1 1 1 1 1 1 1 1
10	2 2 2 2 2 2 2 2 1 2 2 1
11	1 2 2 1 1 1 1 1 2 1 1 1 1 1
12	1 1 1 1 1 2 1 1 1 1 2 2 1 1 1 1 1 1 1 1 1 1
13	2 2 2 2 2 1 1 1 1 1 2 2 2 1 2 2
14	2 1 1 1 1 2 2 1 1 2
15	1 1 1 1 1 1 1 1 1 2 2 2 2 2 2 2
16	2 2 2 2 1 1 1 1 2 2 2 2 2 2 2 2 2 2 2 2 1 2 1 2 2
17	2 2 1 1 2 2 2 2 2 1 1 1 1 2 2 2 2
18	1 1 1 1 1 1 1 2 2 2 2
19	1 1 1 1 1 2 2 1 1 1 1 1 1 1 2 2 2 2 2 2 2
20	1 1 1 1 1 2 2 1 2 1 1 1 1 1 1 2 1 1 2 2 2 2 2 1
21	1 1 1 1 1 1 1 1 1 1
22	2 2 2 2 2 2 2 2 2 2 2 2
23	2 2 2 2 2 2 2 2 2 2 1 1 1 1 1
24	1 2 1 1 1 1 1 1 1 1 1 1 1 1 1 1 2 1
25	2 1 2 2 1 1 2 2 2 2 2
26	2 2 2 2 2 2 2 1 2 2 2 2 2
27	1 1 1 1 1 1 2 2 2 1 1 1 2 2 2 2 2 2 2 2 2 1
28	1 1 2 2 2 2 2 2 2 2 2 1 2
29	2 1 1 1 2 1 1 1 2 2 2
30	1 2 1 1 1 1 1 1 1 1 1
31	1 1 1 1 1 2 2 1 2 2
32	2 2 2 2 2 2 2 2 1 2
33	1 2 2 1 1 1 1 1 1 1 1 1 2 2 2 2 2 1 2
34	1 1 1 1 1 1 1 1 1 1 1 1 1 2 1 1 1
35	1 1 1 1 1 2 2 2 1 1 1 1 1 2
36	1 1 1 1 1 1 1 1 1 1 1 2 1 1 1 1 1 1 1
37	1 1 1 1 2 1 1 2 1 1 1 2
38	1 1 1 1 1 1 1 1 1 1 1 1 1 1 1 2 1 1 1 1 1 1 2 1
39	1 2 2 2 2 2 2 1 1 2 1 2 1 1 2 1 1 1 2 1 1 1 1 1 2 2 2 1 1 2 2 2 1 1 1 2 1 1 1 2 2 2 1 1
40	1 1 2 2 2 2 1 2 2 1
41	2 2 2 2 2 1 2 2 1 1 2 2
42	2 1 2 2 2 2 2 2 2 2 1 2 2 2 2 1 2 2 2 2 2 2 2 2
43	2 1 2 2 1 1 1 1 1 1 1 1 1 2 1 1 1 1 1 1 1 1 1 2 2 1 2 1 2 2 2 1 1 1 1 1 2 2 1 1
44	1 1 1 1 1 1 1 1 1 1 1 1
45	1 1 1 1 1 1 1 1 1 1 1 2 1 1 2
46	2 1 1 1 1 2 2 1 2 2 2 2 1 1 2 2
47	2 2 2 2 2 2 2 1 1 1 1 1 2 1 1 1 1 1 1 1 1 1 1 2 2
48	1 1 1 1 2 2 2 2 2 2 2 2 1 2 2
49	2 2 2 2 2 2 2 2 2 2 2
50	2 2 2 2 2 2 2 2 2 1 2 1 1 1 2 2 1 2 2 1 1 2 2 2 2 1 1 1 1 2
51	1 2 1 2 1 1 1 1 1 1 1 1 1 1 1 1 2
52	2 2 2 2 1 2 1 1 1 1 1 1 2 1 2 2 2 2 2 2 2 2 2 1 1
53	1 1 2 2 2 2 2 2 2 2 2
54	1 1 1 1 2 2 2 2 2 2 2
55	1 1 2 1 1 1 1 1 1 1 1 2 1
56	1 2 1 1 1 1 1 1 1 1 2 1
57	2 2 2 2 2 2 1 1 1 1 2 1 1 1
58	1 1 1 2 2 2 2 2 2 2 2 1 2 2 2 2 2 2 2 2 2 2 2 2 2 2 2 2 2 2 2 1

D.7   The Development of Arthritis Mutilans in Psoriatic Arthritis

The risk of arthritis mutilans in patients with psoriatic arthritis, a condition defined as the presence of at least five severely damaged joints, was discussed in Section 5.4.5. Here we define a joint as severely damaged if it has grade or 4 or 5 damage as determined by radiographic examination and according to the modified Steinbrocker score. This scoring system assigns an integer to each joint according to the following scale: 0 is normal, 1 reflects juxta-articular osteopenia or soft tissue swelling, 2 is the presence of erosion, 3 is presence of erosion and joint space narrowing, and 4 is total joint destruction, either lysis or ankylosis; in this dataset a possible score of 5 is added to reflect a joint that has undergone surgery. We consider data from 613 individuals from the University of Toronto Psoriatic Arthritis Clinic and take the time origin as the time of disease diagnosis, at which point we assumed all 42 joints of the hands (14 per hand), feet (6 per foot) and hips (2 total) were normal (i.e. in the category 0 of the modified Steinbrocker score). For each follow-up assessment, occurring at the times of radiographic examination, we observe the total number of joints of Grades 0 to 5, as well as the state occupied in the 6-state model with state space diagram in Figure 5.5 of Section 5.4.5, with the states reflecting the number of severely damaged joints $(0, 1, 2, 3, 4, \geq 5)$. Fixed covariates include sex, age at diagnosis of PsA, presence of the HLA B27, and presence of the HLA C3. The variable names and descriptions are as follows.

Variable	Description
id	patient ID
sex.female	sex: 0 if male; 1 if female
age.psa	age at diagnosis of PsA (years)
b27	HLA B27: 0 if no; 1 if yes
c3	HLA C3: 0 if no; 1 if yes
enum	number of lines in each patient
times	x-ray assessment time (years)
G0	number of Grade 0 damaged joints
G1	number of Grade 1 damaged joints
G2	number of Grade 2 damaged joints
G3	number of Grade 3 damaged joints
G4	number of Grade 4 damaged joints
G5	number of Grade 5 damaged joints
states	state 1, 2, ..., 6
status	1 if an x-ray assessment time; 0 if end of study period

A few lines of the dataframe follow.

```
> mutilans[mutilans$id == 116,]
```

id	sex.female	age.psa	b27	c3	enum	times	G0	G1	G2	G3	G4	G5	state	status
116	0	32	1	1	1	0.000	42	0	0	0	0	0	1	1
116	0	32	1	1	2	7.587	34	0	2	4	2	0	3	1
116	0	32	1	1	3	8.764	31	0	5	4	2	0	3	1

116	0	32	1	1	4	11.699	30	0	6	4	2	0	3	1
116	0	32	1	1	5	13.832	29	0	6	5	2	0	3	1
116	0	32	1	1	6	15.138	26	0	7	7	2	0	3	1
116	0	32	1	1	7	16.583	25	0	8	6	3	0	4	1
116	0	32	1	1	8	18.968	23	0	9	5	5	0	6	1
116	0	32	1	1	9	21.101	21	0	11	5	5	0	6	1
116	0	32	1	1	10	23.463	20	0	11	5	6	0	6	1
116	0	32	1	1	11	28.586	17	0	14	5	6	0	6	1
116	0	32	1	1	12	30.790	16	1	13	6	6	0	6	1
116	0	32	1	1	13	35.039	18	0	12	6	6	0	6	1
116	0	32	1	1	14	35.959	NA	NA	NA	NA	NA	NA	999	0

D.8 Damage of the Sacroiliac Joints in Psoriatic Arthritis

Here we consider a sub-cohort of 538 individuals from the University of Toronto
Psoriatic Arthritis Clinic. The degree of damage of the sacroiliac (SI) joints was
scored according to the New York Criteria (Bennet and Wood, 1968) with the
following categories: 0 for a normal joint, 1 if the presence of damage is equivocal,
2 if the joint is abnormal due to erosions of the bone surface or sclerosis, 3 if the
joint is unequivocally abnormal and 4 if there is evidence of ankylosis (abnormal
stiffening and immobility due to bone fusion) of the joint. Figure 6.6 displays a pair
of 3-state models used as a basis of some analyses of damage of the SI joints with
states 1 to 3 corresponding to a NYC score ≤ 1, a NYC score 2, and a NYC score of
3 or 4, respectively. Here we take the time origin as the time of the first radiograph
in the University of Toronto Psoriatic Arthritis Clinic, and restrict attention to
individuals whose sacroiliac joints were both in state 1 of Figure 6.6 at this time. This
severity score for the left and right SI joint, is then updated upon each radiographic
assessment and recorded in siL and siR, respectively; the states occupied for process
1 (left) and process 2 (right) of Figure 6.6 are recorded in stateL and stateR,
respectively. The state occupied in the joint 9-state model depicted in Figure 6.8
is in jstate, where the meaning of the states labeled 1 to 9 is given in the table
below.

Variable	Description
id	patient ID
enum	number of lines for each patient
times	x-ray assessment time in years
siL	left SI joint NYC Score (0–4)
siR	right SI joint NYC Score (0–4)
stateL	state for left SI Joint (1–3)
stateR	state for right SI Joint (1–3)
jstate	state for the joint 9-state model:
	1 = state (1, 1), 1 = state (1, 2), 3 = state (1, 3),
	4 = state (2, 1), 5 = state (2, 2), 6 = state (2, 3),
	7 = state (3, 1), 8 = state (3, 2), 9 = state (3, 3)

A few lines of dataframe follow.

```
> sijoints[sijoints$id %in% c(45,70),]

  id enum   etime siL siR stateL stateR jstate
  45    1  0.000   0   0      1      1      1
  45    2  0.008   1   0      1      1      1
  45    3  0.227   1   0      1      1      1
  45    4  3.302   1   0      1      1      1
  45    5  5.218   1   0      1      1      1
  45    6  9.205   3   1      3      1      7
  45    7 10.215   3   3      3      3      9
  45    8 12.096   3   3      3      3      9
  45    9 14.905   3   3      3      3      9
  45   10 18.015   3   3      3      3      9
  45   11 21.120   3   3      3      3      9
  45   12 24.016   3   3      3      3      9
  45   13 28.495   3   3      3      3      9
  45   14 31.751   3   3      3      3      9
  70    1  0.000   1   0      1      1      1
  70    2  9.736   2   0      2      1      4
  70    3 11.863   2   0      2      1      4
  70    4 13.311   2   1      2      1      4
  70    5 15.751   2   1      2      1      4
  70    6 22.037   2   1      2      1      4
  70    7 24.838   2   1      2      1      4
```

D.9   The Incidence of PsA in Individuals with Psoriasis

Here we provide data from 637 individuals who were recruited to the University of
Toronto Psoriasis Clinic (UTPC) and 1378 individuals recruited to the University
of Toronto Psoriatic Arthritis Clinic (UTPAC). The individuals recruited to the
UTPAC reported their age of onset of psoriasis and psoriatic arthritis at the point
of recruitment, while individuals recruited to the UTPC reported their age of onset
of psoriasis; they were then followed over time and some were observed to develop
psoriatic arthritis. The times of last contact or death are also recorded for all patients
along with a status indicator. Variable names and descriptions are as follows. These
data were used to fit a 4-state model depicted in Figure 7.2 in Example 7.1.3 of
Section 7.1.3.

Variable	Description
id	patient ID
clinic	clinic of recruitment:
	1 = psoriasis clinic; 2 = PsA clinic
age.entry	age at recruitment (years)
age.ps	age at onset of psoriasis (years)
age.psa	age at onset of psoriatic arthritis (years)
age.death	age of death (years)
age.last.contact	age of last contact: min(age of death, age on December 5, 2016)
psa.status	censoring status for psoriatic arthritis:
	1 = diagnosed with PsA; 0 = none

death.status   censoring status for death: 1 = died; 0 = alive

---

The dataframe below displays a few patients by type of clinic of recruitment.

```
> incPsA[incPsA$id %in% c(1,2,3,1379,1380,1381,1427,1428,1442),]
```

id	clinic	age.entry	age.ps	age.psa	age.death	age.last.contact	psa.status	death.status
1	2	39.461	18	33	NA	74.853	1	0
2	2	78.689	22	31	86.746	86.746	1	1
3	2	15.540	15	15	NA	52.107	1	0
1379	1	58.097	30	59	NA	68.778	1	0
1380	1	42.360	20	43	NA	53.210	1	0
1381	1	59.817	30	61	NA	69.930	1	0
1427	1	64.808	25	NA	NA	75.699	0	0
1428	1	28.493	26	NA	NA	39.411	0	0
1442	1	63.784	30	NA	66.861	66.861	0	1

Bibliography

Aalen, O. (1975). *Statistical Inference for a Family of Counting Processes.* Ph.D. dissertation, University of California, Berkeley, CA.

Aalen, O. (1976). Nonparametric inference in connection with multiple decrement models. *Scandinavian Journal of Statistics*, 3(1):15–27.

Aalen, O. (1978). Nonparametric inference for a family of counting processes. *Annals of Statistics*, 6(4):701–726.

Aalen, O. (1980). A model for nonparametric regression analysis of counting processes. In Klonecki, W., Kozek, A., and Rosiński, J., editors, *Mathematical Statistics and Probability Theory, Lecture Notes in Statistics, Volume 2*, pages 1–25. Springer-Verlag, New York, NY.

Aalen, O. (1987). Mixing distributions on a Markov chain. *Scandinavian Journal of Statistics*, 14(4):281–289.

Aalen, O. (1988). Heterogeneity in survival analysis. *Statistics in Medicine*, 7(11):1121–1137.

Aalen, O. (1989). A linear regression model for the analysis of life times. *Statistics in Medicine*, 8(8):907–925.

Aalen, O. (1992). Modelling heterogeneity in survival analysis by the compound Poisson distribution. *Annals of Applied Probability*, 2(4):951–972.

Aalen, O. (2012). Armitage lecture 2010: understanding treatment effects: the value of integrating longitudinal data and survival analysis. *Statistics in Medicine*, 31(18):1903–1917.

Aalen, O., Borgan, Ø., and Fekjaer, H. (2001). Covariate adjustment of event histories estimated from Markov chains: the additive approach. *Biometrics*, 57(4):993–1001.

Aalen, O., Borgan, Ø., and Gjessing, H. (2008). *Survival and Event History Analysis: A Process Point of View.* Springer Science + Business Media, New York, NY.

Aalen, O., Cook, R., and Røysland, K. (2015a). Does Cox analysis of a randomized survival study yield a causal treatment effect? *Lifetime Data Analysis*, 21(4):579–593.

Aalen, O., Gran, J., Røysland, K., Stensrud, M., and Strohmaier, S. (2018). Feedback and medication in causal inference illustrated by stochastic process models. *Scandinavian Journal of Statistics*, 45(1):62–86.

Aalen, O. and Husebye, E. (1991). Statistical analysis of repeated events forming renewal processes. *Statistics in Medicine*, 10(8):1227–1240.

Aalen, O. and Johansen, S. (1978). An empirical transition matrix for non-homogeneous Markov chains based on censored observations. *Scandinavian Journal of Statistics*, 5(3):141–150.

Aalen, O., Røysland, K., Gran, J., Kouyos, R., and Lange, T. (2016). Can we believe the DAGs? A comment on the relationship between causal DAGs and mechanisms. *Statistical Methods in Medical Research*, 25(5):2294–2314.

Aalen, O., Valberg, M., Grotmol, T., and Tretli, S. (2015b). Understanding variation in disease risk: the elusive concept of frailty. *International Journal of Epidemiology*, 44(4):1408–1421.

Age-Related Eye Disease Study Research Group (2005). The age-related eye disease study severity scale for age-related macular degeneration: AREDS report no. 17. *Archives of Ophthalmology*, 123(11):1484–1498.

Alioum, A. and Commenges, D. (1996). A proportional hazards model for arbitrarily censored and truncated data. *Biometrics*, 52(2):512–524.

Allignol, A., Beyersmann, J., Gerds, T., and Latouche, A. (2014). A competing risks approach for nonparametric estimation of transition probabilities in a non-Markov illness-death model. *Lifetime Data Analysis*, 20(4):495–513.

Altshuler, B. (1970). Theory for the measurement of competing risks in animal experiments. *Mathematical Biosciences*, 6:1–11.

Andersen, P., Borgan, Ø., Gill, R., and Keiding, N. (1993). *Statistical Models Based on Counting Processes*. Springer-Verlag, New York, NY.

Andersen, P. and Gill, R. (1982). Cox's regression model for counting processes: a large sample study. *Annals of Statistics*, 10(4):1100–1120.

Andersen, P. and Green, A. (1985). Evaluation of estimation bias in an illness-death-emigration model. *Scandinavian Journal of Statistics*, 12(1):63–68.

Andersen, P., Hansen, M., and Klein, J. (2004). Regression analysis of restricted mean survival time based on pseudo-observations. *Lifetime Data Analysis*, 10(4):335–350.

Andersen, P. and Keiding, N. (2002). Multi-state models for event history analysis. *Statistical Methods in Medical Research*, 11(2):91–115.

Andersen, P. and Keiding, N. (2012). Interpretability and importance of functionals in competing risks and multistate models. *Statistics in Medicine*, 31(11–12):1074–1088.

Andersen, P. and Klein, J. (2007). Regression analysis for multistate models based on a pseudo-value approach, with applications to bone marrow transplantation studies. *Scandinavian Journal of Statistics*, 34(1):3–16.

Andersen, P., Klein, J., and Rosthøj, S. (2003). Generalised linear models for correlated pseudo-observations, with applications to multi-state models. *Biometrika*, 90(1):15–27.

Andersen, P. and Listøl, K. (2003). Attenuation caused by infrequently updated covariates in survival analysis. *Biostatistics*, 4(4):633–649.

Andersen, P. and Perme, M. (2008). Inference for outcome probabilities in multistate models. *Lifetime Data Analysis*, 14(4):405–431.

Andersen, P. and Perme, M. (2010). Pseudo-observations in survival analysis. *Statistical Methods in Medical Research*, 19(1):71–99.

Asgharian, M., Wolfson, C., and Wolfson, D. (2014). Analysis of biased survival data: the Canadian study of health and aging and beyond. In Lawless, J., editor, *Statistics in Action: A Canadian Outlook*, chapter 12, pages 193–208. CRC Press, Boca Raton, FL.

Austin, P. (2014). The use of propensity score methods with survival or time-to-event outcomes: reporting measures of effect similar to those used in randomized experiments. *Statistics in Medicine*, 33(7):1242–1258.

Baker, S., Wax, Y., and Patterson, B. (1993). Regression analysis of grouped survival data: informative censoring and double sampling. *Biometrics*, 49(2):379–389.

Barrett, J., Siannis, F., and Farewell, V. (2011). A semi-competing risks model for data with interval-censoring and informative observation: an application to the MRC cognitive function and ageing study. *Statistics in Medicine*, 30(1):1–10.

Bartlett, M. (1978). *An Introduction to Stochastic Processes: with Special Reference to Methods and Applications, Third Edition*. Cambridge University Press, Cambridge, U.K.

Bennet, P. and Wood, P., editors (1968). *Population Studies of the Rheumatic Diseases*, pages 456–457. Excerpta Medica Foundation, Amsterdam, The Netherlands.

Betensky, R. and Finkelstein, D. (1999). A nonparametric maximum likelihood estimator for bivariate interval censored data. *Statistics in Medicine*, 18(22):3089–3100.

Beyersmann, J., Allignol, A., and Schumacher, M. (2012). *Competing Risks and Multistate Models with R*. Springer Science + Business Media, New York, NY.

Binder, N., Gerds, T., and Andersen, P. (2014). Pseudo-observations for competing risks with covariate dependent censoring. *Lifetime Data Analysis*, 20(2):303–315.

Borgan, Ø. and Samuelsen, S. (2014). Nested case-control and case-cohort studies. In Klein, J., van Houwelingen, H., Ibrahim, J., and Scheike, T., editors, *Handbook of Survival Analysis*, chapter 17, pages 343–364. Chapman & Hall/CRC Press, Boca Raton, FL.

Boruvka, A. and Cook, R. (2016). Sieve estimation in a Markov illness-death process under dual censoring. *Biostatistics*, 17(2):350–363.

Boudreau, C. and Lawless, J. (2006). Survival analysis based on the proportional hazards model and survey data. *Canadian Journal of Statistics*, 34(2):203–216.

Breslow, N., Lumley, T., Ballantyne, C., Chambless, L., and Kulich, M. (2009). Using the whole cohort in the analysis of case-cohort data. *American Journal of Epidemiology*, 169(11):1398–1405.

Brookmeyer, R., Corrada, M., Curriero, F., and Kawas, C. (2002). Survival following a diagnosis of Alzheimer disease. *Archives of Neurology*, 59(11):1764–1767.

Brookmeyer, R., Johnson, E., Ziegler-Graham, K., and Arrighi, H. (2007). Forecasting the global burden of Alzheimer's disease. *Alzheimer's & Dementia*, 3(3):186–191.

Broström, G. (2012). *Event History Analysis in R*. CRC Press, Boca Raton, FL.

Bryant, J. and Dignam, J. (2004). Semiparametric models for cumulative incidence functions. *Biometrics*, 60(1):182–190.

Buzkova, P. (2010). Panel count data regression with informative observation times. *International Journal of Biostatistics*, 6(1):1–22. Article 30.

Buzkova, P. and Lumley, T. (2007). Longitudinal data analysis for generalized linear models with follow-up dependent on outcome-related variables. *Canadian Journal of Statistics*, 35(4):485–500.

Buzkova, P. and Lumley, T. (2008). Semiparametric log-linear regression for longitudinal measurements subject to outcome-dependent follow-up. *Journal of Statistical Planning and Inference*, 138(8):2450–2461.

Buzkova, P. and Lumley, T. (2009). Semiparametric modeling of repeated measurements under outcome-dependent follow-up. *Statistics in Medicine*, 28(6):987–1003.

Cai, J. and Zeng, D. (2004). Sample size/power calculation for case-cohort studies. *Biometrics*, 60(4):1015–1024.

Cai, J. and Zeng, D. (2007). Power calculation for case-control studies with nonrare events. *Biometrics*, 63(4):1288–1295.

Carstensen, B. and Plummer, M. (2011). Using Lexis objects to multi-state models in R. *Journal of Statistical Software*, 38(6):1–18.

Carstensen, B., Plummer, M., Läärä, E., and Hills, M. (2016). `Epi`: a Package for Statistical Analysis in Epidemiology. R package version 2.7. `https://CRAN.R-project.org/package=Epi`.

Chaieb, L., Rivest, L., and Abdous, B. (2006). Estimating survival under a dependent truncation. *Biometrika*, 93(3):655–669.

Chen, D., Sun, J., and Peace, K. (2013). *Interval-Censored Time-to-Event Data: Methods and Applications*. CRC Press, Boca Raton, FL.

Chen, P. and Tsiatis, A. (2001). Causal inference on the difference of the restricted mean lifetime between two groups. *Biometrics*, 57(4):1030–1038.

Chi, Y. and Ibrahim, J. (2006). Joint models for multivariate longitudinal and multivariate survival data. *Biometrics*, 62(2):432–445.

Coleman, R., Purohit, O., Vinholes, J., and Zekri, J. (1997). High dose pamidronate: clinical and biochemical effects in metastatic bone disease. *Cancer*, 80(S8):1686–1690.

Collett, D. (2015). *Modelling Survival Data in Medical Research, Third Edition*. CRC Press, Boca Raton, FL.

Commenges, D. and Jacqmin-Gadda, H. (2016). *Dynamical Biostatistical Models*. CRC Press, Boca Raton, FL.

Conlon, A., Taylor, J., and Sargent, D. (2015). Improving efficiency in clinical trials using auxiliary information: application of a multi-state cure model. *Biometrics*,

71(2):460–468.

Cook, R. (1999). A mixed model for two-state Markov processes under panel observation. *Biometrics*, 55(3):915–920.

Cook, R. (2000). Information and efficiency considerations in planning studies based on two-state Markov process. *Journal of Statistical Research*, 34:161–178.

Cook, R., Kalbfleisch, J., and Yi, G. (2002). A generalized mover-stayer model for panel data. *Biostatistics*, 3(3):407–420.

Cook, R. and Lawless, J. (2007). *The Statistical Analysis of Recurrent Events*. Springer Science + Business Media, New York, NY.

Cook, R. and Lawless, J. (2014). Statistical issues in modeling chronic disease in cohort studies. *Statistics in Biosciences*, 6(1):127–161.

Cook, R., Lawless, J., Lakhal-Chaieb, L., and Lee, K. (2009). Robust estimation of mean functions and treatment effects for recurrent events under event-dependent censoring and termination: application to skeletal complications in cancer metastatic to bone. *Journal of the American Statistical Association*, 104(485):60–75.

Cook, R., Lawless, J., and Lee, K. (2003). Cumulative processes related to event histories. *SORT*, 27:13–30.

Cook, R., Ng, E., Mukherjee, J., and Vaughan, D. (1999). Two-state mixed renewal processes for chronic disease. *Statistics in Medicine*, 18(2):175–188.

Cook, R., Yi, G., Lee, K., and Gladman, D. (2004). A conditional Markov model for clustered progressive multistate processes under incomplete observation. *Biometrics*, 60(2):436–443.

Cook, R., Zeng, L., and Lee, K. (2008). A multistate model for bivariate interval-censored failure time data. *Biometrics*, 64(4):1100–1109.

Cornfield, J. (1957). Estimation of the probability of developing a disease in the presence of competing risks. *American Journal of Public Health and the Nation's Health*, 47(5):601–607.

Cortese, G., Gerds, T., and Andersen, P. (2013). Comparing predictions among competing risks models with time-dependent covariates. *Statistics in Medicine*, 32(18):3089–3101.

Couper, D. and Pepe, M. (1997). Modelling prevalence of a condition: chronic graft-versus-host disease after bone marrow transplantation. *Statistics in Medicine*, 16(14):1551–1571.

Cox, D. (1972). Regression models and life tables (with discussion). *Journal of the Royal Statistical Society: Series B (Methodological)*, 34:187–220.

Cox, D. (1975). Partial likelihood. *Biometrika*, 62(2):269–276.

Cox, D. and Hinkley, D. (1979). *Theoretical Statistics*. Chapman & Hall/CRC Press, Boca Raton, FL.

Cox, D. and Miller, H. (1965). *The Theory of Stochastic Processes*. CRC Press, Boca Raton, FL.

Cox, D. and Oakes, D. (1984). *Analysis of Survival Data*. CRC Press, Boca Raton,

FL.

Crowder, M. (2001). *Classical Competing Risks*. Chapman & Hall/CRC Press, Boca Raton, FL.

Crowder, M. (2012). *Multivariate Survival Analysis and Competing Risks*. CRC Press, Boca Raton, FL.

Dabrowska, D., Sun, G., and Horowitz, M. (1994). Cox regression in a Markov renewal model: an application to the analysis of bone marrow transplant data. *Journal of the American Statistical Association*, 89(427):867–877.

Dantan, E., Joly, P., Dartigues, J., and Jacqmin-Gadda, H. (2001). Joint model with latent state for longitudinal and multistate data. *Biostatistics*, 12(4):723–736.

Datta, S. and Satten, G. (2001). Validity of the Aalen-Johansen estimators of stage occupation probabilities and Nelson-Aalen estimators of integrated transition hazards for non-Markov models. *Statistics and Probability Letters*, 55(4):403–411.

Datta, S. and Satten, G. (2002). Estimation of integrated transition hazards and stage occupation probabilities for non-Markov systems under dependent censoring. *Biometrics*, 58(4):792–802.

Datta, S., Satten, G., and Datta, S. (2000). Nonparametric estimation for the three-stage irreversible illness-death model. *Biometrics*, 56(3):841–847.

Datta, S. and Sundaram, R. (2006). Nonparametric estimation of stage occupation probabilities in a multistage model with current status data. *Biometrics*, 62(3):829–837.

de Bruijne, M., le Cessie, S., Kluin-Nelemans, H., and van Houwelingen, H. (2001). On the use of Cox regression in the presence of an irregularly observed time-dependent covariate. *Statistics in Medicine*, 20(24):3817–3829.

de Stavola, B. (1988). Testing departures from time homogeneity in multistate Markov processes. *Journal of the Royal Statistical Society: Series C (Applied Statistics)*, 37(2):242–250.

de Wreede, L., Fiocco, M., and Putter, H. (2011). mstate: an R package for the analysis of competing risks and multi-state models. *Journal of Statistical Software*, 38(7):1–30.

Dewanji, A. and Kalbfleisch, J. (1986). Nonparametric methods for survival/sacrifice experiments. *Biometrics*, 42(2):325–341.

Diabetes Control and Complications Trial Research Group (1993). The effect of intensive treatment of diabetes on the development and progression of long-term complications in insulin-dependent diabetes mellitus. *New England Journal of Medicine*, 329(14):977–986.

Diabetes Control and Complications Trial Research Group (1995). Progression of retinopathy with intensive versus conventional treatment in the Diabetes Control and Complications Trial. *Ophthalmology*, 102(4):647–661.

Diamond, I. and McDonald, J. (1992). The analysis of current status data. In Trussell, J., Hankinson, R., and Tilton, J., editors, *Demographic Applications of Event History Analysis*, chapter 9, pages 231–252. Oxford University Press,

Oxford, U.K.

Diao, L. (2013). *Copula Models for Multi-type Life History Processes*. Ph.D. dissertation, University of Waterloo, Waterloo, ON.

Diao, L. and Cook, R. (2014). Composite likelihood for joint analysis of multiple multistate processes via copulas. *Biostatistics*, 15(4):690–705.

Ding, J., Lu, T., Cai, J., and Zhou, H. (2017). Recent progresses in outcome-dependent sampling with failure time data. *Lifetime Data Analysis*, 23(1):57–82.

Dubin, J., Maller, H., and Wang, J. (2001). Event history graphs for censored survival data. *Statistics in Medicine*, 20(19):2951–2964.

Duchateau, L. and Janssen, P. (2008). *The Frailty Model*. Springer Science + Business Media, New York, NY.

Early Treatment Diabetic Retinopathy Study Research Group (1991). Early photocoagulation for diabetic retinopathy: ETDRS report number 9. *Ophthalmology*, 98(5):766–785.

Ebrahimi, N. (1996). The effects of misclassification of the actual cause of death in competing risks analysis. *Statistics in Medicine*, 15(14):1557–1566.

Ederer, F., Myers, M., and Mantel, N. (1964). A statistical problem in space and time. Do leukemia cases come in clusters? *Biometrics*, 20(3):626–638.

Fahrmeir, L. and Tutz, G. (2001). *Multivariate Statistical Modelling Based on Generalized Linear Models, Second Edition*. Springer Science + Business Media, New York, NY.

Farewell, D., Huang, C., and Didelez, V. (2017). Ignorability for general longitudinal data. *Biometrika*, 104(2):317–326.

Farewell, V. (1982). The use of mixture models for the analysis of survival data with long-term survivors. *Biometrics*, 38(4):1041–1046.

Farewell, V. (1986). Mixture models in survival analysis: are they worth the risk. *Canadian Journal of Statistics*, 14(3):257–262.

Farewell, V., Lawless, J., and Gladman, D. (2003). Tracing studies and analysis of the effect of loss to follow-up on mortality estimation from patient registry data. *Journal of the Royal Statistical Society: Series C (Applied Statistics)*, 52(4):445–456.

Ferguson, N., Datta, S., and Brock, G. (2012). `msSurv`, an R package for nonparametric estimation of multistate models. *Journal of Statistical Software*, 50(14):1–24.

Fine, J. (1999). Analysing competing risks data with transformation models. *Journal of the Royal Statistical Society: Series B (Methodological)*, 61(4):817–830.

Fine, J. (2001). Regression modeling of competing crude failure probabilities. *Biostatistics*, 2(1):85–97.

Fine, J. and Gray, R. (1999). A proportional hazards model for the subdistribution of a competing risk. *Journal of the American Statistical Association*, 94(446):496–509.

Fine, J., Yan, J., and Kosorok, M. (2004). Temporal process regression. *Biometrika*,

91(3):683–703.

Finkelstein, D., Goggins, W., and Schoenfeld, D. (2002). Analysis of failure time data with dependent interval censoring. *Biometrics*, 58(2):298–304.

Fiocco, M., Putter, H., and van Houwelingen, H. (2008). Reduced-rank proportional hazards regression and simulation-based prediction for multi-state models. *Statistics in Medicine*, 27(21):4340–4358.

Fitzpatrick, A., Kuller, L., Lopez, O., Kawas, C., and Jagust, W. (2005). Survival following dementia onset: Alzheimer's disease and vascular dementia. *Journal of the Neurological Sciences*, 229–230:43–49.

Fix, E. and Neyman, J. (1951). A simple stochastic model of recovery, relapse, death and loss of patients. *Human Biology*, 23(3):205–241.

Fleming, T. (1978a). Asymptotic distribution results in competing risks estimation. *Annals of Statistics*, 6(5):1071–1079.

Fleming, T. (1978b). Nonparametric estimation for nonhomogeneous Markov processes in the problem of competing risks. *Annals of Statistics*, 6(5):1057–1070.

Fleming, T. and Harrington, D. (1991). *Counting Processes and Survival Analysis*. John Wiley & Sons, Hoboken, NJ.

Fosen, J., Borgan, Ø., Weedon-Fekjær, H., and Aalen, O. (2006). Dynamic analysis of recurrent event data using the additive hazard model. *Biometrical Journal*, 48(3):381–398.

Frangakis, C. and Rubin, D. (2001). Addressing an idiosyncrasy in estimating survival curves using double sampling in the presence of self-selected right censoring. *Biometrics*, 57(2):333–342.

Frydman, H. (1984). Maximum likelihood estimation in the mover-stayer model. *Journal of the American Statistical Association*, 79(387):632–638.

Frydman, H. (1992). A nonparametric estimation procedure for a periodically observed three-state Markov process, with application to aids. *Journal of the Royal Statistical Society: Series B (Methodological)*, 54(3):853–866.

Frydman, H. (1994). A note on nonparametric estimation of the distribution function from interval-censored and truncated observations. *Journal of the Royal Statistical Society: Series B (Methodological)*, 56(1):71–74.

Frydman, H. (1995). Nonparametric estimation of a Markov "illness-death" process from interval-censored observations, with application to diabetes survival data. *Biometrika*, 82(4):773–789.

Frydman, H., Gerds, T., Grøn, R., and Keiding, N. (2013). Nonparametric estimation in an "illness-death" model when all transition times are interval censored. *Biometrical Journal*, 55(6):823–843.

Frydman, H. and Liu, J. (2013). Nonparametric estimation of the cumulative intensities in an interval censored competing risks model. *Lifetime Data Analysis*, 19(1):79–99.

Frydman, H. and Szarek, M. (2009). Nonparametric estimation in a Markov "illness-death" process from interval censored observations with missing intermediate transition status. *Biometrics*, 65(1):143–151.

Fuchs, C. and Greenhouse, J. (1988). The EM algorithm for maximum likelihood estimation in the mover-stayer model. *Biometrics*, 44(2):605–613.

Gail, M., Santner, T., and Brown, C. (1980). An analysis of comparative carcinogenesis experiments based on multiple times to tumor. *Biometrics*, 36(2):255–266.

Gandy, A. and Jensen, U. (2005a). Checking a semiparametric additive risk model. *Lifetime Data Analysis*, 11(4):451–472.

Gandy, A. and Jensen, U. (2005b). On goodness-of-fit tests for Aalen's additive risk model. *Scandinavian Journal of Statistics*, 32(3):425–445.

Gaston, R., Danovitch, G., Adams, P., Wynn, J., Merion, R., Deierhoi, M., Metzger, R., Cecka, J., Harmon, W., Leichtman, A., Spital, A., Blumberg, E., Herzog, C., Wolfe, R., Tyan, D., Roberts, J., Rohrer, R., Port, F., and Delmonico, F. (2003). The report of a national conference on the wait list for kidney transplantation. *American Journal of Transplantation*, 3(7):775–785.

Gelber, R., Cole, B., Gelber, S., and Goldhirsch, A. (1995). Comparing treatments using quality-adjusted survival: the Q-TWiST method. *American Statistician*, 49(2):161–169.

Gelber, R., Gelman, R., and Goldhirsch, A. (1989). A quality-of-life-oriented endpoint for comparing therapies. *Biometrics*, 45(3):781–795.

Gentleman, R. and Vandal, A. (2002). Nonparametric estimation of the bivariate cdf for arbitrarily censored data. *Canadian Journal of Statistics*, 30(4):557–571.

Gerds, T., Cai, T., and Schumacher, M. (2008). The performance of risk prediction models. *Biometrical Journal*, 50(4):457–479.

Gerds, T., Scheike, T., and Andersen, P. (2012). Absolute risk regression for competing risks: interpretation, link functions and prediction. *Statistics in Medicine*, 31(29):3921–3930.

Geskus, R. (2011). Cause-specific cumulative incidence estimation and the Fine and Gray model under both left truncation and right censoring. *Biometrics*, 67(1):39–49.

Geskus, R. (2016). *Data Analysis with Competing Risks and Intermediate States*. Chapman & Hall/CRC Press, Boca Raton, FL.

Gill, R. (1980). Censoring and stochastic integrals. *Statistica Neerlandica*, 34(2):124.

Gill, R. and Johansen, S. (1990). A survey of product-integration with a view toward application in survival analysis. *Annals of Statistics*, 18(4):1501–1555.

Gill, R. and Keiding, N. (2010). Product-limit estimators of the gap time distribution of a renewal process under different sampling patterns. *Lifetime Data Analysis*, 16(4):571–579.

Gladman, D. and Chandran, V. (2011). Observational cohort studies: lessons learnt from the University of Toronto Psoriatic Arthritis Program. *Rheumatology*, 50(1):25–31.

Gladman, D., Ibañez, D., and Urowitz, M. (2002). Systemic lupus erythematosus disease activity index 2000. *Journal of Rheumatology*, 29(2):288–291.

Glasziou, P., Simes, R., and Gelber, R. (1990). Quality adjusted survival analysis. *Statistics in Medicine*, 9(11):1259–1276.

Glidden, D. (2002). Robust inference for event probabilities with non-Markov event data. *Biometrics*, 58(2):361–368.

Gneiting, T. (2014). Calibration of medium-range weather forecasts. *European Centre for Medium-Range Weather Forecasts*.

Gneiting, T., Balabdaoui, F., and Raftery, A. (2007). Probability forecasts, calibration and sharpness. *Journal of the Royal Statistical Society: Series B (Methodological)*, 69(2):243–268.

Gneiting, T. and Katzfuss, M. (2014). Probabilistic forecasting. *Annual Review of Statistics and Its Application*, 1:125–151.

Gneiting, T. and Raftery, A. (2007). Strictly proper scoring rules, prediction, and estimation. *Journal of the American Statistical Association*, 102(477):359–378.

Goetghebeur, E. and Ryan, L. (1995). Analysis of competing risks survival data when some failure types are missing. *Biometrika*, 4(1):821–833.

Goggins, W. and Finkelstein, D. (2000). A proportional hazards model for multivariate interval-censored failure time data. *Biometrics*, 56(3):940–943.

Goldhirsch, A., Gelber, R., Simes, R., Glasziou, P., and Coates, A. (1989). Costs and benefits of adjuvant therapy in breast cancer: a quality-adjusted survival analysis. *Journal of Clinical Oncology*, 7(1):36–44.

Gómez, G., Calle, M., Oller, R., and Langohr, K. (2009). Tutorial on methods for interval-censored data and their implementation in R. *Statistical Modelling*, 9(4):259–297.

Goodman, L. (1961). Statistical methods for the mover-stayer model. *Journal of the American Statistical Association*, 56(296):841–868.

Grambsch, P. and Therneau, T. (1994). Proportional hazards tests and diagnostics based on weighted residuals. *Biometrika*, 81(3):515–526.

Gran, J., Lie, S., Øyeflaten, I., Borgan, Ø., and Aalen, O. (2015). Causal inference in multi-state models-sickness absence and work for 1145 participants after work rehabilitation. *BMC Public Health*, 15:1082.

Graw, F., Gerds, T., and Schumacher, M. (2009). On pseudo-values for regression analysis in competing risks models. *Lifetime Data Analysis*, 15(2):241–255.

Gray, R. (1988). A class of K-sample tests for comparing the cumulative incidence of a competing risk. *Annals of Statistics*, 16(3):1141–1154.

Griffin, B. and Lagakos, S. (2010). Nonparametric inference and uniqueness for periodically observed progressive disease models. *Lifetime Data Analysis*, 16(2):157–175.

Grøn, R. and Gerds, T. (2014). Binomial regression models. In Klein, J., van Houwelingen, H., Ibrahim, J., and Scheike, T., editors, *Handbook of Survival Analysis*, chapter 11, pages 221–239. Chapman & Hall/CRC Press, Boca Raton, FL.

Grønnesby, J. and Borgan, Ø. (1996). A method for checking regression models in survival analysis based on the risk score. *Lifetime Data Analysis*, 2(4):315–328.

Grossman, R., Mukherjee, J., Vaughan, D., Eastwood, C., Cook, R., LaForge, J., and Lampron, N. (1998). A 1-year community-based health economic study of ciprofloxacin vs usual antibiotic treatment in acute exacerbations of chronic bronchitis: The Canadian Ciprofloxacin Health Economic Study Group. *CHEST*, 113(1):131–141.

Grüger, J., Kay, R., and Schumacher, M. (1991). The validity of inferences based on incomplete observations in disease state models. *Biometrics*, 47(2):595–605.

Grundmann, H., Bärwolff, S., Tami, A., Behnke, M., Schwab, F., Geffers, C., Halle, E., Göbel, U., Schiller, R., Jonas, D., Klare, I., Weist, K., Witte, W., Beck-Beilecke, K., Schumacher, M., Rüden, H., and Gastmeier, P. (2005). How many infections are caused by patient-to-patient transmission in intensive care units? *Critical Care Medicine*, 33(5):946–951.

Gunnes, N., Borgan, Ø., and Aalen, O. (2007). Estimating stage occupation probabilities in non-Markov models. *Lifetime Data Analysis*, 13(2):211–240.

Hajducek, D. and Lawless, J. (2012). Duration analysis in longitudinal studies with intermittent observation times and losses to followup. *Canadian Journal of Statistics*, 40(1):1–21.

Hajducek, D. and Lawless, J. (2013). Estimation of finite population duration distributions with longitudinal survey panels with intermittent followup. *Lifetime Data Analysis*, 19(3):371–392.

Hamerle, A. (1991). On the treatment of interrupted spells and initial conditions in event history analysis. *Sociological Methods & Research*, 19(3):388–414.

Heckman, J. and Walker, J. (1987). Using goodness of fit and other criteria to choose among competing duration models: a case study to Hutterite data. *Sociological Methodology*, 17:247–307.

Heddle, N., Cook, R., Tinmouth, A., Kouroukis, T., Hervig, T., Klapper, E., Brandwein, J., Szczepiorkowski, Z., AuBuchon, J., Barty, R., Lee, K., and for the SToP investigators of the BEST Collaborative (2009). A randomized controlled trial comparing standard- and low-dose strategies for transfusion of platelets (SToP) to patients with thrombocytopenia. *Blood*, 113(7):1564–1573.

Hernán, M. (2010). The hazards of hazard ratios. *Epidemiology*, 21(1):13–15.

Hernán, M. and Robins, J. (2015). *Causal Inference*. Chapman & Hall/CRC Press, Boca Raton, FL.

Hjort, N. (1990). Nonparametric Bayes estimators based on beta processes in models for life history data. *Annals of Statistics*, 18(3):1259–1294.

Hoel, D. (1972). A representation of mortality data by competing risks. *Biometrics*, 28(2):475–488.

Hoem, J. (1971). Point estimation of forces of transition in demographic models. *Journal of the Royal Statistical Society: Series B (Methodological)*, 33(2):275–289.

Hoem, J. (1977). A Markov chain model of working life tables. *Scandinavian Actuarial Journal*, 1977(1):1–20.

Hogan, J., Roy, J., and Korkontzelou, C. (2004). Handling dropouts in longitudinal studies. *Statistics in Medicine*, 23(9):1455–1497.

Hogg, R., Strathdee, S., Craib, K., O'Shaughnessay, M., Montaner, J., and Schechter, M. (1994). Lower socioeconomic status and shorter survival following HIV infection. *Lancet*, 344(8930):1120–1124.

Hortobagyi, G., Theriault, R., Porter, L., Blayney, D., Lipton, A., Sinoff, C., Wheeler, H., Simeone, J., Seaman, J., Knight, R., Heffernan, M., Reitsma, D., Kennedy, I., Allan, S., Mellars, K., and for the Protocol 19 Aredia Breast Cancer Study Group (1996). Efficacy of pamidronate in reducing skeletal complications in patients with breast cancer and lytic bone metastases. Protocol 19 Aredia Breast Cancer Study Group. *New England Journal of Medicine*, 335(24):1785–1791.

Hougaard, P. (1984). Life table methods for heterogeneous populations: distributions describing the heterogeneity. *Biometrika*, 71(1):75–83.

Hougaard, P. (1986). A class of multivariate failure time distributions. *Biometrika*, 73(3):671–678.

Hougaard, P. (1999). Multi-state models: a review. *Lifetime Data Analysis*, 5(3):239–264.

Hougaard, P. (2000). *Analysis of Multivariate Survival Data*. Springer Science + Business Media, New York, NY.

Hudgens, M., Li, C., and Fine, J. (2014). Parametric likelihood inference for interval censored competing risks data. *Biometrics*, 70(1):1–9.

Hudgens, M., Satten, G., and Longini Jr, I. (2001). Nonparametric maximum likelihood estimation for competing risks survival data subject to interval censoring and truncation. *Biometrics*, 57(1):74–80.

Hvidtfeldt, U., Gunter, M., Lange, T., Chlebowski, R., Lane, D., Farhat, G., Freiberg, M., Keiding, N., Lee, J., Prentice, R., Tjønneland, A., Vitolins, M., Wassertheil-Smoller, S., Strickler, H., and Rod, N. (2012). Quantifying mediating effects of endogenous estrogen and insulin in the relation between obesity, alcohol consumption, and breast cancer. *Cancer Epidemiology and Prevention Biomarkers*, 21(7):1203–1212.

Hwang, W. and Brookmeyer, R. (2003). Design of panel studies for disease progression with multiple stages. *Lifetime Data Analysis*, 9(3):261–274.

Ibrahim, J., Chu, H., and Chen, L. (2010). Basic concepts and methods for joint models of longitudinal and survival data. *Journal of Clinical Oncology*, 28(16):2796–2801.

Jackson, C. (2011). Multi-state models for panel data: the `msm` package for R. *Journal of Statistical Software*, 38(8):1–29.

Jackson, C. (2016). `msm`: multi-state Markov and hidden Markov models in continuous time. R package version 1.6.4. `https://CRAN.R-project.org/package=msm`.

Jackson, C. and Sharples, L. (2002). Hidden Markov models for the onset and progression of bronchiolitis obliterans syndrome in lung transplant recipients. *Statistics in Medicine*, 21(1):113–128.

Jackson, C., Sharples, L., Thompson, S., Duffy, S., and Couto, E. (2003). Multistate

Markov models for disease progression with classification error. *Journal of the Royal Statistical Society: Series D (The Statistician)*, 52(2):193–209.

Jensen, A., Ravn, H., Sørup, S., and Andersen, P. (2016). A marginal structural model for recurrent events in the presence of time-dependent confounding: non-specific effects of vaccines on child hospitalisations. *Statistics in Medicine*, 35(27):5051–5069.

Jewell, N. and Kalbfleisch, J. (1996). Markov processes in survival analysis. *Lifetime Data Analysis*, 2(1):15–29.

Joe, H. (1997). *Multivariate Models and Multivariate Dependence Concepts*. Chapman & Hall/CRC Press, Boca Raton, FL.

Joly, P., Commenges, D., Helmer, C., and Letenneur, L. (2002). A penalized likelihood approach for an illness-death model with interval-censored data: application to age-specific incidence of dementia. *Biostatistics*, 3(3):433–443.

Kalbfleisch, J., Krewski, D., and Ryzin, O. (1983). Dose-response models for time-to-response toxicity data. *Canadian Journal of Statistics*, 11(1):25–46.

Kalbfleisch, J. and Lawless, J. (1985). The analysis of panel data under a Markov assumption. *Journal of the American Statistical Association*, 80(392):863–871.

Kalbfleisch, J. and Lawless, J. (1989). Inference based on retrospective ascertainment: an analysis of the data on transfusion-related AIDS. *Journal of the American Statistical Association*, 84(406):360–372.

Kalbfleisch, J. and Prentice, R. (2002). *The Statistical Analysis of Failure Time Data, Second Edition*. John Wiley & Sons, Hoboken, NJ.

Kang, M. and Lagakos, S. (2007). Statistical methods for panel data from a semi-Markov process, with application to HPV. *Biostatistics*, 8(2):252–264.

Keiding, N. (1991). Age-specific incidence and prevalence: a statistical perspective. *Journal of the Royal Statistical Society: Series A*, 154(3):371–412.

Keiding, N. (2006). Event history analysis and the cross-section. *Statistics in Medicine*, 25(14):2343–2364.

Keiding, N. (2011). Age-period-cohort analysis in the 1870s: diagrams, stereograms, and the basic differential equation. *Canadian Journal of Statistics*, 39(3):405–420.

Keiding, N. (2014). Event history analysis. *Annual Review of Statistics and Its Application*, 1(1):333–360.

Keiding, N., Holst, C., and Green, A. (1989). Retrospective estimation of diabetes incidence from information in a prevalent population and historical mortality. *American Journal of Epidemiology*, 130(3):588–600.

Keiding, N., Klein, J., and Horowitz, M. (2001). Multi-state models and outcome prediction in bone marrow transplantation. *Statistics in Medicine*, 20(12):1871–1885.

Kessing, L., Hansen, M., and Andersen, P. (2004). Course of illness in depressive and bipolar disorders. *British Journal of Psychiatry*, 185(5):372–377.

Kim, J., Sit, T., and Ying, Z. (2016). Accelerated failure time model under general biased sampling scheme. *Biostatistics*, 17(3):576–588.

Klein, J. and Andersen, P. (2005). Regression modeling of competing risks data based on pseudovalues of the cumulative incidence function. *Biometrics*, 61(1):223–229.

Klein, J. and Moeschberger, M. (2003). *Survival Analysis: Techniques for Censored and Truncated Data, Second Edition*. Springer-Verlag, New York, NY.

Klein, J., van Houwelingen, H., Ibrahim, J., and Scheike, T. (2014). *Handbook of Survival Analysis*. Chapman & Hall/CRC Press, Boca Raton, FL.

Król, A., Mauguen, A., Mazroui, Y., Laurent, A., Michiels, S., and Rondeau, V. (2017). Tutorial in joint modeling and prediction: a statistical software for correlated longitudinal outcomes, recurrent events and a terminal event. *Journal of Statistical Software*, 81(3):1–52.

Krzanowski, W. and Hand, D. (2009). *ROC Curves for Continuous Data*. CRC Press, Boca Raton, FL.

Kvist, K., Andersen, P., Angst, J., and Kessing, L. (2010). Event dependent sampling of recurrent events. *Lifetime Data Analysis*, 16(4):580–598.

Lagakos, S., Sommer, C., and Zelen, M. (1978). Semi-Markov models for partially censored data. *Biometrika*, 65(2):311–317.

Lange, J., Hubbard, R., Inoue, L., and Minin, V. (2015). A joint model for multistate disease processes and random informative observation times, with application to electronic medical records data. *Biometrics*, 71(1):90–101.

Lange, J. and Minin, V. (2013). Fitting and interpreting continuous-time latent Markov models for panel data. *Statistics in Medicine*, 32(26):4581–4595.

Lawless, J. (2003). *Statistical Models and Methods for Lifetime Data, Second Edition*. John Wiley & Sons, Hoboken, NJ.

Lawless, J. (2013). Armitage Lecture 2011: the design and analysis of life history studies. *Statistics in Medicine*, 32(13):2155–2172.

Lawless, J. and Cook, R. (2017a). Independence conditions and intermittent observation in life history studies. Manuscript in preparation.

Lawless, J. and Cook, R. (2017b). Independence conditions on loss to followup in life history studies. Manuscript in preparation.

Lawless, J. and Fong, D. (1999). State duration models in clinical and observational studies. *Statistics in Medicine*, 18(17–18):2365–2376.

Lawless, J., Kalbfleisch, J., and Wild, C. (1999). Semiparametric methods for response-elective and missing data problems in regression. *Journal of the Royal Statistical Society: Series B (Methodological)*, 61(2):413–438.

Lawless, J. and Nazeri Rad, N. (2015). Estimation and assessment of Markov multistate models with intermittent observations on individuals. *Lifetime Data Analysis*, 21(2):160–179.

Lawless, J., Wigg, M., Tuli, S., Drake, J., and Lamberti-Pasculli, M. (2001). Analysis of repeated failures or durations, with application to shunt failures for patients with paediatric hydrocephalus. *Journal of the Royal Statistical Society: Series C (Applied Statistics)*, 50(4):449–465.

Lawless, J. and Yuan, Y. (2010). Estimation of prediction error for survival models. *Statistics in Medicine*, 29(2):262–274.

Lee, E. and Kim, M. (1998). The analysis of correlated panel data using a continuous-time Markov model. *Biometrics*, 54(4):1638–1644.

Lee, J., Hess, K., and Dubin, J. (2000). Extensions and applications of event charts. *American Statistician*, 54(1):63–70.

Lemaire, M., Islam, Q., Shen, H., Khan, M., Parveen, M., Abedin, F., Haseen, F., Hyder, Z., Cook, R., and Zlotkin, S. (2011). Iron-containing micronutrient powder provided to children with moderate-to-severe malnutrition increases hemoglobin concentrations but not the risk of infectious morbidity: a randomized, double-blind, placebo-controlled, noninferiority safety trial. *American Journal of Clinical Nutrition*, 94(2):585–593.

Liang, K. and Self, S. (1996). On the asymptotic behaviour of the pseudolikelihood ratio test statistic. *Journal of the Royal Statistical Society: Series B (Methodological)*, 58(4):785–796.

Liang, K. and Zeger, S. (1986). Longitudinal data analysis using generalized linear models. *Biometrika*, 73(1):13–22.

Lin, D. and Wei, L. (1989). The robust inference for the Cox proportional hazards model. *Journal of the American Statistical Association*, 84(408):1074–1078.

Lin, D. and Ying, Z. (1994). Semiparametric analysis of the additive risk model. *Biometrika*, 81(1):61–71.

Lin, H., Scharfstein, D., and Rosenheck, R. (2004). Analysis of longitudinal data with irregular, outcome-dependent follow-up. *Journal of the Royal Statistical Society: Series B (Methodological)*, 66(3):791–813.

Lindsey, J. (1996). *Parametric Statistical Inference*. Oxford University Press, New York, NY.

Lindsey, J. and Ryan, L. (1993). A three-state multiplicative model for rodent tumorigenicity experiments. *Journal of the Royal Statistical Society: Series C (Applied Statistics)*, 42(2):283–300.

Lipton, A., Theriault, R., Hortobagyi, G., Simeone, J., Knight, R., Mellars, K., Reitsma, D., Heffernan, M., and Seaman, J. (2000). Pamidronate prevents skeletal complications and is effective palliative treatment in women with breast carcinoma and osteolytic bone metastases. *Cancer*, 88(5):1082–1090.

Lloyd-Jones, D., Martin, D., Larson, M., and Levy, D. (1998). Accuracy of death certificates for coding coronary heart disease as the cause of death. *Annals of Internal Medicine*, 129(12):1020–1026.

Logan, B. and Wang, T. (2014). Pseudo-value regression models. In Klein, J., van Houwelingen, H., Ibrahim, J., and Scheike, T., editors, *Handbook of Survival Analysis*, chapter 10, pages 199–218. Chapman & Hall/CRC Press, Boca Raton, FL.

Lumley, T., Shaw, P., and Dai, J. (2011). Connections between survey calibration estimators and semiparametric models for incomplete data. *International*

Statistical Review, 79(2):200–220.

Lund, J. (2000). Sampling bias in population studies - how to use the Lexis diagram. *Scandinavian Journal of Statistics*, 27(4):589–604.

Lymer, A., Schofield, D., Lee, C., and Colagiuri, S. (2016). NCDMod: a microsimulation model projecting chronic disease and risk factors for Australian adults. *International Journal of Microsimulation*, 9(3):103–139.

Marmot, M., Stansfeld, S., Patel, C., North, F., Head, J., White, I., Brunner, E., Feeney, A., Marmot, M., and Smith, G. (1991). Health inequalities among British civil servants: the Whitehall II study. *Lancet*, 337(8754):1387–1393.

Martinussen, T. and Scheike, T. (2006). *Dynamic Regression Models for Survival Data*. Springer Science + Business Media, New York, NY.

Martinussen, T. and Vansteelandt, S. (2013). On collapsibility and confounding bias in Cox and Aalen regression models. *Lifetime Data Analysis*, 19(3):279–296.

Martinussen, T., Vansteelandt, S., Gerster, M., and Hjelmborg, J. (2011). Estimation of direct effects for survival data by using the Aalen additive hazards model. *Journal of the Royal Statistical Society: Series B (Methodological)*, 73(5):773–788.

Matthews, D. (1984). Some observations on semi-Markov models for partially censored data. *Canadian Journal of Statistics*, 12(3):201–205.

May, S. and Hosmer, D. (2004). A cautionary note on the use of the Grønnesby and Borgan goodness-of-fit test for the Cox proportional hazards model. *Lifetime Data Analysis*, 10(3):283–291.

Maziarz, M., Heagerty, P., Cai, T., and Zheng, Y. (2017). On longitudinal prediction with time-to-event outcome: comparison of modeling options. *Biometrics*, 73(1):83–93.

McCullagh, P. and Nelder, J. (1989). *Generalized Linear Models, Second Edition*. Chapman & Hall, London, U.K.

McKeague, I. and Sasieni, P. (1994). A partly parametric additive risk model. *Biometrika*, 81(3):501–514.

McKnight, B. and Crowley, J. (1984). Tests for differences in tumor incidence based on animal carcinogenesis experiments. *Journal of the American Statistical Association*, 79(387):639–648.

Meade, M., Cook, R., Guyatt, G., Groll, R., Kachura, J., Bedard, M., Cook, D., Slutsky, A., and Stewart, T. (2000). Interobserver variation in interpreting chest radiographs for the diagnosis of acute respiratory distress syndrome. *American Journal of Respiratory and Critical Care Medicine*, 161(1):85–90.

Mealli, F. and Pudney, S. (1999). Specification tests for random effects transition models: an application to the British Youth Training Scheme. *Lifetime Data Analysis*, 5(3):213–237.

Mehtala, J., Auranen, K., and Kulathinal, S. (2015). Optimal designs for epidemiologic longitudinal studies with binary outcomes. *Statistical Methods in Medical Research*, 24(6):803–818.

Mostajabi, F. and Datta, S. (2013). Nonparametric regression of state occupation, entry, exit, and waiting times with multistate right-censored data. *Statistics in Medicine*, 32(17):3006–3019.

Munda, M., Rotolo, F., and Legrand, C. (2012). `parfm`: Parametric frailty models in R. *Journal of Statistical Software*, 51(1):1–20.

Nagelkerke, N., Chunge, R., and Kinoti, S. (1990). Estimation of parasitic infection dynamics when detectability is imperfect. *Statistics in Medicine*, 9(10):1211–1219.

Nazeri Rad, N. (2014). *Multistate Models for Biomarker Processes*. Ph.D. dissertation, University of Waterloo, Waterloo, ON.

Nazeri Rad, N. and Lawless, J. (2017). Estimation of state occupancy probabilities in multistate models with dependent intermittent observation, with application to HIV viral rebounds. *Statistics in Medicine*, 36(8):1256–1271.

Nelsen, R. (2006). *An Introduction to Copulas, Second Edition*. Springer Science + Business Media, New York, NY.

Nelson, W. (1972). Theory and applications of hazard plotting for censored failure data. *Technometrics*, 14(4):945–966.

Ng, E. and Cook, R. (1997). Modeling two-state disease processes with random effects. *Lifetime Data Analysis*, 3(4):315–335.

O'Keeffe, A., Tom, B., and Farewell, V. (2013). Mixture distributions in multistate modelling: some considerations in a study of psoriatic arthritis. *Statistics in Medicine*, 32(4):600–619.

O'Quigley, J. and Xu, R. (2014). Robustness of proportional hazards regression. In Klein, J., van Houwelingen, H., Ibrahim, J., and Scheike, T., editors, *Handbook of Survival Analysis*, chapter 16, pages 323–339. Chapman & Hall/CRC Press, Boca Raton, FL.

Palmer, A., Klein, M., Reboud, J., Cooper, C., Hosein, S., Loutfy, M., Machouf, N., Montaner, J., Rourke, S., Smieja, M., Tsoukas, C., Yip, B., Milan, D., Hogg, R., and the CANOC Collaboration (2011). Cohort Profile: The Canadian Observational Cohort Collaboration. *International Journal of Epidemiology*, 40(1):25–32.

Pan, W. and Chappell, R. (1998). Computation of the NPMLE of distribution functions for interval censored and truncated data with applications to the Cox model. *Computational Statistics and Data Analysis*, 28(1):33–50.

Pena, E. (1998). Smooth goodness-of-fit tests for composite hypothesis in hazard based models. *Annals of Statistics*, 26(5):1935–1971.

Peng, Y. and Dear, K. (2000). A nonparametric mixture model for cure rate estimation. *Biometrics*, 56(1):237–243.

Peng, Y., Dear, K., and Denham, J. (1998). A generalized F mixture model for cure rate estimation. *Statistics in Medicine*, 17(8):813–830.

Pepe, M. (1991). Inference for events with dependent risks in multiple endpoint studies. *Journal of the American Statistical Association*, 86(415):770–778.

Pepe, M. (2003). *The Statistical Evaluation of Medical Tests for Classification and Prediction*. Oxford University Press, Oxford, U.K.

Pepe, M. and Fleming, T. (1991). A nonparametric method for dealing with mismeasured covariate data. *Journal of the American Statistical Association*, 86(413):108–113.

Pepe, M. and Mori, M. (1993). Kaplan-Meier, marginal or conditional probability curves in summarizing competing risks failure time data? *Statistics in Medicine*, 12(8):737–751.

Plummer, M. and Carstensen, B. (2011). `Lexis`: an R class for epidemiological studies with long-term follow-up. *Journal of Statistical Software*, 38(5):1–12.

Prentice, R. (1986). A case-cohort design for epidemiologic cohort studies and disease prevention trials. *Biometrika*, 73(1):1–11.

Prentice, R., Kalbfleisch, J., Peterson, A., Flournoy, N., Farewell, V., and Breslow, N. (1978). The analysis of failure times in the presence of competing risks. *Biometrics*, 34(4):541–554.

Prentice, R., Williams, B., and Peterson, A. (1981). On the regression analysis of multivariate failure time data. *Biometrika*, 68(2):373–379.

Proust-Lima, C., Sane, M., Taylor, J., and Jacqmin-Gadda, H. (2014). Joint latent class models for longitudinal and time-to-event data: a review. *Statistical Methods in Medical Research*, 23(1):74–90.

Proust-Lima, C. and Taylor, J. (2009). Development and validation of a dynamic prognostic tool for prostate cancer recurrence using repeated measures of post-treatment PSA: a joint modeling approach. *Biostatistics*, 10(3):535–549.

Pullenayegum, E. and Feldman, B. (2013). Doubly robust estimation, optimally truncated inverse-intensity weighting and increment-based methods for the analysis of irregularly observed longitudinal data. *Statistics in Medicine*, 32(6):1054–1072.

Pullenayegum, E. and Lim, L. (2016). Longitudinal data subject to irregular observation: a review of methods with a focus on visit processes, assumptions, and study design. *Statistical Methods in Medical Research*, 25(6):2992–3014.

Putter, H., Fiocco, M., and Geskus, R. (2007). Tutorial in biostatistics: competing risks and multi-state models. *Statistics in Medicine*, 26(11):2389–2430.

Putter, H. and van Houwelingen, H. (2015). Frailties in multi-state models: are they identifiable? Do we need them? *Statistical Methods in Medical Research*, 24(6):675–692.

Rahman, P., Gladman, D., Cook, R., Zhou, Y., Young, G., and Salonen, D. (1998). Radiological assessment in psoriatic arthritis. *British Journal of Rheumatology*, 37(7):760–765.

Raina, P., Wolfson, C., Kirkland, S., Griffith, L., Oremus, M., Patterson, C., Tuokko, H., Penning, M., Balion, C., Hogan, D., Wister, A., Payette, H., Shannon, H., and Brazil, K. (2009). The Canadian Longitudinal Study on Aging (CLSA). *Canadian Journal on Aging/La Revue Canadienne Du Vieillissement*, 28(3):221–229.

RECORD Trial Group (2005). Oral vitamin D3 and calcium for secondary prevention of low-trauma fractures in elderly people (Randomised Evaluation of Calcium Or vitamin D, RECORD): a randomised placebo-controlled trial. *Lancet*,

365(9471):1621–1628.

Rizopoulos, D. (2012). *Joint Models for Longitudinal and Time-to-Event Data: With Applications in R*. CRC Press, Boca Raton, FL.

Rizopoulos, D. and Ghosh, P. (2011). A Bayesian semiparametric multivariate joint model for multiple longitudinal outcomes and a time-to-event. *Statistics in Medicine*, 30(12):1366–1380.

Robins, J. and Finkelstein, D. (2000). Correcting for noncompliance and dependent censoring in an AIDS clinical trial with inverse probability of censoring weighted (IPCW) log-rank tests. *Biometrics*, 56(3):779–788.

Robins, J., Hernán, M., and Brumback, B. (2000). Marginal structural models and causal inference in epidemiology. *Epidemiology*, 11(5):550–560.

Robins, J. and Rotnitzky, A. (1992). Recovery of information and adjustment for dependent censoring using surrogate markers. In *AIDS Epidemiology: Methodological Issues*, pages 297–331. Springer Science + Business Media, New York, NY.

Romanowski, B., Marina, R., and Roberts, J. (2003). Patients' preference of valacyclovir once-daily suppressive therapy versus twice-daily episodic therapy for recurrent genital herpes: a randomized study. *Sexually Transmitted Diseases*, 30(3):226–231.

Rondeau, V., Mazroui, Y., and Gonzalez, R. (2012). `frailtypack`: an R package for the analysis of correlated survival data with frailty models using penalized likelihood estimation of parametrical estimation. *Journal of Statistical Software*, 47(4):1–28.

Rosen, L., Gordon, D., Tchekmedyian, S., Yanagihara, R., Hirsh, V., Krzakowski, M., Pawlicki, M., de Souza, P., Zheng, M., Urbanowitz, G., Reitsma, D., and Seaman, J. (2003). Zoledronic acid versus placebo in the treatment of skeletal metastases in patients with lung cancer and other solid tumors: a phase III, double-blind, randomized trial – the zoledronic acid lung cancer and other solid tumors study group. *Journal of Clinical Oncology*, 21(16):3150–3157.

Ross, S. (1996). *Stochastic Processes*. John Wiley & Sons, New York, NY.

Rosychuk, R. and Thompson, M. (2001). A semi-Markov model for binary longitudinal responses subject to misclassification. *Canadian Journal of Statistics*, 29(3):395–404.

Rotnitzky, A. (2009). Inverse probability weighted methods. In Fitzmaurice, G., Davidian, M., Verbeeke, G., and Molenberghs, G., editors, *Longitudinal Data Analysis*, chapter 20, pages 453–476. Chapman & Hall/CRC Press, Boca Raton, FL.

Rubin, D. (1974). Estimating causal effects of treatment in randomized and non-randomized studies. *Journal of Educational Psychology*, 66(5):688–701.

Rubin, D. (1976). Inference and missing data. *Biometrika*, 63(3):581–592.

Saad, F., Gleason, D., Murray, R., Tchekmedyian, S., Venner, P., Lacombe, L., Chin, J., Vinholes, J., Goas, J., Zheng, M., and for the Zoledronic Acid Prostate Cancer

Study Group (2004). Long-term efficacy of zoledronic acid for the prevention of skeletal complications in patients with metastatic hormone-refractory prostate cancer. *Journal of the National Cancer Institute*, 96(11):879–882.

Saarela, O., Kulathinal, S., Arjas, E., and Läärä, E. (2008). Nested case-control data utilized for multiple outcomes: a likelihood approach and alternatives. *Statistics in Medicine*, 27(28):5991–6008.

Samuelsen, S., Ånestad, H., and Skrondal, A. (2007). Stratified case–cohort analysis of general cohort sampling designs. *Scandinavian Journal of Statistics*, 34(1):103–119.

Satten, G. (1999). Estimating the extent of tracking in interval-censored chain-of-events data. *Biometrics*, 55(4):1228–1231.

Satten, G. and Datta, S. (2002). Marginal estimation for multi-stage models: waiting time distributions and competing risks analyses. *Statistics in Medicine*, 21(1):3–19.

Satten, G. and Longini Jr, I. (1996). Markov chains with measurement error: estimating the true course of a marker of the progression of human immunodeficiency virus disease. *Journal of the Royal Statistical Society: Series C (Applied Statistics)*, 45(3):275–309.

Satten, G. and Sternberg, M. (1999). Fitting semi-Markov models to interval-censored data with unknown initiation times. *Biometrics*, 55(2):507–513.

Scharfstein, D. and Robins, J. (2002). Estimation of the failure time distribution in the presence of informative censoring. *Biometrika*, 89(3):617–634.

Scheike, T. and Zhang, M. (2007). Direct modelling of regression effects for transition probabilities in multistate models. *Scandinavian Journal of Statistics*, 34(1):17–32.

Scheike, T. and Zhang, M. (2008). Flexible competing risks regression modeling and goodness-of-fit. *Lifetime Data Analysis*, 14(4):464–483.

Scheike, T. and Zhang, M. (2011). Analyzing competing risk data using the R `timereg` package. *Journal of Statistical Software*, 38(2):i02.

Scheike, T., Zhang, M., and Gerds, T. (2008). Predicting cumulative incidence probability by direct binomial regression. *Biometrika*, 95(1):205–220.

Shen, Y., Ning, J., and Qin, J. (2017). Nonparametric and semiparametric estimation for length-biased survival data. *Lifetime Data Analysis*, 23:3–24.

Siannis, F. (2011). Sensitivity analysis for multiple right censoring processes: investigating mortality in psoriatic arthritis. *Statistics in Medicine*, 30(4):356–367.

Slud, E. and Rubinstein, L. (1983). Dependent competing risks and summary survival curves. *Biometrika*, 70(3):643–649.

Smith, M., Cook, R., Lee, K., and Nelson, J. (2011). Disease and host characteristics as predictors of time to first bone metastasis and death in men with progressive castration-resistant nonmetastatic prostate cancer. *Cancer*, 117(10):2077–2085.

Sparling, Y., Younes, N., Lachin, J., and Bautista, O. (2006). Parametric survival models for interval-censored data with time-dependent covariates. *Biostatistics*,

4(1):599–614.

Stampfer, M., Colditz, G., Willett, W., Manson, J., Rosner, B., Speizer, F., and Hennekens, C. (1991). Postmenopausal estrogen therapy and cardiovascular disease: ten-year follow-up from the Nurses' Health Study. *New England Journal of Medicine*, 325(11):756–762.

Sternberg, M. and Satten, G. (1999). Discrete-time nonparametric estimation for semi-Markov models of chain-of-events data subject to interval censoring and truncation. *Biometrics*, 55(2):514–522.

Stewart, T., Meade, M., Cook, D., Granton, J., Hodder, R., Lapinsky, S., Mazer, C., McLean, R., Rogovein, T., Schouten, B., Todd, T., Slutsky, A., and the Pressure- and Volume-Limited Ventilation Strategy Group (1998). Evaluation of a ventilation strategy to prevent barotrauma in patients at high risk for acute respiratory distress syndrome. *New England Journal of Medicine*, 338(6):355–361.

Steyerberg, E. (2009). *Clinical Prediction Models: A Practical Approach to Development, Validation and Updating*. Springer Science + Business Media, New York, NY.

Struthers, C. and Kalbfleisch, J. (1986). Misspecified proportional hazards models. *Biometrika*, 74(2):363–369.

Sun, J. (2006). *The Statistical Analysis of Interval-Censored Failure Time Data*. Springer Science + Business Media, New York, NY.

Sun, J. and Zhao, X. (2013). *Statistical Analysis of Panel Count Data*. Springer Science + Business Media, New York, NY.

Sutradhar, R. and Cook, R. (2008). Analysis of interval-censored data from clustered multistate processes: application to joint damage in psoriatic arthritis. *Journal of the Royal Statistical Society: Series C (Applied Statistics)*, 57(5):553–566.

Sverdrup, E. (1965). Estimates and test procedures in connection with stochastic models for deaths, recoveries and transfers between different states of health. *Scandinavian Actuarial Journal*, 3–4:184–211.

Taylor, J. (1995). Semi-parametric estimation in failure time mixture models. *Biometrics*, 51(3):899–907.

Taylor, J., Yu, M., and Sandler, H. (2005). Individualized predictions of disease progression following radiation therapy for prostate cancer. *Journal of Clinical Oncology*, 23(4):816–825.

Temkin, N. (1978). An analysis for transient states with application to tumor shrinkage. *Biometrics*, 34(4):571–580.

Therneau, T. (2012). `coxme`: mixed effect Cox models. R package version 2.2-3. `http://CRAN.R-project.org/package=coxme`.

Therneau, T. and Grambsch, P. (2000). *Modeling Survival Data: Extending the Cox Model*. Springer Science + Business Media, New York, NY.

Therneau, T. and Hamilton, S. (1997). rhDNase as an example of recurrent event analysis. *Statistics in Medicine*, 16(18):2029–2047.

TIBCO Spotfire S+® 8.2 (2010). *Guide to Statistics, Volume 2*. TIBCO Software

Inc., Palo Alto, CA.

Titman, A. (2011). Flexible nonhomogeneous Markov models for panel observed data. *Biometrics*, 67(3):780–787.

Titman, A. and Sharples, L. (2010a). Model diagnostics for multi-state models. *Statistical Methods in Medical Research*, 19(6):621–651.

Titman, A. and Sharples, L. (2010b). Semi-Markov models with phase-type sojourn distributions. *Biometrics*, 66(3):742–752.

Tom, B. and Farewell, V. (2011). Intermittent observation of time-dependent explanatory variables: a multistate modelling approach. *Statistics in Medicine*, 30(30):3520–3531.

Touraine, C., Gerds, T., and Joly, P. (2017). SmoothHazard: an R package for fitting regression models to interval-censored observations of illness-death models. *Journal of Statistical Software*, 79(7):1–22.

Tsiatis, A. and Davidian, M. (2004). Joint modeling of longitudinal and time-to-event data: an overview. *Statistica Sinica*, 14(3):809–834.

Tuli, S., Drake, J., Lawless, J., Wigg, M., and Lamberti-Pasculli, M. (2000). Risk factors for repeated cerebrospinal shunt failures in pediatric patients with hydrocephalus. *Journal of Neurosurgery*, 92(1):31–38.

Turnbull, B. (1976). The empirical distribution function with arbitrarily grouped, censored and truncated data. *Journal of the Royal Statistical Society: Series B (Methodological)*, 38(3):290–295.

Vakulenko-Lagun, B., Mandel, M., and Goldberg, Y. (2017). Nonparametric estimation in the illness-death model using prevalent data. *Lifetime Data Analysis*, 23:25–56.

Van den Berg, G. (2001). Duration models: specification, identification and multiple durations. In Heckman, J. and Leamer, E., editors, *Handbook of Econometrics*, chapter 55, pages 3381–3460. North-Holland, Amsterdam, The Netherlands.

Van den Hout, A. (2017). *Multi-State Survival Models for Interval-Censored Data*. CRC Press, Boca Raton, FL.

van der Heijde, D., Dijkmans, B., Geusens, P., Sieper, J., DeWoody, K., Williamson, P., and Braun, J. (2005). Efficacy and safety of infliximab in patients with ankylosing spondylitic: results of a randomized, placebo-controlled trial (ASSERT). *Arthritis & Rheumatology*, 52(2):585–591.

van Houwelingen, H. and Putter, H. (2012). *Dynamic Prediction in Clinical Survival Analysis*. CRC Press, Boca Raton, FL.

Varin, C., Reid, N., and Firth, D. (2011). An overview of composite likelihood methods. *Statistica Sinica*, 21(1):5–42.

Vaupel, J., Manton, K., and Stallard, E. (1979). The impact of heterogeneity in individual frailty on the dynamics of mortality. *Demography*, 16(3):439–454.

Vaupel, J. and Yashin, A. (1985). Heterogeneity's ruses: some surprising effects of selection on population dynamics. *American Statistician*, 39(3):176–185.

von Cube, M., Schumacher, M., Palomar-Martinez, M., Olaechea-Astigarraga, P.,

Alvarez-Lerma, P., and Wolkewitz, M. (2017). A case-cohort approach of multi-state models in hospital epidemiology. *Statistics in Medicine*, 36(3):481–495.

Wei, L., Lin, D., and Weissfeld, L. (1989). Regression analysis of multivariate incomplete failure time data by modeling marginal distributions. *Journal of the American Statistical Association*, 84(408):1065–1073.

White, H. (1982). Maximum likelihood estimation of misspecified models. *Econometrica*, 50(1):1–25.

Whittemore, A. (1997). Multistage sampling designs and estimating equations. *Journal of the Royal Statistical Society: Series B (Methodological)*, 59(3):589–602.

Wienke, A. (2011). *Frailty Models in Survival Analysis*. CRC Press, Boca Raton, FL.

Willekens, F. (2014). *Multistate Analysis of Life Histories with R*. Springer International, Cham, Switzerland.

Williamson, P., Kolamunnage-Dona, R., Philipson, P., and Marson, A. (2008). Joint modelling of longitudinal and competing risks data. *Statistics in Medicine*, 27(30):6426–6438.

Wolfe, R., McCullough, K., Schaubel, D., Kalbfleisch, J., Murray, S., Stegall, M., and Leichtman, A. (2008). Calculating life years from transplant (LYFT): methods for kidney and kidney-pancreas candidates. *American Journal of Transplantation*, 8(4p2):997–1011.

Wulfsohn, M. and Tsiatis, A. (1997). A joint model for survival and longitudinal data measured with error. *Biometrics*, 53(1):330–339.

Xu, J., Kalbfleisch, J., and Tai, B. (2010). Statistical analysis of illness-death processes and semicompeting risks data. *Biometrics*, 66(3):716–725.

Xue, X. and Brookmeyer, R. (1996). Bivariate frailty model for the analysis of multivariate survival time. *Lifetime Data Analysis*, 2(3):277–289.

Yamaguchi, K. (1994). Some accelerated failure-time regression models derived from diffusion process models: an application to a network diffusion analysis. *Sociological Methodology*, 24:267–300.

Yamaguchi, K. (1998). Mover-stayer models for analyzing event nonoccurrence and event timing with time-dependent covariates: an application to an analysis of remarriage. *Sociological Methodology*, 28(1):327–361.

Yamaguchi, K. (2003). Accelerated failure-time mover-stayer regression models for the analysis of last-episode data. *Sociological Methodology*, 33(1):81–110.

Yang, G. (2013). Neyman, Markov processes and survival analysis. *Lifetime Data Analysis*, 19(3):393–411.

Yang, Y. and Nair, V. (2011). Parametric inference for time-to-failure in multi-state semi-Markov models: a comparison of marginal and process approaches. *Canadian Journal of Statistics*, 39(3):537–555.

Yashin, A., Vaupel, J., and Iachine, I. (1995). Correlated individual frailty: an advantageous approach to survival analysis of bivariate data. *Mathematical Pop-

ulation Studies, 5(2):145–159.

Zeng, D. and Lin, D. (2014). Efficient estimation of semiparametric transformation models for two-phase cohort studies. *Journal of the American Statistical Association*, 109(505):371–383.

Zhang, Z. and Rockette, H. (2005). On maximum likelihood estimation in parametric regression with missing covariates. *Journal of the Statistical Planning and Inference*, 134(1):206–223.

Zhao, L., Claggett, B., Tian, L., Uno, H., Pfeffer, M., Solomon, S., Trippa, L., and Wei, L. (2016). On the restricted mean survival time curve in survival analysis. *Biometrics*, 72(1):215–221.

Zhao, Y., Lawless, J., and McLeish, D. (2009). Likelihood methods for regression models with expensive variables missing by design. *Biometrical Journal*, 51(1):123–136.

Zhu, Y., Lawless, J., and Cotton, C. (2017). Estimation of parametric failure time distributions based on interval-censored data with irregular dependent follow-up. *Statistics in Medicine*, 36(10):1548–1567.

Zinn, S. (2014). The `MicSim` package of R: an entry-level toolkit for continuous-time microsimulation. *International Journal of Microsimulation*, 7(3):3–32.

Author Index

Subject Index